Lecture Notes in Physics

Springer
Berlin
Heidelberg
New York
Barcelona
Hong Kong
London
Milan
Paris
Singapore
Tokyo

Physics and Astronomy

ONLINE LIBRARY

http://www.springer.de/phys/

The Editorial Policy for Proceedings

The series Lecture Notes in Physics reports new developments in physical research and teaching – quickly, informally, and at a high level. The proceedings to be considered for publication in this series should be limited to only a few areas of research, and these should be closely related to each other. The contributions should be of a high standard and should avoid lengthy redraftings of papers already published or about to be published elsewhere. As a whole, the proceedings should aim for a balanced presentation of the theme of the conference including a description of the techniques used and enough motivation for a broad readership. It should not be assumed that the published proceedings must reflect the conference in its entirety. (A listing or abstracts of papers presented at the meeting but not included in the proceedings could be added as an appendix.)

When applying for publication in the series Lecture Notes in Physics the volume's editor(s) should submit sufficient material to enable the series editors and their referees to make a fairly accurate evaluation (e.g. a complete list of speakers and titles of papers to be presented and abstracts). If, based on this information, the proceedings are (tentatively) accepted, the volume's editor(s), whose name(s) will appear on the title pages, should select the papers suitable for publication and have them refereed (as for a journal) when appropriate. As a rule discussions will not be accepted. The series editors and Springer-Verlag will normally not interfere with the detailed editing except in fairly obvious cases or on technical matters.

Final acceptance is expressed by the series editor in charge, in consultation with Springer-Verlag only after receiving the complete manuscript. It might help to send a copy of the authors' manuscripts in advance to the editor in charge to discuss possible revisions with him. As a general rule, the series editor will confirm his tentative acceptance if the final manuscript corresponds to the original concept discussed, if the quality of the contribution meets the requirements of the series, and if the final size of the manuscript does not greatly exceed the number of pages originally agreed upon. The manuscript should be forwarded to Springer-Verlag shortly after the meeting. In cases of extreme delay (more than six months after the conference) the series editors will check once more the timeliness of the papers. Therefore, the volume's editor(s) should establish strict deadlines, or collect the articles during the conference and have them revised on the spot. If a delay is unavoidable, one should encourage the authors to update their contributions if appropriate. The editors of proceedings are strongly advised to inform contributors about these points at an early stage.

The final manuscript should contain a table of contents and an informative introduction accessible also to readers not particularly familiar with the topic of the conference. The contributions should be in English. The volume's editor(s) should check the contributions for the correct use of language. At Springer-Verlag only the prefaces will be checked by a copy-editor for language and style. Grave linguistic or technical shortcomings may lead to the rejection of contributions by the series editors. A conference report should not exceed a total of 500 pages. Keeping the size within this bound should be achieved by a stricter selection of articles and not by imposing an upper limit to the length of the individual papers. Editors receive jointly 30 complimentary copies of their book. They are entitled to purchase further copies of their book at a reduced rate. As a rule no reprints of individual contributions can be supplied. No royalty is paid on Lecture Notes in Physics volumes. Commitment to publish is made by letter of interest rather than by signing a formal contract. Springer-Verlag secures the copyright for each volume.

The Production Process

The books are hardbound, and the publisher will select quality paper appropriate to the needs of the author(s). Publication time is about ten weeks. More than twenty years of experience guarantee authors the best possible service. To reach the goal of rapid publication at a low price the technique of photographic reproduction from a camera-ready manuscript was chosen. This process shifts the main responsibility for the technical quality considerably from the publisher to the authors. We therefore urge all authors and editors of proceedings to observe very carefully the essentials for the preparation of camera-ready manuscripts, which we will supply on request. This applies especially to the quality of figures and halftones submitted for publication. In addition, it might be useful to look at some of the volumes already published. As a special service, we offer free of charge LaTeX and TeX macro packages to format the text according to Springer-Verlag's quality requirements. We strongly recommend that you make use of this offer, since the result will be a book of considerably improved technical quality. To avoid mistakes and time-consuming correspondence during the production period the conference editors should request special instructions from the publisher well before the beginning of the conference. Manuscripts not meeting the technical standard of the series will have to be returned for improvement.

For further information please contact Springer-Verlag, Physics Editorial Department II, Tiergartenstrasse 17, D-69121 Heidelberg, Germany

Thierry Passot   Pierre-Louis Sulem   (Eds.)

Nonlinear MHD Waves and Turbulence

Proceedings of the Workshop Held in Nice, France, 1-4 December 1998

 Springer

Editors

Thierry Passot
Pierre-Louis Sulem
Laboratoire Cassini, CNRS UMR 6529
Bd. de L'Observatoire, BP 4229
06304 Nice Cedex 4, France

Library of Congress Cataloging-in-Publication Data.

Die Deutsche Bibliothek - CIP-Einheitsaufnahme

Nonlinear MHD waves and turbulence : proceedings of the
workshop, held in Nice, France, 1 - 4. December 1998 / Thierry
Passot ; Pierre-Louis Sulem (ed.). - Berlin ; Heidelberg ; New York ;
Barcelona ; Hong Kong ; London ; Milan ; Paris ; Singapore ; Tokyo
: Springer, 1999
 (Lecture notes in physics ; Vol. 536)

ISSN 0075-8450
ISBN 978-3-642-08598-7   e-ISBN 978-3-540-47038-0

© Springer-Verlag Berlin Heidelberg 2010
Printed in Germany

The use of general descriptive names, registered names, trademarks, etc. in this publication does not imply, even in the absence of a specific statement, that such names are exempt from the relevant protective laws and regulations and therefore free for general use.

Cover design: *design & production*, Heidelberg

Printed on acid-free paper

Preface

The workshop "Nonlinear MHD Waves and Turbulence" was held at the Observatoire de Nice, December 1–4, 1998 and brought together an international group of experts in plasma physics, fluid dynamics and applied mathematics. The aim of the meeting was to survey the current knowledge on two main topics: (i) propagation of plasma waves (like Alfvén, whistler or ion-acoustic waves), their instabilities and the development of a nonlinear dynamics leading to solitonic structures, wave collapse or weak turbulence; (ii) turbulence in magnetohydrodynamic flows and its reduced description in the presence of a strong ambient magnetic field. As is well known, both aspects play an important role in various geophysical or astrophysical media such as the magnetospheres of planets, the heliosphere, the solar wind, the solar corona, the interplanetary and interstellar media, etc.

This volume, which includes expanded versions of oral contributions presented at this meeting, should be of interest for a large community of researchers in space plasmas and nonlinear sciences. Special effort was made to put the new results into perspective and to provide a detailed literature review. A main motivation was the attempt to relate more closely the theoretical understanding of MHD waves and turbulence (both weak and strong) with the most recent observations in space plasmas. Some papers also bring interesting new insights into the evolution of hydrodynamic or magnetohydrodynamic structures, based on systematic asymptotic methods.

We wish to express our special thanks to the lecturers for their stimulating presentations and to all the participants who contributed to the success of this meeting. We also gratefully acknowledge the support of the staff of the Laboratoire Cassini (CNRS UMR 6529) and of the Observatoire de la Côte d'Azur whose contribution was capital in the organization. The workshop benefited from support from CNRS through the Groupes de Recherche "Propagation des Ondes en Milieux Aléatoires ou Nonlinéaires" and "Mécanique des Fluides Géophysiques et Astrophysiques" and from INTAS Project 96-413.

Nice   *T. Passot*
June 1999   *P.L. Sulem*

Contents

X

Nonlinear Phenomena Involving Dispersive Alfvén Waves

P.K. Shukla and L. Stenflo

Fakultät für Physik und Astronomie, Ruhr-Universität Bochum, D-44780
Bochum, Germany, and Department of Plasma Physics, Umeå University,
SE-90187 Umeå, Sweden

Abstract. This paper presents a comprehensive review of linear and nonlinear
dispersive Alfvén waves in a magnetized plasma. For illustrative purposes, we start
with small amplitude Alfvén waves and their relation with other low-frequency
plasma modes in a uniform plasma. We then show that dispersive Alfvén waves can
be excited either by electron beams or by high-frequency external pump waves.
Finite amplitude dispersive Alfvén waves interact with the background plasma
as well as with themselves, giving rise to a number of interesting nonlinear phenomena, which are described here. The nonlinear effects produced by dispersive
Alfvén waves include three-wave decay interactions, wave-amplitude modulations
and wave-filamentation, density profile modification, as well as self-organization
in the form of vortical structures, and routes to chaos. The nonlinear effects, as
discussed here, can be of relevance to the generation of low-frequency turbulence
and the modification of the background plasma density in the Earth's ionosphere
and magnetosphere, as well as to the large amplitude coherent nonlinear structures
which are frequently observed in auroral plasmas.

1 Introduction

The Alfvén wave is classic in plasma physics. In an Alfvén wave, the
restoring force comes from the pressure of the equilibrium magnetic field,
and the ion mass provides the inertia. The propagation of low-frequency (in
comparison with the ion gyrofrequency ω_{ci}) nondispersive Alfvén waves is
governed by the ideal magnetohydrodynamic (MHD) equations. The inclusion
of finite particle inertia, Larmor radius effects, etc. provides the dispersion
of the Alfvén waves and their coupling to other plasma modes. In space
and laboratory plasmas, finite amplitude Alfvén waves are excited by many
sources such as energetic charged particle beams, nonuniform background
plasma parameters, electrostatic and electromagnetic waves. The dispersive
Alfvén waves have wide ranging applications (e.g. Hasegawa and Uberoi,
1982) in space, fusion, and laboratory plasmas. Leneman, Gekelman and
Maggs (1999) have recently observed shear Alfvén waves in the Large Plasma
Device at UCLA.

Finite amplitude Alfvén waves can cause a number of nonlinear effects.
The latter include parametric processes such as three-wave decay interactions
(Sagdeev and Galeev, 1969; Brodin and Stenflo, 1989), stimulated Compton

scattering instabilities (Shukla and Dawson, 1984), modulational and filamentational interactions (Shukla and Stenflo, 1985; Stenflo, Yu, and Shukla, 1988; Shukla and Stenflo, 1989a; Shukla and Stenflo, 1989b; Shukla *et al.*, 1999a), as well as the modification of the background plasma number density by the Alfvén wave ponderomotive force (Shukla and Stenflo, 1985; 1995; Shukla *et al.*, 1999a). On the other hand, multi- dimensional dispersive Alfvén waves are also subjected to a spatial collapse (Passot and Sulem, 1993; Shukla *et al.*, 1998; Champeaux *et al.*, 1997, 1999), which may cause dissipation of the Alfvén wave energy in magnetoplasmas.

Recently, there has been a great deal of interest in understanding the nonlinear properties (Shukla and Stenflo, 1985; Stenflo and Shukla, 1988, Hada, 1993; Shevchenko *et al.*, 1995; Mishin and Föster, 1995; Shukla *et al.*, 1996; Champeaux *et al.*, 1999) of dispersive Alfvén waves. This is a necessary prerequisite in interpreting the numerous observations of large amplitude low-frequency dispersive electromagnetic waves in space plasmas. In this paper, we shall focus our attention on some of the most important nonlinear phenomena involving dispersive Alfvén waves. Specifically, we shall consider the excitation of dispersive Alfvén waves and the nonlinear effects that are introduced by these waves. Special attention shall be paid to the parametric processes as well as to the formation of coherent nonlinear structures and the transition to chaos, which are all of great importance in many branches of physics. The pattern formation and chaotic motion thus describe the nonlinear dynamics of various scale size disturbances which are nonlinearly interacting in a driven dissipative system. The appropriate dynamical equations for studying the amplitude modulation of a dispersive wave packet as well as vortical and chaotic motions are the nonlinear Schrödinger equation, the Ginzburg-Landau equation, the Henon-Heiles equation, and the Lorenz equations that have been widely employed to investigate the nonlinear dynamics of wave motions in nonlinear optics, in fluids, in superconductors, as well as in astrophysical plasmas and in atmospheric physics.

The manuscript is organized in the following fashion. In section 2, we briefly review the linear properties of Alfvén waves in a homogeneous magnetoplasma. In section 3, we show that dispersive Alfvén waves are excited by electron beams as well as by electron whistlers and lower-hybrid (LH) waves. Finite amplitude dispersive Alfvén waves are subjected to a class of parametric instabilities including the three-wave decay interaction and the modulational/filamentation instabilities, which are discussed in sections 4 and 5. Furthermore, we have found that large scale plasma density perturbations can be created by the ponderomotive force of dispersive Alfvén waves. The self-interaction of Alfvén waves and the formation of vortices in a multi-component magnetoplasma are considered in section 6. Section 7 presents an investigation of chaotic Alfvén waves in a nonlinear dissipative medium. Finally, section 8 contains a summary of our findings and points out possible applications to space plasmas.

2 Dispersive Alfvén waves in a uniform plasma

We shall now shed some light on the linear properties of dispersive Alfvén waves, which are either governed by the Hall-MHD or by the two-fluid model of a low-β ($\beta \ll 1$) uniform magnetoplasma, where β is the ratio between the kinetic and magnetic pressures. The dispersion relation is then (e.g. Brodin and Stenflo, 1990)

$$\left(\omega^2 - k_z^2 v_A^2\right) D_m(\omega, \mathbf{k}) = \omega^2 k_z^2 k^2 \left(\omega^2 - k^2 v_s^2\right) v_A^4 / \omega_{ci}^2 , \tag{1}$$

where $D_m(\omega, \mathbf{k}) = \omega^4 - \omega^2 k^2 (v_A^2 + v_s^2) + k_z^2 k^2 v_A^2 v_s^2$, ω is the wave frequency, $\mathbf{k}\,(= \mathbf{k}_\perp + k_z \hat{z})$ is the wave vector, $v_A = B_0/(4\pi n_0 m_i)^{1/2}$ is the Alfvén velocity, B_0 is the strength of the external magnetic field which is directed along the z-axis, n_0 is the unperturbed plasma number density, m_i is the ion mass, and v_s the ion sound velocity.

In the low-frequency limit ($\omega \ll \omega_{ci} = eB_0/m_i c$, where e is the magnitude of the electron charge and c is the speed of light), Eq. (1) reproduces the standard dispersion relation for Alfvén and magnetosonic waves. On the other hand, for wave propagation along the external magnetic field (i.e. $\mathbf{k}_\perp = 0$), we obtain (Shukla and Stenflo, 1995) from (1) for $v_s \ll v_A$,

$$k_z^2 v_A^2 = \frac{\omega^2 \omega_{ci}}{\omega_{ci} \mp \omega} , \tag{2}$$

which describes the propagation of electromagnetic ion-cyclotron-Alfvén (EM-ICA) waves. The $+(-)$ sign corresponds to the right- (left-) hand polarization.

The finite magnetic-field-aligned wave electric fields, the ion polarization drift, the Hall effect, and the finite Larmor radius and parallel (to the unit vector $\hat{z}$) electron inertial effects are responsible for the Alfvén wave dispersion. In a low-β ($\beta \ll 1$) plasma, we have (Shukla et al., 1999b)

$$\omega^2 = \frac{k_z^2 v_A^2}{(1 + k_\perp^2 \lambda_e^2)(1 + k_z^2 \lambda_i^2)} \left[1 + \left(\frac{3}{4} + \frac{T_e}{T_i}\right) k_\perp^2 \rho_i^2\right] , \tag{3}$$

where $\lambda_e = c/\omega_{pe}$ is the electron skin depth, ω_{pe} is the electron plasma frequency, $T_e(T_i)$ is the electron (ion) temperature, $\rho_i = v_{ti}/\omega_{ci}$ is the ion Larmor radius, $v_{tj} = (T_j/m_j)^{1/2}$ is the thermal velocity of the particle species j (j equals e for the electrons and i for the ions), $\lambda_i = c/\omega_{pi}$ is the ion skin depth, and ω_{pi} is the ion plasma frequency. We note that the parallel dispersive term (the $k_z^2 \lambda_i^2 -$ term) is ascribed to the finite ω/ω_{ci}-effect which arises when the ion inertial and ion Lorentz forces are treated on an equal footing, whereas the term involving $3/4$ comes from the ion-finite Larmor radius effect. The term T_e/T_i originates from the parallel electron kinetics. In a plasma with $T_e \gg T_i$, the perpendicular dispersion solely arises from the parallel electron pressure gradient force, and the bracket in (3) is thus replaced by $1 + k_\perp^2 \rho_s^2$, where $\rho_s = (T_e/m_i)^{1/2}/\omega_{ci} \equiv c_s/\omega_{ci}$ is the ion gyroradius at the electron temperature. Thus, the frequency of the dispersive Alfvén waves in a cold ion plasma is

$$\omega^2 = \frac{k_z^2 v_A^2}{(1 + k_\perp^2 \lambda_e^2)(1 + k_z^2 \lambda_i^2)} (1 + k_\perp^2 \rho_s^2) \equiv \omega_H^2 . \tag{4}$$

For $k_\perp \lambda_e \ll 1$ and $k_z \lambda_i \ll 1$, (4) reduces to (Stefant, 1970; Hasegawa and Chen, 1975)

$$\omega = k_z v_A (1 + k_\perp^2 \rho_s^2)^{1/2} , \tag{5}$$

which is the well known frequency of the kinetic Alfvén wave (KAW) in a medium $\beta \, (m_e/m_i \ll \beta \ll 1)$ plasma, where m_e is the electron mass. Its group velocity is $\mathbf{v}_g = v_{gz} \hat{\mathbf{z}} + \mathbf{v}_{g\perp}$, where $v_{gz} = v_A(1 + k_\perp^2 \rho_s^2)^{1/2}$ and $\mathbf{v}_{g\perp} = (k_z^2 \rho_s^2 v_A^2/\omega)\mathbf{k}_\perp$ are the parallel and perpendicular components of the group velocity. It follows that the KAW is a forward wave in the direction parallel (as well as perpendicular) to the equilibrium magnetic field. We note that (5) also follows from (1) in the limits $k_z \ll k$ and $\omega/k_\perp \ll v_s \ll v_A$. The dispersive terms in (5), which can be regarded as corrections, arise from the ion polarization (or the perpendicular ion inertia) drift. The polarization drift is proportional to the particle mass multiplied by the time rate of change of the wave perpendicular electric field, and is therefore dominated by the ions. We see that the KAW dispersion remains finite even when $T_i = 0$.

Furthermore, when the parallel wavelength is shorter than λ_i, and $k_\perp \lambda_e \ll 1$, we obtain from (4)

$$\omega \approx \omega_{ci}(1 + k_\perp^2 \rho_s^2)^{1/2} , \tag{6}$$

which is the frequency of the electromagnetic ion-cyclotron (or the high-frequency dispersive Alfvén wave) in a warm plasma with $T_e \gg T_i$.

In an extremely low-β $(\beta \ll m_e/m_i)$ plasma, when the parallel phase velocity (ω/k_z) is much larger than v_{te} and $k_z \lambda_i \ll 1$, the dispersion relation for dispersive Alfvén waves in a cold plasma reads

$$\omega = \frac{k_z v_A}{(1 + k_\perp^2 \lambda_e^2)^{1/2}} , \tag{7}$$

which is the frequency of the dispersive inertial Alfvén wave (DIAW) (e.g. Shukla et al., 1982; Lysak, 1990).

The DIAWs have positive (negative) perpendicular group dispersion for $k_\perp \lambda_e > 1/2(k_\perp \lambda_e < 1/2)$. The group velocity of the dispersive inertial Alfvén wave is $\mathbf{v}_g = v_{gz} \hat{\mathbf{z}} + \mathbf{v}_{g\perp}$, where $v_{gz} = k_z v_A^2/\omega(1 + k_\perp^2 \lambda_e^2)$ and $\mathbf{v}_{g\perp} = -\omega \lambda_e^2 \mathbf{k}_p erp/(1 + k_\perp^2 \lambda_e^2)$. It follows that the DAW energy propagates in the forward (backward) direction along (across) the magnetic field lines of force.

On the other hand, for $k_\perp \lambda_e \ll 1$, $k_\perp \rho_i \ll 1$, and $T_i \ll T_e$ we obtain from (3)

$$\omega = \frac{k_z v_A}{(1 + k_z^2 \lambda_i^2)^{1/2}} , \tag{8}$$

which is the frequency of the magnetic field-aligned dispersive Alfvén waves.

Finally, for $k_z \lambda_i \ll 1$ (or $\omega \ll \omega_{ci}$) and $k_\perp \lambda_e \gg 1$, (4) gives the frequency of the modified convective cell

$$\omega = \frac{k_z}{k_\perp} \left(\omega_{ce}\omega_{ci}\right)^{1/2} \left(1 + k_\perp^2 \rho_s^2\right)^{1/2} , \tag{9}$$

where $\omega_{ce} = eB_0/m_e c$ is the electron gyrofrequency.

3 Generation of Alfvén waves

The dispersive Alfvén waves can by excited by linear as well as nonlinear processes. The linear process involves magnetic field-aligned electron beams or sheared magnetic fields. On the other hand, the nonlinear processes usually require the presence of large amplitude high-frequency electromagnetic or electrostatic drivers in the plasma.

3.1 The excitation by electron beams

Let us consider Alfvén wave excitation in the presence of an equilibrium electron current $J_0 = -n_0 e u_0$ in a Maxwellian plasma. Here, the parallel component of the perturbed electron current density in the wave electric field is

$$J_{ez} = i \frac{\omega E_z}{8\pi k_z^2 \lambda_{De}^2} Z'(\xi_e) , \tag{10}$$

where E_z is the parallel component of the wave electric field, Z' the derivative of the standard plasma dispersion function with its argument $\xi_e = (\omega - k_z u_0)/\sqrt{2}k_z v_{te}$, and $\lambda_{De} = (T_e/4\pi n_0 e^2)^{1/2}$ the electron Debye radius. For $\xi_e \ll 1$, (10) becomes

$$J_{ez} = -i \frac{\omega E_z}{4\pi k_z^2 \lambda_{De}^2} \left[1 + i\sqrt{\frac{\pi}{2}} \frac{(\omega - k_z u_0)}{k_z v_{te}}\right] . \tag{11}$$

The perpendicular component of the plasma current density is

$$\mathbf{J}_\perp \approx -i \frac{n_0 e c}{B_0} \frac{\omega \omega_{ci}}{\omega_{ci}^2 - \omega^2} \mathbf{E}_\perp . \tag{12}$$

From $\nabla \cdot \mathbf{J} = 0$, we then obtain

$$\frac{k_z c_s^2}{\omega_{ci}^2 - \omega^2} \mathbf{k}_\perp \cdot \mathbf{E}_\perp = -\left[1 + i\sqrt{\frac{\pi}{2}} \frac{(\omega - k_z u_0)}{k_z v_{te}}\right] E_z . \tag{13}$$

On the other hand, from the Maxwell equations we have

$$\left(k_z^2 c^2 - \frac{\omega_{pi}^2 \omega^2}{\omega_{ci}^2}\right) \mathbf{k}_\perp \cdot \mathbf{E}_\perp = k_\perp^2 k_z c^2 E_z . \tag{14}$$

Combining (13) and (14) we obtain the dispersion relation

$$\omega^2 \approx \omega_H^2 \left[1 - i\sqrt{\frac{\pi}{2}} \frac{(\omega - k_z u_0)}{k_z v_{te}}\right] . \tag{15}$$

Letting $\omega = \omega_H + i\gamma_b$ in (15), we obtain for $\gamma_b < \omega_H$ the growth rate

$$\gamma_b = \sqrt{\frac{\pi}{8}} \omega_H \frac{|k_z u_0 - \omega_H|}{k_z v_{te}} , \tag{16}$$

if

$$|J_0| > \frac{n_0 e v_A (1 + k_\perp^2 \rho_s^2)^{1/2}}{(1 + k_z^2 \lambda_i^2)^{1/2}(1 + k_\perp^2 \lambda_e^2)^{1/2}} . \tag{17}$$

The physical mechanism of the above instability is similar to the Cerenkov process in which the beam electrons resonantly interact with the dispersive Alfvén waves to drive the latter at non-thermal levels.

3.2   Parametric excitation

The dispersive Alfvén (DA) waves can also be parametrically excited either by high-frequency electromagnetic or by electrostatic waves. In the following, we first consider the parametric excitation of DAWs by electron whistlers (Chen, 1977; Shukla, 1977, Antani et al., 1983; Stenflo, 1999), and then formulate the problem of the DA wave excitation by lower-hybrid (LH) waves.

Let us consider the presence of a circularly polarized whistler pump of the form

$$\mathbf{E}_0 = \mathbf{E}_{0\perp} \exp(i\mathbf{k}_0 \cdot \mathbf{r} - i\omega_0 t) + \text{compl. conj}, \tag{18}$$

in a uniform plasma with an external magnetic field $B_0\hat{\mathbf{z}}$. Here, $\mathbf{E}_{0\perp}$ is the perpendicular (to $\hat{\mathbf{z}}$) component of the pump electric field, which is supposed to be much larger than the parallel component $\hat{\mathbf{z}}E_{0z}$. The whistler wave frequency for $\omega_{ci}, \omega_{pi} \ll \omega < \omega_{ce}$ and $\omega_{pe} \ll k_0 c$ is

$$\omega_0 = k_0 k_{0z} c^2 \omega_{ce}/\omega_{pe}^2, \tag{19}$$

where $\mathbf{k}_0 = k_{0x}\hat{\mathbf{x}} + k_{0z}\hat{\mathbf{z}}$ is the wavevector.

The nonlinear interaction between the whistler pump and DA waves $(\omega, \mathbf{k})$ generates Stokes $(\omega - \omega_0, \mathbf{k} - \mathbf{k}_0)$ and anti-Stokes $(\omega + \omega_0, \mathbf{k} + \mathbf{k}_0)$ whistlers, where $\omega \ll \omega_0$. The wave equations for the two whistler sidebands are (Chen, 1977; Shukla, 1977)

$$D_{\pm} E_{x\pm} = \mp \frac{4i\pi}{\omega_0} \frac{\delta n_A ec}{B_0} \alpha_{\pm} E_{0x}(\pm), \qquad (20)$$

where

$$D_{\pm} = (\omega_{pe}^4/k_{\pm}^2 c^2 \omega_{ce}^2) - k_{z\pm}^2 c^2/\omega_{\pm}^2,$$

and

$$\alpha_{\pm} = \mp i(\beta_0 \pm \beta_{\pm})\cos\theta_{\pm} + (1 \pm \beta_0\beta_{\pm})\sin\theta_{\pm},$$

with $\beta_{\pm} = \omega_{pe}^2 \omega_0/k_{\pm}^2 c^2 \omega_{ce}$, $\cos\theta_{\pm} = \hat{\mathbf{e}}_x \cdot \hat{\mathbf{e}}_{x\pm}$, $\hat{\mathbf{e}}_{x\pm} = \mathbf{k}_{\perp\pm}/|k_{\perp\pm}|$, $E_0(+) = E_0$ and $E_0(-) = E_0^*$. Furthermore, the subscript x denotes the x component of the corresponding quantities, and $\delta n_A (\ll n_0)$ is the quasi-neutral density perturbation associated with the DA waves. The latter accompany electromagnetic fields of the form $\mathbf{E}_A = -\nabla\phi_A - c^{-1}\partial_t A_z \hat{\mathbf{z}}$ and $\mathbf{B}_A = -\hat{\mathbf{z}} \times \nabla A_z$, where ϕ is the scalar potential, and A_z is the $z-$ component of the vector potential. The compressional magnetic field perturbation of the DA waves has been neglected in view of the low-β approximation.

The dynamics of the DA waves in the presence of the whistler ponderomotive force is governed by

$$\partial_t \delta n_A + \frac{c}{4\pi e}\partial_z \nabla_{\perp}^2 A_z = 0, \qquad (21)$$

$$\partial_t \left(1 - \lambda_e^2 \nabla_{\perp}^2\right) A_z + c\partial_z \left(\phi_A - \frac{T_e}{e}\frac{\delta n_A}{n_0}\right) = -c\partial_z \psi, \qquad (22)$$

and

$$\left(\partial_t^2 + \omega_{ci}^2\right)\delta n_A - \frac{n_0 e}{m_i}\nabla_{\perp}^2 \phi_A = 0, \qquad (23)$$

which are the electron continuity equation, the parallel component of the electron momentum equation, and the ion continuity equation, respectively. In (21) and (22), we have replaced the electron fluid velocity v_{ez} of the DAW by $(c/4\pi n_0 e)\nabla_{\perp}^2 A_z$. Furthermore, the whistler ponderomotive potential is

$$\psi = \frac{ic}{B_0 \omega_0}\left(E_{0x}E_{x-}\alpha_-^* - E_{0x}^* E_{x+}\alpha_+^*\right). \qquad (24)$$

Combining (21) to (23) we obtain

$$\partial_z^2 \left(\partial_t^2 + \omega_{ci}^2 - c_s^2 \nabla_{\perp}^2\right)\delta n_A - \frac{\omega_{pi}^2}{c^2}\partial_t^2(1 - \lambda_e^2 \nabla_{\perp}^2)\delta n_A = -\frac{n_0 e}{m_i}\partial_z^2 \nabla_{\perp}^2 \psi. \qquad (25)$$

Assuming that δn_A is proportional to $\exp(i\mathbf{k} \cdot \mathbf{r} - i\omega t)$, we find from (25)

$$\epsilon_A \delta n_A = -in_0 \frac{\omega_{ci}}{\omega_0} \frac{k_\perp^2 c^2}{B_0^2} \left(E_{0x} E_{x-} \alpha_-^* - E_{0x}^* E_{x+} \alpha_+^* \right), \qquad (26)$$

where

$$\epsilon_A = \omega^2 \left[1 + \left(1 + k_\perp^2 \lambda_e^2 \right) / k_z^2 \lambda_i^2 \right] - \omega_{ci}^2 - k_\perp^2 c_s^2$$

Substituting for $E_{x\pm}$ from (20) into (26), we obtain the desired dispersion relation

$$\epsilon_A = k_\perp^2 c^2 \frac{\omega_{pi}^2}{\omega_0^2} \left| \frac{E_{0x}}{B_0} \right|^2 \sum_{-,+} \frac{|\alpha_\mp|^2}{D_\mp} . \qquad (27)$$

We now examine (27) for the resonant decay instability. For this purpose, we ignore the upper-sideband D_+, as being off-resonant, and treat the low-frequency DA and the lower-sideband (D_-) as resonant normal modes. That is, we let $\omega = \omega_{DA} + i\gamma_A$ and $\omega_- = \omega_{DA} - \omega_0 + i\gamma_A = -\omega_w + i\gamma_A$, where $\omega_{DA} = \omega_{IA}(1 + k_\perp^2 \rho_s^2)^{1/2}/(1 + k_z^2 \lambda_i^2)^{1/2}$, $\omega_w = |k_z - k_-|c^2 \omega_{ce}/\omega_{pe}^2$, and γ_A is the growth rate. The dispersion relation (27) then reduces to

$$(\gamma_A + \Gamma_A)(\gamma_A + \Gamma_w) \left(\frac{\partial \epsilon_A}{\partial \omega} \right)_{\omega = \omega_{DA}} \left(\frac{\partial D_-}{\partial \omega_-} \right)_{\omega_- = -\omega_w} = -k_\perp^2 c^2 \frac{\omega_{pi}^2}{\omega_0^2} \left| \frac{E_{0x}}{B_0} \right|^2 |\alpha_-|^2 ,$$
$$(28)$$

where Γ_A and Γ_w are linear damping rates of the whistler wave and the dispersive Alfvén wave. It is easy to show that

$$\partial \epsilon_A / \partial \omega = 2\omega \left(1 + k_z^2 \lambda_i^2 + + k_\perp^2 \lambda_e^2 \right) / k_z^2 \lambda_i^2$$

and

$$\partial D_- / \partial \omega_- = -2\omega_{pe}^4 / \omega_w \omega_{ce}^2 k_-^2 c^2 .$$

Setting $\gamma_A = 0$ in (28), we obtain the threshold electric field

$$\left| \frac{cE_{0x}}{B_0 c_s} \right|^2 \geq \frac{4\omega_0^2 \omega_{DA} \omega_{pe}^4 \Gamma_A \Gamma_w \left(1 + k_z^2 \lambda_i^2 + k_\perp^2 \lambda_e^2 \right)}{|\alpha_-|^2 \omega_{pi}^2 k_\perp^2 c_s^2 \omega_w \omega_{ce}^2 k_-^2 c^2 k_z^2 \lambda_i^2} . \qquad (29)$$

The growth rate above the threshold is

$$\gamma_A \approx k_\perp c \frac{|k_- c| \omega_{ce}}{2\omega_{pe}^2} \frac{\omega_{pi}}{\omega_0} \left(\frac{\omega_w}{\omega_{DA}} \right)^{1/2} \frac{k_z \lambda_i |\alpha_-| E_{0x}}{(1 + k_z^2 \lambda_i^2 + k_\perp^2 \lambda_e^2)^{1/2} B_0} . \qquad (30)$$

Next, we consider the excitation of long parallel wavelength (in comparison with λ_i) DA waves by large amplitude electrostatic lower-hybrid (LH) waves. For this purpose, we derive the 3D equations for nonlinearly coupled DA and LH waves that are propagating obliquely to an external magnetic field. We note that the phase velocity of the LHW is much larger than the electron and ion thermal velocities, and that the LHW frequency is smaller (larger) than the electron (ion) gyrofrequency. On the other hand, the parallel (to $\hat{z}$) phase velocity of the low-frequency (in comparison with the ion gyrofrequency ω_{ci}) DA waves can be either smaller or larger than the electron thermal velocity.

By substituting the appropriate fluid velocities for the LH and DA waves in the charge density conservation equation, we readily obtain the wave equation for LH waves

$$
\partial_t^2 \nabla_\perp \cdot \left[1 + \frac{\omega_{pe}^2}{\omega_{ce}^2} \left(1 + \frac{\delta n_A}{n_0} \right) \right] \nabla_\perp \phi_L + \partial_z \left[\partial_t^2 + \omega_{pe}^2 \left(1 + \frac{\delta n_A}{n_0} \right) \right] \partial_z \phi_L
$$

$$
+ \nabla \cdot \left[\omega_{pi}^2 \left(1 + \frac{\delta n_A}{n_0} \right) \cdot \nabla \phi_L \right] + \partial_t^2 \left[a_1 \nabla_\perp^4 + a_2 \nabla_\perp^2 \partial_z^2 + a_3 \partial_z^4 \right] \phi_L
$$

$$
= -\frac{\omega_{pe}^2}{\omega_{ce}} \partial_t \left(\hat{z} \times \nabla \phi_L \cdot \nabla \frac{\delta n_A}{n_0} \right) - \frac{c}{B_0} \partial_t \left(\hat{z} \times \nabla \phi_A \cdot \nabla \nabla_\perp^2 \phi_L \right)
$$

$$
- \frac{e}{m_e} \frac{\omega_{pi}^2}{\omega_{ci}^2} \partial_t \nabla_\perp \cdot \left[(\partial_t \nabla_\perp \phi_A) \partial_t^{-2} \nabla^2 \phi_L \right] + \frac{\omega_{pe}^2}{B_0} \partial_t \left[\hat{z} \times \nabla A_z \cdot \nabla \partial_t^{-1} \partial_z \phi_L \right]
$$

$$
- \frac{c\omega_{pe}^2}{B_0 \omega_{ce}^2} \partial_t \nabla \cdot \left(\hat{z} \times \nabla \phi_A \cdot \nabla \nabla_\perp \phi_L + \hat{z} \times \nabla \phi_L \cdot \nabla \nabla_\perp \phi_A \right), \qquad (31)
$$

where ϕ_L is the electrostatic potential of the LHW, $\delta n_A = n_0 (c/B_0 \omega_{ci}) \nabla_\perp^2 \phi_A$ is the quasi-neutral density perturbation associated with the DA waves, and a_1, a_2, a_3 are the coefficients of the thermal dispersion of the LH waves.

On the other hand, in a dense magnetoplasma with $\omega_{pi} \gg \omega_{ci}$, the appropriate equations for the nonlinear DA waves in the presence of the ponderomotive force of the LH waves are

$$
d_t \nabla_\perp^2 \phi_A + \frac{v_A^2}{c} d_z \nabla_\perp^2 A_z = \frac{c m_e}{B_0 m_i} < \hat{z} \times \nabla \phi_L \cdot \nabla \nabla_\perp^2 \phi_L^* > + \text{compl. conj.}
$$

$$
- \frac{e}{m_i \omega_L^2} \partial_t \nabla_\perp^2 |\nabla \phi_L|^2, \qquad (32)
$$

and

$$
d_t \left(1 - \lambda_e^2 \nabla_\perp^2 \right) A_z + c \partial_z \phi_A - c \rho_s^2 d_z \nabla_\perp^2 \phi_A
$$

$$
= i \frac{c^2}{B_0 \omega_L} \left[< \hat{z} \times \nabla \phi_L^* \cdot \nabla \partial_z \phi_L > \right] - \text{compl. conj.} + \frac{ec}{2 m_e \omega_L} \partial_z |\partial_z \phi_L|^2, \qquad (33)
$$

where $d_t = \partial_t + (c/B_0) \hat{z} \times \nabla \phi_A \cdot \nabla$, $d_z = \partial_z - B_0^{-1} \hat{z} \times \nabla A_z \cdot \nabla$, $\omega_L \approx \left[\omega_{pi} / \left(1 + \omega_{pe}^2 k_\perp^2 / \omega_{ce}^2 k^2 \right)^{1/2} \right] \left(1 + m_i k_z^2 / m_e k^2 \right)^{1/2}$ is the LH wave frequency, and the asterisk stands for the complex conjugate. The angular bracket denotes the ensemble average over the LH wave period.

Equations (31) to (33) are the desired equations for studying the parametric excitation of DA waves by LH waves. They can be analytically as well as numerically analyzed in order to investigate three-wave decay and modulational interactions.

4   Nonlinear effects caused by inertial Alfvén waves

Large amplitude dispersive Alfvén waves nonlinearly interact with the background plasma as well as among themselves. In this section, we present an investigation of the nonlinear effects produced by finite amplitude DIA waves interacting with the plasma slow motions in a uniform magnetoplasma. Specifically, here we focus on the ponderomotive force of the DIA waves that creates magnetic field aligned quasi-stationary (supersonic) density humps (cavities), and examine the parametric instabilities of a coherent DIAW. Furthermore, it will also be shown that the nonlinear DIA waves can propagate in the form of a dark (bright) envelope soliton when the DIWs are modulated by quasi-stationary (finite frequency) electrostatic perturbations.

We consider the nonlinear propagation (in the $x-z$ plane) of low-frequency (in comparison with the ion gyrofrequency), long parallel wavelength (in comparison with the collisionless ion-skin depth) finite amplitude DIA waves in a magnetoplasma. In the DIAW fields, the perpendicular (denoted by the subscript $\perp$) and the parallel (denoted by the subscript z) components of the electron and ion fluid velocities are given by, respectively,

$$\mathbf{v}_{e\perp} \approx (c/B_0)\mathbf{E}_\perp \times \hat{\mathbf{z}}, \tag{34}$$

and

$$\mathbf{v}_{i\perp} \approx (c/B_0)\mathbf{E}_\perp \times \hat{\mathbf{z}} + (c/B_0\omega_{ci})\partial_t\mathbf{E}_\perp. \tag{35}$$

On the other hand, the parallel component of the electron and ion fluid velocities in the parallel electric field of the dispersive IAW are determined from

$$\partial_t v_{ez} = -eE_z/m_e, \tag{36}$$

and

$$\partial_t v_{iz} = eE_z/m_i. \tag{37}$$

The y component of the IAW magnetic field B_y and the IAW electric fields are related by Faraday's law

$$\partial_t B_y = c\partial_x E_z - c\partial_z E_x. \tag{38}$$

We now derive the relevant dynamical equation for the DIA waves, taking into account the nonresonant electrostatic density perturbations δn_s that are created by the ponderomotive force of the DIA waves. For this purpose, we combine the parallel component of the Ampère law with (36) to obtain

$$\partial_{tx}^2 B_y = \frac{\omega_{pe}^2}{c}\left(1 + \frac{\delta n_s}{n_0}\right)E_z, \tag{39}$$

On the other hand, by inserting (34) and (35) into the conservation of the current density equation (viz. $\nabla \cdot \mathbf{J} = 0$) and eliminating the parallel component of the plasma current density from the Ampère law, we have

$$\partial_t E_x = -\frac{v_A^2}{c}\left(1 - \frac{\delta n_s}{n_0}\right)\partial_z B_y, \tag{40}$$

where $\delta n_s/n_0 \ll 1$. For large density perturbations, i. e. $\delta n_s/n_0 > 0.2$, one could extend the validity of (40) by replacing $1 - \delta n_s/n_0$ by $(1 + \delta n_s/n_0)^{-1}$. Eliminating E_x from (38) and (40) we find (Shukla $et\ al$, 1999a)

$$\partial_t^2 B_y - c\partial_{tx}^2 E_z = v_A^2\left(1 - \frac{\delta n_s}{n_0}\right)\partial_z^2 B_y. \tag{41}$$

Combining (39) and (41) we finally obtain the DIAW equation

$$\left(\partial_t^2 - v_A^2\partial_z^2 - \lambda_e^2\partial_x^2\partial_t^2\right)B_y + \frac{\delta n_s}{n_0}\left(v_A^2\partial_z^2 + \lambda_e^2\partial_x^2\partial_t^2\right)B_y = 0, \tag{42}$$

where the last term in the left-hand side of (42) arises from the nonlinear coupling between the DIAWs and the slow density perturbations that are produced by the DIAW driving force. We note that (42) accounts for small amplitude density changes that are created by the ponderomotive force of the DIAWs. It is not valid for large density perturbations.

The parallel component of the driving force of the dispersive DIAWs can be calculated by averaging the parallel component of the convective $m_j\mathbf{v}_j \cdot \nabla v_{jz}$ and Lorentz force $(q_j/c)\left(\mathbf{v}_{j\perp} \times \mathbf{B}_\perp\right)_z$ terms over the IAW period $2\pi/\omega$, where $q_i = e$ and $q_e = -e$. The ponderomotive force of the DIAW produces a space charge ambipolar potential φ and density perturbations $\delta n_{js} = n_{js} - n_0$. The magnetic field-aligned force balance equation for non-resonant electrostatic disturbances in the presence of the driving force of the DIAW is

$$m_j\partial_t v_{js} + m_j\left\langle \mathbf{v}_{j\perp}^* \cdot \nabla v_{jz} + v_{jz}^*\partial_z v_{jz}\right\rangle$$

$$-\frac{q_j}{c}\left\langle \hat{\mathbf{z}} \cdot \left(\mathbf{v}_{j\perp}^* \times \mathbf{B}_\perp\right)\right\rangle = -q_j\partial_z\varphi - \frac{T_j}{n_{js}}\partial_z n_{js}, \tag{43}$$

where v_{js} is the magnetic field-aligned particle velocity involved in the nonresonant electrostatic disturbances, and the angular bracket denotes averaging over the DIAW period. The bracket terms in the left-hand side represents the ponderomotive force, and the asterisk the complex conjugate.

For low parallel phase velocity (in comparison with the electron thermal velocity) electrostatic waves, we can neglect the parallel electron inertial force in (43). Adding the inertialess electron and inertial ion versions of (43) under the quasi-neutrality approximation $n_{es} = n_{is} \equiv n_s$, we obtain

$$m_i\partial_t v_{is} + \frac{i}{n_0\omega}\left\langle \mathbf{J}_\perp^* \cdot \nabla E_z\right\rangle - \frac{1}{n_0 c}\left\langle \hat{\mathbf{z}} \cdot \left(\mathbf{J}_\perp^* \times \mathbf{B}_\perp\right)\right\rangle$$

$$+\frac{e^2}{4m_e\omega^2}\partial_z\left\langle|E_z|^2\right\rangle = -T\partial_z\ln\, n_s, \tag{44}$$

where $T = T_e + T_i$ and the perpendicular component of the plasma current density is given by $\mathbf{J}_\perp = -i(n_0ec\omega/B_0\omega_{ci})\mathbf{E}_\perp$. Inserting this expression for $\mathbf{J}_\perp$ into the left-hand side of (44), we obtain the resulting formula for the quasi-stationary $(v_{is} = 0)$ density that is modified by the ponderomotive force of the DIA waves

$$n_s \approx n_0 \exp[\frac{e^2}{4m_iTw_{ci}^2}\left\langle|\mathbf{E}_\perp|^2\right\rangle - \frac{e^2}{4m_eTw^2}\left\langle|E_z|^2\right\rangle]. \tag{45}$$

We note that Bellan and Stasiewicz (1998) overlooked the importance of the density hump contribution [the $|\mathbf{E}_\perp|^2$-term in Eq. (45)], which arises from the combination of the ion advection and Lorentz ion forces, and suggested that the parallel component of the electron ponderomotive potential [the $|E_z|^2$-term in Eq. (45)] is responsible for the magnetic field-aligned density cavities. However, for DIAWs (with $\omega \ll \omega_{ci}$ and $k_z\lambda_i \ll 1$) we have $\mathbf{E}_\perp = (k_z\omega_{ce}\omega_{ci}/\omega^2k_\perp^2)E_z\mathbf{k}_\perp$, which inserted in (45) yields (Shukla et al., 1999a)

$$n_s = n_0 \exp\left(e^2 < |E_z|^2 > /4m_eTw^2k_\perp^2\lambda_e^2\right) \equiv n_0 \exp\left(< |B_y|^2 > /16\pi n_0T\right). \tag{46}$$

Thus, the ponderomotive force of the DIA waves can create only magnetic field-aligned quasi-stationary density humps. When the wave magnetic energy density is much smaller than the thermal energy density of the plasma, we obtain from (46)

$$\delta n_s = < |B_y|^2 > /16\pi T, \tag{47}$$

where $\delta n_s = (n - n_0) \ll n_0$.

On the other hand, inclusion of the parallel ion inertial force in (43) leads to driven ion-acoustic waves

$$\left(\partial_t^2 - c_s^2\partial_z^2\right)\delta n_s = -\frac{c_s^2}{16\pi T}\partial_z^2 < |B_y|^2 >, \tag{48}$$

where the perpendicular wavelength of the driven ion-acoustic waves is assumed to be much larger than the ion gyroradius at the electron temperature. For $\partial_t^2 \ll c_s^2\partial_z^2$, Eq. (48) reduces to (47). Equation (48) in addition can depict supersonic density cavities produced by the DIAW ponderomotive force. Equations (42) and (47) [Eq. (48)] are our desired set for studying the nonlinear propagation of DIAWs in the presence of quasi-stationary [nonstationary] electrostatic density perturbations.

In the following, we consider the parametric instabilities of DIAWs. For this purpose, we decompose the DIAW field as $B_y = B_{y0}\exp(i\mathbf{k}_0\cdot\mathbf{r} - i\omega_0t)$ + compl. conj. + $B_{y\pm}\exp(i\mathbf{k}_\pm\cdot\mathbf{r} - i\omega_\pm t)$, where $B_{y0}(B_{y\pm})$ is the magnetic field of the DIAW pump (sidebands), $\omega_0 = k_{z0}v_A/(1 + k_{x0}^2\lambda_e^2)^{1/2}$ the pump frequency, $\omega_\pm = \Omega\pm\omega_0$, $\mathbf{k}_\pm = \mathbf{K}\pm\mathbf{k}_0$, and $\Omega(\mathbf{K})$ the frequency (wavevector) of the low-frequency electrostatic oscillations. Thus, (42) and (48) can be Fourier transformed and combined to yield the nonlinear dispersion relation

$$\Omega^2 - K_z^2 c_s^2 = K_z^2 c_s^2 \omega_0^2 \frac{|B_{y0}|^2}{16\pi n_0 T} \sum_{+,-} \frac{1}{D_{A\pm}}, \tag{49}$$

where $D_{A\pm} = \omega_\pm^2(1 + k_{x\pm}^2 \lambda_e^2) - k_{z\pm}^2 v_A^2$. It can be shown that for $K_x \ll k_{x0}$, the latter takes the form $\pm 2\omega_0(1 + k_{x0}^2 \lambda_e^2)(\Omega - K_z v_{g0} \mp \delta)$, where $v_{g0} = k_{z0} v_A^2 / \omega_0 (1 + k_{x0}^2 \lambda_e^2)$ is the group velocity of the pump and $\delta = K_z^2 v_A^2 / 2\omega_0 (1 + k_{x0}^2 \lambda_e^2)$ a small frequency shift arising from the nonlinear interaction.

Equation (49) can be analyzed for three-wave decay and modulational interactions. For the three-wave decay interaction, we can assume D_{A-} to be resonant and ignore the upper sideband D_{A+}, which is off-resonant. Letting $\Omega = K_z c_s + i\gamma_i$ and $K_z v_{g0} - \delta \approx K_z c_s$, we obtain the growth rate γ_i from (49)

$$\gamma_i = (K_z c_s \omega_0)^{1/2} |B_{y0}| / 8[(1 + k_{x0}^2 \lambda_e^2)\pi n_0 T]^{1/2}. \tag{50}$$

On the other hand, for the modulational interaction, both the upper and lower sidebands are resonant (viz. $D_{A\pm} = 0$, whereas the low-frequency perturbations are non-resonant. Here, (49) takes the form

$$(\Omega^2 - K_z^2 c_s^2)[(\Omega - K_z v_{g0})^2 - \delta^2] = K_z^4 c_s^2 v_A^2 |B_{y0}|^2 / 32\pi n_0 T (1 + k_{x0}^2 \lambda_e^2)^2. \tag{51}$$

Equation (51) can be analyzed in two limiting cases. First, for $\Omega \ll K_z c_s$, we have

$$\Omega = K_z v_{g0} \pm \left[\delta^2 - \frac{K_z^2 v_A^2 |B_{y0}|^2}{32\pi n_0 T (1 + k_{x0}^2 \lambda_e^2)^2}\right]^{1/2}, \tag{52}$$

which predicts an oscillatory instability when

$$|B_{y0}|^2 > 8\pi n_0 T K_z^2 (1 + k_{x0}^2 \lambda_e^2) / k_{0z}^2$$

. The maximum growth rate of that instability is

$$K_{z0} v_A |B_{y0}| / \sqrt{32\pi n_0 T}(1 + k_{x0}^2 \lambda_e^2)$$

. Second, for $\Omega^2 \gg K_z^2 c_s^2$ and $\Omega \gg K_z v_{g0}$, Eq. (51) reduces to

$$\Omega^2 = (\delta^2/2) \pm \left[(\delta^4/4) + K_z^4 v_A^2 c_s^2 |B_{y0}|^2 / 32\pi n_0 T (1 + k_{x0}^2 \lambda_e^2)^2\right]^{1/2}. \tag{53}$$

Equation (53) also admits an oscillatory instability.

Next, we consider the nonlinear evolution of the modulated DIAW packets. We assume that the DIAW magnetic field changes slowly due to the nonlinear interaction with electrostatic density perturbations. Accordingly, introducing the concept of two time and space scales and writing B_y as $\tilde{B}_y(x, Z) \exp(ik_z z + ik_x x - i\omega t)$, where $\omega = k_z v_A / (1 + k_x^2 \lambda_e^2)^{1/2}$, we obtain from (42) and (47) the cubic nonlinear Schrödinger equation (Shukla *et al.*, 1999a)

$$-2i\frac{k_z v_A^2}{\omega^2}\partial_Z \tilde{B}_y + \lambda_e^2 \partial_x^2 \tilde{B}_y - \frac{<|\tilde{B}_y|^2>}{16\pi n_0 T}\tilde{B}_y = 0, \qquad (54)$$

where $\partial_Z \tilde{B}_y \ll k_z \tilde{B}_y$.

The convective stability of a constant amplitude (B_{y0}) dispersive IAW packet against magnetic field-aligned quasi-stationary density perturbations can be investigated from (54) by standard techniques. The dispersion relation for convective stability, for our purposes, reads

$$K_z^2 = \frac{K_\perp^2}{4}\left(K_\perp^2 + |B_{y0}|^2\right), \qquad (55)$$

where the parallel and perpendicular wavenumbers, which are denoted by K_z and $K_\perp$, respectively, of the quasi-stationary density modulations are normalized by $\omega^2/2k_z v_A^2$. Furthermore, the magnetic energy density of the pump $|B_{y0}|^2/8\pi$ is in units of $n_0 T$. It follows from (55) that the DIA waves are stable.

Let us now discuss the asymptotic state of the spatially modulated DIA wave packet. For this purpose, we seek a solution of (54) in the form (Hasegawa, 1975)

$$\tilde{B}_y = \sqrt{G(x, z_0)}\exp\left(\frac{i}{2\lambda_e}\int_0^x \sigma(x', z_0)dx'\right), \qquad (56)$$

where

$$G(x, z_0) = G_0\left[1 - \gamma_0 \mathrm{sech}^2\left(x/x_0\right)\right], \qquad (57)$$

and $z_0 = Z\omega^2/k_z v_A^2$, $x_0 = 8\sqrt{\pi n_0 T}\lambda_e/\sqrt{\gamma_0 G_0}$. Here, G_0 is the value of G when $|x| \to \infty$, σ is a phasor, and γ_0 is an arbitrary constant. This type of envelope soliton solution, given by (56), is called a dark soliton (referring to nonlinear optics). The latter can be ducted in self-created density channels over long distances.

Furthermore, in order to study the long term spatio-temporal evolution of (42) and (48), we must invoke the slowly varying envelope approximation on (42) and solve the pair by means of a numerical scheme. It is likely that the coupled equations admit stationary solutions in the form of supersonic density cavities which may trap spatially localized DIA magnetic fields.

5   Amplitude modulation of kinetic Alfvén waves

Nearly two decades ago, Mikhailovskii *et al.* (1977) pointed out possible existence of kinetic Alfvén wave (KAW) envelope solitons due to the amplitude modulation of a coherent KAW by quasi-stationary density and magnetic field perturbations across $B_0\hat{z}$ in a plasma in which the electron and ion temperatures are equal. Here, we relax the latter approximation and present a general description of the nonlinear propagation and modulation/filamentation of the KAW (Shukla *et al.*, 1998a). In addition, we discuss self-focusing of the KAW and cylindrically-symmetric kinetic Alfvén envelope solitons.

The nonlinear interaction of a finite amplitude KAW with quasi-stationary density and compressional magnetic field perturbations gives rise to an envelope of waves. Employing the eikonal operator representation, which implies letting $\omega = \omega_0 + i\partial_t$ and $\mathbf{k} = \mathbf{k}_0 - i\nabla$ in Eq. (5), where $\omega_0(\mathbf{k}_0)$ is the pump frequency (wavevector), and applying the WKB approximation, viz. $\partial_t B_\perp \ll \omega_0 B_\perp$, we find that the envelope evolves according to

$$i\left(\frac{\partial}{\partial t} + \mathbf{v}_g \cdot \nabla\right) B_\perp + \frac{\Omega_{A0}^2 \alpha \rho_s^2}{2\omega_0}\nabla_\perp^2 B_\perp - \frac{\Omega_{A0}^2}{2\omega_0}\frac{\delta B_z}{B_0}B_\perp = 0; \qquad (58)$$

here $\mathbf{v}_g = \partial\omega_0/\partial\mathbf{k}_0$ is the group velocity, $\omega_0^2 = \Omega_{A0}^2(1 + \alpha k_{\perp 0}^2\rho_s^2)$, $\Omega_{A0} = k_{0z}v_A$, $\alpha = 1 + 3T_i/4T_e$, and $B_\perp$ the perpendicular (to $\hat{z}$) component of the KAW magnetic field. In Eq. (58) we have used the frozen-in-field concept and have thus set $n_i/B = $ constant. We have also let $n_i = n_0 + \delta n$ and $B = B_0 + \delta B_z$, where $\delta B_z(\ll B_0)$ is the quasi-stationary compressional magnetic field triggered by the KAW magnetic field pressure. Furthermore, we have also employed the relation $\delta n/n_0 \approx \delta B_z/B_0$.

The magnetic field aligned quasi-stationary magnetic field perturbation δB_z is given by

$$\frac{\delta B_z}{B_0} = -\frac{|B_\perp|^2}{2(1 + 2\beta)B_0^2}. \qquad (59)$$

Combining Eqs. (58) and (59) we obtain

$$i\left(\frac{\partial}{\partial t} + \mathbf{v}_g \cdot \nabla\right) b_\perp + \frac{P}{2}\nabla_\perp^2 b_\perp + Q|b_\perp|^2 b_\perp = 0, \qquad (60)$$

where $b_\perp = B_\perp/B_0$, $P = \Omega_{A0}^2 \alpha \rho_s^2/\omega_0$, and $Q = \Omega_{A0}^2/4\omega_0(1 + 2\beta)$.

Equation (60) is the standard cubic nonlinear Schrödinger equation, which predicts modulation/filamentation instability of a constant amplitude KAW pump, as well as self-focusing of the KAW.

The nonlinear dispersion relation for the modulational/filamentation instability of a constant amplitude $(B_{\perp 0})$ KAW pump is derived from Eq. (60) by decomposing $b_\perp$ as the sum of the pump and the two sidebands. The standard procedure gives

$$(\Omega - \mathbf{K}\cdot\mathbf{v}_g)^2 = PK_\perp^2(\frac{1}{4}PK_\perp^2 - Q|b_0|^2), \qquad (61)$$

where $\Omega(\mathbf{K})$ is the frequency (wavenumber) of the quasi-stationary modulation, $\mathbf{K} = \hat{z}K_z + \mathbf{K}_\perp$, and $b_0 = B_{\perp 0}/B_0$. Letting $\Omega = \mathbf{K}\cdot\mathbf{v}_g + i\gamma$ in Eq. (61), we observe that the growth rate γ of the modulational instability is positive for $Q|b_0|^2 > PK_\perp^2/4$.

On the other hand, stationary filamentation instability can be investigated on the basis of the nonlinear dispersion relation

$$K_z^2 = \frac{PK_\perp^2}{4v_A^2\alpha}(PK_\perp^2 - 4Q|b_0|^2),\tag{62}$$

where v_{gz} is the z component of the group velocity of the KAW. Equation (62) follows from Eq. (61) if we set $\Omega = 0$ and assume that $K_z v_{gz} \gg \mathbf{K}_\perp \cdot \mathbf{v}_{g\perp}$. Setting $K_z = -iK_m(K_m > 0)$ in Eq. (62), we see that convective amplification occurs for $|b_0|^2 > PK_\perp^2/4Q$. The mode number of the most unstable wave is $K_m = (2Qb_0/P)^{1/2}$ and the corresponding spatial amplification rate is $K_i = Q|b_0|^2/v_{gz}$. The minimum scalelength over which the KAW filamentation occurs is $2\pi/K_i$.

We now consider the time independent stationary propagation of KAW along the z axis. Thus, we seek the solution of Eq. (60) in the form $b_\perp = b_\perp(r)\exp(i\lambda z)$ and assume the steady state ($\partial_t B_\perp = 0$), where λ is a constant. Thus, for cylindrically-symmetric KAW beams, we have

$$\frac{1}{r}\frac{d}{dr}\left(r\frac{db_\perp}{dr}\right) - \frac{2}{P}(v_{gz}\lambda - Qb_\perp^2)b_\perp = 0.\tag{63}$$

For $\lambda > 0$, Eq. (63) admits cylindrical soliton solutions.

Furthermore, in the steady state Eq. (60) can be cast in the form

$$i\partial_\xi b_\perp + \frac{1}{2\rho}\frac{\partial}{\partial\rho}(\rho\frac{\partial b_\perp}{\partial\rho}) + D|b_\perp|^2 b_\perp = 0,\tag{64}$$

where $\xi = z\Omega_{A0}^2\alpha/v_{gz}\omega_0, \rho = r/\rho_s$ and $D = 1/2(1+2\beta)\alpha$. Equation (64) has been investigated in detail in the context of self-focusing of laser beams. For a Gaussian beam $|b_\perp|^2 = b_{\perp 0}^2\exp(-\rho^2/a^2)$, the threshold for the self-focusing is $Da^2b_{\perp 0}^2 > 2$, where a is the normalized (by ρ_s) beam width.

6   Modulation of EMICA waves

In the past, Rogister (1971) derived a derivative nonlinear Schrödinger (DNLS) equation for a modulated dispersive Alfvén wave packet that is propagating along the external magnetic field. Many works related to the DNLS have appeared in the literature (Mio et al., 1976; Kennel et al., 1988; Verheest, 1990; Hada, 1993; Passot et al., 1994; Shevchenko, 1995; Nocera and Buti, 1996). On the other hand, attempts (Shukla and Stenflo, 1985; Shukla, Feix and Stenflo, 1988) have been made to investigate the modulation of high-frequency electromagnetic ion-cyclotron Alfvén (EMICA) waves by a spectrum of non-resonant very low-frequency electrostatic/electromagnetic perturbations. This interaction gives rise to amplitude modulation of the EMICA waves, which are coupled with low- frequency disturbances due to the EMICA ponderomotive force. In the WKB approximation (viz. $|\partial_t \mathbf{E}| \ll \omega |\mathbf{E}|$), the EMICA amplitudes evolve according to the nonlinear Schrödinger equation (Shukla, Feix and Stenflo, 1988)

$$i\left(\partial_t + v_g \partial_z\right) E + \frac{S_z}{2}\partial_z^2 E + \frac{S_\perp}{2}\nabla_\perp^2 E - \Delta E = 0, \qquad (65a)$$

where $\mathbf{E} = E(\mathbf{r},t)(\hat{\mathbf{x}} - i\hat{\mathbf{y}})\exp(ikz - i\omega t)$, $E(\mathbf{r},t)$ is the electric field envelope, and ω is given by Eq. (2). For EMICA waves, the group velocity v_g and the coefficients of the parallel and perpendicular group dispersions are, respectively,

$$v_g = -\frac{2kv_A^2}{\omega}\frac{(\omega - \omega_{ci})^2}{\omega_{ci}(\omega - 2\omega_{ci})},$$

$$S_z = \left[1 + \frac{v_g^2}{v_A^2}\frac{\omega_{ci}^3}{(\omega - \omega_{ci})^3}\right]\frac{v_g}{k},$$

and

$$S_\perp \approx \frac{v_g}{2k}.$$

The nonlinear frequency shift is

$$\Delta \approx -\frac{kv_g}{2}N + \omega b + k v_{zs}, \qquad (65b)$$

where $N = \delta n_s/n_0$ and $b = \delta B_{zs}/B_0$ are the relative density and compressional magnetic field perturbations of the plasma slow motion, and

$$v_{zs} = -\partial_t \partial_z^{-1}(N - b). \qquad (65c)$$

The dynamics of the plasma slow motion in the presence of the ponderomotive force of the EMICA waves is governed by

$$\partial_t^2(N - b) - v_s^2\partial_z^2 N = \frac{e^2}{m_i^2\omega_{ci}(\omega - \omega_{ci})}\left(\partial_z^2 + \frac{2}{v_g}\partial_t\partial_z\right)|E|^2, \qquad (66)$$

and

$$\left(\partial_t^2 - v_A^2\nabla^2\right)b - v_s^2\nabla_\perp^2 N = \frac{e^2}{m_i^2(\omega - \omega_{ci})^2}\nabla_\perp^2|E|^2. \qquad (67)$$

The one-dimensional electrostatic modulation of EMICA waves has been considered by Shukla and Stenflo (1985). In that case, (65) and (66) with $b = 0$ constitute a coupled set for studying the modulational instability (Shukla et al., 1986a; Brodin and Stenflo, 1988). On the other hand, (65) to (67) are useful for the investigation of time-dependent filamentation instabilities and the dynamical evolution of EMICA waves on account of their

nonlinear coupling with the MHD perturbations. We note that the filamentation instability of EMICA waves by quasi-stationary perturbations has been
considered by Shukla and Stenflo (1989a). The dispersion relation is

$$K_z^2 = S_\perp K_\perp^2 (S_\perp K_\perp^2 - 4Q|E_0|^2)/4v_g^2, \qquad (68)$$

where $\mathbf{K} = (K_x, K_y, K_z)$ is the wave vector of the stationary perturbation,
$Q = (1 + 2\beta)kc^2/\beta B_0^2 v_A$, $\beta = 8\pi n_0(T_e + T_i)/B_0^2$, and $|E_0|$ is the constant
amplitude of the EMICA pump. The spatial amplification rate is obtained
by letting $K_z = -iK_m (K_m > 0)$ in (68). The minimum spatial scale length
$L(= 2\pi/K_m)$ for $K_\perp^2 = (2Q/S_\perp)|E_0|^2$ is $L \approx 2\pi\beta v_A^2 B_0^2/c^2|E_0|^2 k(1 + 2\beta)$.

Let us finally mention that EMICA envelope solitons may propagate in
the z direction, as $S_z < 0$ for EMICA waves near $|\omega_{ci}|$. The stationary mov-
ing solutions of (65) and (66) with $\nabla_\perp = 0$ and $b = 0$, viz. when the one-
dimensional EMICA waves are modulated by one-dimensional ion-acoustic
perturbations, can be represented as supersonic envelope solitons consist-
ing of localized EMICA wave packets accompanied by compressional density
perturbations. Similar conclusions also hold for left-hand circularly polar-
ized dispersive Alfvén waves given by $\omega \approx kv_A(1 - kv_A/2|\omega_{ci}|)$ in which
$S_z = -v_A^2/|\omega_{ci}|$ and $\Delta \approx kv_A|E|^2/2B_0^2(v_A^2 - v_s^2)$, provided that β is sufficiently low.

7   Self-interaction between dispersive Alfvén waves in nonuniform plasmas

In this section, we consider the nonlinear propagation of low-frequency
(in comparison with the ion gyrofrequency), long wavelength electromagnetic
waves in a nonuniform multi-component magnetoplasma with charged dust
impurities. It has been shown (Pokhotelov et al., 1999) that the presence
of charged impurities provides a possibility of linear as well as nonlinear
couplings between the drift-Alfvén and the Shukla-Varma mode (Shukla and
Varma, 1993). Furthermore, the latter modifies the theory of drift-Alfvén
vortices (Petviashvili and Pokhotelov, 1992) in that the dust density gradient
causes a complete localization of the electromagnetic drift-Alfvén vortex, in
addition to introducing a bound on the vortex speed.

7.1   Derivation of the nonlinear equations

We consider a nonuniform multicomponent plasma whose constituents are
electrons, singly charged positive ions and charged dust impurities immersed
in an external magnetic field $B_0\hat{z}$. The plasma density is assumed to be inhomogeneous along the x-axis. The charged dust impurities are treated as
point charges and their sizes, as well as the inter-grain spacing, are assumed
to be much smaller than the characteristic scale lengths (viz. the electron

skin length, gyroradii, etc). The dust particles are also considered as stationary due to the very large dust grain mass. The wave frequency is assumed to be small in comparison to the proton cyclotron frequency ω_{ci}, and the characteristic perpendicular wavelength is larger than the ion gyroradius.

The equilibrium plasma state satisfies the quasi-neutrality condition, i.e. $n_{i0} = n_{e0} + Z_d n_{d0}$, where n_{j0} is the unperturbed number density of particle species j (j equals e for the electrons, i for the ions, and d for the negatively charged dust grains) and Z_d is the constant number of charges residing on the dust grain surface.

In the electromagnetic fields, the electron and ion fluid velocities are given by

$$\mathbf{v}_e \approx \frac{c}{B_0} \mathbf{E}_\perp \times \hat{\mathbf{z}} - \frac{cT_e}{eB_0 n_e} \hat{\mathbf{z}} \times \nabla n_e + v_{ez} \left(\hat{\mathbf{z}} + \frac{\mathbf{B}_\perp}{B_0} \right), \tag{69}$$

and

$$\mathbf{v}_i \approx \frac{c}{B_0} \mathbf{E}_\perp \times \hat{\mathbf{z}} + \frac{cT_i}{eB_0 n_i} \hat{\mathbf{z}} \times \nabla n_i$$

$$+ \frac{c}{B_0 \omega_{ci}} (\partial_t + \mathbf{v}_i \cdot \nabla) \mathbf{E}_\perp, \tag{70}$$

where $\mathbf{E}_\perp = -\nabla_\perp \phi$ is the perpendicular component of the wave electric field, $\mathbf{B}_\perp = \nabla A_z \times \hat{\mathbf{z}}$ the perturbed magnetic field, ϕ the scalar potential, and A_z the parallel (to $B_0\hat{\mathbf{z}}$) component of the vector potential. The parallel component of the electron fluid velocity is given by

$$v_{ez} \approx \frac{c}{4\pi n_e e} \nabla_\perp^2 A_z, \tag{71}$$

where we have ignored the ion motion parallel to $\hat{\mathbf{z}}$, as well as neglected the compressional magnetic field perturbation. Thus, ion- acoustic and magnetosonic waves are decoupled in our low-β ($\beta \ll 1$) system.

Substituting (69) into the electron continuity equation, letting $n_j = n_{j0}(x) + n_{j1}$, where $n_{j1}(\ll n_{j0})$ is the particle number density perturbation, and using (71) we obtain

$$d_t n_{e1} - \frac{c}{B_0} \hat{\mathbf{z}} \times \nabla n_{e0} \cdot \nabla \phi + \frac{c}{4\pi e} d_z \nabla_\perp^2 A_z = 0, \tag{72}$$

where $d_t = \partial_t + (c/B_0)\hat{\mathbf{z}} \times \nabla \phi \cdot \nabla$ and $d_z = \partial_z + B_0^{-1} \nabla A_z \times \hat{\mathbf{z}} \cdot \nabla$. We have assumed that $(\omega_{pe}^2/\omega_{ce})|\hat{\mathbf{z}} \times \nabla \phi \cdot \nabla| \gg c\partial_z \nabla_\perp^2 A_z$.

On the other hand, substitution of the ion fluid velocity (70) into the ion continuity equation yields

$$d_t n_{i1} - \frac{c}{B_0} \hat{\mathbf{z}} \times \nabla n_{i0} \cdot \nabla \phi - \frac{cn_{i0}}{B_0 \omega_{ci}} (d_t + \mathbf{u}_{i*} \cdot \nabla) \nabla_\perp^2 \phi$$

$$- \frac{c^2 T_i}{eB_0^2 \omega_{ci}} \nabla_\perp \cdot [(\hat{\mathbf{z}} \times \nabla n_{i1}) \cdot \nabla \nabla_\perp \phi] = 0, \tag{73}$$

where $\mathbf{u}_{i*} = (cT_i/eB_0 n_{i0})\hat{z} \times \nabla n_{i0}$ is the unperturbed ion diamagnetic drift.

Subtracting (73) from (72) and assuming $n_{i1} = n_{e1}$ we obtain the modified ion vorticity equation

$$(d_t + u_{i*}\partial_y)\nabla_\perp^2 \phi + \frac{v_A^2}{c}d_z \nabla_\perp^2 A_z + \omega_{ci}\delta_d \kappa_d \partial_y \phi$$

$$+\frac{cT_i}{eB_0 n_{i0}}\nabla_\perp \cdot [(\hat{z} \times \nabla n_{e1}) \cdot \nabla\nabla_\perp \phi] = 0, \tag{74}$$

where $u_{i*} = (cT_i/eB_0 n_{i0})\partial n_{i0}/\partial x$, $v_A = B_0/(4\pi n_{i0}m_i)^{1/2}$ is the Alfvén velocity, $\delta_d = Z_d n_{d0}/n_{i0}$, and $\kappa_d = \partial ln\,(Z_d n_{d0}(x))/\partial x$. The term $\omega_{ci}\delta_d \kappa_d \partial_y \phi$ is associated with the Shukla-Varma mode.

By using (69) and (71), the parallel component of the electron momentum equation can be written as

$$(\partial_t + u_{e*}\partial_y)\,A_z - \lambda_e^2 d_t \nabla_\perp^2 A_z + cd_z \left(\phi - \frac{T_e}{en_{e0}}n_{e1}\right) = 0, \tag{75}$$

where $u_{e*} = -(cT_e/eB_0 n_{e0})\partial n_{e0}(x)/\partial x$ is the unperturbed electron diamagnetic drift.

Equations (72), (74) and (75) are the desired nonlinear equations for the coupled drift-Alfvén-Shukla-Varma modes in nonuniform magnetoplasmas with charged dust impurities. In a uniform cold plasma, they reduce to those derived earlier (Shukla et al., 1985).

7.2   The dispersion relation in an inhomogeneous plasma

In order to derive the local dispersion relation, we neglect the nonlinear terms in (72), (74), and (75) and suppose that n_{e1}, ϕ and A_z are proportional to $\exp(ik_y y + ik_z z - i\omega t)$, where $\mathbf{k} = k_y\hat{y} + k_z\hat{z}$ is the wave vector and ω the frequency. Then, in the local approximation, when the wavelength is much smaller than the scale length of the density gradient, we have from (72), (74), and (75)

$$\omega n_{e1} + \frac{cn_{e0}}{B_0}\kappa_e k_y \phi + \frac{c}{4\pi e}k_z k_y^2 A_z = 0, \tag{76}$$

$$(\omega - \omega_{i*} - \omega_{sv})\,\phi - \frac{k_z v_A^2}{c}A_z = 0, \tag{77}$$

and

$$\left[\omega_{e*} - (1 + k_y^2\lambda_e^2)\omega\right]A_z + k_z c\left(\phi - \frac{T_e}{en_{e0}}n_{e1}\right) = 0, \tag{78}$$

where $\kappa_e = \partial ln\, n_{e0}(x)/\partial x$, $\omega_{j*} = k_y u_{j*}$, and $\omega_{sv} = -\omega_{ci}\delta_d \kappa_d/k_y$ is the Shukla- Varma frequency of the dust-convective cells in dusty plasmas.

Combining (76) to (78) we obtain the linear dispersion relation

$$\left(\omega^2 - \omega\omega_m - \omega_{IA}^2 k_y^2\rho_{ss}^2\right)(\omega - \omega_{i*} - \omega_{sv}) = \omega_{IA}^2(\omega - \omega_{e*}), \tag{79}$$

where $\omega_m = \omega_{e*}/(1 + k_y^2\lambda_e^2)$ is the magnetic drift wave frequency, $\omega_{IA} = k_z v_A/(1 + k_y^2\lambda_e^2)^{1/2}$ the frequency of the inertial Alfvén waves, $\rho_{ss} = c_{ss}/\omega_{ci}$ the ion Larmor radius at the electron temperature, and $c_{ss} = (n_{i0}/n_{e0})^{1/2}c_s$ the ion-acoustic velocity in dusty plasmas.

Several comments are in order. First, in a homogeneous plasma, (79) correctly reproduces the spectra of the dispersive Alfvén waves, namely, $\omega = \omega_{IA}(1 + k_y^2\rho_{ss}^2)^{1/2}$. Second, for $\omega \gg \omega_m, \omega_{j*}$, we observe that the dispersive Alfvén waves are linearly coupled with the Shukla-Varma mode $\omega = \omega_{sv}$ (Shukla and Varma, 1993). Specifically, in a cold ($T_j \to 0$) dusty plasma we have from (79) $\omega^2 - \omega\omega_{sv} - \omega_{IA}^2 = 0$. The latter shows that coupling between the Shukla-Varma mode and the inertial Alfvén wave arises due to the consideration of the parallel electron dynamics in the electromagnetic fields. Third, when the perpendicular wavelength is much larger than λ_e, we obtain from (79) for $\omega \gg \omega_{i*}$

$$\left(\omega^2 - \omega\omega_{sv} - k_z^2 v_A^2\right)(\omega - \omega_{e*}) = k_z^2 v_A^2 k_y^2 \rho_{ss}^2 (\omega - \omega_{sv}), \tag{80}$$

which exhibits the coupling between the drift-Alfvén waves and the Shukla-Varma mode due to finite Larmor radius correction of the ions at the electron temperature.

7.3   Quasi-stationary dipolar vortices

Let us now consider stationary solutions of the nonlinear equations (72), (74) and (75), assuming that all the field variables depend on x and $\eta = y + \alpha z - ut$, where u is the translation speed of the vortex along the y-axis, and α the angle between the wave front normal and the (x, y) plane. Two cases are considered. First, in the stationary η-frame, (75) for $\lambda_e^2|\nabla_\perp^2| \ll 1$ can be written as

$$\hat{D}_A\left(\phi - \frac{T_e}{en_{e0}}n_{e1} - \frac{u - u_{e*}}{\alpha c}A_z\right) = 0, \tag{81}$$

where $\hat{D}_A = \partial_\eta + (1/\alpha B_0)[(\partial_\eta A_z)\partial_x - (\partial_x A_z)\partial_\eta]$. A solution of (81) is

$$n_{e1} = \frac{n_{e0}e}{T_e}\phi - \frac{n_{e0}e(u - u_{e*})}{\alpha c T_e}A_z. \tag{82}$$

Writing (72) in the stationary frame, and making use of (82) it can be put in the form

$$\hat{D}_A\left(\lambda_{De}^2\nabla_\perp^2 A_z + \frac{u(u - u_{e*})}{\alpha^2 c^2}A_z - \frac{u - u_{e*}}{\alpha c}\phi\right) = 0, \tag{83}$$

where $\lambda_{De} = v_{te}/\omega_{pe}$ is the electron Debye radius. A solution of (83) is

$$\lambda_{De}^2\nabla_\perp^2 A_z + \frac{u(u - u_{e*})}{\alpha^2 c^2}A_z - \frac{u - u_{e*}}{\alpha c}\phi = 0. \tag{84}$$

The modified ion vorticity equation (74) for cold ions can be expressed as

$$\hat{D}_\phi \left[\nabla_\perp^2 \phi + \frac{u_{c*}}{u}\phi \right] = 0, \tag{85}$$

where $\hat{D}_\phi = \partial_\eta - (c/uB_0)[(\partial_x\phi)\partial_\eta - (\partial_\eta\phi)\partial_x]$ and $u_{c*} = -c_{ss}\delta_d\kappa_d/\rho_{ss}$.
Combining (84) and (85) we obtain

$$\hat{D}_\phi(\nabla_\perp^2 \phi + \frac{p}{\rho_{ss}^2}\phi + \frac{u - u_{e*}}{\alpha c\rho_{ss}^2}A_z) = 0, \tag{86}$$

where $p = (c_{ss}/u)\left[\delta_d\kappa_d\rho_{ss} + (u_{e*} - u)/c_{ss}\right]$. A typical solution of (86) is

$$\nabla_\perp^2 \phi + \frac{p}{\rho_{ss}^2}\phi + \frac{u - u_{e*}}{\alpha c\rho_{ss}^2}A_z = C_1 \left(\phi - \frac{uB_0}{c}x \right), \tag{87}$$

where C_1 is an integration constant.
Eliminating A_z from (84) and (87), we obtain a fourth order inhomogeneous differential equation

$$\nabla_\perp^4 \phi + F_1\nabla_\perp^2 \phi + F_2\phi + C_1\frac{u^2(u - u_{e*})B_0}{\alpha^2 c^3 \lambda_{De}^2}x = 0, \tag{88}$$

where

$$F_1 = (p/\rho_{ss}^2) - C_1 + u(u - u_{e*})/\alpha^2 c^2\lambda_{De}^2, \tag{89a}$$

$$F_2 = (u - u_{e*})^2/\alpha^2 c^2\lambda_{De}^2\rho_{ss}^2 + (p - C_1\rho_{ss}^2)u(u - u_{e*})/\alpha^2 c^2\lambda_{De}^2\rho_{ss}^2. \tag{89b}$$

We note that in the absence of charged dust we have $\delta_d\kappa_d = 0$ and $F_2 = 0$. Accordingly, the outer solution, where $C_1 = 0$, of (88) has a long tail for $(u - u_{e*})(\alpha^2 v_A^2 - u^2) > 0$. On the other hand, inclusion of a small fraction of dust grains would make F_2 finite in the outer region. Here, we have the possibility of well behaved solutions. In fact, (88) admits spatially bounded dipolar vortex solutions. In the outer region $(r > R)$, where R is the vortex radius, we set $C_1 = 0$ and write the solution of (88) as (Liu and Horton, 1986; Shukla et al., 1986b; Mikhailovskii et al., 1987)

$$\phi = [Q_1 K_1(s_1 r) + Q_2 K_1(s_2 r)]\cos\theta, \tag{90}$$

where Q_1 and Q_2 are constants, and $s_{1,2}^2 = -[-\alpha_1 \pm (\alpha_1^2 - 4\alpha_2)^{1/2}/2$ for $\alpha_1 < 0$ and $\alpha_1^2 > 4\alpha_2 > 0$. Here, $\alpha_1 = (p/\rho_{ss}^2) + u(u - u_{e*})/\alpha^2 c^2\lambda_{De}^2$ and $\alpha_2 = [(u - u_{e*})^2 + u(u - u_{e*})p]/\alpha^2 c^2\lambda_{De}^2\rho_{ss}^2$. In the inner region $(r < R)$, the solution reads

$$\phi = \left[Q_3 J_1(s_3 r) + Q_4 I_1(s_4 r) - \frac{C_1}{\lambda_{De}^2}\frac{u^2(u - u_{e*})B_0}{\alpha^2 c^3 F_2}r \right]\cos\theta, \tag{91}$$

where Q_3 and Q_4 are constants. We have defined $s_{3,4} = [(F_1^2 - 4F_2)^{1/2} \pm F_1]/2$ for $F_2 < 0$. Thus, the presence of charged dust grains is responsible for the complete localization of the vortex solutions both in the outer as well as in the inner regions of the vortex core.

Second, we present the double vortex solution of (72), (74) and (75) in the cold plasma approximation. Thus, we set $T_j = 0$ and write (74) and (75) in the stationary frame as

$$\hat{D}_\phi \left(\nabla_\perp^2 \phi - \frac{\omega_{ci} \delta_d \kappa_d}{u} \phi \right) - \frac{v_A^2 \alpha}{uc} \hat{D}_A \nabla_\perp^2 A_z = 0, \tag{92}$$

and

$$\hat{D}_\phi \left[\left(1 - \lambda_e^2 \nabla_\perp^2 \right) A_z - \frac{\alpha c}{u} \phi \right] = 0. \tag{93}$$

It is easy to verify that (93) is satisfied by

$$\left(1 - \lambda_e^2 \nabla_\perp^2 \right) A_z - \frac{\alpha c}{u} \phi = 0. \tag{94}$$

By using (92) one can eliminate $\nabla_\perp^2 A_z$ from (94), yielding

$$\hat{D}_\phi \left[\nabla_\perp^2 \phi - \frac{\omega_{ci} \delta_d \kappa_d}{u} \phi + \frac{\alpha^2 v_A^2}{u^2 \lambda_e^2} \phi - \frac{\alpha v_A^2}{uc \lambda_e^2} A_z \right] = 0. \tag{95}$$

A typical solution of (95) is

$$\nabla_\perp^2 \phi + \beta_1 \phi - \beta_2 A_z = C_2 \left(\phi - \frac{uB_0}{c} x \right), \tag{96}$$

where $\beta_1 = (\alpha^2 v_A^2 / u^2 \lambda_e^2) - \omega_{ci} \delta_d \kappa_d / u$, $\beta_2 = \alpha v_A^2 / uc \lambda_e^2$, and C_2 is an integration constant.

Eliminating A_z from (92) and (96) we obtain

$$\nabla_\perp^4 \phi + F_1 \nabla_\perp^2 \phi + F_2 \phi - \frac{C_2 u B_0}{\lambda_e^2 c} x = 0, \tag{97}$$

where $F_1 = \lambda_e^{-2} \left(\alpha^2 v_A^2 / u^2 - 1 \right) - (\omega_{ci} \delta_d \kappa_d / u) - C_2$, and

$$F_2 = (C_2 + \omega_{ci} \delta_d \kappa_d / u) / \lambda_e^2.$$

Equation (97) is similar to (87) and its bounded solutions [similar to (90) and (91)] exist provided that $u^2 + \lambda_e^2 \omega_{ci} \delta_d \kappa_d u > \alpha^2 v_A^2$ and $\kappa_d > 0$. In the absence of the dust, we have $F_2 = 0$ in the outer region ($C_2 = 0$), and the outer solution (Yu et al., 1986) of the dust-free case has a long tail (decaying as $1/r$).

The analysis presented in this section reveal that the nonlinearly coupled drift-Alfvén-Shukla-Varma modes in a nonuniform warm dusty plasma as well as the nonlinearly coupled inertial Alfvén- Shukla-Varma modes in a nonuniform cold dusty plasma can be represented in the form of dipolar vortices, which are well behaved both in the outer and inner regions of the vortex core. The presence of charged dust impurities also limits the propagation speeds of the dipolar vortices.

8   Chaos in Alfvénic turbulence

Here, we show that the nonlinear equations governing the dynamics of low-frequency, flute-like Alfvénic disturbances in a nonuniform collisional plasma can be written as a set of three coupled nonlinear equations. The latter are a generalization of the Lorenz-Stenflo equations (Lorenz, 1963; Stenflo, 1996), which admit chaotic fluid behavior of electromagnetic turbulence in a nonuniform magnetoplasma.

The nonlinear dynamics of low-frequency, long wavelength (in comparison with the ion gyroradius) electromagnetic fields in a nonuniform magnetized plasma containing sheared plasma flows is governed by the ion vorticity equation

$$(D_t - \mu\nabla_\perp^2)\nabla_\perp^2\phi - \frac{\omega_{ci}\partial_x J_0}{n_0 ec}\partial_y A_z - \frac{v_A^2}{cB_0}(\hat{z} \times \nabla A_z) \cdot \nabla\nabla_\perp^2 A_z = 0, \qquad (98)$$

and the parallel (to $\hat{z}$) component of the electron momentum equation

$$\left[(1 - \lambda_e^2\nabla_\perp^2)D_t - \eta\nabla_\perp^2\right] A_z + \frac{c\partial_x v_{e0}}{\omega_{ce}}\partial_y\phi = 0, \qquad (99)$$

where $D_t = \partial_t + (c/B_0)\hat{z} \times \nabla\phi\cdot\nabla$, $\nabla_\perp^2 = \partial_x^2 + \partial_y^2$, and $J_0 = n_0 e(v_{i0} - v_{e0})$ is the equilibrium plasma current. Furthermore, $\eta = \nu_e\lambda_e^2$ is the plasma resistivity, and $\mu = (3/10)\nu_i\rho_i^2$ is the coefficient of ion gyroviscosity. The electron and ion collision frequencies are denoted by ν_e and ν_i, respectively, and ρ_i is the ion Larmor radius. We have assumed that the phase velocities of the disturbances are much larger than the electron and ion diamagnetic drift velocities.

Equations (98) and (99) are the nonlinear equations governing the dynamics of finite amplitude Alfvén-like disturbances in a nonuniform magnetoplasma containing equilibrium magnetic field-aligned sheared plasma flows. In the absence of the nonlinear interactions, we readily obtain from (98) and (99) the dispersion relation (Shukla, 1987)

$$\omega^2 + i\omega\Gamma - \Omega^2 + \gamma_0^2 = 0, \qquad (100)$$

where $\Gamma = [\mu + \eta/(1 + K^2\lambda_e^2)]K^2$, $\Omega^2 = \mu\eta K^4/(1 + K^2\lambda_e^2)$, $\mathbf{K} = \hat{x}K_x + \hat{y}K_y$ is the wave vector, and $\gamma_0^2 = -m_e(\partial_x J_0)(\partial_x v_{e0})K_y^2/m_i n_0 e(1 + K^2\lambda_e^2)K^2$. In the absence of the equilibrium current and velocity gradients, (100) gives convective cells and magnetostatic modes, which are decoupled. However, when the current and velocity gradients are opposite to each other, we obtain an instability provided that $\gamma_0^2 > \Omega^2 - \Gamma^2/4$. The maximum growth rate of that instability is $|\gamma_0|$.

In the following, we follow Lorenz and Stenflo and derive a set of equations which are appropriate for studying the temporal behavior of chaotic motion involving low-frequency electromagnetic waves in a dissipative magnetoplasma with sources. Accordingly, we introduce the Ansatz

$$\phi = \phi_1(t)\sin(k_x x)\sin(k_y y), \qquad (101)$$

and

$$A_z = A_1(t)\sin(k_x x)\cos(k_y y) - A_2(t)\sin(2k_x x), \tag{102}$$

where k_x and k_y are constant parameters, and ϕ_1, A_1 and A_2 are amplitudes which are only functions of time. By substituting (101) and (102) into (98) and (99), we readily obtain (Mirza and Shukla, 1997)

$$k^2 d_t \phi_1 = -\mu k^4 \phi_1 + \alpha_1 k_y A_1 - \alpha_2(k^2 - 4k_x^2)k_x k_y A_1 A_2, \tag{103}$$

$$\left(1 + k^2 \lambda_e^2\right) d_t A_1 = -\eta k^2 A_1 - \alpha_3 k_y \phi_1$$
$$+ \frac{c}{B_0}\left[1 + k^2 \lambda_e^2 - 6k_x^2 \lambda_e^2\right] k_x k_y A_2 \phi_1, \tag{104}$$

and

$$(1 + 4k_x^2 \lambda_e^2)d_t A_2 = -\frac{c}{2B_0}(1 + 4k_x^2 \lambda_e^2)k_x k_y \phi_1 A_1 - 4\eta k_x^2 A_2, \tag{105}$$

where $\alpha_1 = \omega_{ci}\partial_x J_0/n_0 ec$, $\alpha_2 = v_A^2/cB_0$ and $\alpha_3 = c\partial_x v_{e0}/\omega_{ce}$. We note that the terms proportional to $\sin(3k_x x)$ have been dropped in the derivation of (103) to (105). This approximation is often employed by many authors for deriving the relevant Lorenz-like equations in many branches of physics.

Equations (103) to (105) can be appropriately normalized so that they can be put in a form which is similar to that of Lorenz and Stenflo. We have (Mirza and Shukla, 1997; Shukla, Mirza, and Faria, 1998b)

$$d_\tau X = -\sigma X + \sigma Y + \delta Y Z, \tag{106}$$

$$d_\tau Y = -XZ + \gamma X - Y, \tag{107}$$

and

$$d_\tau Z = XY - \beta Z, \tag{108}$$

which describe the nonlinear coupling between various amplitudes. Here, $\sigma = \mu\left(1 + k^2 \lambda_e^2\right)/\eta$, $\gamma = -\alpha_1 \alpha_3 k_y^2/(\eta\mu k^6)$, $\beta = 4k_x^2\left(1 + k^2 \lambda_e^2\right)/(1 + 4k_x^2\lambda_e^2)k^2$, and the new parameter $\delta = \alpha_2(k_y^2 - 3k_x^2)\left(1 + k^2 \lambda_e^2\right)\mu^2 k^6 B_0/[\alpha_1^2 c k_y^2(1+k^2\lambda_e^2 - 6k_x^2\lambda_e^2)]$, with $k^2 = k_x^2 + k_y^2$ and $\tau = t/t_0$; where $t_0 = \eta k^2/\left(1 + k^2 \lambda_e^2\right)$.

A comment is in order. If we set $\delta = 0$, which happens for $k_y^2 = 3k_x^2$, then (106) to (108) reduce to the Lorenz type equations. However, the normalizations used here (Mirza and Shukla, 1997) are

$$\phi_1 = \left\{\frac{\sqrt{2}\eta k^2 B_0}{k_x k_y\left[\left(1 + k^2 \lambda_e^2\right)\left(1 + k^2 \lambda_e^2 - 6k_x^2\lambda_e^2\right)\right]^{1/2}}\right\} X,$$

$$A_1 = \left\{\frac{\sqrt{2}\eta\mu k^6 B_0}{c\alpha_1 k_x k_y^2\left[\left(1 + k^2 \lambda_e^2\right)\left(1 + k^2 \lambda_e^2 - 6k_x^2\lambda_e^2\right)\right]^{1/2}}\right\} Y,$$

and

$$A_2 = - \left[\frac{\eta \mu k^6 B_0}{c\alpha_1 k_x k_y^2 \left(1 + k^2 \lambda_e^2 - 6k_x^2 \lambda_e^2\right)} \right] Z.$$

Let us now discuss the chaotic fluid behavior of electromagnetic turbulence that is governed by (106) to (108). We observe that the equilibrium points of the dynamical equations are (Mirza and Shukla, 1998)

$$X_0 = \pm \left\{ \beta \left(\gamma - 2 + \delta\gamma^2/\sigma\right) + \frac{1}{2} \left[\left(\gamma - 2 + \delta\gamma^2/\sigma\right)^2 + 4(\gamma - 1)\right]^{1/2} \right\}^{1/2},$$
(109)

$$Y_0 = \frac{\gamma \beta X_0}{(\beta + X_0^2)},$$
(110)

and

$$Z_0 = \frac{X_0 Y_0}{\beta}.$$
(111)

In the absence of the δ-term, we note that for $|\gamma| > 1$, the equilibrium fixed points $[X_0 = Y_0 = \pm\sqrt{\beta}(|\gamma| - 1)^{1/2}]$ and $Z_0 = |\gamma| - 1)$ are unstable, resulting in convective cell motions. Thus, the linear instability should saturate by attracting to one of these new fixed states. Furthermore, it is worth mentioning that a detailed behavior of chaotic motion for $k_y \neq \sqrt{3}k_x$ can studied numerically (Sparrow, 1982) by solving (106) to (108).

9   Summary and conclusions

In this paper, we have presented a review of the linear and nonlinear dispersive Alfvén waves in homogeneous and non-homogeneous magnetoplasmas. We have highlighted the linear properties of Alfvén waves, taking into account those non-ideal effects which are absent in the ideal MHD description. For this purpose, we have utilized dispersion relations, which are derived either from the Hall-MHD or from the two-fluid equations supplemented by Faraday's and Ampère's laws, in order to show the relation of Alfvén waves with other plasma modes. Specifically, it has been noted that the Alfvén wave dispersion comes from the finite ω/ω_{ci} effect, as well as from the ion polarization drift and parallel electron inertial effects. The dispersive Alfvén waves can be excited due to a Cherenkov process involving the resonant interaction between magnetic field-aligned electron beams and DAWs. The latter are also driven at a nonthermal level when large amplitude electron whistlers and electrostatic lower-hybrid waves parametrically couple with the plasma slow motion that involves the DAWs. Linearly or nonlinearly excited DAWs attain finite amplitudes and they produce many interesting nonlinear effects. Of particular interests are the decay and modulational interactions in which coherent dispersive Alfvén or EMICA waves interact with low-frequency resonant or non-resonant perturbations to excite Stokes and anti-Stokes sidebands.

The latter interact with the pump and produce a low-frequency ponderomotive force which reinforces the low-frequency electrostatic or electromagnetic perturbations. We have presented explicit results for the parametric instabilities of dispersive Alfvén and EMICA waves. Due to the three-wave decay interaction, we have the possibility of generating ion-acoustic and daughter DIA/EMICA waves on account of the pump wave energy. There are indications (Wahlund et al., 1994) of ion-acoustic turbulence in association with the dispersive inertial Alfvén waves in the Freja data. We have also discussed the modulational instabilities of DIAWs in which Stokes and anti-Stokes DIAWs as well as non-resonant quasi-stationary and non-stationary density perturbations are involved. It is found that the driving force of dispersive inertial Alfvén waves can generate a quasi-stationary density hump, which, in turn, can act as a waveguide for the propagation of dark envelope DIAW packets. Thus, our theory predicts that the quasi-stationary large scale density depressions (Stasiewicz et al., 1998; Mäkelä et al., 1998), as seen in the data from the Freja and Fast spacecrafts, cannot be explained in terms of the DIAW ponderomotive force density cavitation, in contrast to the suggestion made by Bellan and Stasiewicz (1998). However, nonstationary supersonic density cavities could be created by the DIAW ponderomotive force. The nonlinear DIAW structures comprising bell shaped DIAW wave magnetic fields and supersonic rarefactive density perturbations could then be associated with the coherent nonlinear DIA waves (with scale sizes of the order of the electron skin depth) provided that further data analysis reveals a detailed knowledge of the speed of the nonlinear structures.

We have also considered the effect of the plasma nonuniformity on the dispersive Alfvén waves in a multi-component collisionless magnetoplasma containing charged dust impurities. In the linear limit, the local dispersion relation exhibits a coupling between drift-Alfvén waves and the Shukla-Varma mode. The latter arises because the divergence of the $\mathbf{E}_\perp \times \mathbf{B}_0$ current remains finite when charged stationary dust grains are present. The stationary solutions of the nonlinear equations for weakly interacting long wavelengths (in comparison with the collisionless electron skin depth and the ion gyroradius) drift-Alfvén-Shukla-Varma modes as well as the inertial Alfvén-Shukla-Varma modes can be represented in the form of dipole vortices. It is found that the presence of the Shukla-Varma mode provides the possibility for the localization of the drift-Alfvén/inertial Alfvén vortex solutions in the outer region, which otherwise would have a long tail. Furthermore, the dipolar vortex speeds have bounds when the dust grains are present in the electron-ion plasma.

Finally, we have investigated the long term behavior of nonlinearly interacting low-frequency flute-like electromagnetic modes in a nonuniform collisional magnetoplasma containing equilibrium density and velocity gradients. It was found that the nonlinear dynamics of such an electromagnetic turbulence can be studied in the context of chaotic attractor theory. Specifically, we have demonstrated that the coupled vorticity and magnetic field diffusion equations in the presence of sheared flows can be put in the form of a simple system (three nonlinear equations for the amplitudes), which resembles the Lorenz-Stenflo equations. The stationary points for our generalized mode coupling equations were derived. The properties of the nontrivial steady state attractor were discussed.

In conclusion, we stress that the results of the present investigation should be useful in identifying the spectra of low-frequency dispersive Alfvén waves as well as nonlinear effects related to the density modification and the formation of coherent structures such as envelope solitons and vortices. The latter are indeed observed (e.g. Petviashvili and Pokhotelov, 1992; Stasiewicz and Potemra, 1998) in the Earth's auroral ionosphere and magnetosphere.

We thank Robert Bingham, Gert Brodin, Arshad Mirza, and Oleg Pokhotelov for their valuable collaboration on different parts presented in this paper. This research was partially supported by the Deutsche Forschungsgemeinschaft through the Sonderforschungsbereich 191 and the European Union (Brussels) through the INTAS grant (INTAS-GEORG-52-98), as well as by the Swedish Natural Science Research Council and the International Space Science Institute (ISSI) at Bern (Switzerland). The authors gratefully acknowledge the warm hospitality of Prof. Bengt Hultqvist at the ISSI.

References

1. Antani S. N., Kaup D. J., & Shukla P. K., 1983, Phys. Fluids, 26, 483.
2. Bellan P. M., & Stasiewicz K, 1998, Phys. Rev. Lett. 80, 3523.
3. Brodin G., & Stenflo L., 1988, Physica Scripta, 37, 89.
4. Brodin G., & Stenflo L., 1989, J. Plasma Phys., 41, 199.
5. Brodin G., & Stenflo L, 1990, Contrib. Plasma Phys., 30, 413.
6. Champeaux S., Gazol A, Passot T., & Sulem P. L., 1997, Astrophys. J., 486, 477.
7. Champeaux S., Passot T., & Sulem P. L., 1999, Phys. Plasmas, 6, 413.
8. Chen L, 1977, Plasma Phys., 19, 47.
9. Hada T., 1993, Geophys. Res. Lett., 20, 2415.
10. Hasegawa, A., 1975, *Plasma Instabilities and Nonlinear Effects* (Springer-Verlag, Berlin), Chap. 4.
11. Hasegawa A., & Chen L., 1975, Phys. Rev. Lett., 35, 440.
12. Hasegawa A, & Uberoi, C., 1982, *The Alfvén Wave*, (Tech. Inf. Center, U. S. Department of Energy, Washington, D. C.).
13. Kennel C. F., Buti B., Hada T., & Pellat R., 1988, Phys. Fluids, 31, 1949.
14. D. Leneman, W. Gekelman, & J. Maggs, 1999, Phys. Rev. Lett, 82, 2673.
15. Lysak, R., 1990, Space Sci. Rev., 52, 33.
16. Liu J., & Horton, W., 1986, J. Plasma Phys., 36, 1.
17. Lorenz E. N., 1963, J. Atmos. Sci., 20, 130.
18. Mäkelä J. S., Mälkki A., Koskinen H., Boehm M., Holback B, & Eliasson L., 1998, J. Geophys. Res., 103, 9391.

19. Mikhailovskii A. B., Petviashvili V. I., & Friedman A. M., 1977, JETP Letters, 24, 43.
20. Mikhailovskii A. B., Lakhin V. P., Aburdzhaniya G. D., Mikhailovskaya L. A., Onishchenko O. G., and Smolyakov, 1987, Plasma Phys. Controlled Fusion, 29, 1.
21. Mio K., Ogino T., Minami K., & Takeda S., 1976, J. Phys. Soc. Jpn., 41, 265.
22. Mirza A. M., & Shukla P. K., 1997, Phys. Lett. A, 229, 313.
23. Mishin E. V., & Förster M., 1995, Geophys. Res. Lett., 22, 1745.
24. Nocera L., & Buti B., 1996, Physica Scripta, T63, 186.
25. Passot T., & Sulem P. L., 1993, Phys. Rev. E, 48, 2966.
26. Passot T., Sulem C., & Sulem P-L., 1994, Phys. Rev. E, 50, 1427.
27. Petviashvili V., & Pokhotelov, O., 1992, Solitary Waves in Plasmas and in the Atmosphere, (Gordon and Breach Science Publishers, Reading).
28. Pokhotelov O. A., Pokhotelov D. O., Gokhberg M. B., Feygin F. Z., Stenflo L., & Shukla P. K., 1996, J. Geophys. Res., 101, 7913.
29. Pokhotelov O. A., Onischenko O. G., Shukla P. K., & Stenflo L., 1999, J. Geophys. Res., to be published.
30. Rogister A., 1971, Phys. Fluids, 14, 2733.
31. Sagdeev R. Z. & Galeev, A. A., 1969, Nonlinear Plasma Theory, revised and edited by T. M. O'Neil and D. L. Book (Benjamin, New York).
32. Shevchenko V. I., Galinsky V. L., Ride S. K., & Baine M., 1995, Geophys. Res. Lett., 22, 2997.
33. Shukla P. K., 1977, Phys. Rev. A, 16, 1294.
34. Shukla P. K., Rahman H. U., & Sharma R. P., 1982, J. Plasma Phys., 28, 125,
35. Shukla P. K., & Dawson J. M., 1984, Astrophys. J, 276, L49.
36. Shukla P. K., & Stenflo L., 1985, Phys. Fluids, 28, 1576.
37. Shukla P. K., Anderson D., Lisak M, & Wilhelmsson, H., 1985, Phys. Rev., 31, 1946.
38. Shukla P. K., Yu M. Y., & Stenflo L., 1986a, Physica Scripta, 34, 169.
39. Shukla P. K., Yu M. Y., & Stenflo L., 1986b, Phys. Rev. A, 34, 1582.
40. Shukla P. K., 1987, Phys. Fluids, 30, 1901.
41. Shukla P. K., Feix G., & Stenflo L., 1988, Astrophys. Space Sci., 147, 383.
42. Shukla P. K., & Stenflo L., 1989a, Astrophys. Space Sci., 155, 145.
43. Shukla P. K., & Stenflo, 1989b, Phys. Fluids B, 1, 1926.
44. Shukla P. K., & Varma R. K., 1993, Phys. Fluids B, 5, 236.
45. Shukla P. K., & Stenflo L., 1995, Physica Scripta, T60, 32.
46. Shukla P. K., Birk G. T., Dreher J., & Stenflo L., 1996, Plasma Phys. Rep., 22, 818.
47. Shukla P. K., Bingham R., & Dendy R., 1998a, Phys. Lett. A, 239, 369.
48. Shukla P. K., Mirza A. M., & Faria Jr. R. T., 1998b, Phys. Plasmas, 5, 616.
49. Shukla P. K., Stenflo L., & Bingham, 1999a, Phys. Plasmas, 6, in press.
50. Shukla P. K., Bingham R., McKenzie J. F., & Axford W. I., 1999b, Solar Phys., 184, in press.
51. Sparrow, C., 1982, The Lorenz Equations: Bifurcations, Chaos and Strange Attractors (Springer-Verlag, Berlin).
52. Stasiewicz K., Holmgren G., & Zanetti L, 1998, J. Geophys. Res., 103, 4251.
53. Stasiewicz K., & and Potemra T., 1998, J. Geophys. Res., 103, 4315.
54. Stefant R. Z., 1970, Phys. Fluids, 13, 440.
55. Stenflo L., Yu M. Y., & Shukla P. K., 1988, Planet. Space Sci., 36, 499.

56. Stenflo L., & Shukla P. K., 1988, J. Geophys. Res., 93, 4115.
57. Stenflo L., 1996, Physica Scripta, 53, 83.
58. Stenflo L., 1999, J. Plasma Phys., 61, 129.
59. F. Verheest, 1990, Icarus, 86, 273.
60. Wahlund J.-E., Louarn, P., Chest T. *et al.*, 1994, Geophys. Res. Lett., 21, 1831; *ibid*, 21, 1835.
61. Yu M. Y., Shukla P. K., & Stenflo L., 1986, Astrophys. J, 309, L63.

Whistler Solitons, Their Radiation and the Self-Focusing of Whistler Wave Beams

V.I. Karpman

Racah Institute of Physics, Hebrew University, Jerusalem 91904, Israel

Abstract. A theory of envelope whistler solitons beyond the approximation based on the nonlinear Schrödinger (NLS) equation is developed. It is shown that such solitons must emanate radiation due to the continuos transformation of trapped whistler modes into other modes that cannot be trapped in the duct, produced by the soliton (such modes are not described by the NLS equation). An equation governing the decrease of soliton amplitude due to the loss of trapped radiation is derived. The soliton radiation increases with the decrease of the soliton size and, therefore only weak solitons have sufficiently large lifetime. The theory is extended to the whistler spiral wave beams which, according to the NLS equation, must be liable to the self-focusing. It is shown that when the wave beam becomes sufficiently narrow, the self-focusing is replaced by the defocusing because of big radiation losses. These predictions are confirmed by numerical experiments. Possible generalizations to other gyrotropic media are briefly discussed.

1 Introduction

Among many kinds of waves in space plasmas, the whistler waves play very important role. After their discovery in the Earth ionosphere about fifty years ago by L.R.O. Storey [1], they became an important tool of diagnostics of the ionosphere and then magnetosphere [2]. Now the whistler waves are intensively studied also in the magnetospheres of other planets. Resonantly interacting with electrons, whistlers play an important role in the dynamics of magnetized plasmas both in space and laboratory. The special role of whistlers is connected with their ability to bounce along magnetic field lines of dipole type configurations of magnetic field. However, as follows from the geometrical optics, the curvature of magnetic field causes a drift of a whistler wave from one magnetic line to another. The attachment of whistlers to the magnetic lines is explained by the existence of plasma ducts, the field aligned plasma inhomogeneities with enhanced electron densities . It can be shown that a whistler wave may be trapped in such a duct and this effect can be understood even in the frame of geometrical optics [2,3]. This trapping is very similar to what happens in a whistler envelope soliton that is a density enhancement with trapped whistler wave, self-consistently interacting with plasma by means, e.g., of the ponderomotive force. Therefore, the whistler trapping in ducts has close connection with the physics of whistler solitons and this will be widely exploited below. One should point out, however, that the approach of geometrical optics and some simplified wave theories, such as based on the the simple nonlinear Schrödinger (NLS) equation [4], miss an important effect : it appears, in fact, that the whistler trapping in ducts is not perfect because of a transformation of the trapped wave into another mode that can leave the duct [5]. Such an effect, to some extent similar to the tunneling, was called the tunneling transformation [6]. The tunneling transformation leads, therefore, to the wave leakage from the duct. As the classical tunneling, this effect is small if the duct width is much greater than the wavelength but it becomes large if they are comparable. Thus, as it will be seen below, only the broad (and, therefore, weak) whistler solitons can exist. The present paper is devoted to the nonlinear theory of localized whistler wave structures. Starting with the slab models of ducts and solitons [7], we then extend our theory to the cylindrical (generally, spiral) solitons that are in fact nonlinear wave beams, self- consistently interacting with plasma. It appears that, similar to the slab solitons, the nonlinear interaction leads to the formation of density enhancements that trap whistler waves. However, the two-dimensional wave beams can be unstable which may result in the self-focusing [8,9]. After the wave beam becomes sufficiently narrow, the wave leakage due to the tunneling transformation leads to a significant loss of the intensity of trapped wave and, as it will be seen below, the self-focusing is replaced by the defocusing [10,11]. The whistler solitons are insructive examples of more general wave solitons in gyrotropic media (i.e. media sensitive to the ambient magnetic field) and our theory can be extended to these media [12,13] that include gyrotropic solids, etc. The theory may also be extended to Alfven solitons, which will be considered separately.

2   Basic equations. Slab ducts

Assume that the ambient magnetic field B is constant and directed along the z- axis and the wave electric field is monochromatic and has the form,

$$\frac{1}{2}\mathbf{E}e^{-i\omega t} + \text{c.c.}, \quad \mathbf{E} = \mathbf{E}(\mathbf{r}). \tag{1}$$

The wave field is described by the Maxwell equations

$$\nabla \times \nabla \times \mathbf{E} = \left(\frac{\omega}{c}\right)^2 \mathbf{D}, \tag{2}$$

$$\nabla \cdot \mathbf{D} = 0, \tag{3}$$

where

$$\mathbf{D} = \hat{\epsilon}\mathbf{E} \tag{4}$$

and $\hat{\epsilon}$ is the dielectric permittivity tensor. We neglect the space dispersion and dissipation, assuming that plasma is collisionless and the angular frequency

ω is sufficiently far from resonances. Assume that

$$\omega_p^2 \gg \omega_c^2, \quad \omega_p = \left(\frac{4\pi N e^2}{m}\right)^{\frac{1}{2}}, \quad \omega_c = \frac{eB}{mc} \tag{5}$$

where ω_p and ω_c are electron plasma and cyclotron frequencies, and $N = N(x)$ is the electron density. Due to the condition in (5), the whistler frequencies are in the range

$$\omega_{LH} \ll \omega < \omega_c, \tag{6}$$

where ω_{LH} is the low hybrid frequency. These conditions are violated in polar regions, where the theory should be modified. Due to condition (6), we can neglect the ion oscillations in the whistler wave field.

Now consider a slab model. In the chosen reference frame, we can write

$$\mathbf{E} = \hat{\mathbf{i}}\, E_x + \hat{\mathbf{k}}\, E_z, \quad \mathbf{r} = \hat{\mathbf{i}}\, x + \hat{\mathbf{k}}\, z. \tag{7}$$

The elements of the dielectric tensor are

$$\epsilon_{xx} = \epsilon_{yy} = \epsilon(\omega, N, B), \quad \epsilon_{xy} = \epsilon_{yx} = -ig(\omega, N, B), \quad \epsilon_{zz} = \eta(\omega, N), \tag{8a}$$
$$\epsilon_{xz} = \epsilon_{zx} = \epsilon_{yz} = \epsilon_{zy}. \tag{8b}$$

As far as the contribution from ion oscillations is neglected, we have

$$\epsilon = 1 + \frac{\omega_p^2}{\omega_c^2 - \omega^2} > 0, \quad g = \frac{\omega_p^2 \omega_c}{\omega(\omega_c^2 - \omega^2)}, \quad \eta = 1 - \frac{\omega_p^2}{\omega^2}. \tag{9}$$

Defining

$$F(x, z) = E_x - iE_y, \quad G(x, z) = E_x + iE_y, \tag{10}$$

we transform the (x, y)-projections of Eq. (2) to

$$\frac{\partial^2 F}{\partial z^2} + \frac{\partial^2 F}{\partial x^2} - \frac{\partial}{\partial x}(\nabla \cdot \mathbf{E}) + \frac{\omega^2}{c^2}(\epsilon + g)F = 0, \tag{11a}$$
$$\frac{\partial^2 G}{\partial z^2} + \frac{\partial^2 G}{\partial x^2} - \frac{\partial}{\partial x}(\nabla \cdot \mathbf{E}) + \frac{\omega^2}{c^2}(\epsilon - g)G = 0. \tag{11b}$$

Instead of the z-projection of Eq. (2), we use Eq. (3) that can be written as

$$\frac{\partial}{\partial z}(\eta E_z) + \frac{1}{2}\frac{\partial}{\partial x}[(\epsilon + g)F + (\epsilon - g)G] = 0. \tag{12}$$

This is the third equation, containing E_z that appears in

$$\nabla \cdot \mathbf{E} \equiv \frac{1}{2}\frac{\partial}{\partial x}(F + G) + \frac{\partial E_z}{\partial z}. \tag{13}$$

Equations (11) and (12), together with the given function $N = N(x)$, constitute a complete system.

At $N(x) = $ const, the simplest solution to this system, describes a whistler
wave, propagating parallel to the ambient magnetic field (z-axis) . It reads

$$F = A\exp\left(i\frac{\omega}{c}p_0 z\right), \quad G = E_z = 0, \tag{14}$$

where A is a constant amplitude and p_0 the normalized wave number

$$p_0^2 \equiv \epsilon_0 + g_0 \approx \frac{\omega_{p0}^2}{\omega(\omega_c - \omega)}. \tag{15}$$

The subzero means that it is substituted

$$N = N_0 = \text{const} \tag{16}$$

into the expression for a corresponding quantity. Equation (15) is the dispersive relation for the parallel propagation. A more general solution, describing
the oblique propagation in weakly inhomogeneous plasma with

$$N = N_0[1 + \nu(x)], \quad \partial_x \ln \nu(x) \ll 1, \tag{17}$$

can be approximately found by means of the WKB method, where all unknown functions are proportional to the factor

$$\exp\left[i\int^x k_x(x')\,dx' + ik_z z\right], \quad k_z = \text{const.} \tag{18}$$

Then from the system (11)–(12), we obtain, neglecting the terms with

$$\partial\epsilon/\partial x, \quad \partial g/\partial x, \quad \partial\eta/\partial x, \tag{19}$$

the same dispersion relation as in the homogeneous plasma. Using $\omega_p^2 \gg \omega_c^2$
and eqs.(9) , we can write it in the form

$$\omega \approx c^2\frac{\omega_c kk_z}{\omega_p^2 + c^2k^2}, \quad k^2 = k_x^2 + k_z^2. \tag{20}$$

Denote

$$k_z = \frac{\omega}{c}p, \quad k_x(x) = \frac{\omega}{c}q(x), \quad k_y = 0. \tag{21}$$

Substituting this into Eq. (20), we come to a quartic equation for q, which
has the following solution (at $\alpha \gg 1$)

$$q^2 \approx \frac{1}{2u^2}\left[(1 - 2u^2)p^2 - 2\alpha^2 \pm p\sqrt{p^2 - 4\alpha^2}\right], \tag{22}$$

where

$$u = \frac{\omega}{\omega_c}, \quad \alpha = \frac{\omega_p}{\omega_c}. \tag{23}$$

Equation (22) shows that, generally, there are two wave branches at given p and ω (if, of course, $q^2 > 0$). From formulas (5) and (17), it follows that $\alpha = \alpha(x)$ and

$$\alpha^2(x) = \alpha_0^2[1 + \nu(x)]. \tag{24}$$

Other quantities in Eq. (22) are constant. One can easily check that the parallel propagation ($q = 0$) is possible only in a homogeneous plasma with $\alpha = \alpha_0$ and $p = p_0$, where p_0 is defined in (15).

Now consider a propagation in weakly stratified plasmas when p is close to p_0. Introducing the notations

$$p = p_0 + \Delta p, \quad \Delta \alpha^2 \equiv \alpha^2(x) - \alpha_0^2 = \alpha_0^2 \nu(x), \quad |\Delta p| \ll 1, \quad |\nu(x)| \ll 1, \tag{25}$$

we have from Eq. (22) the following expressions for the two branches of $q(x)$:

$$q_1^2(x) \approx \frac{2\alpha_0^2}{u(1 - 2u)} \left[\nu(x) - \frac{2\Delta p}{p_0} \right], \tag{26}$$

$$q_2^2(x) \approx \frac{p_0^2(1 - 2u)}{u^2} + 2\frac{1 - 2u(1 - u)}{u^2(1 - 2u)} p_0 \Delta p - 2\alpha_0^2 \frac{1 - u}{(1 - 2u)u^2} \nu(x)$$

$$\approx \frac{p_0^2(1 - 2u)}{u^2}, \tag{27}$$

where Δp should be the same in both expressions and all quantities in denominators are assumed to be not too small. Equation (26) describes nearly parallel propagating wave, while (27) describes a branch with comparable k_x and k_z. Respectively, the first branch has almost circular polarization ($G/F \ll 1$), while for the second branch

$$\frac{G}{F} \approx \frac{g_0 + \epsilon_0 + \eta_0}{g_0 - \epsilon_0 + \eta_0} \approx \frac{1 + u}{1 - u}(1 - 2u). \tag{28}$$

From these relations one can derive conditions of the whistler trapping in a weak duct. Assume, for simplicity, that the duct has a bell type shape. Then $\nu(x) \to 0$ at $|x| \to \infty$. The trapping condition reads: $q^2(x) > 0$, inside the duct, and $q^2(x) < 0$, at $|x| \to \infty$. Then the branch, described by $q_1(x)$, may be trapped if

$$\nu(0) > \frac{2\Delta p}{p_0} > 0 \quad (\omega < \frac{\omega_c}{2}). \tag{29}$$

Therefore at $\omega < \omega_c/2$, the first branch can be trapped only into a desity *hump*; it is easy to see that the second branch, $q_2(x)$, cannot be trapped at all. At $\omega > \omega_c/2$, the first branch can be trapped only into a density trough with

$$\nu(0) < \frac{2\Delta p}{p_0} \quad (\omega > \frac{\omega_c}{2});\tag{30}$$

and the second branch does not exist because $q_2^2(x) < 0$ at $\omega > \omega_c/2$. These results, depicted in Fig.1, are in agreement with the geometrical optics [2].

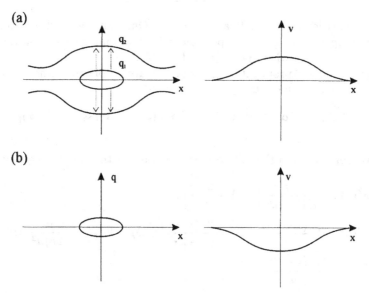

Fig. 1. Behavior of $q(x)$ in the WKB approximation. (a): $\omega/\omega_c < 1/2$: the ducts are density humps. Transformation is symbolically shown by broken lines. (b): $\omega/\omega_c > 1/2$: ; the ducts are density troughs. There is only one real branch of q, in this case, and, respectively, no tunneling transformation.

Consider now what follows from the full wave approach at $\omega < \omega_c/2$. In this case the wave function of the trapped whistler wave should be a superposition of the two above branches that cannot be considered separately. In the case when the WKB approximation is legitimate, this leads to the conclusion that if, initially, the second branch is not present in the duct, it would be generated because of continuous transformation of the first branch into the second one. (In the inhomogeneous plasmas, such transformations are well-known.) As far as the second branch can not be trapped, we conclude that the whistler wave trapping in a duct at $\omega < \omega_c/2$ is not perfect: even if conditions (29) are fulfilled, there should be a wave leakage from the duct. On the other hand, at $\omega > \omega_c/2$, the wave trapping is perfect, because the second branch does not exist. The transformation coefficient for the process $q_1(x) \mapsto q_2(x)$ at $\omega < \omega_c/2$ and the corresponding leakage rate from the duct were calculated in Ref. [5] by means of a modified WKB approach. In the weak ducts, both the transformation coefficient and the leakage rate are exponentially small. The expression for the leakage rate looks very similar to the tunneling coefficient in quantum mechanics [14] and therefore this effect was called tunneling transformation [5,6]. Further developments of this approach were described in Refs. [15,16,11]. One should point out that in the ducts produced by solitons the conditions of the WKB approximation are not fulfilled . It appears, nevertheless, that the above reasoning leads to qualitatively correct predictions also in this case. To obtain quantitative results, we shall use below another approach [7], based on the full wave asymptotic solutions to Eqs. (11) and (12), which are beyond the WKB approximation.

3   Full wave solutions for the slab ducts

Following [7], we first consider weak ducts with the soliton type profile

$$\nu(\xi) = \frac{2\mu^2 b}{p_0^2}\text{sech}^2\xi = \frac{2\mu^2 b}{\epsilon_0 + \eta_0}\text{sech}^2\xi, \tag{31}$$

where ξ is a stretched variable, defined by

$$\xi = \mu\sqrt{\frac{2\eta_0}{\epsilon_0 + \eta_0 + g_0}}\frac{\omega}{c}x. \tag{32}$$

Here b is an arbitrary factor and μ a small parameter, $\mu \ll 1$. The solutions to Eqs. (11) and (12) can be looked in the form

$$F(x,z) = \bar{F}(\xi,\zeta)\exp\left(i\frac{\omega}{c}p_0 z\right), \tag{33a}$$

$$G(x,z) = \bar{G}(\xi,\zeta)\exp\left(i\frac{\omega}{c}p_0 z\right), \tag{33b}$$

$$E_z(x,z) = \bar{E}_z(\xi,\zeta)\exp\left(i\frac{\omega}{c}p_0 z\right), \tag{34}$$

where

$$\zeta = \frac{\mu^2}{p_0}\frac{\omega}{c}z \tag{35}$$

and p_0 is defined in Eq. (15). It will be shown below that in the whistler solitons $\nu(\xi) > 0$. Then the factor b in (31) should be positive and the trapping in the duct $\nu(\xi)$ is possible only at $\omega < \omega_c/2$. (This will be seen also from further results.) Taking into account that

$$\frac{2\eta_0}{\epsilon_0 + \eta_0 + g_0} \approx \frac{2(\omega_c - \omega)}{\omega_c - 2\omega} > 0, \tag{36}$$

we also see that ξ is real at $\omega < \omega_c/2$. Substituting (34) into Eq. (12) and neglecting the terms (19), we have

$$\bar{E}_z = \frac{i\mu}{2\eta_0 p}\sqrt{\frac{2\eta_0}{\epsilon_0 + \eta_0 + g_0}}\partial_\xi[(\epsilon_0 + g_0)\bar{F} + (\epsilon_0 - g_0)\bar{G}] + O(\mu^3). \tag{37}$$

Substituting Eqs. (33), (34) and (37) into (11) and (13), we arrive at the following approximate system of equations for $\bar{F}$ and $\bar{G}$:

$$2i\partial_\zeta \bar{F} + \partial_\xi^2 \bar{F} + \frac{\epsilon_0 - \eta_0 - g_0}{\epsilon_0 + \eta_0 + g_0}\partial_\xi^2 \bar{G} + 2b(\mathrm{sech}^2\xi)\bar{F} = 0, \tag{38}$$

$$2i\partial_\zeta \bar{G} + \frac{\epsilon_0 + \eta_0 - g_0}{\epsilon_0 + \eta_0 + g_0}\partial_\xi^2 \bar{G} + \frac{\epsilon_0 - \eta_0 + g_0}{\epsilon_0 + \eta_0 + g_0}\partial_\xi^2 \bar{F} = \frac{2g_0}{\mu^2}\bar{G}. \tag{39}$$

From Eq. (39), it follows that inside the duct $\bar{G} \sim \mu^2\bar{F}$. Neglecting the term with $\bar{G}$ in Eq. (38), we arrive at the Schrödinger equation [7]

$$2i\partial_\zeta \bar{F} + \partial_\xi^2 \bar{F} + 2b(\mathrm{sech}^2\xi)\bar{F} = 0 \tag{40}$$

which approximately describes the eigenmodes, trapped in the duct. The solutions of Eq. (40) for the trapped eigenmodes have the form

$$\bar{F}_n(\xi,\zeta) = \exp\left(i\frac{\beta_n}{2}\zeta\right)\Psi_n(\xi), \tag{41}$$

where $\Psi_n(\xi)$ satisfies the equation

$$\Psi_n''(\xi) + (2b\mathrm{sech}^2\xi - \beta_n)\Psi_n(\xi) = 0 \tag{42}$$

which has well known analytical solutions . Functions $\Psi_n(\xi)$ must be regular and vanish at $|\xi| \to \infty$. This is possible only at (e.g. Ref. [14], Sec. 23)

$$\beta_n = \frac{1}{4}\left[\sqrt{1 + 8b} - (1 + 2n)\right]^2, \tag{43}$$

where

$$n = 0, 1, 2, \ldots, \quad n < \sigma = \frac{1}{2}(\sqrt{1 + 8b} - 1) \tag{44}$$

(n is the number of nodes) . At large b , there are many eigenmodes and those with large n satisfy the WKB conditions. In fact, such solutions were used in [5,15,16] for the description of trapped modes in the WKB approximation. As we shall see , for the soliton $b = 1$. Then from (44) it follows that there is only one eigenmode with $n = 0$. From Eq. (43) we then have $\beta_0 = 1$ and therefore the solution (41) for $b = 1$ reads

$$\bar{F}_0 = \exp\left(i\frac{\zeta}{2}\right)\Psi_0(\xi), \quad \Psi_0(\xi) = A\mathrm{sech}\xi. \tag{45}$$

However, as we shall see below, omitting the term with G in Eq. (38) we loose the wave leakage from the duct which has been qualitatively discussed above in the WKB approximation. To include this effect, we define

$$\bar{F}(\xi,\zeta) = \Psi(\xi,\zeta)\exp\left(i\frac{\zeta}{2}\right), \quad \bar{G}(\xi,\zeta) = \Phi(\xi,\zeta)\exp\left(i\frac{\zeta}{2}\right) \tag{46}$$

and substitute this into the full system (38), (39). This gives the equations

$$2i\partial_\zeta\Psi + \partial_\xi^2\Psi + \frac{\epsilon_0 - \eta_0 - g_0}{\epsilon_0 + \eta_0 + g_0}\partial_\xi^2\Phi + (2\mathrm{sech}^2\xi - 1)\Psi = 0 \tag{47}$$

$$2i\partial_\zeta\Phi + \frac{\epsilon_0 + \eta_0 - g_0}{\epsilon_0 + \eta_0 + g_0}\partial_\xi^2\Phi + \frac{\epsilon_0 - \eta_0 + g_0}{\epsilon_0 + \eta_0 + g_0}\partial_\xi^2\Psi \approx \frac{2g_0}{\mu^2}\Phi. \tag{48}$$

Substitute here

$$\Psi(\xi,\zeta) = \Psi_0(\xi) + \psi(\xi,\zeta), \quad \Phi(\xi,\zeta) = \Phi_0(\xi) + \varphi(\xi,\zeta), \tag{49}$$

where $\Psi_0(\xi)$ is written in (45) and $\Phi_0(\xi)$ is the solution to Eq. (48) with $\Psi = \Psi_0(\xi)$. Then we have the following equations for $\psi(\xi,\zeta)$ and $\varphi(\xi,\zeta)$ [7]

$$2i\partial_\zeta\psi + \partial_\xi^2\psi + \frac{\epsilon_0 - \eta_0 - g_0}{\epsilon_0 + \eta_0 + g_0}\partial_\xi^2\varphi - \psi + 2\mathrm{sech}^2\xi\,\psi$$
$$= \gamma\frac{(\epsilon_0 - \eta_0)^2 - g_0^2}{4\epsilon_0\eta_0}\widehat{L}^{-1}\partial_\xi^4\Psi_0, \tag{50}$$

$$2i\partial_\zeta\varphi + \frac{\epsilon_0 + \eta_0 - g_0}{\epsilon_0 + \eta_0 + g_0}\partial_\xi^2\varphi + \frac{\epsilon_0 - \eta_0 + g_0}{\epsilon_0 + \eta_0 + g_0}\partial_\xi^2\psi \approx \frac{2g_0}{\mu^2}\varphi. \tag{51}$$

where $\widehat{L}$ is the differential operator

$$\widehat{L} = 1 + \gamma\frac{(\epsilon_0 + \eta_0)^2 - g_0^2}{4\epsilon_0\eta_0}\partial_\xi^2 \tag{52}$$

with

$$\gamma = -\frac{2\epsilon_0\eta_0}{g_0(\epsilon_0 + \eta_0 + g_0)^2}\mu^2. \tag{53}$$

The small parameter γ plays, as it will be seen below, an important role in the description of the wave leakage. From formulas (9), it follows that $\gamma > 0$. We solve eqs. (50)–(51) at the initial conditions

$$\psi(\xi,0) = 0, \quad \varphi(\xi,0) = 0. \tag{54}$$

Thus, we assume that at $\zeta = 0$, there is only the trapped wave described by $\Psi_0(\xi)$ from (45) and associated function $\Phi_0(\xi)$ defined above. The system (50)–(51) can be solved by means of the Fourier (with respect to ξ) and Laplace (with respect to ζ) transforms. The approximate solution reads [7]

$$\psi(\xi,\zeta) \approx CK\left[\Theta(\xi)\Theta(\zeta - \chi\xi)\exp\left(-i\frac{\xi}{\sqrt{\gamma}}\right) + \Theta(-\xi)\Theta(\zeta + \chi\xi)\exp\left(i\frac{\xi}{\sqrt{\gamma}}\right)\right] \tag{55}$$

where C is a numerical constant of order 10,

$$K = \frac{\pi A}{i\sqrt{\gamma}} \frac{(\epsilon_0 - \eta_0)^2 - g_0^2}{4\epsilon_0\eta_0} \exp\left(-\frac{\pi}{2\sqrt{\gamma}}\right), \tag{56}$$

A is a constant factor from (45), $\Theta(Z)$ is a step function defined by

$$\Theta(Z) = 1 \text{ for } Z > 0, \quad \Theta(Z) = 0 \text{ for } Z < 0, \tag{57}$$

and

$$\chi = -\frac{\epsilon_0^2 + \eta_0^2 + (\epsilon_0 + \eta_0)g_0}{2\epsilon_0\eta_0}\sqrt{\gamma} \tag{58}$$

Expression (55) is valid at

$$\zeta > 0, \quad |\xi| \gg 1, \quad |\zeta \pm \chi\xi| \gg 1. \tag{59}$$

As for $\varphi(\xi,\zeta)$, it is given by

$$\varphi(\xi,\zeta) \approx \frac{g_0 + \epsilon_0 + \eta_0}{g_0 - \epsilon_0 + \eta_0}\psi(\xi,\zeta). \tag{60}$$

Expressions (55) and (60) asymptotically determine the radiation outgoing from the duct at initial conditions (54). It is easy to check that $\xi/\sqrt{\gamma} = q_2(\omega/c)x$, where q_2 is given by Eq. (27). This shows again that the radiated wave is indeed the second branch. Also, we see that Eq. (60) agrees with (28), i.e., it correctly determines the polarization of the wave out of the duct. The region, occupied by the radiation, is shown in Fig.2. At conditions (59) the width of the duct can be neglected (in Fig.2, the duct corresponds to the origin of the domain occupied by radiation) and the transient layers are reduced to jumps, described by the step functions in Eq. (55) . (The fine structure of these layers is briefly discussed in [7].) From formula (55) it also follows that $\tan \alpha = \chi$. This can be expressed through the components of the radiation group velocity $\mathbf{U}_g$. Using dispersion equation (20), one can find that

$$\chi = -(U_{gz}/U_{gx})_{q_2} = (U_{gz}/U_{gx})_{-q_2} \tag{61}$$

(note that $\chi > 0$, $\mathrm{sgn}U_{gx}(q_2) = -\mathrm{sgn}q_2$, $U_{gz} > 0$.

As we have already mentioned, the wave leakage from the duct is a result of the transformation of the trapped branch into the nontrapped. Therefore, the amplitude of the trapped wave must decrease with the increase of ζ . Deriving (55), we used an adiabatic approach, assuming that the amplitude A in (45) is constant. This was justified because expression (55) is exponentially small and, therefore, the radiation flux is also small. Now one can obtain an equation for $A(\zeta)$, using (55) and conservation laws. This equation reads [7]

$$\partial_\zeta \ln A \approx -\frac{\pi^2 \left[\epsilon_0^2 + (\eta_0 + g_0)^2\right](\epsilon_0 - \eta_0 + g_0)^2}{8\epsilon_0^2\eta_0^2\chi\gamma}\exp\left(-\frac{\pi}{\sqrt{\gamma}}\right) \tag{62}$$

where, according to (58), $\chi \sim \sqrt{\gamma}$. We see that $\partial_\zeta A$ decreases beyond all powers of γ.

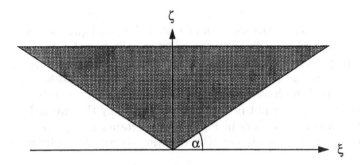

Fig. 2. Region occupied by the radiation at initial conditions (54); $\tan\alpha = |\chi|$

4   Slab whistler solitons

Extending the above results to the solitons, we first assume that the nonlinearity originates because of the ponderomotive interaction of the whistler
wave with plasma. Considering a collisionless plasma with hot electrons and
cold ions, we can neglect the wave damping. Then the density variation produced by the ponderomotive force can be written as [17,11]

$$\nu = \left(|F|^2 - \frac{1-u}{1+u}|G|^2 - \frac{2(1-u)}{u}|E_z|^2\right)E_0^{-2} \tag{63}$$

where

$$E_0^2 \approx 32\pi u(1-u)\alpha_0^{-2}N_0T_e, \tag{64}$$

T_e is the electron temperature and other notations are defined in (23). Now
we use Eqs. (11), (13) and (37) with

$$\epsilon + g = \epsilon_0 + g_0 + \frac{\alpha_0^2\nu}{u(1-u)}, \quad \epsilon - g = \epsilon_0 - g_0 - \frac{\alpha_0^2\nu}{u(1+u)}. \tag{65}$$

[These formulas follow from (9), (23) and (24)]. Then we arrive at the approximate system of equations

$$2i\partial_\zeta\bar{F} + \partial_\xi^2\bar{F} + \frac{\epsilon_0 - \eta_0 - g_0}{\epsilon_0 + \eta_0 + g_0}\partial_\xi^2\bar{G} + \frac{p_0^2}{\mu^2}\nu(\xi)\bar{F} = 0 \tag{66}$$

$$2i\partial_\zeta\bar{G} + \frac{\epsilon_0 + \eta_0 - g_0}{\epsilon_0 + \eta_0 + g_0}\partial_\xi^2\bar{G} + \frac{\epsilon_0 - \eta_0 + g_0}{\epsilon_0 + \eta_0 + g_0}\partial_\xi^2\bar{F} - \frac{p_0^2(1-u)}{\mu^2(1+u)}\nu(\xi)\bar{G}$$
$$= \frac{2g_0}{\mu^2}\bar{G}, \tag{67}$$

where we used the same variables as in eqs. (32)–(35). Assuming that $\nu \sim \mu^2$, we can neglect the last term on the left hand side of Eq. (67) and we arrive at the equations looking like (38) and (39) [Now Eq. (66)] is in fact nonlinear]. Basing on the solution of system (38) and (39) we can guess [7] that eqs.(66) and (67) describe the whistler soliton and a tunneling wave. [The density variation (63), produced by the ponderomotive force, serves as a duct, trapping the whistler wave]. Neglecting also the terms with derivatives of $\bar{G}$ in Eq. (67), we have $\bar{G} \sim \mu^2 \partial_\xi^2 \bar{F}$. This rough estimate, however, is valid inside the soliton, but not in the region occupied by the emitted radiation . Neglecting the term with G in Eq. (66) and retaining only the first term in (63), we reduce Eq. (66) to the NLS equation [7] with the soliton solution

$$\bar{F}(\xi,\zeta) = \left(\sqrt{2}\mu \frac{E_0}{p_0}\right)\Psi_0(\xi)\exp(i\zeta/2), \tag{68}$$

$$\Psi_0(\xi) = \operatorname{sech}\xi. \tag{69}$$

From eqs. (63), (68) an (69) we then obtain the density variation produced by the soliton

$$\nu(\xi) \approx \left(\frac{2\mu^2}{p_0^2}\right)\operatorname{sech}^2\xi. \tag{70}$$

This indeed describes a density hump which traps the whistler wave at $\omega < \omega_c/2$. Equation (70) coincides with (31) for $b = 1$.

Now we take into account the terms with $\bar{G}$, neglecting the term with $\nu(\xi)\bar{G}$ in Eq. (67) and retaining only the first term in Eq. (63). Then the solution of eqs. (66) and (67) can be written in the form

$$\bar{F}(\xi,\zeta) = \left(\sqrt{2}\mu \frac{E_0}{p_0}\right)\Psi(\xi,\zeta)\exp(i\zeta/2), \tag{71a}$$

$$\bar{G}(\xi,\zeta) = \left(\sqrt{2}\mu \frac{E_0}{p_0}\right)\Phi(\xi,\zeta)\exp(i\zeta/2) \tag{71b}$$

where

$$\Psi(\xi,\zeta) = \Psi_0(\xi) + \psi(\xi,\zeta), \tag{72a}$$
$$\Phi(\xi,\zeta) = \Phi_0(\xi) + \varphi(\xi,\zeta), \tag{72b}$$

with $\Psi_0(\xi)$ from (69) and $\Phi_0(\xi)$ satisfying the equation

$$\frac{\epsilon_0 + \eta_0 - g_0}{\epsilon_0 + \eta_0 + g_0}\partial_\xi^2\Phi_0 + \frac{\epsilon_0 - \eta_0 + g_0}{\epsilon_0 + \eta_0 + g_0}\partial_\xi^2\Psi_0 = \frac{2g_0}{\mu^2}\Phi_0. \tag{73}$$

Then we arrive at the following equations for $\psi(\xi,\zeta)$ and $\phi(\xi,\zeta)$

$$2i\partial_\zeta\psi + \partial_\xi^2\psi - \psi + 2\operatorname{sech}\xi(2\psi + \psi^*) + \frac{\epsilon_0 - \eta_0 - g_0}{\epsilon_0 + \eta_0 + g_0}\partial_\xi^2\varphi$$
$$= \gamma\frac{[(\epsilon_0 - \eta_0)^2 - g_0^2]}{4\epsilon_0\eta_0}\widehat{L}^{-1}\partial_\xi^4\Psi_0, \tag{74}$$

$$2i\partial_\zeta\varphi + \frac{\epsilon_0 + \eta_0 - g_0}{\epsilon_0 + \eta_0 + g_0}\partial_\xi^2\varphi + \frac{\epsilon_0 - \eta_0 + g_0}{\epsilon_0 + \eta_0 + g_0}\partial_\xi^2\psi = -\frac{4\epsilon_0\eta_0}{\gamma(\epsilon_0 + \eta_0 + g_0)^2}\varphi \tag{75}$$

where γ is defined in (53) . This system, like eqs. (50) and (51), can be solved by means of the Fourier-Laplace transforms and the solution looks similar. Then we again arrive at eqs. (55), (56) and (60) with somewhat different constant C (but of the same order) and $A = 1$ [7]. This is not surprising in view of the described above similar mechanisms of the wave trapping in ducts and solitons.

As it is seen from Eq. (55), the amplitude of the radiated wave is exponentially small provided $\sqrt{\gamma} \ll 1$ (which is also a condition of the validity of the asymptotic approximation used in [7]). This condition can be expressed in terms of the soliton parameters as follows. Indeed, from (69) and (39) it follows that the soliton width in dimensional units is

$$\delta_s = \frac{c}{\omega}\sqrt{\frac{2(1-2u)}{1-u}}\mu^{-1}. \tag{76}$$

The x-projection of the wave vector of the radiated wave reads

$$k_x^{(2)} = \frac{\omega}{c}q_2 \approx \alpha_0\sqrt{\frac{1-2u}{u^3(1-u)}}\frac{\omega}{c}, \tag{77}$$

where we used Eqs. (27) and (15). From these expressions and Eq. (53), it follows that

$$\delta_s k_x^{(2)} \approx 2\gamma^{-\frac{1}{2}}. \tag{78}$$

Therefore, the condition $\sqrt{\gamma} \ll 1$ can be written as

$$\delta_s k_x^{(2)} \gg 2, \tag{79}$$

i.e., the soliton width must be much larger than $1/k_x^{(2)}$. This condition is violated if ω is sufficiently close to $\omega_c/2$. Though at $\omega \to \omega_c/2$, our theory is not applicable, one can guess that in this case the radiation becomes rather strong and, in fact, the whistler solitons have short lifetimes. This conclusion is confirmed by numerical experiments which show that the soliton lifetimes decrease together with $\omega_c/2 - \omega$; it also naturally agrees with the existence of the whistler solitons only at $\omega < \omega_c/2$.

The wave leakage from the soliton leads to the decrease of its amplitude and, therefore, of the parameter γ . The equation for γ reads [7]

$$\partial_\zeta \gamma \approx \frac{(\epsilon_0 - \eta_0 + g_0)^2[\epsilon_0^2 + (\eta_0 + g_0)^2]}{2\epsilon_0\eta_0[\epsilon_0^2 + \eta_0^2 + \epsilon_0(\epsilon_0 + \eta_0)]}\frac{|C|^2\pi^2}{\sqrt{\gamma}}\exp\left(-\frac{\pi}{\sqrt{\gamma}}\right). \tag{80}$$

The right hand side of (80) is negative because $\epsilon_0\eta_0 < 0$ and the decrease of the soliton amplitude is exponentially small at condition (79). Eq. (80) is analogous to (62) and can be solved by an approach developed in [18] with similar results.

The above approach can be readily extended to the Kerr nonlinearities. As a starting equation we may take

$$\text{curl curl}\mathbf{E} = \frac{\omega^2}{c^2}(\hat{\epsilon}_0\mathbf{E} + \mathbf{D}_N), \tag{81}$$

where $\mathbf{E}(x, z)$ is the electric field amplitude, defined by Eq. (1), $\hat{\epsilon}_0$ the linear dielectric permitivity tensor and $\mathbf{D}_N = \hat{\epsilon}_0\mathbf{E}$ is the nonlinear part of the electric displacement vector. For the ponderomotive nonlinearity, we can write,

$$\hat{\epsilon}_N = \frac{\partial\hat{\epsilon}}{\partial\omega_p}\frac{\partial\omega_p}{\partial\nu}\nu \tag{82}$$

where $\hat{\epsilon}$ is defined by formulas (8) and (9) and ν is given by Eqs. (63) and (64). This leads to Eqs. (66) and (67) and the above results. If

$$\mathbf{D}_N = \kappa\Pi\mathbf{E}, \quad \Pi = \frac{1}{2}(|F|^2 + |G|^2 + 2|E_z|^2), \tag{83}$$

we have an isotropic Kerr nonlinearity $\mathbf{D}_N = \kappa|\mathbf{E}|^2\mathbf{E}$. We shall assume that $\kappa > 0$. One can write

$$\kappa = p_0^2/E_0^2 \tag{84}$$

where E_0 is a constant with the dimension of electric field. As before, we assume that the nonlinear term is small and introduce a parameter μ, such that $|\mathbf{E}|^2/E_0 \sim \mu^2/p_0^2 \ll 1$. Then one can use the approximation $\Pi \approx (1/2)|F|^2$. Using variables (32)–(35), (71) and (72) we arrive at Eqs. (74) and (75) with $\Psi_0(\xi)$ from Eq. (69). The solution of these equations are expressed by formulas (55)–(60) with $A = 1$; it describes the radiation outgoing from the soliton [see Fig.2 and explanations after Eqs. (60) and (75)]. The decrease of the soliton amplitude due to the radiation is again determined by Eq. (80). Therefore, Eqs. (81) and $\mathbf{D}_N = \kappa\Pi\mathbf{E}$ with $\Pi \approx (1/2)|F|^2$ lead to the same results as the ponderomotive nolinearities with the density variation (63). This is in agreement with the results obtained (by means of another approach) for the slab solitons in gyrotropic media, which are not necessarily plasmas [12]. The gyrotropic properties for such matherials are determinened by the parameter g that is called "the gyrotropic parameter". {The basic equations of the linear theory [19] can be reduced to the linear part of Eq. (81) with the elements of $\hat{\epsilon}$ written in Eqs. (8)}. Generally, the gyrotropic parameter g may be both positive, as for whistlers, and negative. At $g < 0$, the solitons do not radiate (in [12], this parameter was used with the sign opposite to the present paper).

5   Self-focusing of whistler wave beams

The wave leakage from nonlinear structures plays an important role in the collapse type phenomena of nonlinear waves. A typical example is the self-focusing of whistler waves [10,11] which will be considered in this section. We confine ourselves to the ponderomotive nonlinearity. Introduce cylindrical coordinates r, ϕ, z and assume that the density variation in the wave beam is axially symmetric, i.e. $\nu = \nu(r, z)$. Then Eq. (2) with (4) and (8) has solutions of the form

$$F \equiv E_r - iE_\phi = F_m(r, z)e^{im\phi}, \ G \equiv E_r + iE_\phi = G_m(r, z)e^{im\phi}, \ E_z = E_{mz}e^{im\phi},$$
$$(85)$$

where $m = 0, \pm 1, \pm 2, \ldots$. Substituting (85) into (2), we arrive at the equations

$$\frac{\partial^2 F_m}{\partial z^2} + \widehat{\Lambda}_{(r)}^{(m-1)} F_m - \left(\frac{\partial}{\partial r} + \frac{m}{r}\right)(\nabla \cdot \mathbf{E}_m) + \frac{\omega^2}{c^2}(\epsilon + g)F_m = 0, \quad (86a)$$
$$\frac{\partial^2 G_m}{\partial z^2} + \widehat{\Lambda}_{(r)}^{(m+1)} G_m - \left(\frac{\partial}{\partial r} - \frac{m}{r}\right)(\nabla \cdot \mathbf{E}_m) + \frac{\omega^2}{c^2}(\epsilon - g)G_m = 0, \quad (86b)$$

$$\nabla \cdot \mathbf{E}_m = \frac{1}{2r}\frac{\partial}{\partial r}[r(F_m + G_m)] - \frac{m}{2r}(F_m - G_m) + \frac{\partial E_{mz}}{\partial z}, \qquad (87)$$

$$\frac{\partial}{\partial z}(\eta E_{mz}) + \frac{1}{2r}\frac{\partial}{\partial r}\left\{[(\epsilon + g)F_m - (\epsilon - g)G_m]\right\}$$
$$+ \frac{m}{2r}[(\epsilon + g)F_m - (\epsilon - g)G_m] = 0, \qquad (88)$$

where

$$\widehat{\Lambda}_{(r)}^{(n)} = \frac{1}{r}\frac{\partial}{\partial r}\left(r\frac{\partial}{\partial r}\right) - \frac{n^2}{r^2}. \qquad (89)$$

At large r these equations turn into (11)–(13).

Now assume, as in the slab problem, that $\nu(r, z)$ is a slow function of its arguments. Introduce stretched variables

$$\rho = \mu\frac{\omega}{c}p_0 r, \quad \zeta = \mu^2\frac{\omega}{c}p_0 z \qquad (90)$$

($\mu \ll 1$) and neglect the terms with the derivatives of $\nu(r, z)$ (and, respectively, the derivatives of ϵ, η, g). Similar to formulas (33) and (34), we also write

$$F_m = \bar{F}_m(\rho, \zeta)\exp[i(\omega/c)p_0 z], \quad G_m = \bar{G}_m(\rho, \zeta)\exp[i(\omega/c)p_0 z] \quad (91a)$$
$$E_{mz} = \bar{E}_{mz}(\rho, \zeta)\exp[i(\omega/c)p_0 z] \qquad (91b)$$

and neglect terms with $\mu^4 \partial_\zeta^2 \bar{F}_m$ and $\mu^4 \partial_\zeta^2 \bar{G}_m$. Then we obtain from Eqs. (86a)–(88) and (65)

$$i\frac{\partial \bar{F}_m}{\partial \zeta} + \frac{\epsilon + g + \eta}{4\eta} \widehat{\Lambda}_{(\rho)}^{(m-1)} \bar{F}_m + \frac{\nu}{2\mu^2} \bar{F}_m$$
$$+ \frac{\epsilon - g - \eta}{4\eta}\left[\widehat{\Lambda}_{(\rho)}^{(m+1)} \bar{G}_m + \frac{2m^2}{\rho^2}\bar{G}_m + \frac{2m}{\rho^2}\frac{\partial}{\partial\rho}(\rho\bar{G}_m)\right] = 0, \quad (92)$$

$$i\frac{\partial \bar{G}_m}{\partial \zeta} - \frac{g}{\mu^2 p_0^2}\bar{G}_m + \frac{\epsilon - g + \eta}{4\eta}\Lambda_{(\rho)}^{(m+1)} \bar{G}_m$$
$$+ \frac{\epsilon + g - \eta}{4\eta}\left[\widehat{\Lambda}_{(\rho)}^{(m-1)} \bar{F}_m + \frac{2m^2}{\rho^2}\bar{F}_m - \frac{2m}{\rho^2}\frac{\partial}{\partial\rho}(\rho\bar{F}_m)\right] = 0, \quad (93)$$

where operator $\widehat{\Lambda}_{(\rho)}^{(n)}$ is given by Eq. (89) with $r \to \rho$, and the terms with $\mu^2 \partial_\zeta^2 \bar{F}_m$ and $\mu^2 \partial_\zeta^2 \bar{G}_m$ are neglected. The additional terms, containing derivatives of ϵ, η, g, are derived in [11]; a qualitative investigation, supported by numerical calculations, shows that they are less important then those retained in Eqs. (92) and (93). We require that all terms in Eq. (92), containing $\bar{F}_m$, should be of the same order. Therefore $\nu \sim \mu^2$,

$$\epsilon = \epsilon_0 + O(\mu^2), \quad g = g_0 + O(\mu^2), \quad \eta = \eta_0 + O(\mu^2), \quad (94)$$

and we shall consider Eqs. (92) and (93) with

$$\epsilon \to \epsilon_0, \quad \eta \to \eta_0, \quad g \to g_0. \quad (95)$$

To these equations we add

$$\bar{E}_{mz} \approx \frac{i\mu}{2p_0\eta_0\rho}\left\{(\epsilon_0 + g_0)\left[\frac{\partial}{\partial\rho}(\rho\bar{F}_m) - m\bar{F}_m\right] + (\epsilon_0 - g_0)\left[\frac{\partial}{\partial\rho}(\rho\bar{G}_m) + m\bar{G}_m\right]\right\}$$
$$(96)$$

which follows from Eq. (88), after similar simplifications. From Eq. (63), we see that $\nu \sim |\bar{F}_m|/E_0$. Therefore μ is equal, by the order of magnitude, to the normalized electric field strength (as in the slab case).

It is instructive to apply, first, these equations to the homogeneous plasma [11]. Then $\nu = 0$ and it is reasonable to return to the original variables r and z. The solutions to Eqs. (92), (93) with $\nu = 0$ are

$$\bar{F}_m(r, z) = A_1 Z_{m-1}\left(\frac{\omega}{c}qr\right)\exp\left(i\frac{\omega}{c}\mu p_1 z\right), \quad (97a)$$

$$\bar{G}_m(r, z) = A_1 Z_{m+1}\left(\frac{\omega}{c}qr\right)\exp\left(i\frac{\omega}{c}\mu p_1 z\right), \quad (97b)$$

where p_1 is a free parameter, $A_{1,2}$ are constant coefficients and $Z_n(w)$ are cylindrical functions, satisfying the Bessel equation [20]

$$\widehat{\Lambda}_{(r)}^{(n)} Z_n\left(\frac{\omega}{c}qr\right) = -\left(\frac{\omega}{c}q\right)^2 Z_n\left(\frac{\omega}{c}qr\right). \quad (98)$$

If $Z_n(w)$ are Hankel functions $H_n^{(1)}(w)$, expressions (97) and (91) describe a spiral outgoing cylindrical wave with the parallel wave number $(\omega/c)(p_0+\mu p_1)$ and perpendicular wave number $(\omega/c)q$. Using recursive formulas for the cylindrical functions [20] and Eq. (98), we arrive at the system of equations for $A_{1,2}$

$$[4\eta_0\mu p_1 p_0 + (\epsilon_0 + g_0 + \eta_0)q^2]A_1 - (\epsilon_0 - g_0 - \eta_0)q^2]A_2 = 0, \quad (99a)$$
$$(\epsilon_0 + g_0 - \eta_0)q^2]A_1 - [4\eta_0\mu p_1 p_0 + 4g_0\eta_0 + (\epsilon_0 - g_0 + \eta_0)q^2]A_2 = 0. \quad (99b)$$

The condition of its solvability approximately reads

$$\epsilon_0\eta q^4 + g_0(\epsilon_0 + g_0 + \eta_0)q^2 + 4\eta_0 g_0\mu p_1 p_0 = 0. \quad (100)$$

This gives two branches of q at fixed p_1,

$$q_1^2 \approx -\frac{4\eta_0 p_0^2}{\epsilon_0 + g_0 + \eta_0}\frac{\mu p_1}{p_0} \approx -4\frac{\alpha_0^2}{u(1 - 2u)}\frac{\mu p_1}{p_0}, \quad (101)$$

$$\left(\frac{A_2}{A_1}\right)_1 = \frac{\epsilon_0 + g_0 - \eta_0}{4\eta_0 g_0}q_1^2 \approx -(1 + u)\frac{\mu p_1}{p_0}, \quad (102)$$

and

$$q_2^2 \approx -g_0\frac{\epsilon_0 + g_0 + \eta_0}{\epsilon_0} \approx \frac{1 - 2u}{u^2}p_0^2, \quad (103)$$

$$\left(\frac{A_2}{A_1}\right)_2 = \frac{\epsilon_0 + g_0 + \eta_0}{\epsilon_0 - g_0 - \eta_0} \approx \frac{(2u - 1)(1 + u)}{1 - u}. \quad (104)$$

Evidently, the existence of the first branch is indispensable, because otherwise there will not be a continuous transition to the longitudinal propagation at $\mu \to 0$. From formula (101) it then follows that $p_1 < 0$ at $u < 1/2$ ($\omega < \omega_c/2$) and there are two real branches of q, while at $\omega > \omega_c/2$ there is only one branch, q_1 with $p_1 > 0$ (q_2 is imaginary). Thus we, naturally, arrive at the same physical conclusions as in the plane geometry. Comparing formulas (101)–(104) with the plane wave expressions (26)–(28) at $\nu(x) = 0$ we see that they are in agreement and $\mu p_1 = \Delta p$.

In fact, the only simplification made in the derivation of Eqs. (92) and (93) from (86)–(88) for the homogeneous plasma, was the neglect of terms with the second derivatives and (the derivatives $\partial_z^2\bar{F}_m \sim \mu^2\partial_\zeta\bar{F}_m$ and $\partial_z^2\bar{G}_m \sim \mu^2\partial_\zeta\bar{G}_m$ (the derivatives of $\bar{F}_m, \bar{G}_m$ over r were not neglected). This leads to the restriction $\mu p_1 \ll 1$ and this is the only reason for the appearance of a small parameter μ in applications of system (92), (93) to the homogeneous plasma. The full solutions to Eqs. (86)–(88) describing cylindrical waves in homogeneous plasma [21,11] with $|p - p_0| \sim p_0$ lead to the dispersion equation (22) obtained for slab geometry; as we know, at $|p - p_0| \ll p_0$, it gives approximate formulas (26)–(28) for plane and (101)–(104) for cylindrical whistler waves. It is remarkable that Eq. (22) as well as Eqs. (101)–(104) do not contain the azimuth number m. Now we apply Eqs. (92) and (93) to inhomogeneous plasmas and to nonlinear whistler waves with the ponderomotive nonlinearity. Then a small parameter μ originates in the assumptions characterizing density variations ($\nu \sim \mu^2$, $\partial_r \ln \nu \sim \mu$). Assume that the wave beam and, therefore, ν vanish at ∞. Then we can argue in a way similar to that one in the plane case. From Eq. (93) we find that inside the beam $\bar{G}_m \sim \mu^2 \partial_\rho^2 \bar{F}_m$. Due to that, we consider the terms with $\bar{G}_m$ in Eq. (92) as small *inside* the beam and neglect them (in the lowest approximation). Likewise, in Eq. (63) we can neglect both terms with G and E_z. Then Eq. (92), with account of (95), approximately turns into

$$i\frac{\partial \bar{F}_m}{\partial \zeta} + \frac{\epsilon_0 + g_0 + \eta_0}{4\eta_0}\widehat{\Lambda}_{(\rho)}^{(m-1)}\bar{F}_m + \frac{1}{2\mu^2 E_0^2}|\bar{F}_m|^2\bar{F}_m = 0, \qquad (105)$$

which plays a role of the NLS equation for $\bar{F}_m$ for spiral beams [11]. Instead of the operator $\widehat{\Lambda}_{(\rho)}^{(m)}$, that is the radial part of the Laplacian for the m th angular mode, it contains $\widehat{\Lambda}_{(\rho)}^{(m-1)}$. The reason is that in addition to the angular momentum of the m th harmonic there is a "spin" angular momentum, associated with the right-handed polarization of the field described by $\bar{F}_m$. Solving Eq. (105), we then can calculate $\bar{G}_m$ and $\bar{E}_z$ by means of Eqs. (93) and (96).

Equation (105), in the dimensional units, can be written as [11]

$$iV_g\frac{\partial \bar{F}_m}{\partial z} + \frac{1}{2}S\widehat{\Lambda}_{(r)}^{(m-1)}\bar{F}_m - \delta\omega \bar{F}_m = 0, \qquad (106)$$

where

$$V_g = \left(\frac{\partial \omega}{\partial k_z}\right)_0, \quad S = \left(\frac{\partial^2 \omega}{\partial k_\perp^2}\right)_0, \quad \delta\omega = N_0\left(\frac{\partial \omega}{\partial N}\right)_0\nu. \qquad (107)$$

Here $\omega = \omega(k, k_z, N)$ is the function (20) with $k^2 = k_z^2 + k_\perp^2$ and $(\ )_0$ means that after differention one should put $k_\perp = 0$, $N = N_0$. The parameter $\delta\omega$ is the nonlinear frequency shift. Equation (106) describes the stationary self-focusing if $S\delta\omega < 0$ (which is a necessary condition). From the explicit expressions

$$V_g \approx \frac{2c(1-u)}{p_0}, \quad S \approx \frac{c^2(1-2u)}{\omega p_0^2}, \quad \delta\omega \approx -\omega(1-u)\nu, \qquad (108)$$

it follows that the self-focusing is possible only at $\omega < \omega_c/2$, because according to Eq. (63), $\nu > 0$ coincides with the conditions of the existence of plane solitons) . Another condition of the self-focusing is [11],

$$J_m = \int_0^\infty \left(\left|\frac{\partial \bar{F}_m}{\partial r}\right|^2 + \left(\frac{m-1}{r}\right)^2|\bar{F}_m|^2 - \frac{\omega p_0}{2c}|\bar{F}_m|^4\right) r\,dr < 0 \qquad (109)$$

which follows from the virial theorem [22]. We also observe that the asymptotic behavior of the wave beam at sufficiently large r and suffuciently far from the collapse point can be described by a Fourier integral of the functions (97) with the variable of integration p_1, $A_1 = A_1(p_1)$, $Z_{m-1}(w) = H_{m-1}^{(1)}(w)$ and q_1 from Eq. (101). Now consider all terms in Eqs. (92) and (93). We can use a reasoning similar to that one applied to Eqs. (66) and (67), describing slab solitons. At sufficiently large r, the solutions to Eqs. (92), (93) should turn into a sum of two Fourier integrals containing functions (97) with two branches described by (101), (102) and (103), (104). The first of them, containing q_1, represents the asymptotics of the wave beam and the second integral, containing q_2, describes the radiation emanated by the wave beam. The radiation increases with the narrowing of the beam and losses due to the radiation may cause the violation of the self-focusing condition (109). Then should come a defocusing stage and the radiation will be less intensive, etc. Finally the wave beam should become defocused because of continual radiation, contrary to the prediction that follows from the NLS equation (105). A confirmation of this scenario was obtained by means of numerical experiments.

Here we describe relatively simple experiments based on the numerical solutions of the system (92), (93), (63) with simplifications (95) and (96). Of course, it is a model system based on the assumption of the smallness of parameter μ which is correct only for sufficiently broad beams. But the NLS equation, predicting self-focusing, must also be considered as a model equation formally valid at small μ. [In fact, it is less accurate than the system (92), (93)]. Thus, consider the solution of the system in question at $m = 1$ for the initial conditions,

$$\bar{F}_1(\rho, 0) = F_{10}\exp(-\rho^2/\rho_0^2), \quad \bar{G}_1(\rho, 0) = 0 \tag{110}$$

(in all computations we have taken $\rho_0 = 20\mu$) and with boundary conditions

$$\partial \bar{F}_1/\partial \rho = 0, \quad \bar{G}_1 = 0 \quad (\rho = 0). \tag{111}$$

At the end of the computational region ($\rho = \rho_{max}$ we have required the total absorption of the field. The results are presented in Figs. 3 and 4 for $u = \omega/\omega_c = 0.4$ and Fig. 5 for $u = 0.45$. Fig. 3 shows the evolution of the normalized field intensity versus ζ in the center of the wave beam ($\rho = 0$). The full line is $\bar{F}_1(0, \zeta)|^2$, obtained from the system (92) and (93) and the dotted line is the solution of NLS equation (105). The solution of Eqs. (92) and (93) shows, in accordance with the above analytical theory, the initial increase of the intensity and then decrease after a peak at $\zeta \sim 10^3\mu$. The decrease is caused by the radiation. At the same time, the numerical solution of the NLS equation, presented by the dotted line, shows an unlimited increase of the intensity which is nothing than collapse. The third line depicts the behavior

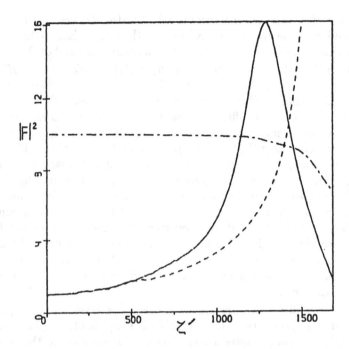

Fig. 3. Numerical solutions of the system (92), (93) and the NLS equation (105) at $\rho = 0$ for $u = 0.4$, $m = 1$ and $|F_{10}|^2 = 0.004$. (———) Plot of $|\bar{F}_1/\bar{F}_{10}|^2$ versus $\zeta' = \zeta/\mu^2$, obtained from Eqs. (92) and (93). (— — —) Plot of $|\bar{F}_1/\bar{F}_{10}|^2$, obtained from Eq. (105). (— · —) Plot of $I/10$, where I is the integral (112).

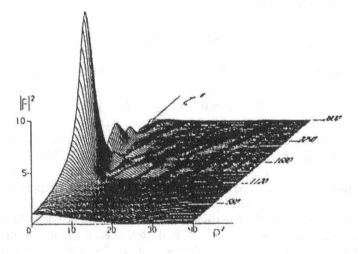

Fig. 4. Surface $|\bar{F}_1/\bar{F}_{10}|^2$ for $u = 0.4$, $m = 1$ and $|\bar{F}_{10}|^2 = 0.004$, obtained from Eqs. (92) and (93). $\rho' = \rho/\mu$ and $\zeta' = \zeta/\mu$.

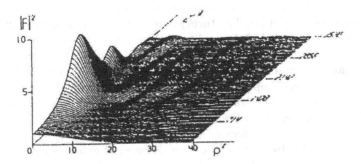

Fig. 5. The same as Fig.4 for $u = 0.45$ and $|\bar{F}_{10}|^2 = 0.0035$

of the integral

$$I(\zeta) = \int_0^{\rho_{max}} |\bar{F}_1(\rho, \zeta)|^2 \rho \, d\rho. \tag{112}$$

This function becomes noticeably decreasing after the radiation, emanating from the beam, reaches the end of the computational region ρ_{max}, where the condition of full absorption is imposed. The full relief of $|\bar{F}_1(\rho, \zeta)|^2$ at $u = 0.4$ is shown in Fig. 4. Here, one can clearly see the oscillatory regime of the beam defocusing as well as radiated waves. A similar relief for $u = 0.45$ is shown in Fig. 5. Here, the peak of intensity is lower than at u=0.4 while the wavelength of the radiation is larger. This figure confirms the analytical predictions that the intensity of the resonant radiation increases and the perpendicular wavenumber decreases if the whistler frequency approaches $\omega_c/2$.

Very similar results were obtained for other m, e.g., for $m = 0$ [10] and $m \neq 0, 1$ (unpublished). Numerical studies of other, more complicated, models with softer restrictions, which are now in progress, seem to confirm the described analytical predictions.

6 Concluding remarks

Considering plasma density duct, aligned parallel to the ambient magnetic field, with trapped whistler waves we have shown that at $\omega < \omega_c/2$, when the ducts are density humps, the trapping is not complete and therefore waves are continuously emanated from the duct. This process takes place because of the transformation of the trapped whistler wave into another mode that cannot be trapped and therefore escaping from the duct. Then we exploited a substantial similarity between envelope whistler solitons (which exist only at $\omega < \omega_c/2$) and plasma ducts and showed that the whistler solitons must also emanate radiation and therefore loose their energy. A full-wave analytical theory of the whistler radiation from the plain (one-dimensional) ducts and solitons was developed [7]. The theory was then extended to the cylindrical geometry where, instead of plain solitons, there are spiral wave beams parallel to the ambient magnetic field. The equations, self-consistently describing such beams that ineract with plasma by means of the ponderomotive force, are reduced in the lowest approximation to the NLS equation which predicts the beams self-focusing. The leakage of the trapped radiation from the beam, which has many common features with the tunneling [11,23], increases with the decrease of the beam effective diameter and this causes the seasing of the self-focusing and subsequent defocusing. Numerical experiments confirming these processes [10,11] are discussed.

The developed theory can be extended to other gyrotropic media with Kerr nonlinearities where the gyrotropic constant g can be not only positive (as for whistlers) but also negative [12,13]. The solitons can exist at both signs of g. However, at $g < 0$, the soliton radiation does not take place [12]. A study of wave beams with cylindrical symmetry in gyrotropic media shows that they are defocusing at $g > 0$ (in oscillatory regime) and weakly oscillate around stable cylindrical beams at $g < 0$ [13].

In this paper we have considered only steady structures. One can show, adding to some basic equations the appropriate terms with time derivatives, that the considered structures are stable with respect to the time dependent perturbations (within the range of the validity of used approximations).

References

1. L.R.O. Storey, Phil.Trans.Roy.Soc. A, 246, 113 (1953).
2. R.A.Helliwell, *Whistlers and Related Ionospheric Phenomena* (Stanford University Press, Stanford, 1965).
3. R.L.Smith, R.A. Helliwell and I.W. Yabroff, J. Geophys. Res. 65, 815 (1960).
4. A.G. Litvak, Zh.Eksp.Teor.Fiz. 57, 629 (1969) [Sov. Phys. JETP 30, 344 (1970)].
5. V.I. Karpman and R.N. Kaufman, Zh. Eksp. Teor. Fiz. Pis'ma Red. 33, 266 (1981) [Sov. Phys. JETP Lett. 33, 252 (1981)].
6. V.I. Karpman and R.N. Kaufman, Zh. Exp. Teor. Fiz. 80, 1845 (1981) [Sov. Phys. JETP, 53, 956 (1981)] .
7. V.I.Karpman, Phys. Plasmas 5 , 156 (1998).
8. E. A. Kusnetsov, A.M. Rubenchik and V.E. Zakharov, Phys.Rep. 142, 103 (1986).
9. J.J. Rasmussen and K. Rypdal, Phys. Scr. 33, 481 (1986).
10. V.I. Karpman and A.G. Shagalov, Zh.Eksp.Teor.Fiz. 87, 422(1984) [Sov. Phys. JETP 60, 242 (1984)].
11. V.I. Karpman, R.N. Kaufman and A.G. Shagalov, Phys. Fluids B 4 3087 (1992).
12. V.I. Karpman, Phys. Rev. Lett. 74, 2475 (1995).
13. V.I. Karpman and A.G. Shagalov, Phys. Rev. A 46, 518 (1992).
14. L.D. Landau and E.M. Lifshitz, *Quantum Mechanics. Non- relativistic Theory* (Pergamon Press, Oxford, 1977).

15. V.I.Karpman and R.N. Kaufman, Plan. Space Sci. 32, 1505 (1984).
16. V.I. Karpman and R.N. Kaufman, Radio Sci. 22 ,1026 (1987).
17. Yu. S. Barash and V.I. Karpman,Zh. Exp. Teor. Fiz. 85 , 1942 (1983) [Sov. Phys. JETP 58. 1139 (1983).
18. V.I. Karpman , Phys. Lett.A 244,397 (1998).
19. L.D. Landau and E.M. Lifshitz, *The Electrodynamics of Continuous Media* (Pergamon, Oxford, 1983).
20. I.S.Gradsteyn and I.M.Ryzhik *Table of Integrals, Series and Products*, edited by A.Jefrey (Academic Press, New York,1994).
21. R.N. Kaufmam, Izv. Vysh. Uchebn. Zaved. Radiofiz. 28, 566 (1985) [Radiophys. Quantum Electron. 28, 390 (1985)].
22. S.N.Vlasov, V.A. Petrishchev and V.I. Talanov, Izv. Vysh. Uchebn. Zaved. Radiofiz. 14, 1353 (1972) [Radiophys. Quantum Electron. 14, 1062 (1985)].
23. V.I. Karpman, Phys.Rev.E 47, 2073 (1993).

Alfvén Wave Filamentation
and Plasma Heating

S. Champeaux[1], A. Gazol[2], T. Passot[3], and P.L. Sulem[3]

[1] Physics Department, University of California San Diego,
    La Jolla CA 92093-0319
[2] Instituto de Astronomía, J. J. Tablada 1006, col. Balcones de Sta. María,
    Morelia, Michoacán, 58090 México
[3] CNRS UMR 6529, Observatoire de la Côte d'Azur, BP 4229,
    06304 Nice Cedex 4, France

Abstract. Alfvén wave filamentation is an important instability as it can lead
to wave collapse and thus to the formation of small scales. Different asymptotic
equations are here derived to describe this phenomenon. They apply in different
regimes, depending on the level of the dispersion with respect to the nonlinearity.
The (scalar) nonlinear Schrödinger equation, valid when the wave is strongly dispersive, allows the study of the influence of the coupling to magneto-sonic waves on
the development of the instability. This equation generalizes to a vector nonlinear
Schrödinger equation when the dispersion is decreased. The amount of dissipated
energy that results from the wave collapse when damping processes are retained,
is also estimated in these two cases. When the dispersion is weak and comparable
to the effects of the nonlinearities, a reductive perturbation expansion can be used
to derive long-wave equations that generalize the DNLS equations and also contain
the reduced MHD for the dynamics in the plane transverse to the propagation.

1   Introduction

It is believed that Alfvén waves play a central role in the dynamics of
magnetized plasmas. These waves have been widely observed in astro and
geophysical plasmas such as the solar wind [1], planetary and interplanetary shocks and near comets ([22] and references therein). The interaction
of Alfvén waves with particles can lead to particle heating, acceleration and
wave damping. Waves of finite amplitude are also subject to several important nonlinear processes and under their effect can undergo a transformation from an undamped state to a state where due to the formation of small
scales, dissipative processes can act. The equations governing the dynamics of
a three-dimensional plasma are in fact of a considerable complexity for both
analytic developments and numerical simulations on the present-day com-
puters, even when considered in the MHD approximation. As a consequence,
strong interest has been paid to the development of asymptotic approaches
leading to drastic simplifications of the problem in specific limits.

Neglecting kinetic effects, we concentrate here on the so-called Hall-MHD,
where dispersive effects associated with ion inertia originate from the Hall

term in the generalized Ohm's law. When written in a non-dimensional form where the density, the magnetic field and the velocities are measured respectively in terms of the unperturbed density ρ_0, the ambient magnetic field $\mathcal{B}$ and the Alfvén velocity $v_A = \mathcal{B}/\sqrt{4\pi\rho_0}$, the equations for a polytropic gas read

$$\partial_t \rho + \mathbf{\nabla} \cdot (\rho\mathbf{u}) = 0 \tag{1}$$

$$\rho(\partial_t \mathbf{u} + \mathbf{u} \cdot \mathbf{\nabla}\mathbf{u}) = -\frac{\beta}{\gamma}\mathbf{\nabla}\rho^\gamma + (\mathbf{\nabla}\times\mathbf{b})\times\mathbf{b} \tag{2}$$

$$\partial_t\mathbf{b} - \mathbf{\nabla}\times(\mathbf{u}\times\mathbf{b}) = -\frac{1}{R_i}\mathbf{\nabla}\times\left(\frac{1}{\rho}(\mathbf{\nabla}\times\mathbf{b})\times\mathbf{b}\right) \tag{3}$$

$$\mathbf{\nabla}\cdot\mathbf{b} = 0. \tag{4}$$

The quantity R_i denotes the nondimensional ion-gyromagnetic frequency, the parameter β the square ratio of the sound speed c_s and the Alfvén velocity, while γ is the polytropic gas constant.

In the case of weakly nonlinear dispersive waves, two different asymptotics are usually considered. When the dispersion is kept finite, the modulation of a monochromatic Alfvén wave with a small (but finite) amplitude is governed by a nonlinear Schrödinger (NLS) equation for the complex amplitude of the wave with, in some instance, a coupling to low frequency magnetosonic waves driven by the modulation. In contrast, a reductive perturbative expansion deals with long wavelength waves for which the dispersion is comparable to the nonlinearity. In the context of Alfvén waves propagating along a uniform ambient magnetic field and in the one- dimensional case, this asymptotics leads to the so-called "derivative nonlinear Schrödinger" (DNLS) equation ([20], [28], [21] and [22] for a recent review), because of the equality between the phase velocity of the Alfvén wave and the sound speed in the zero-dispersion limit. As noted in [24], this equation is to be generalized by retaining the coupling to magneto-sonic waves, to describe the filamentation instability.

Note that the DNLS formalism does not retain the decay instability which involves contrapropagating waves. This longitudinal instability which occurs for $\beta < 1$ affects wavenumbers larger than that of the carrier and may lead to the development of strongly nonlinear effects [4]. In higher dimensions, the linear problem involves additional instabilities [33] and the analysis of their competition in the nonlinear regime requires the direct formulation of the primitive equations. As a first step, we choose to isolate the various processes and to concentrate in this paper on the filamentation instability, using the modulation and long wavelength asymptotics associated respectively with situations where the nonlinearity is small or comparable to the dispersion at the scale of the carrier.

A well-known nonlinear phenomenon that takes place in the presence of a finite dispersion is the collapse of waves of small but finite amplitude, which collects energy over large distances and leads to an intense focusing of the wave. Section 2 discusses recent results on the transverse collapse (filamentation) of dispersive Alfvén waves propagating along the ambient magnetic field. The envelope dynamics of a weakly-nonlinear circularly polarized wave-train propagating along the ambient magnetic field then obeys a NLS equation with a potential resulting from the coupling to low-frequency magneto-sonic waves that are themselves driven by ponderomotive forces [5]. Special attention is paid to the quasi-stationary convective filamentation instability which develops in a strongly magnetized plasma (small β), a regime where kinetic effects are negligible. In this regime, on the scales where the modulation can be viewed as time-independent, the dispersion is negligible compared to the diffraction, the magneto-sonic waves become slaved to the Alfvén wave and the Alfvén-wave amplitude is governed by the usual two-dimensional NLS equation. The nonlinear development of the filamentation instability and its contribution to the heating of the plasmas are discussed with a possible application to the quiet regions of the solar corona [3]. When deviation from purely stationary regime is made relevant by reducing the duration of the incident wave-packet, the coupling to the magneto-sonic waves is to be retained. This coupling leads to the formation of sharp magneto-sonic fronts [26], [5], which provides an additional mechanism for small-scale generation and, in some instances, can also arrest the Alfvén wave collapse. Note that the envelope dynamics are significantly different when the direction of propagation of the Alfvén wave makes a finite angle θ with the ambient magnetic field [9].

Section 3 deals with the regime where the dispersion is too weak to enforce circular polarization of the wave on the modulation scales. In this case, the filamentation dynamics was shown to be governed by a vector NLS equation [6] and to produce different small-scale structures. The consequences on the wave dissipation are also considered.

Section 4 addresses the long-wavelength regime where nonlinearity and dispersion can balance. Generalized DNLS equations are presented. They take into account the coupling between the nonlinear dynamics of the Alfvén waves and the two-dimensional hydrodynamics that develops in the planes transverse to the mean magnetic field and also include the coupling with nonlinear magneto-sonic waves, thus extending the formalisms presented in [25] and [10]. They form a good starting point for the numerical exploration of the filamentation process beyond the framework of the envelope equations.

2   Filamentation of dispersive Alfvén waves

2.1   Modulation analysis

Consider a monochromatic Alfvén wave-train of small amplitude, propagating along the ambient magnetic field in an homogeneous and dispersive plasma. The modulation analysis is performed by introducing the slow variables $T = \mu t$, $X = \mu x$, $Y = \mu y$ and $Z = \mu z$ and by expanding the various fields in the form

$$b_y = \mu b_{y_1} + \mu^2 b_{y_2} + \cdots,$$
$$b_z = \mu b_{z_1} + \mu^2 b_{z_2} + \cdots,$$
$$b_x = 1 + \mu^2 b_{x_2} + \mu^3 b_{x_3} + \cdots,$$
$$\rho = 1 + \mu^2 \rho_2 + \mu^3 \rho_3 + \cdots,$$

$$u_y = \mu u_{y_1} + \mu^2 u_{y_2} + \cdots,$$
$$u_z = \mu u_{z_1} + \mu^2 u_{z_2} + \cdots,$$
$$u_x = \mu^2 u_{x_2} + \mu^3 u_{x_3} + \cdots,$$

with μ a small parameter. The different magnitudes of the transverse and longitudinal components select the Alfvén wave eigenmode at the level of the linear problem.

To leading order, one gets the linear system

$$\partial_t u_{y_1} - \partial_x b_{y_1} = 0, \qquad \partial_t b_{y_1} - \partial_x u_{y_1} - \frac{1}{R_i}\partial_{xx} b_{z_1} = 0, \tag{5}$$

$$\partial_t u_{z_1} - \partial_x b_{z_1} = 0, \qquad \partial_t b_{z_1} - \partial_x u_{z_1} + \frac{1}{R_i}\partial_{xx} b_{y_1} = 0 \tag{6}$$

which admits monochromatic solutions in the form

$$b_{y_1} = B_{y_1}(X, Y, Z, T)e^{i(kx - \omega t)} + \text{c.c.}$$

with complex amplitudes B_{y_1}, U_{y_1}, B_{z_1}, U_{z_1} related by the characteristic relations of Alfvén modes $U_{y_1} = -\frac{k}{\omega}B_{y_1}$ and $U_{z_1} = -\frac{k}{\omega}B_{z_1}$. The presence of dispersion at the scale of the wavelength when the parameter R_i is kept finite, prescribes that the amplitudes of the transverse magnetic field components then obey $B_z = i\sigma B_y$ with $\sigma = \pm 1$. The Alfvén wave is then right-hand ($\sigma = 1$) or left-hand ($\sigma = -1$) circularly polarized, with a dispersion relation $(\omega/k)^2 = 1 + \sigma\omega/R_i$ (where ω assumed positive).

At order μ^2 of the expansion, the equations for the density and the longitudinal velocity and magnetic field components read

$$\partial_t \rho_2 + \partial_x u_{x_2} = -(\partial_Y u_{y_1} + \partial_Z u_{z_1}) \tag{7}$$

$$\partial_t u_{x_2} + \beta \partial_x \rho_2 = -\frac{1}{2}\partial_x(b_{y_1}^2 + b_{z_1}^2) \tag{8}$$

$$\partial_t b_{x_2} = -(\partial_Y u_{y_1} + \partial_Z u_{z_1}) + \frac{1}{R_i}\partial_x(-\partial_Z b_{y_1} + \partial_Y b_{z_1}) \tag{9}$$

$$\partial_x b_{x_2} = -(\partial_Y b_{y_1} + \partial_Z b_{z_1}) \tag{10}$$

and are solved as

$$\rho_2 = \tilde{\rho}_2 e^{i(kx-\omega t)} + \text{c.c.} + \bar{\rho}_2 \tag{11}$$

$$u_{x_2} = \tilde{u}_{x_2} e^{i(kx-\omega t)} + \text{c.c} + \bar{u}_{x_2} \tag{12}$$

$$b_{x_2} = \tilde{b}_{x_2} e^{i(kx-\omega t)} + \text{c.c.} + \bar{b}_{x_2}, \tag{13}$$

where overbars refer to non-oscillating contributions which appear to be driven when considering the next order of the expansion. Due to the circular polarization of the Alfvén wave, no second harmonics arise. The amplitudes of the oscillating parts associated with the first harmonics are given by

$$\tilde{u}_{x_2} = \frac{\beta k}{\omega}\tilde{\rho}_2 \tag{14}$$

$$\tilde{\rho}_2 = -\frac{ik}{(\beta k^2 - \omega^2)}(\partial_Y + i\sigma\partial_Z)B_{y_1} \tag{15}$$

$$\tilde{b}_{x_2} = \frac{i}{k}(\partial_Y + i\sigma\partial_Z)B_{y_1}, \tag{16}$$

far from the resonance $\omega/k = \beta^{1/2}$ where the phase velocity of the Alfvén wave identifies with the sound speed.

In a formal multiple-scale analysis, the envelope equations arise as the solvability conditions which eliminate the resonant terms belonging to the null space of the adjoint of the linear operator. These terms include both terms proportional to $e^{i(kx-\omega t)}$ and non-oscillating contributions. The presence of dispersion in the linear problem prevent harmonics of the carrier to be resonant. Details of the analysis are given in [5]. It turns out that, on time and length scales of order μ^{-1} (compared with the period and wavelength of the carrier), the Alfvén wave envelope $B = B_y - i\sigma B_z = 2B_y = 2(B_{y_1} + \mu B_{y_2})$, the long wavelength longitudinal components $\bar{u}_x = \bar{u}_{x_2} + \mu\bar{u}_{x_3}$, $\bar{b}_x = \bar{b}_{x_2} + \mu\bar{b}_{x_3}$ and the mean density $\bar{\rho} = \delta + \bar{b}_x$ obey

$$i\left(\partial_T B + v_g\partial_X B\right) + \mu\alpha\Delta_\perp B$$

$$+\mu\frac{\omega}{k^2+\omega^2}\left(\frac{\omega^2}{k^2}\partial_{XX}B + \frac{k^2}{\omega^2}\partial_{TT}B + \frac{2k}{\omega}\partial_{XT}B\right)$$

$$-\mu k v_g\left(\frac{1}{v_g}\bar{u}_x + \frac{k^2}{2\omega^2}\bar{b}_x - \frac{1}{2}\delta\right)B = 0 \tag{17}$$

$$\partial_T\delta + \partial_X\bar{u}_x = 0 \tag{18}$$

$$\partial_T\bar{u}_x + \beta\partial_X(\delta + \bar{b}_x) + \partial_X\frac{|B|^2}{2} = -i\frac{\alpha}{v_g}\mu(B\Delta_\perp B^* - B^*\Delta_\perp B) \tag{19}$$

$$\partial_{TT}\bar{b}_x - \partial_{XX}\bar{b}_x - \Delta_\perp\left(\beta\delta + (\beta+1)\bar{b}_x + \frac{k^2}{2\omega^2}|B|^2\right) = 0. \tag{20}$$

In the above equations, $v_g = \omega' = \frac{2\omega^3}{k(k^2+\omega^2)}$ denotes the group velocity of the Alfvén wave. The coefficient $\alpha = \frac{k\omega}{k^2+\omega^2}\left(\frac{\omega^2}{2k^3} - \frac{\beta k}{2(\beta k^2 - \omega^2)}\right)$ identifies with the diffraction coefficient $\frac{1}{2}\frac{\partial^2\Omega}{\partial k_y^2}$ or $\frac{1}{2}\frac{\partial^2\Omega}{\partial k_z^2}$, evaluated for $\Omega = \omega$, $k_x = k$, $k_y = k_z = 0$.

Equations (17)–(20) can be viewed as an initial value problem either in time, in the context of the so-called absolute modulational instability [15] or in space, for the convective modulational instability [16]. Note that the concept of absolute/convective modulational instabilities referred to here should not be confused with the one used in the context of open flows where it characterizes the response of the flow to a localized perturbation [13].

Using that, to leading order, $\partial_T B + v_g \partial_X B = 0$, the longitudinal and temporal dispersive terms arising in (17) rewrite in the usual form $\frac{\omega''}{2}\partial_{XX}B$ for the absolute regime and $\frac{\omega''}{2v_g^2}\partial_{TT}B$ for the convective one, with $\omega'' = 2(\frac{\omega}{k^2+\omega^2})(\frac{\omega^2}{k^2} + \frac{k^2}{\omega^2}v_g^2 - \frac{2k}{\omega}v_g)$. Furthermore, the right-hand side of eq. (19) becomes relevant in the case of quasi-transverse (absolute) or quasi-static (convective) modulation, preventing from the degeneracy displayed by eqs. (18)-(19). Indeed in such a regime, at the scale of the envelope modulation the dispersive effects can be neglected compared with the diffraction in eq. (17) and in this case $-i\alpha\mu(B\Delta_\perp B^* - B^*\Delta_\perp B)$ can be replaced by $(\partial_T + v_g\partial_X)|B|^2$. The ponderomotive force with both a time and a space derivative contribution, given in [29] is thus recovered. Equations (18)-(19) can be viewed as the linearized fluid equations for the longitudinal dynamics, forced by a ponderomotive force whose expression can be obtained from the potential of the NLS equation using the Hamiltonian character of the system. Equation (17) indicate that, as usual, to leading order on time scales of order μ^{-1}, the Alfvén-wave packet is advected at the group velocity. To eliminate this trivial effect, in the absolute regime, the equations are rewritten in the reference frame moving at the Alfvén group velocity by defining $\xi = X - v_gT$ and introducing a slower time scale $\tau = \mu T$ typical of the envelope dynamics, while in the convective regime a delayed time $\tau = \mu(T - X/v_g)$ and a longer longitudinal length scale $\xi = \mu X$ is considered. It follows that the magnetosonic waves evolve on a time scale (absolute regime) or a longitudinal length scale (convective regime) shorter by a factor μ than the Alfvén wave envelope.

2.2 Convective filamentation

When in the convective regime, the wave-train modulation is assumed stationary ($\partial_\tau=0$), the description strongly simplifies and the envelope equation for the Alfvén wave reduces to a two-dimensional NLS equation

$$i\partial_\xi B + \frac{\alpha}{v_g}\Delta_\perp B + \frac{k}{4}\left(\frac{1}{\beta} + \frac{(k^2+\omega^2)^2}{\omega^4}\right)|B|^2 B = 0 \tag{21}$$

where the cubic nonlinearity reflects the interactions of the Alfvén wave with the mean fields $\bar{u}_x$, $\bar{b}_x$ and $\bar{\rho}$ which in such regime are slaved to the wave intensity and given by

$$\bar{u}_x = 0, \quad \bar{b}_x = -\frac{k^2+\omega^2}{2\omega^2}|B|^2, \quad \bar{\rho} = \frac{|B|^2}{2\beta}. \tag{22}$$

It follows that whatever its polarization, a plane Alfvén wave of amplitude B_0 is modulationally unstable relatively to large-scale perturbations when

$\beta < (\frac{\omega}{k})^2$ in agreement with numerical observations [16]. Considering perturbations in the form $e^{i(\mathbf{K}_\perp \cdot \mathbf{r} + K_x \xi)}$, the most unstable transverse scale associated with this instability is given by $K_{\perp max} = \left(\frac{q v_g}{\alpha}\right) B_0$ and its spatial growth rate is $\Im(K_{\parallel max}) = q B_0^2$ with $q = \frac{k}{4}\left(\frac{1}{\beta} + \frac{(k^2 + \omega^2)^2}{\omega^4}\right)$ reproducing in the case of long waves and small β, the analysis of Shukla and Stenflo [30]. Note the divergence in the above formulas as $\beta \to 0$, a regime where decay instabilities are known to play an important role. In a plasma with small but finite β, the convective filamentation instability is shown to be dominant compared with the longitudinal modulational instability leading to solitonic structures [5]. The nonlinear development of the filamentation instability can lead to the transverse collapse of the wave in a quasi-stationary way as this one propagates away from its source. This convective instability can be understood in the following way. A local increase of the plasma density reduces the Alfvén velocity $v_A = (\mathcal{B} + \bar{b}_x)/\sqrt{4\pi(\rho_0 + \bar{\rho})}$, where $\mathcal{B}$ denotes the ambient magnetic field and ρ_0 the unperturbed density. This produces a bending of the wave fronts which in turn leads to an enhancement of the transverse magnetic field. The plasma being magnetically dominated (small β), the counter effect of the thermal pressure is inefficient to balance the resulting transverse Lorentz force which confines the particles within a filament around which the magnetic field spirals. Both the density increase and the reduction of the longitudinal magnetic field component lead to the further decrease of the Alfvén velocity, and thus to a wave collapse. In the longitudinal direction, the Lorentz force is canceled by the longitudinal pressure gradient, preventing the development of a longitudinal velocity. This scenario is intrinsically three-dimensional and requires circular polarization to produce spiraling magnetic field lines, ensuring plasma confinement.

A main property of the multidimensional cubic NLS equation is the possible existence of a singularity on finite propagation distance, associated with a blow-up of the wave amplitude ($\sup_{\mathbf{x}_\perp} |B|^2 \to \infty$ and $\int |\nabla_\perp B|^2 d\mathbf{x}_\perp \to \infty$). Dimension two being critical for the existence of a finite-distance singularity, the collapse requires as a necessary condition that the wave energy $N = \frac{1}{2}\int |B|^2 d\mathbf{x}_\perp$ in each transverse plane exceeds a critical value N_T associated with the energy of the so-called Townes soliton [34]. When the wave collapses, N_T is indeed the amount of energy captured in the focus and thus the largest amount of energy available for dissipation (see [32], chapter 5, for a review).

The collapse results in a violent transfer of energy from large to small scales. In reality, the singularity is never reached since dissipation processes become relevant near collapse. Indeed, when the filament diameter becomes of order of the characteristic length $2\pi/k_{diss}$ for dissipation to become relevant, the energy carried by the Fourier wavenumbers exceeding k_{diss} is dissipated. As a consequence, the energy becomes smaller than N_T, the collapse is arrested and the remaining energy is dispersed. In the limit of large k_{diss}, the amount of energy dissipated in one such event scales like $\Delta N \sim (\ln \ln k_{diss})^{-1}$ [8]. This variation with k_{diss}, although very weak, appears to be relevant when dealing with large scale separation such as those present in the solar corona.

2.3 Energy dissipation and application to the solar corona

In a turbulent regime, two main processes dissipate energy: the resonant wave-wave interactions which produce a gradual energy transfer towards small scales and wave collapse. The latter process is dominant when dealing with low amplitude waves [23]. Dissipation due to collapses thus provides a lower bound for the total dissipation. A statistical estimate of the dissipation rate due to collapses is given in [23]. It is written $\langle \gamma \rangle = \omega_c \langle r \rangle N_T$ where ω_c denotes the occurrence frequency of collapses and $\langle r \rangle$ the fraction of energy burned out in the mean in one collapse event. As already mentioned, $\langle r \rangle$, which scales like $(\ln \ln k_{diss})^{-1}$, was estimated numerically.

An estimate of the global energy dissipation also requires the determination of the event frequency ω_c which depends on the turbulent level and the various dynamical scales. The issue was considered in [3] in the context of the solar corona, to argue that Alfvén wave filamentation could provide an efficient heating mechanism in quiet regions. In this medium, the dispersion is weak and the filamentation dynamics was considered in the long-wavelength regime by taking the limit of small-k in eq. (21). The associated dissipated power is thus given by $P = \langle r \rangle N_T v_A$ where N_T appears to be an energy per unit length. Taking into account the coefficients entering the two-dimensional NLS equation, it reads (in c.g.s. units) $N_T = 5.84\beta B^2/(4\pi k^2)$. It follows that waves of wavelength λ emitted from a transverse surface larger than $L_c^2 = \frac{5.84\beta\lambda^2 B^2}{4\pi^2(\delta B)^2}$ can possibly develop a collapse. Assuming that the turbulent fluctuations are space filling, the number of collapses contained in a volume limited by a transverse area S and a distance L along the ambient magnetic field is estimated by $\nu_c = SL/L_\perp^2 L_\parallel$, where $L_\perp$ denotes the transverse scale of the most unstable mode for the convective modulational instability and $L_\parallel$ its e-folding length, which are given by $L_\perp = \lambda\sqrt{\beta}/B_0$ and $L_\parallel = 4\beta\lambda/B_0^2$ in terms of the Alfvén wavelength and the relative amplitude $B_0 = \sqrt{2}\frac{\delta B}{B}$. Note that the energy contained in one period of the most unstable mode for the linear instability is sufficient to trigger a nonlinear collapse, since L_T is larger than the critical transverse scale L_c. The number ν_c is in fact underestimated since the longitudinal length for convective collapse is expected to be shorter than the e-folding length for the linear instability. The number of individual collapses required to significantly reduce the Alfvén wave energy flux $F_e = v_A(\delta B)^2/4\pi$ being large, a differential equation for F_e is written in the form

$$\frac{dF_e}{dL} = \frac{4\pi}{B^2 l_0 v_A} F_e^2 \tag{23}$$

with $l_0 = \frac{4\pi^2 \beta \lambda}{5.84\langle r\rangle}$. Here δB is the dimensional Alfvén wave amplitude. Assuming constant values of β and $\mathcal{B}$ over the propagation distance of Alfvén waves, it follows that $F_e(L) = \frac{F_e(0)}{1 + L/L_d}$, where $L_d = \frac{4\pi^2 \beta \lambda}{5.84\langle r\rangle (\frac{\delta B}{\mathcal{B}})^2}$ defines the dissipation length associated with the Alfvén wave filamentation.

At the base of solar coronal holes where $n \approx 10^8 \mathrm{cm}^{-3}$, $\mathcal{B} \approx 10\mathrm{G}$ and $T \approx 10^6\mathrm{K}$, the velocity fluctuations are $v_{\mathrm{r.m.s}} \approx 30\mathrm{km\ s}^{-1}$, the β of the plasma is $\beta = \frac{c_s^2}{v_A^2} \approx 5.79 \times 10^{-3}$ and the relative amplitude is $B_0 = \sqrt{2}\frac{\delta B}{\mathcal{B}} \approx 1.95 \times 10^{-2}$. For Alfvén waves having a typical wavelength $\lambda = 2\pi v_A/\omega \approx 1.37 \times 10^4 (\omega/\mathrm{rad\ s}^{-1})^{-1}\mathrm{km}$, the transverse wavelength of the most unstable mode and its e-folding length are $L_\perp \approx 5.36 \times 10^4 (\omega/\mathrm{rad\ s}^{-1})^{-1}\mathrm{km}$ and $L_\parallel \approx 8.38 \times 10^5 (\omega/\mathrm{rad\ s}^{-1})^{-1}\mathrm{km}$. In such region, the wave energy loss in the mean during the burnout of a filament has been estimated at about 5% [3]. The corresponding dissipation length then becomes

$$L_d \approx 5.35 \times 10^7 \left(\frac{T}{10^6\mathrm{K}}\right)\left(\frac{n}{10^8\mathrm{cm}^{-3}}\right)^{-1/2}\left(\frac{\mathcal{B}}{10\mathrm{G}}\right)\left(\frac{\omega}{\mathrm{rad\ s}^{-1}}\right)^{-1}\left(\frac{v_{\mathrm{r.m.s}}}{30\mathrm{km\ s}^{-1}}\right)^{-2}\mathrm{km}.$$

The above expression provides a condition on the Alfvén waves frequency to contribute to the coronal heating. For example, for a value of L_d of order of $0.5R_\odot$ required to accelerate the fast solar wind up to its observed velocity [19], a frequency $f = \frac{\omega}{2\pi} > 27Hz$ is needed in open coronal holes.

2.4   Coupling to low-frequency magneto-sonic waves

The effect of a small deviation from exact stationarity on the convective filamentation instability was addressed in [5]. It turns out that when the duration of the incident Alfvén wave is reduced with the dispersive effect remaining nevertheless subdominant compared with the diffraction, the magneto-sonic waves become relevant, and in some instances can affect the Alfvén wave filamentation. This regime is conveniently analyzed by using a frame of reference moving at the group velocity of the Alfvén wave and introducing a slower time scale. It is also convenient to replace the field δ by $d = \frac{\delta}{\beta+1} - \frac{1}{2}(\frac{1}{\beta} + \frac{k^2}{(\beta+1)\omega^2})|B|^2$. After introducing a parameter $\tilde{\mu}$ and rescaling the variables in the form

$$\tilde{\xi} = |k|\xi, \quad \tilde{\tau} = \tilde{\mu}|k||v_g|\tau \quad \tilde{\mu} \ll 1, \quad \tilde{Y} = \sqrt{\frac{|kv_g|}{|\alpha|}}Y, \quad \tilde{Z} = \sqrt{\frac{|kv_g|}{|\alpha|}}Z, \quad (24)$$

$$|B|^2 = \frac{|\tilde{B}|^2}{Q}, \quad \bar{u}_x = \frac{A|v_g|}{Q}\tilde{u}_x, \quad d = \frac{Av_g^2}{\beta Q}\tilde{d}, \quad (25)$$

with $A = \frac{\beta+1}{2\beta}(1 + \frac{\beta k^2}{(\beta+1)\omega^2})$ and $Q = \frac{1}{4}(\frac{1}{\beta} + (\frac{k^2+\omega^2}{\omega^2})^2)$, the system is written in an Hamiltonian form which reads after dropping the tildes

$$i\partial_\xi B = -\sigma_1\sigma_3\Delta_\perp B - \sigma_2\sigma_3\left[W(d + Du_x) + |B|^2\right]B \qquad (26)$$

$$\eta\partial_\xi d - \sigma_3\partial_\tau d = -\partial_\tau u_x - D\partial_\tau|B|^2 \qquad (27)$$

$$\eta\partial_\xi u_x - \sigma_3\partial_\tau u_x = -\frac{1}{M^2}\partial_\tau d - \partial_\tau|B|^2, \qquad (28)$$

where $M = \frac{1}{|v_g|}\sqrt{\frac{\beta}{\beta+1}}$, $D = -\frac{\sigma_3}{|v_g|^2A}$, $W = \frac{|v_g|^2A^2}{Q}$ and $\sigma_1 = \text{sign}(\alpha)$, $\sigma_2 = \text{sign}(kv_g)$, $\sigma_3 = \text{sign}(v_g)$. The parameter η denotes the ratio $\mu/\tilde{\mu}$. The stationary convective regime addressed in Section 2 is recovered as $\eta \to \infty$. This system can be viewed as a special case of the Zakharov-Rubenchik [36] equations describing the coupling of a high-frequency wave to low-frequency acoustic type waves. In the present case the acoustic wave equations include only longitudinal derivatives, since, to leading order, the transverse velocity is slaved to the transverse magnetic field. Note that the above system is similar to that derived [27] with canonical variables which are nevertheless different, since the source term in the density equations does not come from the interaction between density and velocity fluctuations but arises from the time and space derivative in the ponderomotive force (see Section 2.1) which, in the convective regime, leads to redefine the "density" and to introduce a term of self-interaction in the potential of the NLS equation.

On short propagation distances, the magneto-sonic waves are negligible in eq. (26). Assuming that $\sigma_1\sigma_2 > 0$, the carrying wave first focuses in the transverse planes for which the initial L^2 norm $|B_0(.,\tau)|_{L^2(R^2)}$ exceeds the critical value for collapse. Then strong gradients develop, producing a significant ponderomotive force. The latter drives magneto-sonic waves, which in turn react on the focusing process. The resulting dynamics was investigated numerically in [5] for radially symmetric initial conditions corresponding to a localized wave packet $B_0(r) = 4e^{-(\tau^2+r^2)}$ with $d = u = 0$ (here r denoting the radial coordinate). It turns out that regardless of the Alfvén wave polarization, three different regimes can be encountered according to the value of the parameter η which measures the duration of the incident Alfvén pulse. In the case of short pulses, illustrated by Fig. 1 with $\eta = 0.1$, $k/R_i = 0.1$, $\beta = 0.5$ and a left-hand polarization, the filamentation is rapidly inhibited under the effect of the developing magneto-sonic waves. These waves first display an antisymmetric profile whose left part is rapidly advected away from the Alfvén wave packet. The other part stays beneath the pulse, and reaches its adiabatic limit, producing a caviton that in the small k-limit, cancels out the nonlinearity in the equation for the Alfvén wave envelope. When the duration of the initial pulse is slightly increased, as in Fig. 2 for which $\eta = 0.3$, $k/R_i = 1$, $\beta = 0.1$ and the Alfvén wave is left-hand polarized, the Alfvén wave focusing is also arrested. In this case, however, magneto-sonic waves develop an antisymmetric front whose strength increases with η and

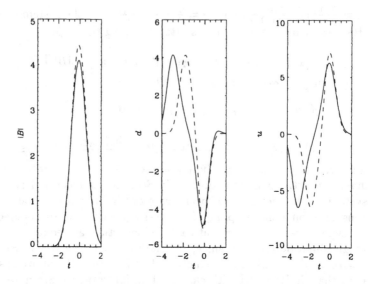

Fig. 1. Arrest of convective filamentation with adiabatic dynamics of the magneto-sonic waves, for $\eta = 0.1$, $k/R_i = 0.1$, $\beta = 0.5$ and left-hand polarization. The time evolution of $|B|$, d and u_x is shown on the $r = 0$ axis, at points $\xi = 0.0734$ (dashed line) and $\xi = 0.119$ (solid line) behind and beyond the location of the filamentation arrest respectively.

which induces strong nonlinear effects in the longitudinal direction not included in the envelope formalism. In the case of longer pulses displayed by Fig.3 for $\eta = 1$, $k/R_i = 0.65$, $\beta = 0.5$ and a right-hand polarization, the filamentation is not arrested. A sharp antisymmetric magneto-sonic front develops, and the Alfvén wave collapses at a finite distance. Note that the above regimes are not restricted to the convective case but were also obtained for the absolute filamentation instability which develop when $\beta > 1$ [26],[5]. We conclude that except for very short Alfvén pulses, for which the collapse is arrested, the coupling to magneto-sonic waves which leads the formation of sharp magneto-sonic fronts, provides an additional mechanism for the generation of small scales and the burn-out of the wave energy.

3   Filamentation of weakly dispersive wave trains

When the carrier wavelength is large compared with the ion gyromagnetic radius, the dispersion is small at the scale of the carrying wave. This long-wavelength ($k \ll 1$) or small dispersion ($R_i \gg 1$) limit can formally be taken on the amplitude equations discussed in the previous sections. In this regime, the amplitude of the Alfvén wave is assumed to be reduced at the same rate as its wavenumber, and the polarization of the Alfvén wave remains

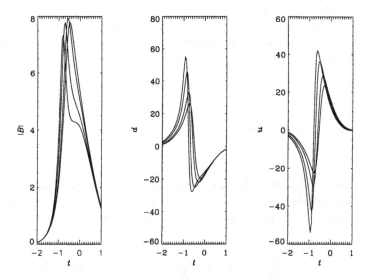

Fig. 2. Arrest of the filamentation with formation of strong acoustic fronts for $\eta = 0.3$, $k/R_i = 1$, $\beta = 0.1$ and left-hand polarization. The time evolution of $|B|$, d and u_x is shown on the $r = 0$ axis, at point $x = 0.139$ behind the filamentation arrest and at points $\xi = 0.149$, $\xi = 0.166$, $\xi = 0.177$ beyond it.

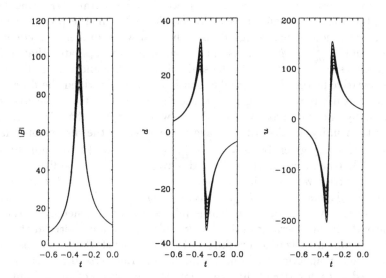

Fig. 3. Finite distance wave collapse for $\eta = 1$, $k/R_i = 0.65$, $\beta = 0.5$ and right-hand polarization. The time evolution of $|B|$, d and u_x is shown at various points of the $r = 0$ axis, located between $\xi = 0.156$ and $\xi = 0.16$, near the collapse.

enforced on the typical scales of the modulation. We consider in this section a regime where the wave amplitude is still small enough to permit a modulation analysis but not sufficiently to enable the dispersion to enforce the circular polarization.

3.1   Envelope dynamics

It is well known that the amplitude equation for a monochromatic wave of small amplitude ψ can be viewed as the Taylor expansion of the dispersion relation made amplitude-dependent by the nonlinear couplings

$$(\omega + \mu i \frac{\partial}{\partial \tau} - \Omega(k - \mu i \frac{\partial}{\partial X}, \mu^2 \Delta_\perp, \zeta, \mu^2 |\psi|^2))\psi = 0. \tag{29}$$

The space and time derivatives acting on the slowly varying complex amplitude ψ result from the broadening of the wavevector and frequency of the modulated carrier. Furthermore, the dependence in terms of the dispersion ζ, was also explicited in eq. (29) because this parameter being taken small, the usual μ-expansion is to be supplemented in the present context by a ζ-expansion, which leads to an additional wavenumber dependent frequency shift $\zeta \frac{\partial \Omega}{\partial \zeta}|_{\zeta=0}\psi$, reminiscent of the dispersion operator that, due to the smallness of ζ, now acts at the scale of the carrier and thus reduces to a multiplicative factor. This formal expansion is valid provided the differential frequency shift remains larger than the broadening of the wave frequency induced by the modulation, a condition which implies that ζ/μ^2 must be larger than unity. In the Alfvén wave context where the dispersion relation is asymptotically given by $\omega = k + \sigma \frac{k^2}{2R_i}$ (where $\sigma = +1$ for right-hand or $\sigma = -1$ left-hand circularly polarized wave), the above condition requires that the carrier wavelength λ be sufficiently small compared with l_{disp}/μ^2 where the dispersion scale l_{disp} is given by $l_{disp}/\lambda = k/2R_i$. An additional aspect of the Alfvén wave problem is related to the degeneracy of the zero mode of the linear operator in the dispersionless limit, a property at the origin of the circular polarization breaking. At a formal level, the equations for the velocity and magnetic field components in the transverse directions become decoupled at each order of the expansion, the null space of the adjoint linear operator becomes two-dimensional and the resonant terms proportional to $e^{i(kx-\omega t)}$ must be eliminated in two sub-systems. As a consequence, the solvability conditions governing the (vector) Alfvén wave amplitude denoted by $\mathbf{B} = (0, B_y, B_z)$ takes the form of a vector NLS equation coupled to the low-frequency density and longitudinal components of the velocity and magnetic field

$$i(\partial_\tau + \partial_\xi)\mathbf{B} + \sigma\nu k\mathbf{B} + i\nu k(\mathbf{B}\times\mathbf{e}_1) + \frac{1}{2k(1-\beta)}\nabla(\nabla\cdot\mathbf{B})$$

$$-k(\bar{b}_x + \bar{u}_x - \frac{1}{2}\bar{\rho})\mathbf{B} - \frac{k}{4(1-\beta)}(\mathbf{B}\cdot\mathbf{B})\mathbf{B}^* = 0 \tag{30}$$

$$\partial_\tau(\bar{\rho} - \bar{b}_x) + \partial_\xi\bar{u}_x = 0 \tag{31}$$

$$\partial_\tau\bar{u}_x + \partial_\xi(\beta\bar{\rho} + |\mathbf{B}|^2) = 2(\partial_\tau + \partial_\xi)|\mathbf{B}|^2, \tag{32}$$

with $\bar{b}_x = -\beta\bar{\rho} - |\mathbf{B}|^2$ and where the parameter $\nu = \frac{k}{2R_i}\mu^{-2} > 1$ is assumed significantly larger than one.

Here $\mathbf{e}_1$ denotes the unit vector along the propagation axis, and $\sigma = +1$ when the emitted Alfvén wave is right hand (or $\sigma = -1$ when it is left hand) circularly polarized. The presence of the dispersive terms proportional to ν, arises from the expansion of the linear operator with respect to the dispersion and prevents the harmonics of the carrying wave to become resonant. The diffraction term in the above vector NLS equation originates from the total (thermal and magnetic) pressure gradient. The pressure $\beta\rho + b_x$ is indeed proportional to $\nabla\cdot\mathbf{B}$. As shown below, when the dispersion is relatively strong, due to the circular polarization of the wave, the envelope equation becomes scalar, the diffraction term reduces to a Laplacian and the coupling $(\mathbf{B}\cdot\mathbf{B})\mathbf{B}^*$ disappears.

For stationary convective modulation,

$$\bar{u}_x = 0, \quad \bar{b}_x = -2|\mathbf{B}|^2 \quad \bar{\rho} = \frac{1}{\beta}|\mathbf{B}|^2, \tag{33}$$

which yields

$$i\partial_\xi\mathbf{B} + \sigma\nu k\mathbf{B} + i\nu k(\mathbf{B}\times\mathbf{e}_1) + \frac{1}{2k(1-\beta)}\nabla(\nabla\cdot\mathbf{B})$$

$$+\frac{k}{4\beta}\left(\frac{2(1-\beta)(4\beta+1)-\beta}{1-\beta}\right)|\mathbf{B}|^2\mathbf{B} + \frac{k}{4(1-\beta)}\mathbf{B}\times(\mathbf{B}\times\mathbf{B}^*) = 0. \tag{34}$$

Note that the presence of the two nonlinear terms is canonical in the case of a vector NLS equation. They also occur in the context of Langmuir waves (where $\nu = 0$) when the induced magnetic field is retained [2], [14]. In this case a $\nabla\times\nabla\times$ operator originating from Maxwell equations is also present.

Equation (34) governs the coupling between right and left-hand Alfvén modes. Defining $B_{\pm} = B_y \pm iB_z$, the vector NLS equation rewrites

$$i\partial_\xi B_- + (\sigma - 1)\nu k B_- + \frac{1}{4k(1-\beta)}(\Delta_\perp B_- + \partial_\perp^{*2} B_+)$$

$$+ k\frac{4\beta + 1}{4\beta}(|B_+|^2 + |B_-|^2)B_- + \frac{k}{4(\beta - 1)}|B_+|^2 B_- = 0 \quad (35)$$

$$i\partial_\xi B_+ + (\sigma + 1)\nu k B_+ + \frac{1}{4k(1-\beta)}(\Delta_\perp B_+ + \partial_\perp^2 B_-)$$

$$+ k\frac{4\beta + 1}{4\beta}(|B_+|^2 + |B_-|^2)B_+ + \frac{k}{4(\beta - 1)}|B_-|^2 B_+ = 0, \quad (36)$$

with $\partial_\perp = \partial_Y + i\partial_Z$. In the limit $\nu \to \infty$, the constraint of circular polarization is restored and the equation for $B = 2B_y = 2i\sigma B_z$ reproduces the scalar NLS equation (21) where the coefficients are taken in the long-wavelength limit $k \to 0$. In contrast, the regime where ν, although much larger than one, is kept finite, is associated with the scaling $l_{disp}/\lambda \sim \mu^2$, $L_M/\lambda \sim \mu^2$ and $\delta B/B \sim \mu$, where L_M is the scale of the modulation. It follows that, assuming the same level of dispersion, the vector NLS equation corresponds to a stronger nonlinear regime, than the scalar NLS equation taken in the weak dispersion limit, and the parameter $\nu = \frac{l_{disp}}{\lambda}(\frac{\delta B}{B})^{-2}$ measures the strength of the dispersion compared with the nonlinearity at the scale of the carrier wavelength.

The nonlinear regime was numerically investigated in [6] in the case of a circularly polarized incident plane wave subject to either a monochromatic perturbation at large-scale ($B_{y0} = (1+0.1\sin y)(1+0.2\sin z)$ with $B_{z0} = B_{z0}$ or to a low amplitude noise. A right-hand ($B_+ = 0$ when $\sigma = 1$) or a left-hand ($B_- = 0$ when $\sigma = -1$) circularly polarized incident wave does not remain so when modulationally unstable, because of the anisotropic diffraction. A specific dynamics then develops. No blow up is observed for the wave amplitude which remains moderate but small scales are nevertheless created in the form of strong magnetic field gradients in the transverse directions (see Fig. 4).

Figure 5 shows that the filamentation dynamics leads to the formation of thin layers of intense gradients. The resulting structures arise from successive splittings of the initial focusing structure where the local maximum amplitude saturates while new structures emerge which undergo a similar process. The increase of the dispersion at the scale of the carrier wavelength associated with an increase of the parameter ν, leads to a progressive transformation of the small scale structures from sheets to foci as displayed on Fig. 6. The saturation level of the wave-amplitude increases with the dispersion (see Fig 7a), and the circular polarization of the emitted wave is enforced on larger propagation distances (see Fig. 7b). Figure 7c shows that at short distance, the typical size of the smallest transverse scales (measured by the logarithmic decrement δ_z of the transverse energy spectrum) is weakly sensitive to ν, up to a distance where the maximum amplitude begins to sat-

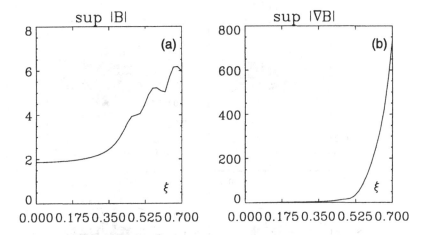

Fig. 4. Variation with the propagation distance of the maximum of the Alfvén wave amplitude (a) and of its transverse gradient norm (b) in the zero dispersion limit, for a circularly polarized initial condition resulting from a large-scale perturbation of a plane wave.

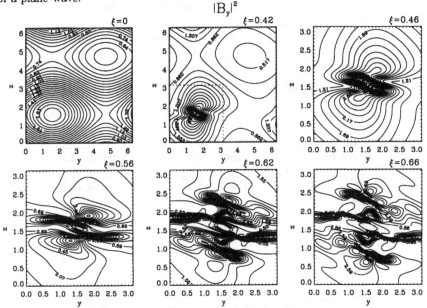

Fig. 5. Contours of $|B_y^2|$ at increasing propagation distances, in the same conditions as in Fig. 4. For each field, the four last panels correspond to the active sub-domain $[0, \pi] \times [0, \pi]$ limited by dotted lines on the second panel. Similar structures are obtained for $|B_z|^2$.

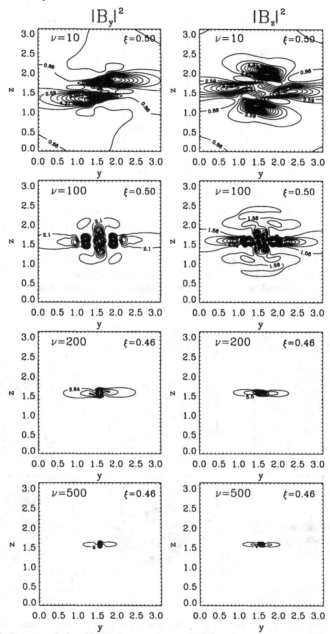

Fig. 6. Influence of the dispersion, as measured by the parameter ν (see text) on the contours of $|B_y|^2$ and $|B_z|^2$ in the sub-domain $[0, \pi] \times [0, \pi]$, for the same initial condition as in Fig. 4.

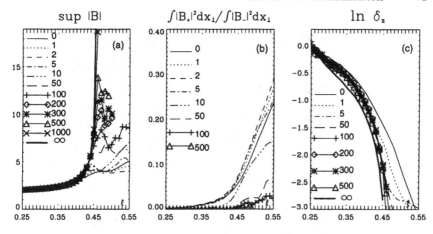

Fig. 7. Variation with the propagation distance of : (a) the maximum Alfvén wave amplitude, (b) the deviation from the initial circular polarization corresponding to $\int |B_+|^2 dx_\perp / \int |B_-|^2 dx_\perp = 0$, (c) the logarithmic decrement δ_z of the transverse energy spectrum, for the same initial condition as in Fig. 4 and various values of ν indicated with each panel.

urate and deviation from the circular polarization becomes significant. The rate of small scale formation then slows down as the dispersion magnitude decreases with ν. The critical transverse scale up to which the collapse proceeds with a roughly circular polarization can be estimated, for large ν, as $l_\perp / \lambda \approx (8(1 - \beta)\nu)^{-1/2}$. The corresponding maximum amplitude of the wave B_M, given by $B_M^2 = \sup_{y,z}(|B_y|^2 + |B_z|^2)$ scales like $\nu(\frac{4\beta}{4\beta+1})B^2$. Beyond this point, the dynamics characteristic of the weak dispersive regime (moderate values of ν) is recovered with the formation of sheets of strong gradients.

As illustrated in Fig. 8 and 9, similar structures are produced in the case of a circularly polarized Alfvén wave perturbed by a weak random noise.

3.2   Dissipation of weakly dispersive Alfvén wave

An interesting question concerns the estimate of the energy dissipated by weakly dispersive Alfvén waves and its comparison with that arising in the usual dispersive regime. A quantitative approach to this estimate, similar to that presented in section 2, would require the development of a detailed theory on the vector NLS equation. The influence of reducing the dispersion on the wave energy decay was thus investigated numerically [7] in the situation where a linear dissipation is retained. Following Refs. [23] and [8], a dissipative term is included in the right-hand side of eq. (34), in the form $-i \int \gamma(k)\hat{\mathbf{B}}(\mathbf{k}, x)e^{i\mathbf{k}\cdot\mathbf{r}}d\mathbf{k}$ with $\gamma(k) = \eta k^2 h(k/k_{diss})$, where η is a positive number and h denotes a cutoff function $h(x) = \frac{1}{6x^5}e^{5(1-1/x^2)}$ for $x \leq 1$ and $1 - \frac{5}{6}e^{\frac{1}{2}(1-x^2)}$ for $x > 1$, making a smooth transition from zero to a constant value as the wavenumber $k = |\mathbf{k}|$ increases through k_{diss}.

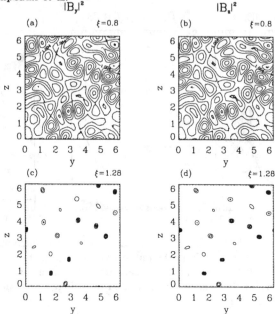

Fig. 8. Contours of $|B_y|^2$ and $|B_z^2|$ at increasing propagation distances, for $\nu = 200$, in the case of a right-hand circularly polarized plane wave perturbed by a noise of amplitude 10^{-6} and random phases, equidistributed on all the retained Fourier modes.

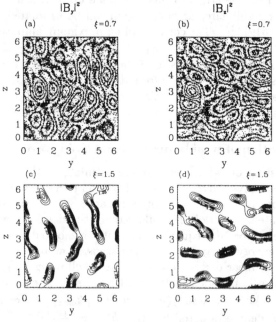

Fig. 9. Contours of $|B_y|^2$ and $|B_z^2|$ at increasing propagation distances, for $\nu = 10$ and the same initial conditions as in Fig. 8.

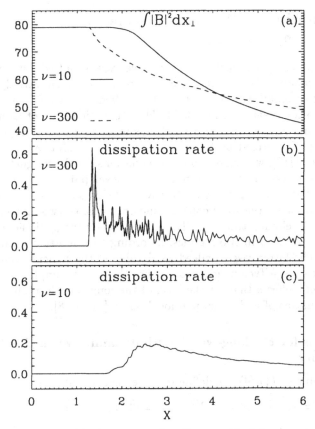

Fig. 10. Variation with the propagation distance X of the wave energy $\int |\mathbf{B}|^2 dx_\perp$ for $\nu = 300$ and $\nu = 10$ (a), of the dissipation rate $\sum_k \gamma(k)|\mathbf{B_k}|^2$ for $\nu = 300$ (b) and for $\nu = 10$ at resolution$(256)^2$ (solid line) and $(1024)^2$ (dotted line) (c), in the case of a circularly polarized emitted wave $B_z = iB_y$ and $B_y = 1$ subject to a noise with a random phase and a low amplitude 10^{-6}, equally distributed on the Fourier modes (k_y, k_z) inside the disk $(k_y^2 + k_z^2)^{1/2} \leq 60$.

These simulations reveal that the dissipative structures which, for large ν, correspond to localized quasi-isotropic foci scattered throughout the transverse plane, become less distinct and more space filling as the dispersion decreases with the parameter ν. Figure 10 shows that although a larger distance is required before dissipation starts to play a role, this process remains efficient on longer distances leading to an eventually larger total amount of dissipated energy. Note that the burst of dissipation associated with the dissipation of foci, disappear in the weakly dispersive case. A main conclusion is that even in the weakly dispersive regime, Alfvén wave filamentation, can lead to a significant energy dissipation and thus contribute to the plasma heating.

4   Filamentation in the weakly dispersive longwavelength regime

In this section we use a reductive perturbative expansion to derive asymptotic equations capable of describing both the nonlinear Alfvén wave dynamics in the limit of small wavenumbers and the "reduced MHD" in planes perpendicular to the mean magnetic field. These equations do not assume a quasi-monochromatic wave and their validity extends to waves of larger amplitude than those described by the envelope formalism. Moreover, extending the analysis of [10], we derive nonlinear equations for the magneto-sonic waves, a necessary ingredient when strong fronts develop due to the filamentation process. The coupling to mean transverse fields also allows to study the interaction between the wave collapse and the possible two-dimensional hydrodynamics developing transversally to the ambient magnetic field. The equations will be derived in two regimes, according to whether β is close or far from unity. The latter case is indeed special due to the coincidence of the sound speed with the Alfvén phase velocity. For the sake of simplicity, we do not include kinetic effects that, at the level of the long-wave formalism, can be modeled by means of additional non-local terms [21,31,18].

4.1   Equations for the long-wavelength dynamics when β is far from unity

Starting from eqs. (1)-(4), we define the stretched variables

$$\xi = \epsilon(x - t), \quad \eta = \epsilon^{3/2}y, \quad \zeta = \epsilon^{3/2}z, \quad \tau = \epsilon^2 t$$

and expand

$$\rho = 1 + \epsilon\rho_1 + \epsilon^2\rho_2 + \cdots, \quad u = \epsilon^{1/2}(u_1 + \epsilon u_2 + \cdots),$$
$$u_x = \epsilon u_{x_1} + \epsilon^2 u_{x_2} + \cdots, \quad b = \epsilon^{1/2}(b_1 + \epsilon b_2 + \cdots),$$
$$b_x = 1 + \epsilon b_{x_1} + \epsilon^2 b_{x_2} + \cdots,$$

where we have defined

$$u = u_y + i u_z, \quad b = b_y + i b_z. \tag{37}$$

At order $\epsilon^{3/2}$, we have

$$\partial_\xi u_1 + \partial_\xi b_1 = 0. \tag{38}$$

In order to include a coupling between the Alfvén waves and the hydrodynamics transverse to the ambient magnetic field, we include a mean contribution corresponding to an average over the ξ variable and denoted by an overline, in the transverse components of the velocity and magnetic fields

$$b_1 = \widetilde{b}_1(\xi, \eta, \zeta, \tau) + \bar{b}_1(\eta, \zeta, \tau), \tag{39}$$
$$u_1 = \widetilde{u}_1(\xi, \eta, \zeta, \tau) + \bar{u}_1(\eta, \zeta, \tau). \tag{40}$$

Equation (38) implies that the fluctuating parts (denoted by tildes) satisfy

$$\tilde{u}_1 = -\tilde{b}_1. \tag{41}$$

Separating the mean and fluctuating parts, we obtain, after some algebra (details are given in [10]),

$$\partial_\tau \bar{\mathbf{u}} + \bar{\mathbf{u}} \cdot \nabla \bar{\mathbf{u}} = -\nabla p + \bar{\mathbf{b}} \cdot \nabla \bar{\mathbf{b}} \tag{42}$$

$$\partial_\tau \bar{\mathbf{b}} - \nabla \times (\bar{\mathbf{u}} \times \bar{\mathbf{b}}) = 0 \tag{43}$$

$$\nabla \cdot \bar{\mathbf{u}} = 0; \qquad \nabla \cdot \bar{\mathbf{b}} = 0 \tag{44}$$

$$\partial_\tau \tilde{b} + \partial_\xi \left(\frac{1}{2}(\bar{b} + \tilde{b})\tilde{P} + (\bar{u}_x + \frac{1}{2}\bar{b}_x - \frac{1}{2}\bar{\delta})\tilde{b} \right) - \frac{1}{2}\partial_\perp \tilde{P}$$

$$+(\bar{\mathbf{u}} + \bar{\mathbf{b}}) \cdot \nabla \tilde{b} + \frac{i}{2R_i}\partial_{\xi\xi}\tilde{b} = 0 \tag{45}$$

$$\partial_\xi \tilde{b}_x + \frac{1}{2}\left(\partial_\perp^* \tilde{b} + \text{c.c.}\right) = 0 \tag{46}$$

$$\partial_\tau \bar{\delta} + \nabla \cdot (\bar{\delta}\bar{\mathbf{u}} + \bar{u}_x \bar{\mathbf{b}}) = 0 \tag{47}$$

$$\partial_\tau \bar{u}_x + \nabla \cdot (\bar{u}_x \bar{\mathbf{u}} - \bar{b}_x \bar{\mathbf{b}}) = \frac{1}{2}\left(\partial_\perp^* \langle \tilde{b}\tilde{P} \rangle + \text{c.c.}\right) \tag{48}$$

$$\Delta_\perp \left((1 + \beta)\bar{b}_x + \beta\bar{\delta} + \frac{\langle|\tilde{b}|^2\rangle}{2}\right) + \nabla \cdot (\bar{\mathbf{u}} \cdot \nabla \bar{\mathbf{u}} - \bar{\mathbf{b}} \cdot \nabla \bar{\mathbf{b}}) = 0. \tag{49}$$

Here $\langle \cdot \rangle$ denotes the average on the x-variable. The fluctuations of magnetic pressure are given by

$$\tilde{P} = \frac{1}{2(1 - \beta)}(2\tilde{b}_x + |\tilde{b} + \bar{b}|^2 - \langle|\tilde{b} + \bar{b}|^2\rangle). \tag{50}$$

It is convenient to introduce the field

$$\delta = \rho - b_x \tag{51}$$

that includes a mean part $\bar{\delta}$ and fluctuations $\tilde{\delta}$.

We also used the complex representation $\partial_\perp = \partial_\eta + i\partial_\zeta$ for the transverse gradient $\nabla = (\partial_\eta, \partial_\zeta)$, and introduced the vectorial notation $\bar{\mathbf{u}}$ and $\bar{\mathbf{b}}$ for the mean transverse fields. Note that the equations for the mean transverse fields are the usual two-dimensional MHD equations. They are not forced by the Alfvén waves.

Pushing to the next order the asymptotic expansion for the equations governing the magneto-sonic waves, summing fluctuating and mean components and combining the two orders of the expansion, we get

$$\partial_\tau \tilde{b} + \partial_\xi \left(\tilde{b}(u_x + b_x - \frac{\rho}{2}) + \frac{1}{2}\bar{b}(u_x + b_x) \right) - \frac{1}{2}\partial_\perp \left(\beta\tilde{\rho} + (1-\beta)\tilde{P} \right)$$
$$+(\bar{u} + \bar{b}) \cdot \boldsymbol{\nabla}\tilde{b} + \frac{i}{2R_i}\partial_{\xi\xi}\tilde{b} = 0 \tag{52}$$

$$\partial_\tau \delta + \frac{1}{\epsilon}\partial_\xi(u_x - \delta) + \partial_\xi(\rho u_x) + \frac{1}{2}\left(\partial_\perp^*(\tilde{b}(u_x - \delta)) + \text{c.c.} \right)$$
$$+\boldsymbol{\nabla} \cdot (\bar{u}\delta + \bar{b}u_x) - \frac{i}{2R_i}\partial_\xi(\partial_\perp^*\tilde{b} - \partial_\perp \tilde{b}^*) = 0 \tag{53}$$

$$\partial_\tau u_x - \frac{1}{\epsilon}\partial_\xi\left(u_x - \beta\rho - \frac{|\tilde{b} + \bar{b}|^2}{2} \right)$$
$$+\partial_\xi\left(\frac{u_x^2}{2} + (\beta(\gamma-1)-1)\frac{\rho^2}{2} - b_x(b_x + u_x - \rho) \right)$$
$$-\frac{1}{2}\left(\partial_\perp^*(\tilde{b}(b_x + u_x)) + \text{c.c.} \right) + \boldsymbol{\nabla} \cdot (\bar{u}u_x - \bar{b}b_x) = 0. \tag{54}$$

$$\partial_\tau b_x + \frac{1}{\epsilon}\left(- \partial_\xi\bar{b}_x + \boldsymbol{\nabla} \cdot \bar{u} \right)$$
$$-\frac{1}{2}\left(\partial_\perp^*\tilde{b}(u_x + b_x) + \text{c.c.} \right) + \boldsymbol{\nabla} \cdot (\bar{u}b_x - \bar{b}u_x) = 0 \tag{55}$$

$$\partial_\tau \bar{u} - \frac{1}{\epsilon}\partial_\xi(\bar{u} + \bar{b}) + \boldsymbol{\nabla}(\beta\bar{\rho} + \bar{b}_x + \langle\frac{|\tilde{b}|^2}{2}\rangle) + \bar{u}\cdot\boldsymbol{\nabla}\bar{u} = (\boldsymbol{\nabla}\times\bar{b})\times\bar{b} \tag{56}$$

$$\partial_\tau \bar{b} - \frac{1}{\epsilon}\partial_\xi(\bar{u} + \bar{b}) - \boldsymbol{\nabla}\times(\bar{u}\times\bar{b}) = 0. \tag{57}$$

We also have

$$\partial_\xi \bar{b}_x + \frac{1}{2}(\partial_\perp(\bar{b}^* + \tilde{b}^*) + \partial_\perp^*(\bar{b} + \tilde{b})) = 0. \tag{58}$$

In these equations, the mean (longitudinal and transverse) fields are implicitly assumed to depend on a large longitudinal scale $X = \epsilon\xi$. The partial derivative ∂/∂_ξ should thus be viewed as also including a term of the form $\epsilon\partial/\partial_X$.

It is easily seen that if the transverse turbulence is neutral (i.e. $\bar{b} = 0$), the above system admits solutions in the form of circularly polarized waves. When their amplitude is small, their modulation is governed by the scalar NLS equation with an additional advection by the mean transverse velocity. In the presence of a mean transverse magnetic field, in contrast, the wave cannot keep its circular polarization on the time scale of the modulation. One is then led to consider an Alfvén wave of arbitrary polarization. The analysis described in [10] then shows that the vector amplitude of the wave obeys a system of equations which generalizes that obtained in the weak dispersion limit, by the inclusion of the couplings to the mean transverse fields. One can thus expect that in the present context, the transverse collapse does not lead to the blow up of the wave amplitude at local foci, but rather to the formation of strong gradients localized on thin sheets in the planes transverse to the propagation, the wave amplitude remaining moderate. Note that the equations include nonlinearities for the magneto-sonic waves. These terms are usually subdominant, except when violent phenomena such as filamentation occur (see Section 2.4). In the latter case, the transverse MHD flow can also develop a compressible component under the effect of the gradient of the wave intensity in eq. (56).

4.2   Long-wavelength dynamics for $\beta \approx 1$

Equations derived in Section 4.1 are only valid when β is significantly different from 1. To study the dynamics for $\beta \approx 1$ we must consider stronger fluctuations in the longitudinal fields and we shall see that the magneto-sonic waves then obey nonlinear equations at dominant order [12]. We thus define

$$\xi = \epsilon^{1/2}(x - t), \quad \eta = \epsilon y, \quad \zeta = \epsilon z, \quad \tau = \epsilon t,$$

write $\beta = 1 + \alpha \epsilon^{1/2}$ and expand

$$\rho = 1 + \epsilon^{1/2}\rho_1 + \epsilon\rho_2 + \cdots, \qquad u = \epsilon^{1/2}(u_1 + \epsilon^{1/2}u_2 + \cdots),$$
$$u_x = \epsilon^{1/2}u_{x_1} + \epsilon u_{x_2} + \cdots, \qquad b = \epsilon^{1/2}(b_1 + \epsilon^{1/2}b_2 + \cdots),$$
$$b_x = 1 + \epsilon^{1/2}b_{x_1} + \epsilon b_{x_2} + \cdots.$$

As previously, $b = b_y + ib_z$ and $u = u_y + iu_z$.

At order ϵ, we obtain

$$\tilde{u}_1 = -\tilde{b}_1, \quad \tilde{u}_{x1} = \tilde{\rho}_1, \quad \tilde{b}_{x1} = 0. \tag{59}$$

We get at order $\epsilon^{3/2}$, after separating mean and fluctuating parts

$$2\partial_\tau \tilde{\rho}_1 + \partial_\xi(\alpha\tilde{\rho}_1 + \frac{1+\gamma}{2}\tilde{\rho}_1^2 + \frac{|\tilde{b}_1 + \bar{b}_1|^2}{2}) - \frac{1}{2}(\partial_\perp^* \tilde{b}_1 + \partial_\perp \tilde{b}_1^*)$$
$$+[(\gamma - 1)\bar{\rho}_1 + 2\bar{u}_{x1}]\partial_\xi \tilde{\rho}_1 = 0 \tag{60}$$

$$\partial_\tau \tilde{b}_1 + \frac{1}{2}\partial_\xi(\tilde{\rho}_1 \tilde{b}_1) - \frac{1}{2}\partial_\perp \tilde{\rho}_1 + \frac{i}{2R_i}\partial_{\xi\xi}\tilde{b}_1$$
$$+\partial_\xi[\tilde{b}_1(\bar{u}_{x1} + \bar{b}_{x1} - \frac{\bar{\rho}_1}{2})] = 0 \tag{61}$$

$$\partial_\perp^* \bar{b}_1 + \partial_\perp \bar{b}_1^* = 0 \tag{62}$$

$$\partial_\tau \bar{\rho}_1 + \frac{1}{2}(\partial_\perp^* \bar{u}_1 + \partial_\perp \bar{u}_1^*) = 0 \tag{63}$$

$$\partial_\tau \bar{b}_{x1} + \frac{1}{2}(\partial_\perp^* \bar{u}_1 + \partial_\perp \bar{u}_1^*) = 0 \tag{64}$$

$$\partial_\tau \bar{u}_1 + \partial_\perp(\bar{\rho}_1 + \bar{b}_{x1}) = 0 \tag{65}$$

$$\partial_\tau \bar{b}_1 = \partial_\tau \bar{u}_{x1} = 0. \tag{66}$$

The above system describes the coupling between the nonlinear longitudinal dynamics of Alfvén and sound waves with the linear magneto-sound propagating transversally to the ambient magnetic field with a speed equal to $(V_A^2 + c_s^2)^{1/2} = (1 + \beta)^{1/2} = \sqrt{2}$. A nonlinear dynamics in the transverse directions occurs on the longer time scale $T = \epsilon^{3/2} t$. We can push the expansion for the mean fields to order ϵ^2. We thus obtain equations which include two time scales. These equations are only useful when all the mean fields (of order $\epsilon^{1/2}$) are taken independent of the shorter time scale τ, a condition that is consistent with the fact that the mean field equations arising at order $\epsilon^{3/2}$ are unforced. In this case, the transverse flow $\bar{u}_1$ becomes incompressible and the constraint of total pressure balance as expressed by eq. (65) prescribes the next order component $\bar{u}_2$ in the form

$$\partial_\perp^* \bar{u}_2 + \text{c.c.} = \langle \partial_\perp^* (\tilde{\rho}_1 \tilde{b}_1) \rangle + \frac{1}{2} \partial_\perp^* (\bar{u}_{x1} \bar{b}_1) + \text{c.c.} \ . \tag{67}$$

Averaging on the time τ and denoting by $\langle\langle . \rangle\rangle$ the mean value on both the ξ and τ variables and by $\bar{p}$ the total pressure we obtain, after dropping the subscripts,

$$\partial_T \bar{\mathbf{u}} + \bar{\mathbf{u}} \cdot \nabla \bar{\mathbf{u}} = -\nabla \bar{p} + \bar{\mathbf{b}} \cdot \nabla \bar{\mathbf{b}} \tag{68}$$

$$\partial_T \bar{\mathbf{b}} - \nabla \times (\bar{\mathbf{u}} \times \bar{\mathbf{b}}) = 0 \tag{69}$$

$$\nabla \cdot \bar{\mathbf{u}} = 0; \qquad \nabla \cdot \bar{\mathbf{b}} = 0 \tag{70}$$

$$\partial_T \bar{u}_x + \nabla \cdot (\bar{u}_x \bar{\mathbf{u}} - \bar{b}_x \bar{\mathbf{b}}) = \frac{1}{2} \langle\langle \partial_\perp^* (\tilde{b} \tilde{\rho}) \rangle\rangle + \text{c.c.} \tag{71}$$

$$\partial_T \bar{b}_x + \nabla \cdot (\bar{b}_x \bar{\mathbf{u}} - \frac{1}{2} \bar{u}_x \bar{\mathbf{b}}) = 0 \tag{72}$$

$$\bar{p} = -\bar{b}_x, \tag{73}$$

together with

$$\partial_\tau \tilde{b} + \frac{1}{2} \partial_\xi (\tilde{\rho} b) - \frac{1}{2} \partial_\perp \tilde{\rho} + \frac{i}{2R_i} \partial_{\xi\xi} \tilde{b} + \partial_\xi [\tilde{b}(\bar{u}_x + \bar{b}_x - \frac{\bar{\rho}}{2})] = 0 \tag{74}$$

$$2\partial_\tau \tilde{\rho} + \partial_\xi (\alpha \tilde{\rho} + \frac{1+\gamma}{2} \tilde{\rho}^2 + \frac{|\tilde{b} + \bar{b}|^2}{2}) - \frac{1}{2}(\partial_\perp^* \tilde{b} + \partial_\perp \tilde{b}^*)$$
$$+ [(\gamma - 1)\bar{\rho} + 2\bar{u}_x] \partial_\xi \tilde{\rho} = 0. \tag{75}$$

Note that, although decoupled, the longitudinal velocity and magnetic field components are of the same order of magnitude as the transverse ones. This situation is similar to turbulence in $2\frac{1}{2}$ dimensions discussed by Zank and Matthaeus [37] in the context of the reduced MHD for $\beta \approx 1$. In the present analysis however, the mean longitudinal velocity is driven by the nonlinear Alfvén waves governed by eqs. (74)–(75). Furthermore, the coefficient $\frac{1}{2}$ in eq. (72) resulting from the contribution of the term $\bar{u}_2$ is absent in Zank and Matthaeus' approach which is not based on a systematic asymptotic expansion but rather uses a prescribed form of slowly varying solutions. In the absence of mean transverse fields, the equations derived by Hada [12] are recovered.

4.3   Transverse modulational analysis of the generalized DNLS equations

The long-wave equations derived in Sections 4.1 and 4.2 using a reductive perturbation expansion, allow to consider wave amplitudes much larger than those permitted by the envelope formalism. It is thus of interest to study the filamentation instability using these systems of equations. This problem was investigated in [10]. It is indeed often the case that a modulational-type instability disappears when the carrying wave has a large amplitude, a situation not captured by the NLS equation and usually addressed within the framework of the Whitham equations [35,20]. Furthermore, (52)–(55) allow to study the effect of a finite compressibility of the transverse mean velocity $\bar{u}$ on the filamentation (non-zero value of ϵ).

As a first step we consider in the absence of dispersion, the transverse modulational stability (for β not close to unity) of the exact stationary solution $\widetilde{b}^{(0)} = B_0 e^{ik\xi}$. The phase and amplitude of this carrying wave are perturbed as $\widetilde{b} = B_0(1+A)e^{i(k\xi+\phi)}$, where B_0 is real, and A and ϕ are taken independent on the longitudinal variable X. The other physical quantities are written as $p = p^{(0)} + p^{(1)}$, (where the unperturbed value $p^{(0)} = 0$). The perturbation $p^{(1)}$ separates into a mean contribution $\bar{p}^{(1)} = \bar{p}(\eta, \zeta, \tau)$ and an oscillating part $\widetilde{p}^{(1)} = \widetilde{p}e^{i(k\xi+\phi)} + \text{c.c.}$ where $\widetilde{p} = \widetilde{p}^R(\eta, \zeta, \tau) + i\widetilde{p}^I(\eta, \zeta, \tau)$.

Neglecting nonlinear terms and projecting on the first Fourier mode, we obtain the following set of equations for the evolution of perturbations

$$B_0(i\partial_\tau\phi + \partial_\tau A) + ikB_0(\bar{u}_x + \frac{\bar{b}_x}{2} - \frac{\bar{\delta}}{2}) - \frac{1}{2}\partial_\perp(\beta\widetilde{\rho} + \widetilde{b}_x) = 0 \qquad (76)$$

$$\widetilde{u}_x - \widetilde{\delta} = 0 \qquad (77)$$

$$\widetilde{u}_x - \beta\widetilde{\delta} - \beta\widetilde{b}_x = 0 \qquad (78)$$

$$ik\widetilde{b}_x + \frac{B_0}{2}(\partial_\perp^* A + i\partial_\perp^* \phi) = 0 \qquad (79)$$

$$\partial_\tau\bar{u} + \partial_\perp(B_0^2 A + \beta\bar{\delta} + (\beta+1)\bar{b}_x) = 0 \qquad (80)$$

$$\partial_\tau\bar{\delta} + \frac{B_0}{2}(\partial_\perp(\widetilde{u}_x - \widetilde{\delta}) + \partial_\perp^*(\widetilde{u}_x^* - \widetilde{\delta}^*)) = 0 \qquad (81)$$

$$\partial_\tau\bar{u}_x - \frac{B_0}{2}(\partial_\perp(\widetilde{b}_x + \widetilde{u}_x) + \partial_\perp^*(\widetilde{b}_x^* + \widetilde{u}_x^*)) = 0 \qquad (82)$$

$$\partial_\tau\bar{b}_x + \frac{1}{2\epsilon}(\partial_\perp\bar{u}^* + \partial_\perp^* \bar{u}) - \frac{B_0}{2}(\partial_\perp(\widetilde{b}_x + \widetilde{u}_x) + \partial_\perp^*(\widetilde{b}_x^* + \widetilde{u}_x^*)) = 0. \qquad (83)$$

We have $\widetilde{\delta} = \widetilde{u}_x$, $\widetilde{b}_x = \frac{(1-\beta)}{\beta}\widetilde{u}_x$, and $\bar{\delta} = 0$. Separating real and imaginary parts and assuming that oscillating and non-oscillating perturbations are proportional to $e^{i(K_\eta \eta + K_\zeta \zeta - \Omega \tau)}$, we obtain the following dispersion relation

$$\Omega^4 - \Omega^2 K_\perp^2 \left(\frac{K_\perp^2}{16 k^2 (1-\beta)^2} + \frac{3 B_0^2}{4(1-\beta)} + \frac{(\beta+1)}{\epsilon} \right)$$
$$+ \frac{K_\perp^6 (\beta+1)}{16 \epsilon k^2 (1-\beta)^2} + B_0^2 \frac{K_\perp^4 (4\beta+3)}{8\epsilon(1-\beta)} = 0. \tag{84}$$

As $\epsilon \to 0$ this dispersion relation identifies with that obtained from envelope equations taken in the limit $\mu \to 0$. The instability criterion for eq. (84) is not affected when ϵ is taken finite : perturbations with wave numbers satisfying the condition $K_\perp^2 < 2 B_0^2 k^2 (\beta-1)(3+4\beta)/(\beta+1)$ are unstable. In particular, as B_0 is increased, no restabilization is obtained. This contrasts with the case of longitudinal perturbations. Note however that for non-vanishing values of ϵ, there exists a finite range of unstable transverse wavenumbers around a value proportional to $\epsilon^{-1/2}$. The question arises of the nature of this extra instability which affects wavenumbers outside the range of accuracy of the asymptotics [4].

5   Conclusions

The filamentation of dispersive Alfvén waves propagating along the ambient magnetic field, provides an efficient mechanism for small scale formation and a way to heat the plasma. This process does not require waves of initially high-amplitude nor plasma inhomogeneities. Furthermore, the dissipation resulting from the wave filamentation can be supplemented by an additional small-scale mechanism when the coupling to magneto-sonic waves become relevant. Indeed, except for very short Alfvén pulses, sharp magneto-sonic fronts develop, where viscous dissipation is efficient, possibly leading to a total burnout of the filaments. In a regime where the dispersion is weaker, the circular polarization of the wave is not enforced anymore, and the nonlinear dynamics is significantly different from that of the strongly dispersive case. The wave amplitude remains moderate, but small scales are generated in the form of thin layers of intense gradients. The resulting dissipation appears to be more homogeneous than in the strongly dispersive regime and globally enhanced. The present investigation of filamentation dynamics is restricted to monochromatic waves. A first step towards an analysis in a more realistic situations consists in using the generalized DNLS equations presented in Section 4. In this context, the dispersion is at the same level as the nonlinearities of the carrier wave. It becomes possible to study the effect of non-quasi-monochromatic waves as well the nonlinear dynamics of the magneto-sonic waves generated by the filamentation process. A numerical integration of these DNLS equations, in absence of mean transverse fields, is under way. Preliminary simulations [17] show that, in the condition where the dispersion dominates the nonlinearity by a factor of order 30, a quasi-monochromatic plane wave undergoes an isotropic transverse collapse which saturates after the amplitude has increased by an order of magnitude. In contrast, when the dispersion is reduced by a factor 10, the observed dynamics is qualitatively similar to that predicted by eq. (34), with the formation of elongated structures of strong gradients but moderate amplitude (see Fig. 9). It still remains to evaluate the role of kinetics effects as well as the competition with decay instabilities ([11] and references therein). In the one-dimensional problem, depending on the β, they were observed to possibly lead to the rapid development of a strongly nonlinear regime, even in the case of small-amplitude waves [4].

Acknowledgements

This work benefited from partial support from the "Programme National Soleil Terre" of CNRS. Part of the computations were performed on the CRAY94 and CRAY98 machines of the Institut du Développement et des Ressources en Informatique Scientifique.

References

1. J.W. Belcher and L.Jr. Davis, J. Geophys. Res. **76**, 3534 (1971).
2. S.A. Bel'kov and S.A. Tsytovich, Sov. Phys. JETP **49**, 656 (1979).
3. S. Champeaux, A. Gazol, T. Passot and P.L. Sulem, ApJ **486**, 477 (1997).
4. S. Champeaux, D. Laveder, T. Passot and P.L. Sulem, Remarks on the parallel propagation of small-amplitude dispersive Alfvén waves, Nonlinear Processes in Geophys., in press.
5. S. Champeaux, T. Passot and P.L. Sulem, J. Plasma Phys. **58**, 665 (1997).
6. S. Champeaux, T. Passot and P.L. Sulem, Phys. Plasmas **5**, 100 (1998).
7. S. Champeaux, T. Passot and P.L. Sulem, Phys. Plasmas **6**, 413 (1999).
8. S. Dyachenko, A. C. Newell, A Pushkarev and V. E. Zakharov, Physica D **57**, 210 (1992).
9. A. Gazol, T. Passot, and P.L. Sulem, J. Plasma Phys. **60**, 95 (1998).
10. A. Gazol, T. Passot, and P.L. Sulem, Phys. Plasmas, **6**, 3114 (1999).
11. S.Ghosh, A.F. Viñas and M.L. Golstein, J. Geophys. Res., **98**, 15561 (1993).
12. T. Hada, J. Geophys. Res. **20**, 2415 (1993).
13. P. Huerre and P.A. Monkewitz, Ann. Rev. Fluid Mech. **22**, 473-537 (1990).
14. M. Kono, M.M. Škorić and D. ter Harr, J. Plasma Phys. **26**, 123 (1981).
15. S.P.P. Kuo, M. H. Wang and M. C. Lee, J. Geophys. Res. **93**, 9621 (1988).
16. S.P.P. Kuo, M. H. Wang and G. Schmidt, Phys. Fluids B **4**, 734 (1988).
17. D. Laveder, T. Passot and P.L. Sulem, Alfvén wave filamentation beyond the envelope formalism, Proceedings of ITCPP99 (Faro, Portugal, 1999), Physica Scripta, in press.
18. M. V. Medvedev and P. H. Diamond, Phys. Plasmas **3**, 863 (1995).
19. J. F. McKenzie, M. Banaszkiewicz and W. I. Axford, Astron. Astrophys. **303**, L45 (1995).
20. E. Mjølhus, J. Plasma Phys. **16**, 321, (1976).
21. E. Mjølhus and J. Wyller, J. Plasma Phys. **40**, 299 (1988).

22. E. Mjølhus and T. Hada, in *Nonlinear Waves in Space Plasmas*, T. Hada and H. Matsumoto eds., 121, (TERRAPUB, Tokyo, 1997).

23. A. C. Newell, D. A. Rand and D. Russell, Physica D **33**, 281 (1988).

24. T. Passot and P.L. Sulem, Phys. Rev. E **48**, 2966 (1993).

25. T. Passot and P.L. Sulem, in *Small-scale structures in three-dimensional hydro-dynamic and magnetohydrodynamic turbulence*, M. Meneguzzi, A. Pouquet and P.L. Sulem eds., Lecture Notes in Physics **462**, 405 (Springer Verlag, 1995).

26. T. Passot, C. Sulem and P.L. Sulem, Phys. Rev. E **50**, 1427 (1994).

27. I. V. Relke and A. M. Rubenchik, J. Plasma. Phys. **39**, 369 (1988).

28. M.S. Ruderman, Fluid Dyn. **22**, 299 (1987) (Isv. Akad. Nauk SSSR, Mekh. Zhid. i Gaza, **2**, 159).

29. P. K. Shukla, G. Feix and L. Stenflo, Astrophys. Space Sci. **147**, 383 (1988).

30. P.K. Shukla and L. Stenflo, Astrophys. Space Sci. **155**, 145 (1989).

31. Spangler, S.R. Phys. Fluids B **2**, 407 (1990).

32. C. Sulem and P.L. Sulem, *The nonlinear Schrödinger equation: self-focusing and wave collapse*, Applied Mathematics Sciences **139**, (Springer Verlag 1999).

33. Viñas, A.F. and Goldstein, M.L. J. Plasma Phys. **46**, 129 1991.

34. M. I. Weinstein, Comm. Math. Phys. **87**, 567 (1983).

35. G.B. Whitham, *Linear and nonlinear waves* (Wiley, New York, 1974).

36. Zakharov, V.E. and Rubenchik, A.M., Prikl. Mat. Techn. Fiz. **5**, 84 (1972) (in Russian), see also Zakharov, V.E. and Kuznetsov, E.A., Sov. Sci. Rev., Section C: Math. Phys. Rev. **4**, 167 (1984), or Zakharov, V.E. and Schulman, E.I., in *What is integrability ?*, V.E. Zakharov, ed., 185, Springer Series on Nonlinear Dynamics, (Springer Verlag 1991).

37. G.P. Zank and W.H. Matthaeus, J. Plasma Physics, **48**, 85 (1992).

Nonlinear Quasiresonant Alfvén Oscillations in a One-Dimensional Magnetic Cavity

L. Nocera[1] and M.S. Ruderman[2]

[1] Institute of Atomic and Molecular Physics, National Research Council, Via Giardino 7, I-56127 Pisa, Italy
[2] School of Mathematical and Computational Sciences, University of St Andrews, St Andrews, Fife KY16 9SS, Scotland, UK

Abstract. The steady state of nonlinear, small-amplitude, one-dimensional quasiresonant Alfvén oscillations in a homogeneous dissipative hydromagnetic cavity forced by the shear motion of its boundaries is studied. It is shown that, even in the case of strong nonlinearity, these oscillations can be represented, to leading order, by a sum of two solutions in the form of oppositely propagating waves with permanent shapes. An infinite set of nonlinear algebraic equations for the Fourier coefficients of these solutions is derived. It is then reduced to a finite set of equations by trancation and solved analytically in the one-mode approximation and numerically in the general case. The comparison of the analytical and numerical results is carried out.

1 Introduction

Study of driven Alfvén oscillations in magnetic cavities are important in the context of magnetospheric and solar physics. In magnetospheric physics this problem arises when considering the excitation of shear Alfvén waves in the magnetospheric cavity or in the high altitute Alfvénic resonator (see [1] and references therein). The well-known examples of magnetic cavities are coronal magnetic loops. Resonant absorption of Alfvén waves has been suggested in [2] as a mechanism of heating of coronal loops. This mechanism has then received much attention by solar physicists (see, e.g., [3–9]). There are also evidencies that magnetospheric cavities exist in the solar chromosphere (see, e.g., [10–12]).

When the driver frequency is close to the fundamental frequency of the cavity or to the frequency of one of the overtones, the driven oscillations are quasiresonant and their amplitudes can be large even when the driver amplitude is small. Driven quasiresonant Alfvén oscillations in magnetic cavities have been studied either analytically in the low-mode approximation (e.g. [1,13–17]), or by means of numerical integration of the full set of MHD equations (e.g. [1,15]).

In this paper we aim to suggest a new method of studying driven quasiresonant Alfvén oscillations in homogeneous magnetic cavities. This method is based on the selection of resonant modes in the double Fourier series. We consider only the steady state of driven oscillations. However the method can be easily generalized to include the slow time variation. The method results in enormous saving of the computational time when the multi-mode description is used. It also allows to outline the scope of applicability of the low-mode approximations.

The paper is organized as follows. In the next section we give the setup of the problem and discuss main assumptions. In Section 3 we derive the governing equation for the nonlinear quasi-resonant driven Alfvén waves written in the form of an infinite set of algebraic equations for the Fourier coefficients. In Section 4 we derive expressions for the energy of Alfvén oscillations in the cavity and for the energy flux into the cavity. In Section 5 we consider the bifurcations of the system in the parameter range where multiple solutions exist. In Section 6 we study two particular ranges of parameters, one where solutions with very large spatial gradients (shocks) exist, and another where the saddle-node bifurcation occurs for large values of the Reynolds number. In Section 7 we present the discussion of our results and conclusions.

2 Basic equations and assumptions

We consider a homogeneous hydromagnetic cavity with an equilibrium magnetic field in the x-direction (see Fig. 1). At one end of the cavity the plasma is at rest, while it is harmonicly driven in the y-direction at the other end. Throughout the cavity the z-components of the velocity and the magnetic field are zero, and perturbations of all quantities depend on x and t only.

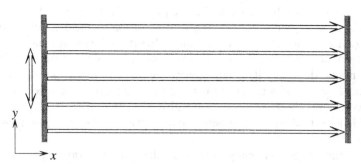

Fig. 1. Sketch of the cavity with left moving (vertical arrow) and right fixed boundary. The long horizontal arrows show the equilibrium magnetic field.

The plasma is assumed to be viscous and resistive, viscosity and resistivity being isotropic. The last assumption is not satisfied for the solar corona where viscosity is strongly anisotropic. However in what follows we consider Alfvén oscillations only. Although we shall develop a nonlinear theory of such oscillations, dissipative terms in the momentum equation and in the induction equation are taken in the linearized form. In the case of Alfvén oscillations, the leading term in the linearized tensorial expression for viscosity (proportional to η_0, Braginskii [18]) is zero. This fact enables us to use isotropic viscosity in what follows. Furthermore, in the corona, the Hall term, which appears in the generalized Ohm's law due to anisotropy of electrical conductivity, is larger than the resistive term. However, we need non-ideal terms in the momentum and induction equations to provide dissipation only, whereas the Hall term in the induction equation provides dispersion. We thus neglect this term.

Let us introduce the viscous and magnetic Reynolds numbers

$$R_e = \frac{LV_A}{\bar{\nu}}, \quad R_m = \frac{LV_A}{\bar{\lambda}}, \tag{1}$$

where L is the length of the magnetic cavity, $\bar{\nu}$ kinematic coefficient of the shear viscosity, and $\bar{\lambda}$ coefficient of magnetic diffusion. The square of the Alfvén speed V_A is given by

$$V_A^2 = \frac{B_0^2}{\mu\rho_0}, \tag{2}$$

where B_0 is the equilibrium magnetic field, ρ_0 the equilibrium density, and μ the magnetic permeability. To characterise the total dissipation due to both viscosity and resistivity we introduce the total Reynolds number R_t given by

$$\frac{1}{R_t} = \frac{1}{R_e} + \frac{1}{R_m}. \tag{3}$$

The total Reynolds number is large in the solar atmosphere. We introduce the small parameter ϵ such that $R_t = \mathcal{O}(\epsilon^{-2})$. Although at present this scaling seems to be artificial, we shall see in what follows that it is convenient and natural. To explicitly show this scaling in the MHD equations we introduce scaled coefficients of viscosity and magnetic diffusion

$$\nu = \epsilon^{-2}\bar{\nu}, \quad \lambda = \epsilon^{-2}\bar{\lambda}. \tag{4}$$

Then we write the equations of viscous resistive MHD in the form

$$\frac{\partial \rho}{\partial t} + \frac{\partial(\rho u)}{\partial x} = 0, \tag{5a}$$

$$\frac{\partial u}{\partial t} + u\frac{\partial u}{\partial x} = -\frac{1}{\rho}\frac{\partial p}{\partial x} - \frac{B_y}{\mu\rho}\frac{\partial B_y}{\partial x} + \epsilon^2\nu\frac{\partial^2 u}{\partial x^2}, \tag{5b}$$

$$\frac{\partial v}{\partial t} + u\frac{\partial v}{\partial x} = \frac{B_0}{\mu\rho}\frac{\partial B_y}{\partial x} + \epsilon^2\nu\frac{\partial^2 v}{\partial x^2}, \tag{5c}$$

$$\frac{\partial B_y}{\partial t} = \frac{\partial}{\partial x}(vB_0 - uB_y) + \epsilon^2\lambda\frac{\partial^2 B_y}{\partial x^2}, \tag{5d}$$

$$\frac{\partial}{\partial t}\left(\frac{p}{\rho^\gamma}\right) + u\frac{\partial}{\partial x}\left(\frac{p}{\rho^\gamma}\right) = 0. \tag{5e}$$

Here u and v are the x- and y-components of the velocity, B_y is the y-component of the magnetic field, ρ the density, p the pressure, and γ the adiabatic index. Although the dissipation is present we use the adiabatic equation (5e). The reason is that dissipative terms that should appear on the right-hand side of equation (5e) are nonlinear and would therefore give higher order corrections to the dynamics of the Alfvén wave. In what follows we only use the linearized form of equation (5e).

The magnetic cavity is bounded by the two surfaces perpendicular to the background magnetic field at $x = 0$ and $x = L$. The magnetic field is assumed to be frozen in the dense plasmas beyond these boundaries, so that the magnetic field lines follow plasma motions beyond the boundaries. In application to coronal loops, for instance, the boundaries model the dense highly electrically conducting photosphere. We assume that the plasma at $x = L$ is immovable, while it harmonically oscillates in the y-direction at $x = 0$ (Fig. 1). In accordance with this we have the boundary conditions

$$u = v = 0 \quad \text{at} \quad x = L, \tag{6a}$$

$$u = 0, \quad v = A\sin(\omega t) \quad \text{at} \quad x = 0, \tag{6b}$$

where A and ω are the amplitude and (real) frequency of the driver. The fundamental eigenfrequency of the cavity is $\omega_A = \pi V_A / L$. When $\omega = \omega_A$, linear theory predicts the amplitude of oscillations of the order $AR_t \sim A\epsilon^{-2}$. As for the case of nonlinear dynamical systems with finite numbers of degrees of freedom, the most interesting (i.e. stucturally unstable) behaviour is to be sought in the situation in which nonlinearity, dissipation and driving produce effects which are of the same order (e.g. Guckenheimer and Holmes [19]). This is why, in what follows, we assume that the mistuning is small and oscillations are quasi-resonant: $|\omega/\omega_A - 1| \ll 1$. When there is no dissipation ($R_t \to \infty$) linear theory predicts the amplitude of oscillations of the order $A|\omega/\omega_A - 1|^{-1}$. In what follows we only consider the situation where the two effects, mistuning and dissipation, are of the same order. Therefore we assume that $\omega/\omega_A - 1 = \mathcal{O}(\epsilon^2)$ and write ω in the form

$$\omega = \omega_A(1 + \epsilon^2 \Delta). \tag{7}$$

The quantity Δ is called the mistuning parameter.

Linear theory of quasi-resonant oscillations is only valid when the amplitude of oscillations is small. Since this amplitude is of the order $A\epsilon^{-2}$, this condition is equivalent to $A \ll \epsilon^2$. To make analytical progress when studying nonlinear oscillations we assume that the amplitudes of nonlinear oscillations are also small. In what follows we only study the situation where the effect of nonlinearity is of the same order as the effect of dissipation. This implies that the dissipative terms in equations (5c) and (5d), which are the terms proportional to ν and λ, are of the same order as nonlinear terms. A typical nonlinear term in these equations is $u\partial v/\partial x$. Its ratio to the dissipative term in equation (5c), which is the term proportional to ν, is of the order $u\epsilon^{-2}$.

This ratio is of the order unity when $u = \mathcal{O}(\epsilon^2)$. Let us estimate the order of magnitude of the quantity u in terms of ϵ and A. To do this we analyze how nonlinearity acts. As we have already seen, the amplitude of Alfvén oscillations in the cavity is of the order $A\epsilon^{-2}$. The presence of Alfvén oscillations causes a variation of the total pressure of the order $A^2\epsilon^{-4}$. This variation of the total pressure drives plasma motions which are parallel to the equilibrium magnetic field. The interaction of the longitudinal motion with the basic Alfvén oscillations creates nonlinear corrections to the Alfvén oscillations. Hence, in particular, u is of the order of the total pressure variation, that is $u = \mathcal{O}(A^2\epsilon^{-4})$. Then the condition $u = \mathcal{O}(\epsilon^2)$ leads to $A = \mathcal{O}(\epsilon^3)$. In what follows we take $A = \epsilon^3 V_A$ and rewrite boundary condition (6b) as

$$u = 0, \quad v = \epsilon^3 V_A \sin(\omega t) \quad \text{at} \quad x = 0. \tag{8}$$

In the next section we use the set of equations (5) and boundary conditions (6a) and (8) to derive a nonlinear governing equation for quasiresonant Alfvén oscillations in the magnetic cavity.

3   Derivation of the governing equations

To derive the governing equation for nonlinear resonant oscillations in the magnetic cavity we use the singular perturbation method. Since $v = \mathcal{O}(\epsilon^3)$ at the boundary, while $v = \mathcal{O}(\epsilon)$ in the cavity, we can take $v \approx 0$ at $x = 0$ in the main order approximation. In what follows we only consider the steady state of resonant oscillations, where perturbations of all quantities oscillate with the driver's frequency ω. Hence we can expand the quantity v in a Fourier series with respect to x and t:

$$v = \sum_{n=-\infty}^{\infty} e^{in\omega t} \sum_{m=1}^{\infty} v_{nm} \sin \frac{\pi m x}{L}. \tag{9}$$

The fundamental harmonic in this series, represented by terms

$$v_{11} e^{i\omega t} \sin \frac{\pi x}{L} \quad \text{and} \quad v_{-11} e^{-i\omega t} \sin \frac{\pi x}{L},$$

is excited by the driver. All other harmonics in series (9) are generated by the two-step nonlinear mechanism described in section 2. The driving force that appears due to the interaction of the longitudinal motion with the amplitude of the order ϵ^2, and the Alfvén oscillations with the amplitude of the order of ϵ, has amplitude of the order ϵ^3. Then it is straightforward to see that the amplitudes of the higher resonant terms in (9), which are terms with $m = n$, are of the order ϵ, while the non-resonant terms, which are terms with $m \neq n$, are of the order ϵ^3. Hence, we can take v in the form

$$v = \sum_{n=-\infty}^{\infty} v_{nn} e^{in\omega t} \sin \frac{\pi n x}{L} + \mathcal{O}(\epsilon^3). \tag{10}$$

The sum in this equation can be rewritten as

$$\sum_{n=-\infty}^{\infty} v_{nn} e^{in\omega t} \sin \frac{\pi n x}{L} = \epsilon [f(\xi) - f(\eta)], \tag{11}$$

where

$$\xi = \omega t + \frac{\pi x}{L}, \quad \eta = \omega t - \frac{\pi x}{L}, \tag{12a}$$

$$f = \sum_{n=-\infty}^{\infty} f_n e^{in\xi}, \quad \epsilon f_n = -\frac{i}{2} v_{nn}. \tag{12b}$$

It follows from equation (12b) that $f_{-n} = f_n^*$, where the asterisk indicates a complex conjugate quantity. We see that in the main order approximation the nonlinear resonant Alfvén oscillation is represented by a superposition of two nonlinear waves, one of which propagates in the positive direction of the x-axis and the other in the negative direction. Simple estimates based on the set of equations (5) show that the perturbations of the quantities ρ and p are of the same order as u, that is of the order ϵ^2, while B_y is of the order ϵ and does not contain terms of the order ϵ^2. Consequently, we look for the solution to the set of equations (5) in the form of expansions

$$v = \epsilon [f(\xi) - f(\eta)] + \epsilon^3 v^{(3)} + \dots, \quad B_y = \epsilon B_y^{(1)} + \epsilon^3 B_y^{(3)} + \dots, \tag{13a}$$

$$\rho = \rho_0 + \epsilon^2 \rho^{(2)} + \dots, \quad p = p_0 + \epsilon^2 p^{(2)} + \dots, \quad u = \epsilon^2 u^{(2)} + \dots \tag{13b}$$

In what follows we only outline the derivation of the governing equation. The details of the derivation can be found in [20]. We obtain the equations of the first order approximation collecting terms of the order ϵ in equations (5c) and (5d). It follows from these equations that

$$B_y^{(1)} = \frac{B_0}{V_A} [f(\xi) + f(\eta)]. \tag{14}$$

The equations of the second order approximation are obtained by collecting terms of the order ϵ^2 in equations (5a), (5b) and (5e). It follows from these equations that $p^{(2)} = c_S^2 \rho^{(2)}$ with $c_S^2 = \gamma p_0 / \rho_0$ the square of the sound speed, and $u^{(2)}$ and $\rho^{(2)}$ are given by

$$u^{(2)} = \sum_{n=-\infty}^{\infty} u_n^{(2)}(x) e^{in\omega t}, \quad \rho^{(2)} = \sum_{n=-\infty}^{\infty} \rho_n^{(2)}(x) e^{in\omega t}, \tag{15}$$

$$u_n^{(2)} = -2iV_A \left[\frac{1}{n(V_A^2 - c_S^2)} \sum_{m=-\infty}^{\infty} m f_m f_{n-m} \sin \frac{\pi n x}{L} \right.$$

$$\left. + n \sum_{m=-\infty}^{\infty} \frac{m f_m f_{n-m}}{V_A^2 n^2 - c_S^2 (2m-n)^2} \sin \frac{\pi (2m-n)x}{L} \right], \tag{16}$$

$$\rho_n^{(2)} = -\frac{2\rho_0}{n(V_A^2 - c_S^2)} \sum_{m=-\infty}^{\infty} m f_m f_{n-m} \cos \frac{\pi n x}{L}$$

$$+ 2\rho_0 \sum_{m=-\infty}^{\infty} \frac{m(2m-n) f_m f_{n-m}}{V_A^2 n^2 - c_S^2 (2m-n)^2} \cos \frac{\pi(2m-n)x}{L}. \tag{17}$$

Equation (17) is valid for $n > 0$. The quantity $\rho_0^{(2)}$ is determined by

$$\rho_0^{(2)} = -\frac{\rho_0}{c_S^2} \sum_{m=-\infty}^{\infty} |f_m|^2 \cos \frac{2\pi m x}{L}. \tag{18}$$

In the third order approximation we collect terms of the order ϵ^3 in equations (5c) and (5d) and eliminate $B_y^{(3)}$ from the obtained equations. As a result we arrive at

$$\frac{\partial^2 v^{(3)}}{\partial t^2} - \frac{\omega^2 L^2}{\pi^2 B_0} \frac{\partial^2 v^{(3)}}{\partial x^2} = F, \tag{19}$$

where

$$F = -2\Delta \frac{\partial^2 v^{(1)}}{\partial t^2} - \frac{\partial}{\partial t} \left(u^{(2)} \frac{\partial v^{(1)}}{\partial x} - \frac{\rho^{(2)}}{\rho_0} \frac{\partial v^{(1)}}{\partial t} \right)$$

$$- \frac{V_A^2}{B_0} \frac{\partial^2}{\partial x^2} \left(u^{(2)} B_y^{(1)} \right) + (\nu + \lambda) \frac{\partial^3 v^{(1)}}{\partial t \partial x^2}. \tag{20}$$

The function $v^{(3)}$ satisfies the boundary conditions

$$v^{(3)} = 0 \quad \text{at} \quad x = L, \tag{21a}$$

$$v^{(3)} = V_A \sin(\omega t) \quad \text{at} \quad x = 0. \tag{21b}$$

In order to have homogeneous boundary conditions we make substitution

$$v^{(3)} = g + V_A \left(1 - \frac{x}{L} \right) \sin(\omega t). \tag{22}$$

Then we rewrite equation (19) as

$$\frac{\partial^2 g}{\partial t^2} - \frac{\omega^2 L^2}{\pi^2 B_0} \frac{\partial^2 g}{\partial x^2} = F + V_A \omega^2 \left(1 - \frac{x}{L} \right) \sin(\omega t). \tag{23}$$

The function g satisfies the boundary conditions

$$g = 0 \quad \text{at} \quad x = 0 \quad \text{and} \quad x = L. \tag{24}$$

Now we expand the functions g and F in the Fourier series with respect to time:

$$g = \sum_{n=-\infty}^{\infty} g_n e^{in\omega t}, \quad F = \sum_{n=-\infty}^{\infty} F_n e^{in\omega t}. \tag{25}$$

The substitution of these expansions into (23) yields

$$\frac{\partial^2 g_n}{\partial x^2} + \frac{\pi^2 n^2}{L^2} g_n = -\frac{F_n}{V_A^2} + \frac{i\omega^2}{2V_A}\left(1 - \frac{x}{L}\right)(\delta_{1,n} - \delta_{-1,n}), \qquad (26)$$

where $\delta_{j,n}$ is the Kronecker delta-symbol. Equation (26) is compatible only when its right-hand side is orthogonal to the eigenfunction of the differential operator determined by the left-hand side of equation (26) and boundary conditions (24) that corresponds to the zero eigenvalue. This eigenfunction is $\sin(\pi n x/L)$. Then the orthogonality condition yields

$$2\Delta f_n - \frac{i\omega(\nu + \lambda)}{V_A^2} n f_n - \frac{(f^3)_n}{2(V_A^2 - c_S^2)}$$

$$- f_n \sum_{m=-\infty}^{\infty} \frac{(2n - m)^2 |f_{n-m}|^2}{V_A^2 m^2 - c_S^2 (2n - m)^2} = \frac{V_A \delta_{1,n}}{2\pi}, \qquad (27)$$

where $(f^3)_n$ is the nth Fourier coefficient for the function f^3. This equation is valid for $n > 0$. For $n < 0$ the Fourier coefficients are determined by the relation $f_n = f_{-n}^*$.

Equation (27) is the nonlinear governing equation for the function f written in terms of the Fourier coefficients of this function. In section 5 we analytically study this equation in the one-mode approximation; in section 6 we present results of its numerical solution in the multi-mode case. Here we only note one very important property of equation (27). If we take the approximation of cold plasmas and put $c_S = 0$, we obtain that the sum in (27) becomes singular since its term corresponding to $m = 0$ is infinite. This property implies that in a cold plasma the steady state of driven oscillations cannot be attained.

4   Energetics

In this section we calculate the energy dissipated and the energy stored in the magnetic cavity. Since the cavity has an infinite extension in the y- and z-direction, we calculate the energy averaged over one period dissipated in a volume with the length L in the x-direction and with the unit lengths in the y- and z-direction. In the steady state of driven oscillations this energy is equal to the energy flux through the unit area of the surface $x = 0$ averaged over one period. Since the x-component of the velocity is zero at $x = 0$, the instantaneous energy flux through the unit area of the surface $x = 0$ is given by the Poynting vector $\boldsymbol{E} \times \boldsymbol{B}/\mu$, where $\boldsymbol{E}$ is the electric field. With the use of the Ohm's and Ampere's laws we get

$$\boldsymbol{E} \times \boldsymbol{B} = \boldsymbol{v}B^2 - \boldsymbol{B}(\boldsymbol{v} \cdot \boldsymbol{B}) + \epsilon^2 \lambda(\nabla \times \boldsymbol{B}) \times \boldsymbol{B}. \qquad (28)$$

With the aid of equations (8) and (13a) we find that at $x = 0$

$$(\boldsymbol{E} \times \boldsymbol{B})_x = -\epsilon^4 V_A B_0 B_y^{(1)} \sin(\omega t) - \epsilon^4 \lambda B_y^{(1)} \frac{\partial B_y^{(1)}}{\partial x} + \mathcal{O}(\epsilon^6). \qquad (29)$$

Now we use equation (14) to reduce this expression to

$$(\boldsymbol{E} \times \boldsymbol{B})_x = -2\epsilon^4 B_0^2 f(\omega t) \sin(\omega t) + \mathcal{O}(\epsilon^6). \qquad (30)$$

And, finally, we arrive at the following expression for the period-averaged energy flux $\mathcal{S}$:

$$\mathcal{S} = \frac{\omega}{2\pi\mu} \int_0^{2\pi/\omega} (\boldsymbol{E} \times \boldsymbol{B})_x \, dt = 2\epsilon^4 \rho_0 V_A^2 \Im(f_1) + \mathcal{O}(\epsilon^6), \qquad (31)$$

where the symbol $\Im$ indicates the imaginary part of a quantity.

The period-averaged energy stored in the volume of the magnetic cavity with the length L in the x-direction and with the unit lengths in the y- and z-direction is

$$\mathcal{E} = \frac{\epsilon^2 \omega}{4\pi} \int_0^{2\pi/\omega} dt \int_0^L \left[\rho_0 (v^{(1)})^2 + \frac{1}{\mu}(B_y^{(1)})^2 \right] dx + \mathcal{O}(\epsilon^4), \qquad (32)$$

where $v^{(1)} = f(\xi) - f(\eta)$. When deriving this equation we have used the fact that the contribution of the perturbation of the inner energy of the plasma into the stored energy is of the order $\mathcal{O}(\epsilon^4)$. We use equation (14) and the relation $f_{-n} = f_n^*$ to transform expression (32) to

$$\mathcal{E} = 4\epsilon^2 \rho_0 L \sum_{n=1}^{\infty} |f_n|^2 + \mathcal{O}(\epsilon^4). \qquad (33)$$

It is interesting to note that $\mathcal{S}/\mathcal{E} = \mathcal{O}(\epsilon^2)$.

In the next sections expressions (31) and (33) are used to calculate the energy flux into the cavity and the energy stored in the cavity analytically in the one-mode approximation and numerically in the multi-modal case.

5   Multistability

The nonlinear algebraic system (27) has in general multiple solutions. This fact holds true even in the 1–mode approximation. Let us introduce the dimensionless quantities

$$V = \frac{f}{V_A}, \quad \beta = \frac{c_S^2}{V_A^2}, \quad R = \frac{V_A^2}{\omega(\nu + \lambda)}. \qquad (34)$$

In accordance with the assumption made in Section 3 the quantity $\beta^{1/2}$ is an irrational number. The quantity $R = \epsilon^2 R_t/\pi$ can be called the scaled total Reynolds number.

5.1   One-mode approximation

Let us consider the one-mode approximation where $v_n = 0$ for $|n| > 1$. Then it is straightforward to obtain

$$(V^3)_1 = 3V_1|V_1|^2, \quad \sum_{m=-\infty}^{\infty} \frac{(2-m)^2|V_{1-m}|^2}{m^2 - \beta(2-m)^2} = -\frac{|V_1|^2}{\beta}. \tag{35}$$

With the use of these results we get from equation (27)

$$V_1 \left(2\Delta + \frac{2-5\beta}{2\beta(1-\beta)}|V_1|^2 - iR^{-1} \right) = \frac{1}{2\pi}. \tag{36}$$

The quantity V_{-1} is determined by the relation $V_{-1} = V_1^*$. We multiply equation (36) by the complex conjugate equation to derive the equation for $|V_1|^2$

$$Y^3 + 2Y^2 + (1 + 3\kappa)Y - C = 0, \tag{37}$$

where we have introduced the notation

$$Y = \frac{2-5\beta}{4\Delta\beta(1-\beta)}|V_1|^2, \quad \kappa = \frac{1}{12\Delta^2 R^2}, \quad C = \frac{2-5\beta}{64\pi^2\Delta^3\beta(1-\beta)}. \tag{38}$$

In what follows we assume that $\beta < 2/5$. We also assume that $\Delta \neq 0$. The case where $\Delta = 0$ will be considered separately.

Since $|V_1|^2 > 0$ we are looking for solutions to equation (37) satisfying the condition $Y \mathrm{sign}\Delta > 0$. When $\Delta > 0$ the polynomial on the left-hand side of equation (37) is a monotonically growing function for $Y > 0$. Since $C < 0$ there is exactly one positive root to equation (37).

When $\Delta < 0$ the analysis is more complicated. Now we look for negative roots to equation (37) and it is straightforward to see that all real roots to this equation are negative. The discriminant of this equation is

$$D = \sigma^3 + \sigma^2 + 2d\sigma + d^2, \tag{39}$$

where

$$\sigma = \kappa - \frac{1}{9}, \quad d = \frac{1}{2}C + \frac{4}{27}. \tag{40}$$

Equation (37) has one real root when $D > 0$ and three real roots when $D < 0$. The discriminant of the cubic equation $D(\sigma) = 0$ is $\frac{1}{4}d^3(d - \frac{4}{27})$. Since $C < 0$ the quantity d satisfies the restriction $d < \frac{4}{27}$. We consider two cases:

i) $d < 0$. In this case there is only one real root σ_1 to the equation $D(\sigma) = 0$. $D(\sigma) < 0$ when $\sigma < \sigma_1$, while $D(\sigma) > 0$ when $\sigma > \sigma_1$. Since $D(-\frac{1}{9}) > 0$, the root σ_1 satisfies the inequality $\sigma_1 < -\frac{1}{9}$. The condition $\kappa > 0$ leads to $\sigma > -\frac{1}{9} > \sigma_1$. Hence, for all possible values of σ we have $D(\sigma) > 0$. This implies that there is only one real root to equation (37).

ii) $0 < d < \frac{4}{27}$. Then there are three real roots $\sigma_1 < \sigma_2 < \sigma_3 < 0$ to equation (39). It is obvious that $D < 0$ when either $\sigma < \sigma_1$ or $\sigma_2 < \sigma < \sigma_3$, while $D > 0$ otherwise. Since $\sigma_1 + \sigma_2 + \sigma_3 = -1$, it follows that $\sigma_1 < -\frac{1}{3}$. Since $D(-\frac{1}{9}) < 0$ we obtain that $\sigma_2 < -\frac{1}{9} < \sigma_3$. Since $\sigma > -\frac{1}{9}$ it follows that $D < 0$ when $\sigma < \sigma_3$, while $D > 0$ when $\sigma > \sigma_3$. Hence, equation (37) has one real root when $\sigma < \sigma_3$ and three real roots when $\sigma > \sigma_3$.

Summarizing the analysis we state that there are three real roots to equation (37) when the following two inequalities are satisfied:

$$-\frac{8}{27} < C < 0, \quad 0 < \kappa < \sigma_3 + \frac{1}{9}. \tag{41}$$

If at least one of them is not satisfied, there is only one real root to equation (37). Examples of unique and triple solutions are shown in Figs. 3a and 3b respectively.

In terms of Δ and R inequalities (41) are rewritten as

$$\Delta_R(\beta, R) < \Delta < \Delta_c(\beta) \equiv -\frac{3}{8} \left[\frac{2 - 5\beta}{\pi^2 \beta(1 - \beta)} \right]^{1/3}, \tag{42}$$

where $\Delta_R(\beta, R)$ is obtained inverting the relation

$$R = \frac{1}{2|\Delta_R|} \left(\frac{3}{1 + 9\sigma_3(\Delta_R)} \right)^{1/2}. \tag{43}$$

Note that the dependence Δ_R on R is given by the asymptotic formula

$$\Delta_R \simeq -\frac{R^2}{8\pi^2} \frac{2 - 5\beta}{\beta(1 - \beta)}, \quad \text{for } 1 \ll R < \epsilon^{-1/2} \tag{44}$$

The upper bound on R is needed since the quasiresonant perturbation scheme adopted in equation (7) requires $\max|\Delta| = \mathcal{O}(\epsilon^{-1})$.

It can be seen that the equation for $|V_1|^2$ has only one positive real root for all values of R when $\Delta = 0$. The function $\Delta_c(\beta)$ monotonicly glows from $-\infty$ to 0 when β is changed from 0 to $2/5$. In Fig. 2 the dependence of Δ_R on R for $\beta = 0.2$ is shown.

We use equation (36) to rewrite expressions (31) and (33) for the Poynting flux and the energy in the cavity as

$$\mathcal{S} = \epsilon^4 \rho_0 V_A^3 S, \quad \mathcal{E} = \epsilon^2 \rho_0 L V_A^2 E, \tag{45}$$

where

$$S = \frac{(3\kappa)^{1/2}}{\pi |\Delta| [(1 + Y)^2 + 3\kappa]}, \quad E = \frac{1}{\pi^2 \Delta^2 [(1 + Y)^2 + 3\kappa]}. \tag{46}$$

In Figs. 3b and 4b the dependencies of the quantities S and E on R are shown for $\beta = 0.2$ and $\Delta = 0.3$ and $\Delta = -1$ respectively. It is seen from

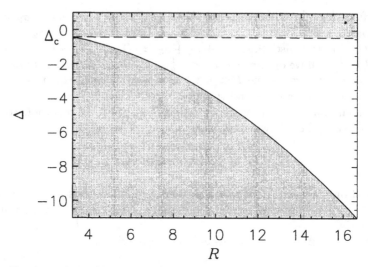

Fig. 2. The dependence of Δ_R on R for $\beta = 0.2$ (solid line). Unique (triple) solutions exist in the shaded (non-shaded) area.

formulae (38) and (46) that $S \propto 1/R$ and $E \to$ const as $R \to \infty$. This trend is in good agreement with Figs. 3b and 4b.

One peculiarity of systems with multiple solutions is their ability to 'bifurcate' from one equilibrium state to another, one classical example being the nonlinear Duffing oscillator. This indeed happens in our hydromagnetic cavity, as it can be seen in Figs. 4b and 5. The transitions the system experiences in going from one 'branch' of the response curve to the other imply an energy dissipation if the arrival branch has a lower energy than the starting branch. When $R \gg 1$ the transition takes place for $|\Delta| = |\Delta_R| \gg 1$, so that, from equations (37), (44) and (46)

$$E = \frac{4R^2}{\pi^2} + \mathcal{O}(R), \quad \text{upper branch,}$$
$$E = \mathcal{O}(R^{-4}), \quad \text{lower branch,}$$

(47)

and the energy jump is

$$\Delta E = \frac{4R^2}{\pi^2} + \mathcal{O}(R). \tag{48}$$

The time in which this energy is dissipated depends on the details of the dynamics in the neighbourhood of the stationary states: its estimate is $\mathcal{O}((R/\epsilon^2)L/V_A)$. Taking equation (48) into account we may write the following estimate for the released power:

$$\mathcal{W} \approx \epsilon^4 \rho_0 V_A^3 W, \quad W = \frac{4}{\pi^2} R. \tag{49}$$

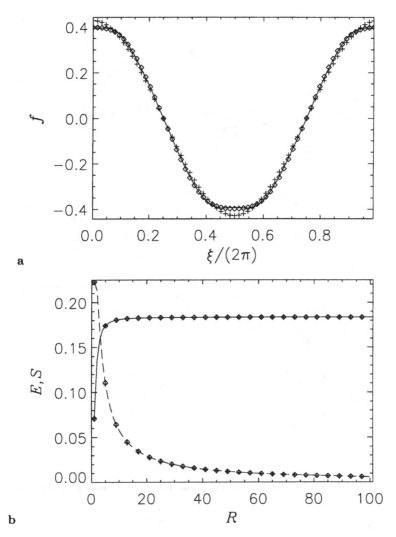

Fig. 3. A unique stationary state for $\beta = 0.2$ and $\Delta = 0.3$. **a.** f vs ξ for $R = 1000$. The solid line shows the numerical solution with the use of 128 modes. **b.** The dimensionless energy, E, (solid line) and Poynting flux, S, (dashed line) vs the normalized Reynolds number R calculated numerically with the use of 128 modes. In both figures $+$ and $\diamond$ shows the 1-mode (analytical) and the 4-mode (numerical) solutions respectively.

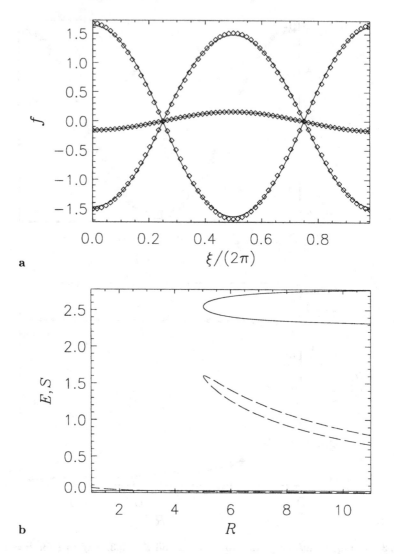

Fig. 4. a. f computed using equation (12b) with 128 modes for $\beta = 0.2$, $\Delta = -1$, and $R = 1000$. The three curves correspond to the three different solutions to equation (27). ◇ shows the 1-mode (analytical) solution. **b.** Bifurcation diagrams for equation (27). The dimensionless energy, E, (solid line) and Poynting flux, S, (dashed line) vs the normalized Reynolds number R for $\beta = 0.2$ and $\Delta = -1$. Note the coalescence of the upper two branches at $R = 5$.

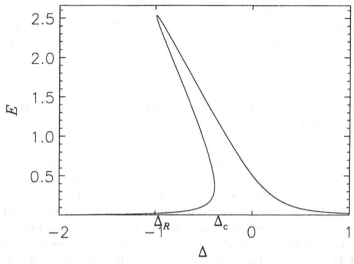

Fig. 5. Energy response of the cavity vs the normalized frequency mistuning at $\beta = 0.2$ and $R = 5$. Note the similarity with the response curve of the Duffing oscillator.

The quantity W is comparable with the normalized Poynting flux (46) at the bifurcation point of the upper branch ($\Delta = \Delta_R$), i.e.

$$S = \frac{2}{\pi}R. \tag{50}$$

We recall that in (49) and (50) $R < \epsilon^{-1/2}$, in accordance with the consistency constraint of equation (44).

5.2   Numerical results

To solve the infinite set of algebraic equations (27) we have used the following method. First we truncated this set of equations and take $f_n = 0$ for $|n| > N$. The relation $f_n = f_{-n}^*$ and the fact that $f_0 = 0$ enabled us to consider $n > 0$ only. As a result we have to solve a set of N algebraic equations. We use the Newton–Raphson method to solve this set of equations [21]. Convergence of this method is highly improved if a 'good' initial guess is given for the root. It turned out that for values of Δ out of the two narrow windows, $[0.20, 0.23]$ and $[0.40, 0.42]$, the one-mode approximation gives a very good initial guess. In Figs. 3 and 4 numerical results for $R = 1000$ and $\Delta = 0.3$ and $\Delta = -1$ are shown. These results were obtained with $N = 128$. We see that the one-mode solution gives an excellent approximation for the multi-modal solution. We have scanned a wide range of R and $\Delta \notin [0.20, 0.23] \cup [0.40, 0.42]$ and obtained the same result: the one-mode and multi-modal solutions practically coincide. Hence, we conclude that the one-mode approximation provides a very good description of the steady state of Alfvén oscillations in the magnetic cavity for $\beta = 0.2$ unless either $0.20 \leq \Delta \leq 0.23$ or $0.40 \leq \Delta \leq 0.42$.

6   Solutions with shocks and high-Reynolds-number bifurcations

When $0.20 \leq \Delta \leq 0.23$ the one-mode approximation does not give a good initial guess for large values of R. To find this initial guess we have used the following procedure. First we found a solution for a small value of R, where the one-mode analytical solution is an acceptable initial guess. Then we increased R and used the solution for the previous value of R as a guess. This procedure was iterated until R reached the desired value. The function $f(\xi)$ for $\beta = 0.2$, $\Delta = 0.2$, and $R = 10$, 100, and 1000 is shown in Fig. 6a. We see that the solution for $R = 1000$ is characterized by very large spatial gradients. As a matter of fact regions with these large spatial gradients correspond to shock waves for $R \to \infty$. The important property of the shock solution is that the corresponding energy flux, S, does not vanish when $R \to \infty$. Instead, it saturates.

Once the solution f_n is found, we can reconstruct the velocity field $v(x,t)$ using equation (10). The time evolution of the velocity $v(x,t)$ is shown in Fig. 6b. It is instructive to compare Fig. 6a with Fig. 7a in [20], where the shock solution for $\beta = 0.1$ is shown. We can see that for $\beta = 0.1$ shocks are much more pronounced than for $\beta = 0.2$. Our numerical study shows that the shocks are strongly pronounced in the shock solution for $\beta < 0.15$. When β is increased beyond this value, the shocks slowly fade away. It is also worth to note the robustness of the shock window for Δ with respect to the variation of β. It is the same, $[0.20, 0.23]$, for $0.1 \leq \beta \leq 0.2$.

It was found in [20] that the shock solution for $\beta = 0.1$ and $\Delta \in [0.20, 0.23]$ is not unique. At $R = R_c \approx 160.5$ the saddle-node bifurcation occurs and two additional low-modal solutions exist for $R > R_c$ (see Fig. 8 in [20]). When $\beta = 0.2$, the situation is different. We did not find bifurcations with respect to R for $\Delta \in [0.20, 0.23]$. So, on the basis of the numerical analysis, we conclude that the shock solution is unique. However, we found that the saddle-node bifurcation occurs when $\Delta \in [0.40, 0.42]$. This bifurcation is shown in Fig. 7. It occurs at $R = R_c \approx 270$. In Fig. 7a the two solutions for $R = 500$ are shown. The solid line shows the solution that also exists for $R < R_c$, while the dashed line shows one of the two solutions that appear as a result of the bifurcation. The second solution is low-modal and it is excellently described by the four-mode approximation. Although more modes are necessary to provide the correct description of the first solution, this solution is also rather smooth and does not contain large gradients. It is interesting to note that the energy

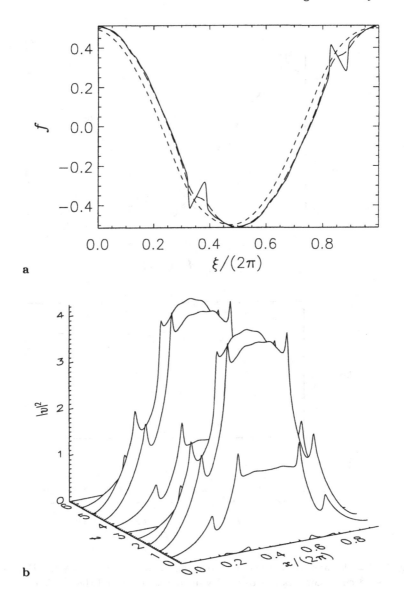

Fig. 6. The solutions with shocks for β = 0.2 and Δ = 0.2. **a.** f vs ξ numerically calculated with N = 512. Note the progressive development of shocks as R is increased from 10 (short-dashed line) to 100 (long-dashed line) and 1000 (solid line). **b.** The time-evolution of the velocity field $v(x, t)$ for R = 1000.

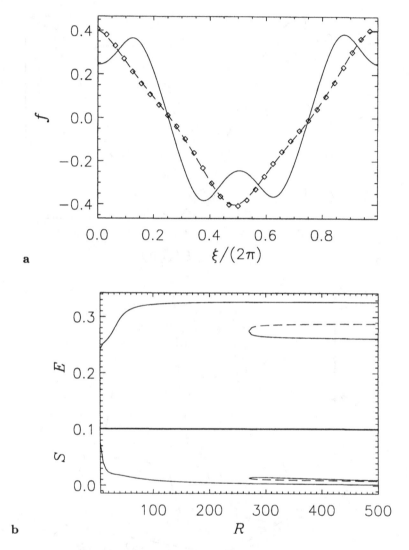

Fig. 7. $\beta = 0.2$ and $\Delta = 0.4$. obtained using $N = 256$. **a.** f vs ξ for $R = 500$. The dashed curve corresponds to the low-modal solution and the ◇'s denote its 4-mode approximation. **b.** upper panel. The dimensionless energy, E, vs normalized Reynolds number R. **b.** lower panel. The Poynting flux, S. The two-branched curves in both panels of Fig. **b** (one branch solid and one branch dashed) correspond to the low-modal solution. Note the saturation of E for both solutions. Although it is not clearly seen in the figure, we checked that $S \propto 1/R$ for both solutions.

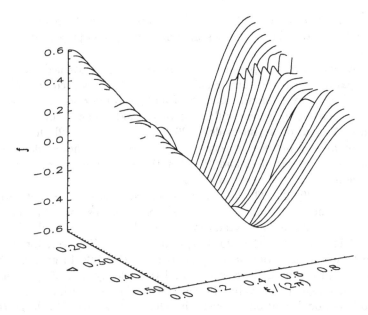

Fig. 8. The dependence of the shock solution on Δ for $\beta = 0.2$ and $R = 1000$.

E saturates when $R \to \infty$ for all three solutions, and the energy flux S is inversely proportional to R for $R \gg 1$.

To better understand how the existence of the shock solution depends on Δ we studied the behaviour of the shock solution when Δ is varied. The results of this study are given in Fig. 8. We see that this solution is rather smooth and can be only conventionally called "shock solution" when Δ is far away from the interval $[0.20, 0.23]$. When Δ approaches one of the boundaries of this interval large gradients start to be built-up. And these large gradients become well-developed when Δ is in the interval. We also can see the bifurcation at $\Delta \approx 0.4$.

7   Conclusions

In the present paper we addressed the problem of nonlinear MHD oscillations in a one-dimensional homogeneous magnetic cavity. We studied the stationary response of the cavity to the excitation of Alfvén waves in its interior by the oscillatory motion of one of its boundaries.

The main results of our work are:

- Even in the strongly nonlinear regime the steady state of the driven oscillations can be represented as a sum of two waves with the same permanent shape propagating in the opposite directions.

- The behaviour of the oscillations is determined by the three dimensionless parameters: the plasma β, the scaled mistuning Δ and the scaled Reynold number R.
- All numerical results presented in our paper were obtained for $\beta = 0.2$. Comparison with the same results for $\beta = 0.1$ presented in [20] shows that, for relatively small values of β and the wide range of variation of Δ, the behaviour of the oscillations only weakly depends on β.
- When Δ is out of the intervals, $[0.20, 0.23]$ and $[0.40, 0.42]$, the one-mode approximation provides a very good description of the oscillations. Depending on values of Δ and R either there is a unique solution to the problem, or there are tree solutions.
- When $\Delta \in [0.20, 0.23]$ the one-mode approximation does not properly describe the oscillations. For large values of R this solution (the shock solution) contains large gradients. Note, that the same behaviour was found in [20] for $\Delta \in [0.20, 0.23]$ and $\beta = 0.1$. However, for $\beta = 0.1$ shocks are much more pronounces than for $\beta = 0.2$. In general, shocks are well pronounced for $\beta < 0.15$ and they slowly fade away when β is increased beyond this value. In contrast to the case with $\beta = 0.1$, where two low-modal solution co-exist with the shock solution for large R, the shock solution is unique for $\beta = 0.2$.
- When $\Delta \in [0.40, 0.42]$, a saddle-node bifurcation occurs for large R. Two low-modal solutions appear as a result of this bifurcation. This result is in contrast to the corresponding result for $\beta = 0.1$, where this bifurcation occurs for $\Delta \in [0.20, 0.23]$, i.e. in the same interval where the shock solution exsts.
- For the low-modal solutions the energy flux into the cavity averaged over the oscillation period tends to zero when R is increased. In contrast, it saturates for the shock solution.

References

1. Tikhonchuk, V. T., Rankin, R. et al. (1995) Nonlinear dynamics of standing Alfvén waves. Phys. Plasmas **2**, 501–515.
2. Ionson, J. A. (1978) Resonant absorption of Alfvénic surface waves and the heating of solar coronal loops. Astrophys. J. **226**, 650–673.
3. Kuperus, M., Ionson, J. A., Spicer, D. (1981) On the theory of coronal heating mechanisms. Ann. Rev. Astron. Astrophys. **19**, 7–40.
4. Davila, J. M. (1987) Heating of the solar corona by the resonant absorption of Alfvén waves. Astrophys. J. **317**, 514–521.
5. Goossens, M. (1991) MHD waves and wave heating in non-uniform plasmas. in: Advances in Solar System Magnetohydrodynamics (ed. E. R. Priest, A. W. Hood), p. 135. Cambridge Univ. Press, Cambridge.
6. Hollweg, J. V. (1991) Alfvén waves. In: Mechanisms of Chromospheric and Coronal Heating (ed. P. Ulmschneider, E. R. Priest, R. Rosner), p. 423. Springer-Verlag, Berlin.

On the Reflection of Alfvén Waves in the Inhomogeneous Solar Wind

L.M.B.C. Campos, N.L. Isaeva, and P.J.S. Gil

Secção de Mecânica Aeroespacial, ISR, Instituto Superior Técnico,
Av. Rovisco Pais, 1049-001 Lisboa, Portugal

Abstract. The linear propagation of Alfvén waves in the solar wind is considered for arbitrary wavelengths, which may exceed the lengthscales of variation of (i) the mass density of the medium, (ii) strength and direction of the external magnetic field, (iii) of the mean flow velocity and (iv) of multiple ion species in the background. The effects (i) to (iv) are studied initially isolated and then in combination, causing the reflection of Alfvén waves, which is shown to lead to a steepening of the wave energy spectrum with distance, which is consistent asymptotically with Kraichnan or Kolmogorov exponents. Particular attention is given to singularities of the wave equation in the presence of mean flow e.g. critical layers where the mean flow velocity equals the Alfvén speed, and localized wave reflection may occur, as well as other changes in wave properties.

1 Introduction

The solar wind is an inhomogeneous medium, with non-uniform mass density and mean flow velocity, under an external magnetic field of varying strength and direction [119,121,27,12,96,68,103]. The inhomogeneity of the solar wind significantly affects Alfvénic perturbations in the range of periods observed at the earth, say from 1 hour to 1 day. Taking $a = 5 \times 10^7 \, \mathrm{cm\,s^{-1}}$ for the average Alfvén speed in the solar wind, the periods $\tau = 3.6 \times 10^3 - 8.6 \times 10^5 \, \mathrm{s}$ would correspond to a wavelength $\lambda = a\tau = 1.8 \times 10^{11} - 4.3 \times 10^{12} \, \mathrm{cm} = 2.6 - 62 \, r_0$ which exceeds the solar radius $r_0 = 7 \times 10^{10} \, \mathrm{cm}$; it is clear that under such conditions the reflection of Alfvén waves by gradients of the background medium is significant, with two consequences. First, for linear Alfvén waves closer to the sun, the process of reflection is frequency dependent, and can lead to Kolmogorov or Kraichnan-type spectra, as will be shown in §2.4. Secondly, further from the sun, as Alfvén waves become non-linear, the waves propagating in opposite directions interact, leading to the energy cascade of hydromagnetic turbulence [80,81,56,58,138,139,142,145,140,141].

Although the process of wave reflection is certainly present in the solar wind, much of the literature on the subject incorporates explicit or implicit assumptions which preclude the occurence of theis process. Three examples are given immediately below. First, much of the literature on Alfvén waves and Alfvénic turbulence in the solar wind uses Elsasser [61] variables:

$$Z^\pm = V \pm B\sqrt{\mu/4\pi\rho}, \tag{1a}$$

7. Ofman, L., Davila, J. M. (1995) Nonlinear resonant absorption of Alfvén waves in 3 dimensions, scaling laws, and coronal heating. J. Geophys. Res. **100**, 23,427–23,441.

8. Ruderman, M. S., Berghmans, D. et al. (1997) Direct excitation of resonant torsional Alfvén waves by footpoint motions. Astron. Astrophys. **320**, 305–318.

9. Tirry, W. J., Poedts, S. (1998) Wave heating of coronal arcades driven by toroidally polarised footpoint motions - Stationary behaviour in dissipative MHD. Astron. Astrophys. **329**, 754–764.

10. Steffens, S., Deubner, F.-L. et al. (1995) Is there a chromospheric mode at 6 μhz? Astron. Astrophys. **302**, 277–284.

11. Deubner, F.-L., Waldschik, Th., Steffen, S. (1996) Dynamics of the solar atmosphere. 6. Resonant oscillations of an atmospheric cavity: Observations. Astron. Astrophys. **307**, 936–946.

12. Steffens, S., Schmitz, F., Deubner, F.-L. (1997) The influence of the solar atmospheric stratification on the structure of the acoustic wave field. Solar Phys. **172**, 85–92.

13. Nocera, L., Priest, E. R. (1991) Bistability of a forced hydromagnetic cavity. J. Plasma Phys. **46**, 153–177.

14. Nocera, L. (1994) Subharmonic oscillations of a forced hydromagnetic cavity. Geophys. Astrophys. Fluid Dynamics **76**, 239–252.

15. Rankin, R., Frycz, P. et al. (1994) Nonlinear standing Alfvén waves in the Earth's magnetosphere. J. Geophys. Res. **99**, 21,291–21,301.

16. Nocera, L., Bologna, M., Califano, F. (1997) Homoclinic chaos in a forced hydromagnetic cavity. Geophys. Astrophys. Fluid Dynamics **86**, 131–148.

17. Bologna, M., Nocera, L. (1998) Self oscillations of a forced inhomogeneous hydromagnetic cavity. Astron. Astrophys. **336**, 735–742.

18. Braginskii, S. I. (1965) Transport Processes in Plasma. In: Review of Plasma Physics (ed. M. A. Leontovitch), Consultants Boureau, New York, vol. 1, 205.

19. Guckenheimer, J., Holmes, P. (1983) Nonlinear Oscillations, Dynamical Systems, and Bifurcations of Vector Fields. Springer-Verlag, Berlin

20. Nocera, L., Ruderman, M. S. (1998) On the steady state of nonlinear quasiresonant Alfvén oscillations in one-dimensional magnetic cavity. Astron. Astrophys. **340**, 287–299.

21. Press, W. H., Teukolsky, S. A. et al. (1992) Numerical Recipes in Fortran. The Art of Scientific Computing. Cambridge Univ. Press, Cambridge.

where V is the velocity, B the magnetic field, μ the magnetic permeability and ρ the mass density; the Elsasser variables are a natural representation of Alfvén waves propagating in opposite directions. The vanishing of one of them, e.g. $Z^+ \neq 0 = Z^-$, implies no reflection; thus for Alfvén waves in the inhomogeneous solar wind, $Z^+ \neq 0 \neq Z^-$ because gradients of background properties imply that an Alfvén waves propagating in one direction give rise to reflected waves propagating in the opposite direction. Second, the assumption of sinusoidal perturbations in space:

$$Z^{\pm}(x,t) = Z_0^{\pm} \exp[i(k \cdot x - \omega t)], \tag{1b}$$

implies that the medium is homogeneous, and there is no reflection; for an inhomogeneous medium the wavevector k does not exist in the direction of stratification, because the MHD equations do not have sinusoidal solutions. Third, the JWKB-approximation assumes there is no wave reflection [79,146,82,23].

There is a fourth way in which wave reflection is excluded, often implicitly, by the use of Elsasser equations in their original form:

$$\partial Z^{\pm}/\partial t + (Z^{\pm} \cdot \nabla) Z^{\pm} + \nabla P = 0, \tag{2a}$$

or linearized versions, where P denotes the total pressure, i.e. gas p plus magnetic pressure, per unit mass:

$$P \equiv (p + \mu B^2/8\pi)/\rho. \tag{2b}$$

The Elsasser equations have been extended to compressible fluids [98] by coupling to an additional equation. In their original form (2a) they assume constant mass density, which excludes (i) compressive MHD modes (slow and fast waves) and (ii) non-uniform mass density of the background medium. If the latter (ii) but not the former (i) is considered, the original Elsasser equations (2a) are replaced [52] by:

$$\partial Z^{\pm}/\partial t + (Z^{\pm} \cdot \nabla) Z^{\pm} + \nabla P$$
$$= (Z^{\pm} - Z^{\mp}) [Z^{\mp} \cdot \nabla (\log \rho)] - P \nabla (\log \rho). \tag{3}$$

The ratio of the terms on the r.h.s of (3) to the terms of the l.h.s. is λ/L where λ in the wavelength of Alfvénic perturbations and $L \equiv |\nabla (\log \rho)|^{-1} = \rho/|\nabla \rho|$ is the lengthscale of change of background mass density. It follows that the original Elsasser equations (2a) apply to Alfvénic perturbations only if $\lambda \ll L$, i.e. the background is homogeneous, in which case there is no reflection. Note that the restriction $\lambda \ll L$ is more severe than the JWKB-approximation $\lambda^2 \ll L^2$ [18,147,104,74,75,102].

In the present paper the MHD equations are taken as the starting point, and none of the preceding assumptions excluding wave reflection is made; the theory to be developed applies to all wavelengths, including the cases of wavelength comparable to $\lambda \sim L$ or longer than $\lambda > L$ the lengthscales of variation of background properties. If the background medium is steady, a sinusoidal dependence on time $e^{i\omega t}$ can be assumed, because the frequency ω is conserved; but no sinusoidal dependence on the coordinate z or r in the direction of stratification can be assumed, because the 'wavenumber' $k = \omega/a(z)$ is not conserved. Wave theories are simplest for linear waves in homogeneous media, but become more complex for (i) non-linear waves or (ii) strongly inhomogeneous media, and almost intractable for (i) and (ii) together. In the present paper the choice is to consider linear waves, and to account fully for inhomogeneity of the background medium. This is justified for Alfvén waves in the solar wind near the sun, since the velocity perturbation (a few tens of km / s) is much smaller than the Alfvén speed (a few hundreds of km / s). The evolution of the wave fields with distance will then determine if, when or where non-linear effects become important.

In order to model Alfvénic perturbations in the solar wind, at least four effects need to be considered [19,21,26,59,60,14,15,97,16,24,122]: (i) an external magnetic field of varying strength and direction; (ii) non-uniform mass density; (iii) a mean flow velocity, which is small near the sun, and increases with distance to become superalfvénic before the earth; (iv) the presence of multiple ion species [65,99,77]. To consider all these effects in combination would be a tall order indeed. Thus for a start these effects are considered (§2) separately, before being combined (§3) into more 'realistic' solar wind models. The main physical ingredients are: (§2.1) the effects of density stratification and Hall currents, which are demonstrated for a plane parallel atmosphere: (§2.2) the effects of an external magnetic field of varying strength and direction, which are demonstrated by considering Alfvén waves in spiral coordinates; (§2.3) the effects of non-uniform mean flow, which are considered first for a plane parallel atmosphere; (§2.4) the effects of spherical divergence, which are considered for Alfvén waves in a radial density stratification and monopole magnetic field, leading to Kolmogorov and Kraichnan type spectra.

The simplest combination of these effects which could claim to include the main physical features of the solar wind is Alfvén waves in a radial mean flow and density stratification, under a monopole external magnetic field (§3). The Alfvén waves can be of two types: (a) one-dimensional, with velocity and magnetic field perturbations along parallels, depending only on radial distance and time; (b) three-dimensional, with velocity and magnetic field perturbations also along meridians, depending on the three spherical coordinates and time. Both for the (a) one- and (b) three-dimensional cases, the radial dependence is specified by the same Alfvén wave equation, for which exact analytical solutions are given, valid for all wave frequencies and radial distances, for four distinct backgrounds. Each background is consistent with momentum balance and mass conservation, and thus is specified by the profile of mean flow velocity: (§3.1) the case of uniform mean flow is the only one to which the cut-off frequency applies, discussed in the literature

[69,17,91,93,117]; (§3.2) in the case of mean flow velocity proportional to the radius, the wave fields are specified by Gaussian hypergeometric functions, with different parameters from case §3.1; (§3.3) the only case without a critical layer is that of a homogeneous medium, or mean flow velocity varying like the inverse square of the radius, which has a transition level; (§3.4) the closest approximation to the solar wind may be a mean flow velocity proportional to the square root of the radius, in which case there is a critical layer, where the mean flow velocity equals the Alfvén speed.

2   Effects of stratification, Hall currents, non-uniform magnetic fields and mean flow

Alfvén waves in the solar wind are affected by (i) density stratification, (ii) mean background flow, (iii) non-uniform external magnetic field and (iv) multiple ion species. Before attempting to study several of these effects in combination (§3), it is desirable to understand the physics of each of them in isolation (§2). This suggests the following sequence, considering: (§2.1) Alfvén waves, with Hall effect, in a plane parallel atmosphere, under an uniform external magnetic field, i.e. effects of density stratification and a two-fluid model with ions and electrons; (§2.2) Alfvén waves propagating in a spiral external magnetic field, i.e. effects of density stratification and external magnetic field of varying strength and direction; (§2.3) Alfvén waves in a plane parallel atmosphere, under an uniform magnetic field, in the presence of a mean flow, which causes the appearance of a critical layer, where the mean flow velocity equals the Alfvén speed, i.e. effects (i) and (iii); (§2.4) spherical Alfvén waves in a radially inhomogeneous medium under a monopole magnetic field, leading to the steepening of the energy spectrum, with Kolmogorov- or Kraichnan- type exponents, as observed in the solar wind.

2.1   Effects of density stratification and Hall currents

For linear magneto-acoustic waves in an inhomogeneous medium under an uniform external magnetic field [76,13,85,57,28,107], the incompressible, shear Alfvén mode [5,7,8] decouples from the compressible slow and fast modes, even in the presence of dissipation [48] and with non-linear terms acting as sources of MHD waves [84,30,132]. This decoupling extends to a plane parallel atmosphere only if gravity, the external magnetic field and the horizontal wavevector lie on the same plane [22,37,55]. In what follows this condition will be assumed to be met, so that it is necessary to consider neither slow and fast magneto-acoustic-gravity waves [149,105,100,135,115,3,4,128,126,136, 151-153,31,32,36-38,41,42,54], nor acoustic-gravity waves [108,90,33]. The present paper will thus concentrate on Alfvén waves, excluding dissipative effects either on incompressible [57,107,71,114,133,113,32,33,39,41,44,45] or compressible modes [148,95,34,29,2], which are relevant to the heating and other phenomena in the solar atmosphere [6,20,129,118,89,74,137,154,35,40,53].

The simplest case of linear decoupling of Alfvén waves from slow and fast modes concerns propagation in a plane parallel isothermal atmosphere under a uniform vertical magnetic field, with zero horizontal wavenumber [63,64,86,72,73,31,9,10,125,130,109,83,110,130]. The extension to oblique magnetic field and non-zero horizontal wavenumber [127,39] and the inclusion of displacement currents [87], lead to solutions which are still specified by Bessel functions; consideration has been given to Alfvén waves in non-isothermal atmospheres [150,120] and to Hall effects [88,43]. Although in the solar atmosphere the density is sufficiently high to justify the use of MHD equations [11,25,155,134,67], Hall currents could become relevant in the distant solar wind, and are a simple example of two-fluid or ion-electron plasma effects. If the ion gyro-frequency varies with altitude it can lead to the appearance of a critical layer [101,51]; since critical layers will be considered subsequently in other contexts (§2.3, §3.1, §3.2, §3.4), only the case of constant gyro frequency will be considered in this section (§2.1).

In order to describe linear Alfvén waves, the equations of continuity, state and energy are not needed, and the only relevant MHD equations are those of momentum:

$$\partial \boldsymbol{V}/\partial t + (\boldsymbol{V} \cdot \nabla) \boldsymbol{V} + \rho^{-1}\nabla p = \boldsymbol{g} - (\mu/4\pi\rho) \, \boldsymbol{H} \wedge (\nabla \wedge \boldsymbol{H}) \,, \qquad (4a)$$

and induction:

$$\partial \boldsymbol{H}/\partial t + \nabla \wedge (\boldsymbol{H} \wedge \boldsymbol{V}) = \nabla \wedge [\zeta \boldsymbol{H} \wedge (\nabla \wedge \boldsymbol{H})] \qquad (4b)$$

where dissipative terms were omitted, but the Hall current was included, and $\boldsymbol{V}$ is the velocity, $\boldsymbol{H}$ the magnetic field, ρ the mass density, p the gas pressure, μ the magnetic permeability; the Hall diffusivity is $\zeta \equiv c/4\pi Ne$ where c is the speed of light in vacuo, e the electron charge and N the electron density. The total state of the fluid is assumed to consist of a steady, non-uniform mean state at rest, with density ρ_0, pressure p_0, and external magnetic field $\boldsymbol{B}$, which may depend on the position vector $\boldsymbol{x}$, upon which unsteady non-uniform velocity $\boldsymbol{v}$, magnetic field $\boldsymbol{h}$, pressure p' and density ρ' perturbations are superimposed, which may depend on the position vector $\boldsymbol{x}$ and time t:

$$\{\boldsymbol{V}, \boldsymbol{H}, p, \rho\}(\boldsymbol{x}, t) = \{\boldsymbol{O}, \boldsymbol{B}, p_0, \rho_0\}(\boldsymbol{x}) + \{\boldsymbol{v}, \boldsymbol{h}, p', \rho'\}(\boldsymbol{x}, t); \qquad (5a\text{-}d)$$

Substitution of (5a-d) in (4a,b) yields the linearized equations:

$$\partial \boldsymbol{v}/\partial t + \rho_0^{-1}\nabla p' - (\rho'/\rho_0) \, \boldsymbol{g} + (\mu/4\pi\rho_0) \, \boldsymbol{B} \wedge (\nabla \wedge \boldsymbol{h}) = 0, \qquad (6a)$$
$$\partial \boldsymbol{h}/\partial t + \nabla \wedge (\boldsymbol{B} \wedge \boldsymbol{v}) = \nabla \wedge [\zeta \boldsymbol{B} \wedge (\nabla \wedge \boldsymbol{h})], \qquad (6b)$$

where the external magnetic field was assumed to be current free.

Elimination of the velocity perturbation between (6a,b), leads to the Alfvén-Hall wave equation for the magnetic field perturbation:

$$\partial^2 h/\partial t^2 - \nabla \wedge \left\{ (A/B^2)\, B \wedge [B \wedge (\nabla \wedge h)] \right\} =$$
$$\nabla \wedge \left\{ B \wedge [(\nabla p' - \rho' g)/\rho_0] \right\} + \nabla \wedge \left\{ (A^2/\omega_g B)\, B \wedge [\nabla \wedge (\partial h/\partial t)] \right\}, \quad (7)$$

where A denotes the Alfvén speed (8a) and ω_g the gyro-frequency (8b):

$$A^2 \equiv \mu B^2/4\pi\rho_0, \quad \omega_g \equiv \mu B M e/m_i c; \qquad (8a,b)$$

in (8b) the mass density $\rho_0 = m_i N_i = m_i N/M$, was calculated for the ions, whose mass is much larger than that of the electrons $m_i \gg m_e$, and the ion density $N_i = N/M$ is related to the electron density $N \gg N_i$ by the charge number M, viz. $\omega_g = A^2/B\zeta = \mu B N_e/\rho_0 c$ leading to (8b). In the case of a vertical uniform external magnetic field, the Alfvén-Hall wave equation (7) shows that only the horizontal components of the external magnetic field propagate:

$$B = B e_z : \quad \ddot{h}_x - \left[A^2 \left(h'_x + h'_y/\omega_g \right) \right]' = 0 = \ddot{h}_y - \left[A^2 \left(h'_y - h'_x/\omega_g \right) \right]', \qquad (9a,b)$$

where dot denotes derivative with regard to time and prime derivative with regard to spatial coordinate (altitude in this case):

$$\left\{ \dot{f}, f' \right\} \equiv \{ \partial f/\partial t, \partial f/\partial z \} : \quad \dot{v}_x - (A^2/B)\, h'_x = 0 = \dot{v}_y - (A^2/B)\, h'_y, \qquad (10a,b)$$

and the velocity perturbation are related to the magnetic field perturbations by (10a,b), which follow from (6a). It follows that the perturbations are incompressible $\nabla \cdot v = 0$, and hence $\rho' = 0 = p'$ the mean state mass density $\rho = \rho_0$ and gas pressure $p = p_0$ can be denoted without the subscript "0".

The cartesian components of the horizontal magnetic field perturbations satisfy two coupled second-order differential equations (9a,b) which, in the case of an homogeneous medium, can be eliminated leading to a fourth-order wave equation [88]. An alternative, which applies as well to an atmosphere, is to consider right- and left- hand polarized waves:

$$h_\pm \equiv h_x \pm i\, h_y, \quad v_\pm \equiv v_x \pm i\, v_y, \qquad (11a,b)$$

which are decoupled:

$$\ddot{h}_\pm - \left[A^2 \left(h'_\pm \pm h'_\pm/\omega_g \right) \right]' = 0 = \dot{v}_\pm - (A^2/B)\, h'_\pm. \qquad (12a,b)$$

Since the background atmosphere does not depend on time, the velocity v and magnetic field h perturbations are represented by Fourier decompositions:

$$h_\pm, v_\pm(z,t) = \int_{-\infty}^{+\infty} H, V(z;\omega) e^{-i\omega t}\, d\omega, \qquad (13a,b)$$

where H is the magnetic field and V the velocity perturbation spectrum of a wave of frequency ω at altitude z, which satisfy (12a,b), viz.:

$$(1 \pm \omega/\omega_g)\left(A^2 H'\right)' + \omega^2 H = 0, \quad V = i\left(A^2/\omega_g B\right) H'; \qquad (14a,b)$$

substitution of (14b) into (14a) yields the Alfvén-Hall wave equation for the velocity perturbation spectrum:

$$(1 \pm \omega/\omega_g)A^2 V'' + \omega^2 V = 0, \qquad (15a)$$

which shows that the Hall effect corresponds to a change of Alfvén speed:

$$A \to A\sqrt{1 \pm \omega/\omega_g} \equiv \bar{A}. \qquad (15b)$$

The implication is that right-polarized waves have a phase speed larger than the Alfvén speed, and left-polarized waves have a smaller propagation speed if $\omega < \omega_g$, and become evanescent above the gyro-frequency $\omega > \omega_g$, because then the phase speed is imaginary. At the gyro-frequency $\omega = \omega_g$ the phase speed is zero $\bar{A} = 0$, corresponding to a critical layer [101,51].

Concerning the effects of density stratification, they are different for inhomogeneities transverse to the direction of propagation [111,112], which cause phase mixing, and for inhomogeneity in the direction of propagation, which is considered next; the velocity (15a) and magnetic field (14a) perturbations satisfy different wave equations:

$$A^2 V'' + \omega^2 V = 0, \qquad \left(A^2 H'\right)' + \omega^2 H = 0, \qquad (16a,b)$$

and thus equipartition of kinetic and magnetic energies $V/A = H/B$ is not possible. In the general case of a non-isothermal atmosphere, as the density tends to zero $\rho \to 0$ at high-altitude $z \to \infty$, the Alfvén speed diverges $A \to \infty$ in (8a), and the magnetic field perturbation $H' \to 0$ in (16b) tends to a constant $H' \to const$, and the velocity perturbation $V'' \to 0$ in (16a) diverges linearly $V \sim z$; this corresponds to propagating waves. For standing modes the velocity perturbation must be bounded $V \to const$ at high-altitude $z \to \infty$, and thus the magnetic field perturbation $H \sim V'$ tends to zero $H \to 0$, like the mass density $\rho \sim e^{-z/L}$, where L is the density scale height $L \equiv -1/(\log \rho)'$. The Table 1 shows that: (i) for standing waves the total energy density $E \equiv E_v + E_h$ decays $E \sim \rho \sim E_v$ and is dominated by the kinetic energy; (ii) for propagating waves the total energy density tends to a constant $E \sim const \sim E_h$ and is dominated by the magnetic energy. Note the property (ii) relates to Alfvén waves in the solar wind, for which the magnetic field perturbation tends to a constant fraction of the background magnetic field $h/B \sim const$.

These results can be confirmed in the case of an isothermal atmosphere, for which the mass density (17a) decays exponentially on altitude on the scale height (17b):

$$\rho(z) = \rho_0 e^{-z/L}, \quad L \equiv RT/g, \qquad (17a,b)$$

Table 1. Alfvén waves in an atmosphere

Type of wave	standing	propagating
Velocity perturbation	$v \sim const$	$v \sim z$
Magnetic field perturbation	$h \sim \rho \to 0$	$h \sim const$
Kinetic energy $E_v \equiv 1/2\rho v^2$	$E_v \sim \rho \to 0$	$E_v \sim z^2 \rho \to 0$
Magnetic energy $E_h \equiv \mu h^2/4\pi$	$E_h \sim \rho^2 \to 0$	$E_h \sim const$

where R is the gas constant, and T the temperature. For a uniform external magnetic field it follows that the Alfvén speed (8a) increases exponentially on twice the scale height (18a) from an initial value (18b) at altitude zero:

$$A(z) = ae^{z/2L}, \quad a \equiv B\sqrt{\mu/4\pi\rho_0}. \tag{18a,b}$$

Note that the profile of Alfvén speed with altitude (18a) implies that an Alfvén wave propagates to infinity in twice the time it would take to cover one scale height at the initial Alfvén speed:

$$t = \int_{-\infty}^{+\infty} \{A(z)\}^{-1} \, dz = a^{-1}\int_{-\infty}^{+\infty} e^{-z/2L} \, dz = 2L/a. \tag{18c}$$

The solution of the Alfvén wave equation (16a) with Alfvén speed profile (18a) is [63,31]:

$$V(z;\omega) = Z_0(\zeta), \quad \zeta \equiv 2\omega L/A(z) = (2\omega L/a)\, e^{-z/2L}, \tag{19a,b}$$

where Z_0 is a linear combination of Bessel J_0, Neumann Y_0 or Hankel H_0 functions [1] of order zero:

$$Z_0(\zeta) \equiv C_1 J_0(\zeta) + C_2 Y_0(\zeta) = C_+ H_0^{(1)}(\zeta) + C_- H_0^{(2)}(\zeta), \tag{20a,b}$$

where C_1, C_2 or $C_\pm$ are arbitrary constants, determined from boundary conditions.

Noting that increasing altitude z corresponds to decreasing ζ in (19b), it may be expected that $H_0^{(1)}(\zeta)$ represents an upward and $H_0^{(2)}(\zeta)$ a downward propagating wave at infinity:

$$z \to \infty, \zeta \to 0: \quad H_0^{(1,2)}(\zeta) \sim \pm(2i\pi)\log\zeta = \pm(2i\pi)\left[-z/2L + \log(2\omega L/a)\right]; \tag{21a,b}$$

it is clear from (21a) that $H^{(1)}$ has a positive spatial phase (upward propagation) and $H^{(2)}$ a negative spatial phase (downward propagation). The phase is asymptotically bounded at high altitude, as should be expected from the finite propagation time (18c); the amplitude grows asymptotically $z \to \infty$ as a linear function of altitude $\sim \pi z/L$ for an isothermal atmosphere, confirming the prediction $V \sim z$ in (Table 1) for propagating waves, without restriction to isothermal atmospheres. The radiation condition, excluding waves coming from infinity, implies $C_- = 0$, and the remaining constant of integration C_+ is determined from the initial velocity perturbation spectrum at amplitude $z = 0$:

$$V(z;\omega) = V(0;\omega) H_0^{(1)} \left((2\omega L/a) \, e^{-z/2L} \right) / H_0^{(1)} (2\omega L/a) . \qquad (22a)$$

Note that $H_0^{(1)}(\zeta) \neq 0$ so that there are no resonances for propagating waves.

For standing waves, perfectly reflected from infinity, the amplitude must be bounded $V < \infty$ asymptotically as $z \to \infty$. Since this condition is not met by $Y_0(\zeta) \sim \log \zeta \sim z$, one must set $C_2 = 0$ in (20a), and the remaining constant of integration C_1 is determined from the initial velocity spectrum:

$$V(z;\omega) = V(0;\omega) J_0 \left((2\omega L/a) \, e^{-z/2L} \right) / J_0 (2\omega L/a) . \qquad (22b)$$

It follows that the standing modes have resonances at the frequencies ω_n corresponding to the roots j_n of the Bessel function J_0:

$$J_0(j_n) = 0 : \omega_n = j_n a/2L . \qquad (23a)$$

Substituting (22b) in the Fourier integral (13b), and using the residues the integrand at the simple poles corresponding to the frequencies (23a), it follows that the wave field:

$$V(z;\omega) = -i\pi (a/L) \sum_n \{ V(0;\omega_n) / [\omega_n J_1(j_n)] \} J_1 \left(j_n e^{-z/2l} \right) , \qquad (23b)$$

consists of a superposition of normal modes; the first three normal modes of standing Alfvén waves in an isothermal atmosphere (Figure 1), show that the increase of Alfvén speed with altitude increases the spacing between nodes of the waveform, and oscillations cease when $\omega L/A(z) < j_1 = 2.408$. This has been interpreted in the literature as a 'cut-off frequency' $\omega_* = 2.4a/L$ for Alfvén waves. In fact, Alfvén waves are not filtered in an atmosphere, and a more accurate statement is that they cease to oscillate beyond an altitude $\zeta < j_1$ in (19b), viz.:

$$z > 2L \log (2\omega L/j_1 a) = 2L [-0.18 + \log (\omega L/a)] \equiv z_* \qquad (24)$$

and thus appear to be 'evanescent' above. This demonstrates that the density gradients in an atmosphere can cause strong reflection of Alfvén waves, as might be expected from the rapid increase of the Alfvén speed with altitude.

In Figure 1 the first four modes (23a,b) of Alfvén waves are shown, standing vertically in an isothermal atmosphere under a uniform external magnetic

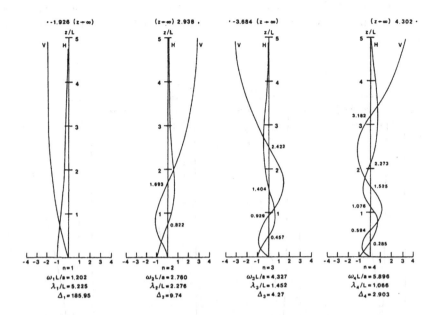

Fig. 1. Velocity $\mathcal{V}$ and magnetic field $\mathcal{H}$ perturbations of first four modes $n = 1, 2, 3, 4$ of Alfvén waves standing vertically in an isothermal atmosphere, perfectly reflected from infinity, versus altitude made dimensionless by dividing by the scale height. For each natural frequency $\Omega_n = \omega_n L/a$, is indicated the corresponding local wavelength $\lambda_n/L = 2\pi/\Omega_n$ and change in mass density over a wavelength $\Delta_n = \exp(\lambda_n/L)$.

field, with perfect reflection from infinity. The density scale height (17b) and initial Alfvén speed (18b) specify through (23a) the eigenvalue or natural frequency ω_n, where j_n is a root of the Bessel function J_0. Clearly the waves are not sinusoidal in space, as one might expect from a "local wavenumber" $k(z) = \omega/A(z) = (\omega/a)e^{-z/2L}$ which depends on altitude. A constant "reference number" $k_n = \omega_n/a$ or wavelength $\lambda_n = 2\pi/k_n = 2\pi a/\omega_n = 4\pi L/j_n$ may be defined from the initial Alfvén speed (18b); the ratio of the "reference wavelength" to the scale height $\lambda_n/L = 4\pi/j_n$ does not satisfy the JWKB-approximation $(\lambda_n/L)^2 \ll 1$ at least for the first four standing modes. The implication is that the atmosphere mass density (17a) changes significantly over a wavelength $\Delta_n = \exp(\lambda_n/L) = \exp(4\pi/j_n)$, so the medium is highly inhomogeneous on the scale of a reference wavelength. Note that the n-th mode has (n-1) nodes, since the node at infinity corresponds to finite, non-zero amplitude. Also the nodes are not equally spaced, due to increase in Alfvén speed with altitude.

In the case of propagating waves the spectrum is continuous, and the frequency is given four values $\omega L/a = 0.5, 1, 2, 5$ of which only the first corresponds to the JWKB approximation. For propagating waves the wave fields (22a) are complex, so the modulus or amplitude are plotted in Figure 2 and

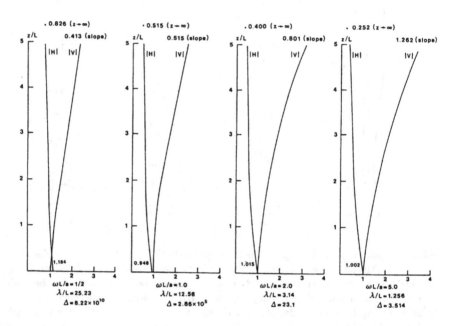

Fig. 2. Modulus of velocity $|\mathcal{V}|$ and magnetic field $|\mathcal{H}|$ perturbation spectra of Alfvén wave propagating vertically in an isothermal atmosphere; the wave field at altitude z is normalized to the value at altitude zero, and plotted versus altitude made dimensionless by dividing by the scale height z/L, for four values of dimensionless frequency $\Omega_0 \equiv \omega L / a = 0.5, 1.0, 2.0, 5.0$, and corresponding local wavelength $\lambda / L \equiv 2\pi / \Omega_0$ and change in mass density over a wavelength $\Delta = \exp(\lambda/L)$

the argument or phase in Figure 3. In the cases plotted, the local "reference wavelength" is larger than the scale height $\lambda/L = 2\pi/kL = 2\pi a/\omega L > 1$, and the change of atmospheric density over a reference wavelength is significant $\Delta = \exp(\lambda/L) = \exp(2\pi a/\omega L) > e$. Both for standing (Figure 1) and propagating (Figures 2 and 3) waves the velocity perturbation (22a,b) and corresponding magnetic field perturbations are plotted, which are related by (13a,b) and (6b, with $\zeta = 0$):

$$H(z;\omega) = i(B/\omega)dV(z;\omega)/dz. \tag{25}$$

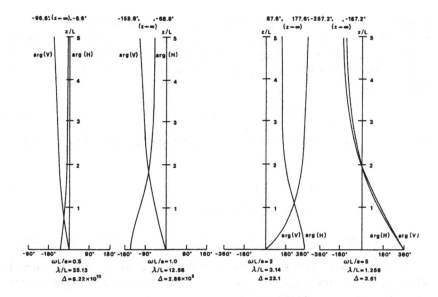

Fig. 3. As figure 2, for phase of velocity $arg(\mathcal{V})$ and magnetic field $arg(\mathcal{H})$ perturbations

Substitution of (22a,b) yields:

$$H(z;\omega) = i(B/a)V(0;\omega)e^{-z/2L}H_1^{(1)}\left((2\omega L/a)\,e^{-z/2L}\right)/H_0^{(1)}(2\omega L/a)\,,$$
(26a)

$$H_n(z;\omega) = i(B/a)V(0;\omega)e^{-z/2L}J_1\left((2\omega_n L/a)\,e^{-z/2L}\right)/J_0(2\omega_n L/a)\,,$$
(26b)

respectively for propagating (26a) and standing (26b) waves. The plots of dimensionless velocity $\mathcal{V} \equiv |\,V(z;\omega)/V(0;\omega)\,|$ and magnetic filed $\mathcal{H} \equiv |\,H(z;\omega)a/V(0;\omega)B\,|$ perturbations in Figures 1 and 2 confirm the results in Table 1 concerning the asymptotic wave fields at high-altitude, and Figure 3 shows the finite asymptotic phase at $z \to \infty$, consistent with the limit (18c) on propagation time.

2.2 Alfvén waves in a spiral external magnetic field

Besides density gradients, another important feature of Alfvén waves in the solar wind is propagation along an external magnetic field of varying strength and direction, along Parker's spiral [119,121]. The Parker's spiral emerges radially from the sun $\psi_0 = 0°$, and its angle with the radial direction increases with distance up to $\psi_1 = 56°$ at the earth. An approximation is to consider Alfvén wave propagation in spiral coordinates [49], which is an example of Alfvén wave propagation in non-uniform magnetic fields [116,46], other examples of such propagation include consideration of flux tubes or magnetic slabs [124,123,131]. Spiral coordinates (α, β) are defined by the complex function

$$\beta + i\alpha = (1 - i\chi) \log \left(re^{i\theta}\right) \tag{27}$$

where χ is constant. Separating the real and imaginary parts of (27) it follows that

$$\alpha = \theta - \chi \log r, \qquad \beta = \log r + \chi\theta, \tag{28a,b}$$

the curves $\alpha, \beta = const$ are logarithmic spirals

$$(rd\theta/dr)_\alpha = \chi = \tan\psi, \quad (rd\theta/dr)_\beta = -1/\chi = \tan\left(\pi/2 - \psi\right), \tag{29a,b}$$

making a constant angle $\psi = \arctan\chi$ and $\pi/2 - \psi$ with all radial lines. It follows that the spiral coordinates are orthogonal, as should be expected from the conformal transformation (27).

Since spiral coordinates are specified by a conformal transformation, the scale factor q is the same along both coordinate curves, i.e. the arc-length is given in polar (r, θ) and spiral (α, β) coordinates by

$$(d\ell)^2 = (dr)^2 + r^2 (d\theta)^2 = q^2 \left[(d\alpha)^2 + (d\beta)^2\right] \tag{30a}$$

where the scale factor is

$$q = r/\sqrt{1 + \chi^2} = \left(1 + \chi^2\right)^{1/2} \exp\left[(\beta - \chi\alpha)/\sqrt{1 + \chi^2}\right] \tag{30b}$$

and the inverse of (28a,b) was used:

$$\left(1 + \chi^2\right)\{\theta, \log r\} = \{\alpha + \chi\beta, \beta - \chi\alpha\}. \tag{31a,b}$$

Spiral coordinates are illustrated in Figure 4a for the equilateral case $\psi = \pi/4$, $\chi = 1$ when both coordinate curves cut all radial lines at the same angle $\psi = 45° = \pi/2 - \psi$ and for the case when one set of curves cuts radial lines at angle $\psi = 15^0$ and the other at $\pi/2 - \psi = 75^0$ is illustrated in Figures 4b. The total magnetic field is assumed to consist (Figure 5) of a background field B along e_β the spirals $\alpha = const$, with transverse and parallel magnetic field $\mathbf{h}$ and velocity $\mathbf{v}$ perturbations, depending on time t and spiral coordinate β, which varies along the coordinate curve $\alpha = const$, viz.:

$$\boldsymbol{H} = B\left(\beta\right) \boldsymbol{e}_\beta + h\left(\beta, t\right) \boldsymbol{e}_\alpha, \quad \boldsymbol{v} = v\left(\beta, t\right) \boldsymbol{e}_\alpha. \tag{32a,b}$$

The linearized induction and momentum equations in spiral coordinates are

$$q\dot{h} = (Bv)' \quad q\dot{v} = (\mu B/4\pi\rho)(qh)' \tag{33a,b}$$

(a)

(b)

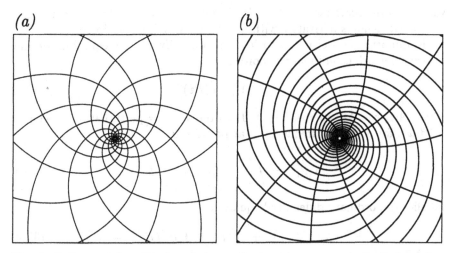

Fig. 4. Orthogonal logarithmic spiral coordinates where one set of spirals makes an angle with radial direction equal to: (a) $\phi = 45°$, (b) $\phi = 15°$.

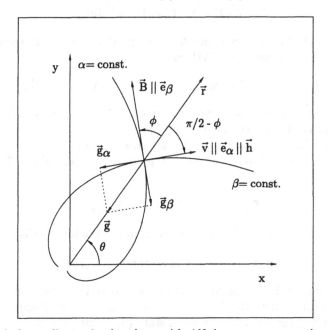

Fig. 5. Spiral coordinates in the plane with Alfvén waves propagating along the external magnetic field, aligned with one spiral, with transverse velocity and magnetic field perturbations, aligned with the orthogonal spiral, in the presence of a radial gravity field

where q is the scale factor (30b), dot denotes derivative with regard to time and prime denotes derivative with regard to the spiral coordinate β.

Elimination between (33a,b) leads to the Alfvén wave equation for the velocity perturbation in spiral coordinates

$$q\ddot{v} = \left(A^2/B\right)\left(Bv\right)'' \tag{34}$$

where A is the Alfvén speed (8a). Thus the velocity perturbation spectrum $V\left(\beta;\omega\right)$, for a wave of frequency ω at position β,

$$v\left(\beta,t\right) = \int_{-\infty}^{+\infty} V\left(\beta;\omega\right)e^{-i\omega t}d\omega, \tag{35a}$$

satisfies the ordinary differential equation

$$V'' + 2\left(B'/B\right)V' + \left(\omega^2 q^2/A^2 + B''/B\right)V = 0. \tag{35b}$$

The Maxwell equation

$$0 = \nabla \cdot \boldsymbol{B} = \nabla \cdot \left(Be_\beta\right) = q^{-2}\left(qB\right)' \tag{36a}$$

implies that the external magnetic field decays along the spiral $\alpha = const$ on the inverse of the scale factor (31), viz.:

$$B\left(\beta\right)/b = q_0/q\left(\beta\right) = r_0/r = \exp\left[-\left(\beta - \beta_0\right)/\left(1+\chi^2\right)\right], \tag{36b}$$

where b is the magnetic field strength at the reference radius, taken as the solar surface r_0. The remaining undetermined coefficient in (35b) is the Alfvén speed (8a), which depends on the density stratification.

The latter is specified by the magnetohydrostatic equilibrium of the mean state:

$$\nabla p + \left(\mu/4\pi\right)\boldsymbol{B}\wedge\left(\nabla\wedge\boldsymbol{B}\right) = \rho\boldsymbol{g}, \tag{37}$$

where the pressure is specified by the equation of state for a perfect gas (38a):

$$p = \rho RT, \quad g = g_0\left(r_0/r\right) = g_0\left(q_0/q\right) \tag{38a,b}$$

and the gravity field g is assumed to be (38b) divergence-free (36a) like the external magnetic field (36b). The latter is force-free $\nabla\wedge\left(B\left(\beta\right)e_\beta\right) = 0$, so the momentum equation for the mean state (37) simplifies to

$$p^{-1}dp = -\left(g/RT\right)dr = -\left(g_0 r_0/RT\right)r^{-1}dr, \tag{39a}$$

where (38a,b) were used. In isothermal conditions this leads to the density or pressure stratification

$$T = const: p/p_0 = \left(r/r_0\right)^{-g_0 r_0/RT} = \left(q/q_0\right)^{-g_0 r_0/RT} = \rho/\rho_0. \tag{39b}$$

Substituting (36b, 39b) in the Alfvén speed (8a) it follows that it varies like a power of radial distance (40a):

$$A = a(q/q_0)^{1-\nu_0}, \quad \nu_0 \equiv 2 - g_0 r_0 / 2RT, \qquad (40\text{a,b})$$

with exponent (40b), from an initial value (18b) at the solar surface.

Substituting the external magnetic field (36b) and Alfvén speed (40a), the differential equation (35b) becomes

$$V'' - 2(q'/q)V' + \left[(q'/q)^2 + (q_0\omega/a)^2 (q/q_0)^{2\nu_0}\right] V = 0 \qquad (41\text{a})$$

whose solution specifies the velocity perturbation spectrum:

$$V(\beta;\omega) = (q/q_0) Z_0 \left((1+\chi^2)(g_0\omega/a\nu)(q/q_0)^{\nu_0}\right), \qquad (41\text{b})$$

where Z_0 is again a linear combination of Bessel functions of order zero (20a,b). Using (30b) the scale factor q can be replaced by the radial distance in the velocity perturbation spectrum (42a):

$$V(r;\omega) = (r/r_0) Z_0 \left(\Omega_0 (r/r_0)^{\nu_0}\right), \quad \Omega_0 \equiv (r_0\omega/a\nu_0)\sec\psi, \qquad (42\text{a,b})$$

which involves one parameter, namely the dimensionless frequency (42b), formed by the Alfvén speed a at the solar radius r_0 and the wave frequency, and involving also the exponent ν_0 and the angle ψ of the logarithmic spirals with all radial lines. Once more, taking $Z_0 \equiv J_0$ to be a Bessel function specifies standing modes, whereas the Hankel functions represent outward $H^{(1)}$ and inward $H^{(2)}$ propagating waves. The outward V_+ and inward V_- propagating waves, normalized to the initial spectra

$$V_\pm(r;\omega) \equiv V(r;\omega)/V(r_0;\omega) = RH_0^{(1,2)}(\Omega_0 R^{\nu_0})/H_0^{(1,2)}(\Omega_0) \qquad (43\text{a,b})$$

are plotted versus radial distance r, made dimensionless by dividing by the solar radius r_0, over a range of 1 AU or 215 solar radii:

$$1 < R \equiv r/r_0 < 215, \ |V(r;\omega)| \equiv |V_\pm(r;\omega)|, \ \arg[V(r;\omega)] \equiv \pm\arg[V_\pm(r;\omega)],$$
$$(44\text{a,b,c})$$

and note that the inward and outward propagating waves have the same amplitude (44b) and opposite phases (44c).

Using $r_0 = 7.0 \times 10^{10}$ cm for the solar radius, $g_0 = 2.74 \times 10^4$ cm s^{-1} for the acceleration of gravity at the solar surface, $R = 8.31 \times 10^7$ cm^2 s^{-2} K^{-1} for the gas constant and $T = 1.8 \times 10^6$ K for the coronal temperature, gives for the exponent (41b) the value $\nu = -4.41$. Near the sun the particle density is taken to be $N_0 = 5 \times 10^8$ cm^{-3}, corresponding, for a proton mass $m_i = 1.67 \times 10^{-24}$ g to a mass density $\rho_0 = N_0 m = 8.35 \times 10^{-16}$ g cm^{-3}; for an average solar magnetic field $b = 12$ G, this leads to an Alfvén speed of $a = b\sqrt{\mu/4\pi\rho_0} = 1.17 \times 10^8$ cm s$^{-1} = 1170$ km / s. At 1 AU, near the earth, the particle density $N_1 = 15\,\mathrm{cm}^{-3}$, mass density $\rho_1 = N_1 m_i = 2.5 \times 10^{-23}\,\mathrm{g\,cm}^{-3}$ and magnetic field strength $B_1 = 5 \times 10^{-5}\,\mathrm{G}$, leads to a smaller Alfvén speed $A_1 = B_1\sqrt{\mu/4\pi\rho_1} = 2.82 \times 10^6\,\mathrm{cm\,s}^{-1} = 28\,\mathrm{km/s}$. The geometric mean of Alfvén speed is thus $A_2 = \sqrt{A_1 a} = 1.82 \times 10^7\,\mathrm{cm\,s}^{-1} = 182\,\mathrm{km/s}$. The dimensionless frequency (42b) is calculated for the reference case of equilateral spirals $\psi = \pi/4$, by $\Omega_0 = \sqrt{2/|\nu|}\,(r_0\omega/a) = 4.23\,(r_0/\tau a) = 2.53 \times 10^3/\tau$, in terms of the period in seconds. Considering periods of Alfvénic perturbations in the solar wind up to one day $\tau \leqslant 24\,\mathrm{h} = 8.64 \times 10^4\,\mathrm{s}$ leads to $\Omega_0 \geqslant 2.93 \times 10^{-2}$. The values chosen for the dimensionless frequency:

$$\Omega_0 = 3 \times 10^{-2}, 3 \times 10^{-1}, 3, 3 \times 10, 3 \times 10^2, 3 \times 10^3, \qquad (45a)$$

correspond to periods from $\tau \leqslant 1$ day to $\tau \geqslant 2.53 \times 10^3/3 \times 10^3 = 0.84\,\mathrm{s}$. Six values are given to the spiral angle

$$\psi = 0, \pi/12, \pi/6, \pi/4, \pi/3, 5\pi/12 = 0°, 15°, 30°, 45°, 60°, 75°, \qquad (45b)$$

in equal steps of $15°$. It follows from Figure 6 (left hand side) that the velocity perturbation of Alfvén waves in a spiral magnetic field increases rapidly with radial distance close to the sun, and then stabilizes, to a larger value for higher frequency and larger spiral angle; the phase in Figure 6 (right hand side) stabilizes after a few solar radii to a value which is larger for higher frequency and larger spiral angle. In figure 6 the velocity perturbation at radius r, is normalized (44a) to the value at the solar surface r_0, for each frequency. The evolution of the spectrum of Alfvénic disturbances from the sun ω^{-1} outward will be considered subsequently (§ 2.4), after discussing another effect, viz. mean flow.

2.3   Effect of vertical mean flow in a plane parallel atmosphere

In addition to stratification and multi-fluid effects (§ 2.1), and external magnetic fields of varying strength and direction (§ 2.2), Alfvén waves in the solar wind are affected by the mean flow, which is super-Alfvénic at the earth. Before combining the mean flow with spherical divergence and non-uniform magnetic field (§ 3), the effect of mean flow is considered for a plane parallel atmosphere, under a uniform vertical magnetic field, in the presence of density stratification. The total state of the fluid is now assumed to consist of a mean state with a steady, non-uniform magnetic field B and mean flow velocity U, upon which are superimposed unsteady, non-uniform magnetic field h and velocity v perturbations:

$$H(x,t) = B(x) + h(x,t), \quad V(x,t) = U(x) + v(x,t). \qquad (46a,b)$$

The momentum (4a) and induction (4b) equations are linearized:

$$\partial v/\partial t + (U \cdot \nabla)v + (v.\nabla)U = -(\mu/4\pi\rho)\,[B \wedge (\nabla \wedge h) + h \wedge (\nabla \wedge B)], \qquad (47a)$$

$$\partial h/\partial t - \nabla \wedge (U \wedge h) = -\nabla \wedge (B \wedge v), \qquad (47b)$$

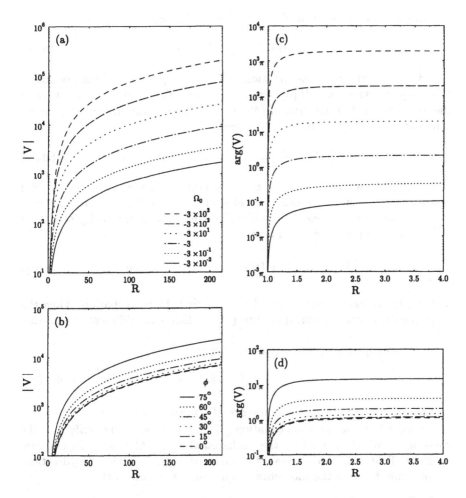

Fig. 6. Modulus (left) and phase (right) of velocity perturbation, normalized to the value at the solar surface, versus radial distance normalized by solar radius, for Alfvén waves, propagating along a logarithmic spiral: (top) making an angle $\phi = 45°$ with the radial direction, for five values of dimensionless frequency; (bottom) for fixed dimensionless frequency $\Omega_0 = -3$ and five values of spiral angle

where the Hall current was omitted in the latter.

Rather than obtaining general forms of the Alfvén wave equation in inhomogeneous flowing media [50] by elimination between (47a,b), one may proceed directly to the case of Alfvén waves in vertical uniform external magnetic field with horizontal and parallel velocity perturbations:

$$\boldsymbol{B} = B\boldsymbol{e}_z, \qquad \boldsymbol{v}, \boldsymbol{h} = v, h\left(z, t\right)\boldsymbol{e}_x, \tag{48a,b}$$

allowing for a vertical mean flow, which must be non-uniform (49a):

$$U = U(z) e_z, \qquad \rho(z) U(z) = \rho_0 u, \qquad (49a,b)$$

in order to satisfy mass conservation (49b) in the presence of density strati-fication $\rho(z)$, where $u = U(0)$ denotes the mean flow velocity at the base of the atmosphere. Substitution of (46a,b;48a,b;49a) in the linearized momen-tum and induction equations (47a,b) leads to

$$\dot{v} + U v' = (A^2/B) h', \qquad \dot{h} + (Uh)' = B v' \qquad (50a,b)$$

where dot, prime have the usual meaning (10a), and the Alfvén speed (8a) was introduced. Careful elimination between (50a,b), taking into account that the external magnetic field B is uniform but the mean flow velocity $U(z)$ and Alfvén speed $A(z)$ are not, leads to

$$\ddot{h} + 2\left(U\dot{h}\right)' + [U(Uh)']' - (A^2 h')' = 0, \qquad (51)$$

as the Alfvén wave equation for the magnetic field perturbation. Thus the magnetic field perturbation spectrum (13a) satisfies the differential equation

$$(A^2 - U^2) H'' + (2i\omega U + 2A'A - 3U'U) H'$$
$$+ \left[\omega^2 + 2i\omega U' - U''U - (U')^2\right] H = 0 \quad (52)$$

which simplifies to (16b) in the absence of a mean flow $U = 0$.

In order to solve the wave equation (52) it is necessary to specify first the coefficients, which depend on the background state. Although the external magnetic field is uniform (48a), the mean state is not one of hydrostatic equilibrium, because the momentum equation for the mean state:

$$dp/dz + \rho U'U = -\rho g, \qquad (53)$$

involves the non-uniform mean flow (49a). For a perfect gas (38a) and uniform gravity $g = const$, the momentum equation (53), viz.:

$$RTp^{-1}dp + U dU = -g dz, \qquad (54)$$

is readily integrated for an isothermal, flowing atmosphere:

$$T(z) = const : \ (U^2 - u^2)/2 + RT\log(p/p_0) = -gz, \qquad (55)$$

leading from (49b) to the implicit density stratification

$$\rho(z)/\rho_0 = e^{-z/L} \exp\left\{(u^2/2RT_0)\left[1 - (\rho_0/\rho(z))^2\right]\right\} \qquad (56)$$

which would become explicit (17a) in the absence of mean flow $u = 0$. The assumption that the mean flow is of low Mach number (57a)

$$1 \gg u^2/2RT = (\gamma/2)\,(u/c)^2 = (\gamma/2)\,M^2, \quad c^2 = \gamma RT \qquad (57a,b)$$

where the sound speed (58b) was used, allows simplification of (56) to

$$\rho\,(z)\,/\rho_0 = e^{-z/L} = u/U\,(z) = [a/A\,(z)]^2 \qquad (58a,b,c)$$

which leads to the profiles of mass density (17a,b), Alfvén speed (18a,b), and mean flow velocity $U\,(z) = ue^{z/L}$ increasing with altitude, to preserve the mass flux, in the presence of decaying density. It is clear from the restriction (57a) that the atmosphere cannot extend to 'infinity', but it can be shown that the condition is met beyond the altitude of the critical layer.

The critical layer occurs at the altitude such that the Alfvén speed (18a) and mean flow velocity(58b) are equal:

$$A(z_*) = U(z_*): \; z_* = -2L \log(a/u) = -2L \log N, \qquad (59a,b)$$

and is a singularity of the wave equation (52), which can be written

$$\left(1 - N^2 e^{z/L}\right) L^2 H'' + \left(2i\Omega N + 1 - 3N^2 e^{z/L}\right) LH'$$
$$+ \left(\Omega_1^2 e^{-z/L} + 2i\Omega N - 2N^2 e^{z/L}\right) H = 0, \quad (60)$$

in terms of two dimensionless parameters, namely the Alfvén number (61a) and dimensionless frequency (61b)

$$N \equiv u/a, \quad \Omega_0 \equiv \omega L/a. \qquad (61a,b)$$

By means of changes of independent (62a) and dependent (62b) variable:

$$\zeta = 1 - N^{-2} e^{-z/L} = 1 - e^{(z_* - z)/L}, \qquad (62a)$$
$$H\,(z;\omega) = \zeta^{-1/2 + iK}\,(1 - \zeta)\,Z_{1-2iK}\left(2iK\sqrt{\zeta}\right), \qquad (62b)$$

the differential equation (60) is transformed to a Bessel equation with imaginary variable $2iK\sqrt{\zeta}$ and complex order $1 - 2iK$, which involves only one dimensionless parameter

$$K_1 \equiv \Omega_0 N = \omega Lu/a^2 \qquad (63)$$

which is a combination of (61a,b).

Since the order of the Bessel function is not an integer, the general wave field is a linear combination:

$$H\,(z;\omega) = C_+ H_+\,(z;\omega) + C_- H_-\,(z;\omega), \qquad (64)$$

where $C_\pm$ are arbitrary constants, of solutions specified by Bessel functions of order $\pm(1-2iK)$, viz.:

$$H^\pm(z;\omega) \equiv \zeta^{-1/2+iK}(1-\zeta)J_{\pm(1-2iK_1)}\left(2iK_1\sqrt{\zeta}\right). \tag{65a,b}$$

The critical layer $z = z_*$ in (59), corresponds to the origin $\zeta = 0$ under the change of variable (62a), and thus the wave fields in its neighbourhood are specified by limit of the Bessel functions in (65a,b) for small variable

$$\zeta \to 0: J_{\pm(1-2iK_1)}\left(2iK_1\sqrt{\zeta}\right) \sim [\Gamma(1\pm(1-2iK_1))]^{-1}\left(iK_1\sqrt{\zeta}\right)^{\pm(1-2iK)}, \tag{66a,b}$$

leading to

$$H^+(z;\omega) \sim \left[(iK_1)^{1-2iK_1}/\Gamma(2-2iK_1)\right](1-\zeta), \tag{67a}$$

$$H^-(z;\omega) \sim \left[(iK_1)^{-1+2iK_1}/\Gamma(2iK_1)\right](1-\zeta)\zeta^{-1+2iK_1}. \tag{67b}$$

The variable ζ (62a) scales like the distance from the critical layer divided by the scale height:

$$\zeta = (z-z_*)/L + O\left((z-z_*)^2/L^2\right), \tag{68}$$

and thus one of the wave fields (67a):

$$H^+(z;\omega)[\Gamma(2-2iK_1)]^{-1}\exp[(1-2iK_1)\log(iK_1)], \tag{69}$$

is finite, non-zero and non-oscillatory at the critical layer.

The other wave field (67b) has a singularity at the critical layer like the inverse of distance:

$$H^-(z;\omega) \sim [(z-z_*)/L]^{-1}\exp\{2iK_1\log[(z-z_*)/L]\}, \tag{70}$$

and in addition, the second factor causes an amplitude jump across the critical layer:

$$H^-(z;\omega) \sim [(z-z_*)/L]^{-1}\exp\{2iK_1\log[|z-z_*|/L]\} \times \begin{cases} 1 & \text{if } z > z_*, \\ e^{2K_1\pi} & \text{if } z < z_*, \end{cases} \tag{71a,b}$$

where for $z < z_*$ the imaginary part of the logarithm was taken to be $-i\pi$ and not $+i\pi$. The justification for this [106] is to give the frequency a small positive imaginary part $\omega = \bar\omega + i\varepsilon$, to simulate the slow triggering of an instability $\exp(-i\omega t) = \exp(-i\bar\omega t)\exp(\varepsilon t)$. Noting that $\omega = ka$, the same transformation may be applied to Alfvén speed in (62a), viz.:

$$a = \bar{a} + i\epsilon: \zeta = 1 - (a/u)^2 e^{-z/L}$$

$$\sim 1 - (\bar{a}/u)^2 e^{-z/L} - (2i\bar{a}\epsilon/u^2) e^{-z/L} + O\left(\epsilon^2\right); \tag{72}$$

as the critical layer is approached $\operatorname{Im}(\zeta) < 0$, justifying $\operatorname{Im}(\log \zeta) = -i\pi$ in the second line of (71b). The implication of (71a,b) is that as the Alfvén wave crosses the critical layer from below its amplitude is reduced from $\exp(2K_1\pi)$ to unity, corresponding to a transmission factor:

$$S_1 = e^{-2K_1\pi} = e^{-\omega/\omega_1}, \quad \omega_1 = a^2/2\pi Lu; \qquad (73\mathrm{a,b})$$

thus waves of low-frequency compared with (73b) are totally transmitted $S_1 \sim 1$ for $\omega \ll \omega_1$, no transmission occurs $S_1 \sim 0$ for $\omega \gg \omega_1$, and partial transmission occurs for $\omega \sim \omega_1$. Note that this transmission coefficients applies to the wave component (70) which is singular at the critical layer, and not to the one (69) that is finite there.

The wave field components which are finite H^+ and singular H^- at the critical layer are complex, and thus the amplitudes $|H^\pm|$ and phases $\arg(H^\pm)$ are plotted separately in Figure 7 versus distance from the critical layer made dimensionless by dividing by the scale height (74a):

$$Z \equiv (z - z_*)/L; \quad \phi_1 = \omega/\omega_1 = 0.1, 0.5, 1, 2, 5, 10, \qquad (74\mathrm{a,b})$$

is a dimensionless parameter, specifying the transmission coefficient $S_1 = e^{-\phi_1}$ in (73a), to which six values are given (74b). The wave field which is finite H^+ at the critical layer, has an amplitude $|H^+|$ which is (Figure 7, top left) smaller for lower frequency, with amplitude oscillations before the critical layer for $\phi \leq 2$; the amplitude decays with distance, with a steeper slope after the critical layer. The phase of the wave field component finite at the critical layer $\arg(H^+)$ tends (Figure 7, bottom left) to a constant asymptotic value after the critical layer; before the critical layer the phase varies more rapidly for higher frequency waves, and is negative, implying downward propagation. In contrast, the phase of the wave field component which is singular at the critical layer (Figure 7, bottom right) is positive $\arg(H^-) > 0$, implying upward propagation; the phase in this case is also finite asymptotically and varies more rapidly for higher frequency. The amplitude of the wave field component which is singular at the critical layer $|H^-|$ is (Figure 7, top right) larger for lower frequencies, both before and after the critical layer; the spread of amplitudes is larger before than after the critical layer.

2.4 Steepening of the energy spectrum with Kolmogorov or Kraichnan exponents

Two processes of reflection of Alfvén waves have been demonstrated: (i) gradual reflection by gradients of Alfvén speed, which may result from density gradients (§ 2.1) or non-uniform external magnetic field (§ 2.2); (ii) localized reflection at a critical layer, which exists in the presence of background mean flow (§ 2.3), where the flow velocity equals the Alfvén speed. It will be shown next that linear process of wave reflection can steepen the energy spectrum of Alfvénic perturbations from $E \sim \omega^{-1}$ near the sun, to Kolmogorov or

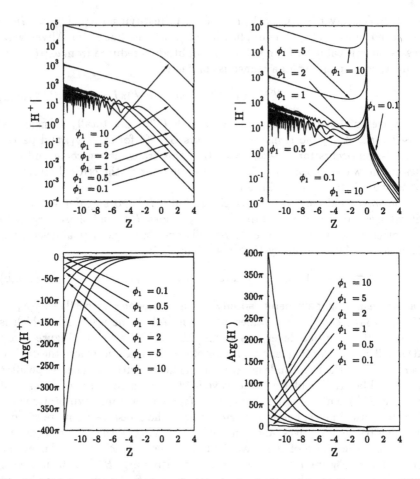

Fig. 7. Modulus (top) and phase (bottom) of velocity perturbation spectrum for vertical Alfvén wave in a flowing isothermal atmosphere; component of wave field which is finite H^+ (l.h.s.) or singular H^- (r.h.s.) at critical layer, versus dimensionless distance from critical layer, for five values of dimensionless frequency ϕ_1

Kraichnan-type spectra $E \sim \omega^{-\alpha}$ with $3/2 < \alpha < 5/3$ farther into the solar wind. The simplest way to obtain this result is to consider an Alfvén speed which varies like a power of radial distance (40a), as was found for Alfvén waves in a spiral magnetic field (§ 2.3).The same kind of dependence can be found for a radial external magnetic field and radial density stratification, with mass density varying like a power of radial distance. Leaving the exponent free in the dependence of the mass density $\rho \sim r^{-2-\nu}$ will allow later some choice in the exponent of the energy spectrum. The neglect of mean flow (which will be considered in §3), means that the model applies to Alfvén waves in the solar wind before the critical layer, where the density gradients are larger, and thus reflection is stronger.

The total state of the fluid is assumed to consist (Figure 8) of a mean state at rest under a radial non-uniform external magnetic field, with transverse

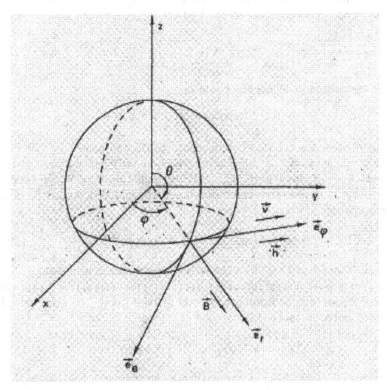

Fig. 8. Spherical Alfvén wave, with velocity v and magnetic field h perturbations along parallels e_ϕ, propagating along a radial B external magnetic field e_r

and parallel velocity and magnetic field perturbations along the parallels, depending only on radial distance and time:

$$H\left(x,t\right) = B\left(r\right)e_r + h\left(r,t\right)e_\varphi, \; V\left(x,t\right) = v\left(r,t\right)e_\varphi. \tag{75a,b}$$

Substitution in the momentum (4a) and induction (4b) equations yields

$$\frac{\partial v}{\partial t} - \left(\frac{A^2}{Br}\right)\frac{\partial}{\partial r}\left(rh\right) = 0, \; \frac{\partial h}{\partial t} - r^{-1}\frac{\partial}{\partial r}\left(Brv\right) = 0, \tag{76a,b}$$

where the Alfvén speed (8a) was introduced. Elimination between (76a,b) leads to the wave equation for the velocity perturbation of radial one-dimensional

Alfvén waves:

$$\frac{\partial^2 v}{\partial t^2} - \left(\frac{A^2}{Br}\right) \frac{\partial^2}{\partial r^2} (Brv) = 0. \tag{77}$$

It follows that the velocity perturbation spectrum $V(r; \omega)$ for a wave of frequency ω at radial distance r:

$$v(r, t) = \int_{-\infty}^{\infty} V(r; \omega) e^{-i\omega t} \, d\omega, \quad h(r, t) = \int_{-\infty}^{\infty} H(r; \omega) e^{-i\omega t} \, d\omega, \tag{78a,b}$$

satisfies the ordinary differential equation

$$(rB)^{-1} (rBV)'' + rB (\omega/A)^2 V = 0 \tag{79}$$

where prime denotes derivative with regard to the radius r.

The coefficients of the differential equation (79) depend on the mean state, which is specified next. The Maxwell equation $0 = \nabla \cdot [B(r) e_r] = r^{-2}d(r^2 B)/dr$, implies that the radial magnetic field decays like the inverse square of distance (80a)

$$B(r) = b(r_0/r)^2, \quad \rho(r) = \rho_0 (r_0/r)^{2+\nu}, \tag{80a,b}$$

and for the mass density (80b) a power law dependence is also assumed, with arbitrary exponent, including values $0 < \nu < 1$ corresponding to the solar wind case. From (80a,b) it follows that the Alfvén speed (8a) also varies like a power of radial distance (81a):

$$A(r) = a(r_0/r)^{1-\nu/2}, \quad a \equiv b\sqrt{\mu/4\pi\rho} \tag{81a,b}$$

from an initial value (81b) calculated from the magnetic field strength $b = B(r_0)$ in (80a) at the solar surface $r = r_0$. The condition of hydrostatic equilibrium (82a) for a radial gravity field (82b):

$$\frac{dp}{dr} = -\rho(r) g(r), \quad g(r) = g_0 (r_0/r)^2, \tag{82a,b}$$

leads to the stratification of gas pressure:

$$p(r) - p_\infty = [\rho_0 g_0 r_0/(\nu + 3)] (r_0/r)^{3+\nu}, \tag{83a}$$

where the pressure is constant at infinity p_∞ for $\nu > -3$; from (83a) and (80b) follows a polytropic law for the mean state:

$$p(r) - p_\infty = [\rho_0 g_0 r_0/(\nu + 3)] [\rho(r)/\rho_0]^{(3+\nu)/(2+\nu)}, \tag{83b}$$

with exponent $4/3 < (3+\nu)/(2+\nu) < 3/2$ for the solar wind case $0 < \nu < 1$.

Substitution of the background (80a, 81a) in the differential equation (79) yields

$$r^2 V'' - 2rV' + (\omega r/a)(r_0/r)^{\nu-2} V = 0 \qquad (84a)$$

whose solution can be expressed in terms of Bessel functions:

$$V(r;\omega) = R^{3/2} Z_{1/(4-\nu)} \left(2\Omega R^{2-\nu/2} / (2-\nu/2) \right), \qquad (84b)$$

where R is the dimensionless radial distance divided by the solar radius (44a)≡(85a) and Ω the dimensionless frequency (85b):

$$R \equiv r/r_0, \quad \Omega \equiv r_0 \omega/a. \qquad (85a,b)$$

The radiation condition, specifying outward propagating waves at infinity, requires the choice of the Hankel function $H^{(1)}$ multiplied by an arbitrary constant, which is determined from the initial velocity perturbation spectrum at the solar surface $V(r_0;\omega)$ viz.:

$$f(r/r_0) \equiv H^{(1)}_{1/(4-\nu)} \left(2\Omega (r/r_0)^{2-\nu/2} / (2-\nu/2) \right), \qquad (86a)$$

$$V(r;\omega) = V(r_0;\omega)(r/r_0)^{3/2} [f(r/r_0)/f(1)]. \qquad (86b)$$

This solution will be analyzed next for large and small radii.

The lengthscale of variation of the background mass density (80b) is proportional to the radius:

$$L = -\rho/(d\rho/dr) = -[d(\log \rho)/dr]^{-1} = (\nu+2)r, \qquad (87)$$

and likewise for the lengthscales of variation of (80a) the strength of external magnetic field $2r$, and of (81a) the Alfvén speed $(1-\nu)r$. Thus the gradients of background properties are larger for small radius, i.e. nearer to the sun, where the mean flow velocity is still sub-Alfvénic, and stronger reflection occurs. The region of stronger reflection of Alfvén waves corresponds to small variable in (84b), viz.:

$$r \ll r_0 [2\Omega/(2-\nu/2)]^{-1/(2-\nu/2)} \equiv r_1, \quad f \sim \Omega^{-1/(4-\nu)} r^{-1/2}. \qquad (88a,b)$$

Note that (88b) corresponds by (86b) to a velocity perturbation (89a):

$$V(r \ll r_1) \sim r, \quad N \equiv V/A \sim r^{\nu/2}, \qquad (89a,b)$$

and hence from (81a) to an Alfvén number (81b) which may increase with distance $0 < \nu/2 < 1/2$, so that non-linear effects could arise at large radius. Assuming that the linear approximation (88a) is still valid at large radius,

$$r \gg r_1 : f \sim \Omega^{-1/2} r^{-1+\nu/4}, \qquad (90)$$

the velocity perturbation (91a):

$$V\left(r \gg r_1\right) \sim r^{1/2+\nu/4}, \; N \equiv V/A \sim r^{-1/2+\nu/4}, \tag{91a,b}$$

leads to an Alfvén number (91b) which scales on the radius with exponent $-1/2 < -1/2 + \nu/4 < -1/4$, so that there is no further growth.

Concerning the dependence of the energy spectrum of Alfvénic perturbations with distance, the kinetic energy scales initially (88b) like

$$E_\nu\left(r \ll r_1; \omega\right) \equiv \frac{1}{2}\rho\left|V\left(r \ll r_*; \omega\right)\right|^2 \sim \omega^{-1/(2-\nu/2)}\left|V\left(r_0; \omega\right)\right|^2 \equiv E_0\left(\omega\right).$$

$$\tag{92}$$

Bearing in mind that the initial energy spectrum of Alfvénic perturbations near the sun has exponent -1 in (93a):

$$E_0\left(\omega\right) \sim \omega^{-1}, \; V\left(r_0; \omega\right) \sim \omega^{-(1-\nu/4)/(4-\nu)}, \tag{93a,b}$$

it follows that the initial velocity perturbation (93b) has exponent $-1/4 < -\left(\nu/2 - 1\right)/\left(4 - \nu\right) < -1/6$. The asymptotic energy spectrum follows from (90), viz.:

$$E_\infty \equiv E_\nu\left(r \gg r_1; \omega\right) \sim \omega^{-1}\left|V\left(r_0; \omega\right)\right|^2 \sim \omega^{-(6-\nu)/(4-\nu)}, \tag{94}$$

and has exponent $-3/2 < -\left(6 - \nu\right)/\left(4 - \nu\right) < -5/3$, which takes: (i) the highest Kolmogorov's exponent $-5/3$ for $\nu = 1$ corresponding to mass density decaying (81b) like $\rho \sim r^{-3}$; (ii) the lowest Kraichnan exponent $-3/2$ for $\nu = 0$ corresponding to mass density decaying like $\rho \sim r^{-2}$. The preceding results are consistent with observations in the solar wind, and show that the Kolmogorov and Kraichnan spectra $E_\infty \sim \omega^{-\alpha}$ with $3/2 < \alpha < 5/3$, can arise from a linear process of reflection of Alfvén waves, starting with an energy spectrum $E_0 \sim \omega^{-1}$ near the sun, as a consequence of a monopole external magnetic field $B \sim r^{-2}$ and a density stratification $\rho \sim r^{-2-\nu}$ with exponent $2 < \nu + 2 < 3$ between the square and cube of the inverse of the radial distance. It could be argued that this result contradicts the JWKB theory that high frequency waves are not reflected, and thus the spectrum cannot become steeper as frequency increases; the flaw of the reasoning lies with the JWKB approximation, which neglects reflection, and does not apply asymptotically over many wavelengths. In fact the JWKB theory cannot be used to predict asymptotic spectra because (i) it does not apply to low frequencies; (ii) it shows that reflection is weak for high frequency waves over one wavelength, but cannot be applied over long distances corresponding to many wavelengths. The asymptotic spectrum of Alfvén waves due to reflection by gradients in the background medium can be determined only from exact solutions, which is exactly what was done in deriving (94).

The Kolmogorov or Kraichnan spectra of Alfvénic perturbations in the solar wind has been interpreted as evidence of MHD or hydromagnetic turbulence, involving a non-linear energy cascade; the preceding result shows that the linear process of reflection of Alfvén waves in the inhomogeneous solar wind is an alternative explanation. The two processes differ not only in the former being non-linear and the latter linear, but also in another respect: the non-linear energy cascade will lead to Kolmogorov or Kraichnan spectra from any initial condition, whereas the linear process of reflection of Alfvén waves will lead to Kolmogorov or Kraichnan spectra (94) with $0 < \nu < 1$ only if the initial energy spectrum near the sun (93a) has exponent -1. The two explanations for the Kolmogorov or Kraichnan spectra of Alfvénic perturbations in the solar wind are physically distinct, but could be complementary [52]: (i) the initial energy spectrum of Alfvén waves near the sun (94a) evolves, as a consequence of the linear process of reflection by gradients in the inhomogeneous solar wind, which are stronger near the sun, into a Kolmogorov or Kraichnan spectrum (95); (ii) this spectrum is then maintained at larger radial distances, when the Alfvén waves may have grown in amplitude, by a 'turbulence-like' energy cascade, resulting from the non-linear interaction of Alfvén waves propagating in opposite directions, as a consequence of reflection. Thus the process of wave reflection can lead to Kolmogorov or Kraichnan type spectra both at linear and non-linear level.

The velocity perturbation spectrum (86a,b) is given, in the case $\nu = 0$ of asymptotic Kraichnan energy spectrum, by:

$$V(R) \equiv V(r;\omega)/V(r_0;\omega_0) = R^{3/2} H_{1/4}^{(1)}\left(\Omega R^2\right)/H_{1/4}^{(1)}(\Omega), \qquad (95)$$

where R is the radial distance made dimensionless by dividing by the solar radius (44a). Substituting (78a,b) in (76b) specifies the magnetic field perturbation spectrum:

$$H(r;\omega) = i(b/\omega)r_0^2 r^{-1}(d/dr)\left[r^{-1}V(r;\omega)\right], \qquad (96a)$$

in terms of the velocity perturbation spectrum (94) leading to:

$$H(R) \equiv H(r;\omega)/H_0(r;\omega) = R^{1/2} H_{-3/4}^{(1)}\left(\Omega R^2\right)/H_{-3/4}^{(1)}\left(\Omega R^2\right). \qquad (96b)$$

Since the velocity (95) and magnetic field (96b) perturbations of outward propagating waves are complex, the amplitudes and phases are plotted separately in Figure 9, versus the radius; the latter is restricted to $R < 10$ due to the neglect of the mean flow in the background state. The phase of the velocity (Figure 9, bottom left) and magnetic field (Figure 9, bottom right) perturbations are larger for larger dimensionless frequency (85b), and increase radially with radial distance (85a), as could be expected from the arguments of ΩR^2 of the Hankel functions in (95,96b). For large radius $\Omega R^2 \gg 1$ in the sense of (88a), the velocity perturbation spectrum (Figure 9, top left) increases like the square root of the radius $|V| \sim R^{1/2}$; since at large radius the JWKB approximation holds $V/A = H/B$, the amplitude of the magnetic field perturbation $|H| \sim |V|\rho^{1/2} \sim R^{1/2}R^{-1} \sim R^{-1/2}$ decays (Figure 9, top right) like the inverse square root of radial distance, as can be checked from (96b).

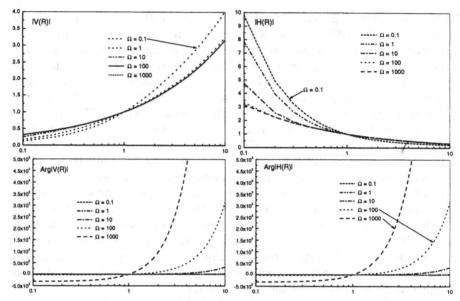

Fig. 9. Modulus (top) and phase (bottom) of velocity (l.h.s.) and magnetic field (r.h.s.) perturbation spectra of spherical Alfvén wave in radially stratified medium at rest normalized to value at reference radius $r = r_0$ (solar surface), versus radial distance r made dimensionless by dividing by reference radius $R \equiv r/r_0$, for five values of dimensionless frequency Ω

3   Models combining radial stratification, mean flow and monopole external magnetic field

Having considered the separate effects of density stratification and an ion-electron two-fluid model (§ 2.1), external magnetic field of varying strength and direction (§ 2.2), non-uniform background mean flow (§ 2.3) and spherical or radial divergence (§ 2.4) on the spectral and spatial evolution of Alfvén waves, the next step is to combine these effects to simulate physical conditions in the solar wind. The simplest model with the essential features could be: (a) a radial monopole external magnetic field, which is consistent with Maxwell's equations; (b) radial non-uniform mean flow velocity, mass density and gas pressure, consistent with momentum balance and conservation of the mass flux. The latter two conditions imply that the background state is uniquely determined by specifying the mean flow velocity as a function of the radius, e.g. like a power law $U(r) \sim r^\nu$. The radial Alfvén wave equation for the velocity perturbation is solved exactly for all frequencies and distances, in four cases: (i) uniform flow $\nu = 0$, leading to the convected Alfvén wave equation (§ 3.1); (ii) mean flow velocity a linear function of the radius $\nu = 1$, which adds extra terms to the convected Alfvén wave equation (§ 3.2); (iii) in both of the preceding cases there is a critical layer, which is excluded (§3.3) in the case of mean flow velocity inversely proportional to the square of the radius $\nu = -2$, and then a transition level is present; (iv) in the case of mean flow velocity proportional to the square root of the radius $\nu = 1/2$, which is probably closest to the solar wind profile, there is (§3.4) both a critical layer and a transition level.

3.1 One- and three- dimensional Alfvén waves in a radial uniform mean flow

The radial wave equation has been solved for a uniform radial flow [69], in terms of Gaussian hypergeometric functions, and a cut-off frequency identified [17,92,93,117]; the availability of numerical results in the literature [94,144,143] suggests that this case is chosen first for exact analytical solution for all distances and frequencies. For one-dimensional spherical Alfvén waves , the total state of the fluid is assumed to consist of a mean state of radial steady non-uniform magnetic field $B(r)$ and a mean flow $U(r)$, upon which transverse and parallel velocity v and magnetic field h perturbations along the parallels are superimposed, depending only on time t and radial distance r, viz.:

$$H(x,t) = B(r)e_r + h(r,t)e_\varphi, \quad V(x,t) = U(r)e_r + v(r,t)e_\varphi. \qquad (97a,b)$$

Substitution in the momentum (4a) and induction (4b) equations yields:

$$\partial\bar{v}/\partial t + U\partial\bar{v}/\partial r = \left(A^2/B\right)\partial\bar{h}/\partial r, \qquad (98a)$$

$$\partial\bar{h}/\partial t + \partial\left(U\bar{h}\right)/\partial r = \partial\left(B\bar{v}\right)/\partial r, \qquad (98b)$$

where $\bar{v}, \bar{h}$ denote the modified velocity and magnetic field perturbations multiplied by the dimensionless radius:

$$\bar{v}, \bar{h}(r,t) \equiv (r/r_0)\, v, h(r,t) \qquad (99a,b)$$

and the Alfvén speed (8a) was introduced.

A more general case is three-dimensional Alfvén waves, for which the mean state is the same as for the one-dimensional case (97a,b) but the transverse velocity and magnetic field perturbations may have components along the parallels and meridians, and may depend on time t and on all three spherical coordinates (r, θ, φ), viz.:

$$H(x,t) = B(r)e_r + h_\varphi(r,\theta,\varphi,t)e_\varphi + h_\theta(r,\theta,\varphi,t)e_\theta, \qquad (100a)$$

$$V(x,t) = U(r)e_r + v_\varphi(r,\theta,\varphi,t)e_\varphi + v_\theta(r,\theta,\varphi,t)e_\theta. \qquad (100b)$$

Substitution in the momentum and induction equations (4a,b) leads again to (98a,b), where $\bar{v}, \bar{h}$ now denote the square of the radius multiplied by the radial component of vorticity $\nabla \wedge \boldsymbol{v}$ and electric current $\nabla \wedge \boldsymbol{h} = (4\pi/c)\,\boldsymbol{J}$, viz.:

$$\bar{v} \equiv r^2 \left(\nabla \wedge \boldsymbol{v}\right)_r = r \csc \theta \left[\partial \left(v_\varphi \sin \theta\right)/\partial \theta - \partial v_\theta/\partial \varphi\right], \tag{101a}$$

$$\bar{h} \equiv r^2 \left(\nabla \wedge \boldsymbol{h}\right)_r = r \csc \theta \left[\partial \left(h_\varphi \sin \theta\right)/\partial \theta - \partial h_\theta/\partial \varphi\right]. \tag{101b}$$

In the case of three-dimensional Alfvén waves (100a,b), the angular dependence is specified by spherical harmonics $Y_l^m\,(\theta, \varphi)$, and the radial dependence is the same (98a,b) as for one-dimensional Alfvén waves (97a,b).

Since the background is steady, the wave frequency ω is conserved, and the spectra of the modified velocity and magnetic field perturbations (99a,b), viz.:

$$\bar{v}, \bar{h}(z,t) = \int_{-\infty}^{+\infty} \bar{V}, \bar{H}(r;\omega)e^{-i\omega t}\,d\omega, \tag{102a,b}$$

relate to the spectra (78a,b) of the unmodified velocity and magnetic field perturbations in the same way:

$$V, H(r;\omega) = (r/r_0)\bar{V}, \bar{H}(r;\omega). \tag{103a,b}$$

Elimination between (98a,b) in the case of uniform mean flow velocity, leads to the convected Alfvén wave equation for the velocity perturbation:

$$U' = 0: \left(\frac{\partial}{\partial t} + U\frac{\partial}{\partial r}\right)\frac{B}{A^2}\left(\frac{\partial}{\partial t} + U\frac{\partial}{\partial r}\right)\bar{v} = \frac{\partial^2}{\partial r^2}(B\bar{v}), \tag{104}$$

and to an ordinary differential equation, for its spectrum:

$$\ddot{\bar{V}} + U\dot{\bar{V}} + (UA^2/B)\left(B\dot{\bar{V}}/A^2\right)' + (UA^2/B)(BU\,\bar{V}'/A^2)' = (A^2/B)(B\,\bar{V})'', \tag{105}$$

where the external magnetic field B, Alfvén speed A and mean flow velocity U appear in the coefficients, and are specified by the background state.

The conservation of the mass flux:

$$\rho(r)\,U(r)\,r^2 = const = \rho_0 u r_0^2, \tag{106}$$

implies that a uniform mean flow (107a) corresponds to a mass density (107b) decaying on the inverse square of the radius:

$$U(r) = u = const, \quad \rho(r) = \rho_0\,(r_0/r)^2, \tag{107a,b}$$

i.e. the $\nu = 0$ in (80b) and the Alfvén speed (81a). Substituting (80a; 107a) and (81a with $\nu = 0$) in (105) leads, for the modified velocity perturbation spectrum (102a, 103a) as a function of the (85) dimensionless radius:

$$\Phi(R) = \Phi(r/r_0) \equiv \bar{V}(r;\omega) = (r/r_0)\,V(r;\omega), \tag{108}$$

to the differential equation:

$$(1 - N^2 R^2)\Phi'' + 2\left(iN\Omega R^2 - 2/R\right)\Phi' + \left(\Omega^2 R^2 + 6/R^2\right)\Phi = 0, \qquad (109)$$

involving two dimensionless parameters, namely the dimensionless frequency (85b) and the Alfvén number (61a). The changes of independent (110a) and dependent (110b) variable:

$$\xi \equiv 1 + NR = 1 + ur/r_0 a, \ \Psi(\xi) \equiv R^{-2}\exp\left(-iK_2 NR\right)\Phi(R), \qquad (110a,b)$$

lead to a differential equation:

$$\xi(1 - \xi)\Psi'' + (2 + iK_2 - 4\xi)\Psi' - (2 + K_2^2)\Psi = 0, \qquad (111)$$

where the Alfvén number and dimensionless frequency appear only in the combination:

$$K_2 \equiv \Omega/N^2 = \omega r_0 a/u^2 = (\omega r_0/u)(a/u). \qquad (112)$$

The differential equation is of the Gaussian hypergeometric type [62]:

$$\xi(1 - \xi)\Psi'' + [\gamma - (\alpha + \beta + 1)\xi]\Psi' - \alpha\beta\Psi = 0, \qquad (113)$$

with parameters specified by:

$$\gamma_1 = 2 + iK_2, \ \alpha_1 + \beta_1 = 3, \ \alpha_1\beta_1 = 2 + (K_2)^2. \qquad (114a,b,c)$$

The roots of (114b,c) are:

$$\alpha_1, \beta_1 = 3/2 \pm \delta, \ 2\delta \equiv \sqrt{1 - (K_2)^2}, \qquad (115a,b)$$

and the change of δ from real to imaginary specifies the cut-off frequency:

$$2\delta = \sqrt{1 - (\omega/\omega_*)^2}, \ \omega_* = \omega/(2K_2) = u^2/(2r_0 a) \qquad (116a,b)$$

The cut-off frequency (116b) has been discussed in the literature [69,17,91,92, 117], and separates imaginary or propagating waves $\omega > \omega_*$ from real or standing modes $\omega < \omega_*$.

The differential equation (109) has a singularity at the critical layer, where the mean flow velocity equals the Alfvén speed (117a):

$$u = A(r_2) = ar_0/r_2, \ r_2 = r_0 R_2 = r_0 a/u = r_0/N, \qquad (117a,b)$$

i.e. at radial distance (117b). Among the six solutions of the hypergeometric equation (111)≡(113), with variables $\xi, 1 - \xi, 1/\xi, 1(1 - \xi), \xi/(1 - \xi)$ and $1 - 1/\xi$, the latter is the most useful, because it applies over all radial distances:

$$V(r;\omega) = (r/r_2)\exp\left(iK_2 r/r_2\right)\{C_1 V_1(r;\omega) + C_2 V_2(r;\omega)\} \qquad (118)$$

is a linear combination of two wave fields, where the coefficients are arbitrary constants of integration. One of the wave fields:

$$V_1(r;\omega) = [(1 + r/r_2)/2]^{-3/2+\delta}$$
$$\times F(3/2 + \delta, 1/2 - iK_2 + \delta; 2 - iK_2; (r - r_2)/(r + r_2)),$$
(119a)

is finite at the critical layer:

$$\bar{s} \equiv r/r_2 - 1: \ V_1(r;\omega) = 1 + O(\bar{s}), \tag{119b}$$

where s is the dimensionless distance from the critical layer; the other wave field:

$$V_2(r;\omega) = [(1 + r/r_2)/2]^{-iK_2+\delta-1/2}(r - r_2)^{-1+iK_2}$$
$$\times F(1/2 + iK_2 - \delta, -\delta - 1/2; iK_2; (r - r_2)/(r + r_2)), \quad (120a)$$

is singular at the critical layer, viz.:

$$\bar{s} \to 0: \ V_2(r;\omega) \sim (1 - r/r_2)^{-1} \exp\{iK_2 \log(1 - r/r_2)\}, \tag{120b}$$

i.e. the amplitude varies like the inverse of the distance from the critical layer s^{-1}, and there is an amplitude jump:

$$\exp\{iK_2 \log(1 - r/r_2)\} = \exp(iK_2|s|) \times \begin{cases} 1 & \text{if} \quad r < r_2, \\ e^{-\pi K_2} & \text{if} \quad r > r_2, \end{cases} \quad (121a,b)$$

where the choice of $\text{Im}\{\log(s)\} = i\pi$ is justified as in (71a,b; 72). The amplitude jump (121a,b) corresponds to a transmission factor (122a):

$$S_2 = \exp(-K_2\pi) = \exp(-\omega/\omega_2), \quad \omega_2 \equiv u^2/(\pi r_0 a), \tag{122a,b}$$

with transition frequency $\omega_2 = (2/\pi)\omega_*$ (122b), close to the cut-off frequency (116b). The transition frequency (122b) is not very significant, because it applies only to the singular component of the wave field (120a), and in that case wave transmission is determined by the way the wave field decays on either side. Thus (116b) is the relevant cut-off frequency, because it applies to all components of the wave fields.

The wave field (119a), which is finite at the critical layer (123b):

$$s \equiv 1 + \bar{s} = r/r_2, \quad Q_0 \equiv V_1(r;\omega) = se^{iK_2 s}[(1 + s)/2]^{-3/2+\delta}$$
$$\times F(3/2 + \delta, 1/2 + iK_2 + \delta; 2 + iK_2; (s - 1)/(s + 1)), \quad (123a,b)$$

is plotted versus (123a) the radial distance r, made dimensionless by dividing by the distance r_2 in (117b) of the critical layer from the center (Figure 10, left-hand side). Concerning the wave field component (120a) which is singular at the critical layer, a factor $(1 - s)/(1 + s)$ is inserted:

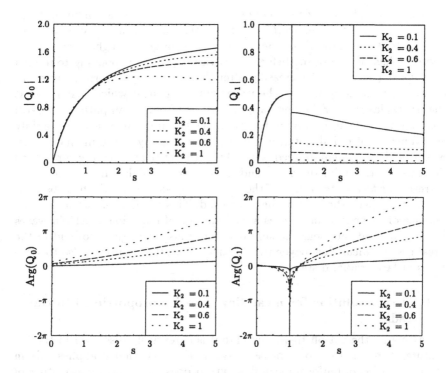

Fig. 10. Modulus (top) and phase (bottom) of velocity perturbation spectrum of Alfvén wave in an uniform radial mean flow; wave field component which is finite (l.h.s) or singular (r.h.s) at critical layer, the latter with singularity removed; both are plotted versus dimensionless distance from critical layer, for four values of dimensionless frequency

$$Q_1(s) \equiv [(1 - s)/1 + s)] \, V_2 \, (r; \omega) = [s/(1 + s)] \, e^{-iK_2 s} \, [(1 + s)\,/2]^{iK_2 - 1/2 - \delta}$$
$$\times F\left(1/2 + \delta - iK_2, -1/2 + \delta; -iK_2; (s - 1)\,/\,(s + 1)\right) \qquad \text{(124a,b)}$$

so as to suppress the amplitude singularity $(1 - s)^{-1}$ at the critical layer (120b), also maintaining $(1 - s)/(1 + s) \to 1$ the asymptotic wave field as $s \to \infty$ or $r \to \infty$; the amplitude jump (121a,b) remains in (124a,b), as can be seen confirmed (Figure 10, right-hand side). Since the wave fields (123b, 124b) are complex, the modulus and phase are plotted separately. The modulus of the wave field component which is finite at the critical layer $|Q_0|$ increases (Figure 10, top left) with radial distance more rapidly before the critical layer in a way, which is nearly independent of frequency (112); after the critical layer the amplitude increases less with distance, or may even decay for larger frequencies. The phase of the wave field component which is finite at the critical layer $\arg(Q_0)$ varies (Figure 10, bottom left) linearly with distance, faster for higher frequency, with no distinction before, after or across the critical layer. The wavefield component whose amplitude is singular at the critical layer has a phase $\arg(Q_1)$ which is also singular at the critical layer (Figure 10, bottom right); the phase variation is larger for higher frequencies, and the phase is decreasing before the critical layer, corresponding to inward propagating waves, and increasing after the critical layer, corresponding to outward propagating waves. The wave field component which is singular at the critical layer, after the singularity is removed from the amplitude $|Q_1|$, still has an amplitude jump (Figure 10, top right); the amplitude increases rapidly as a function of radial distance, before the critical layer in a manner which is almost independent of frequency. After the critical layer the amplitude decreases with radial distance, and is smaller for higher frequencies; this corresponds to the steepening of the wave energy spectrum, which is observed for Alfvénic perturbations in the solar wind. This result was obtained in the absence of mean flow in §2.4, as a consequence of reflection of Alfvén waves by density gradients near the sun; it is confirmed here (§3.1) to hold in the presence of a mean flow beyond the critical layer, thus applying also to wave fields at large radial distances.

3.2   Exact solution for mean flow velocity proportional to the radius

Although the cut-off frequency (116b) is extensively discussed in the literature, it may not be too representative, in the sense that it applies only in the case (107a) of uniform mean flow, viz. it does not occur for any other of the three velocity profiles considered next (§3.2-§3.4). The linearized momentum (98a) and induction (98b) equations may be eliminated for the modified velocity perturbation:

$$\left(\frac{\partial}{\partial t} + \frac{\partial}{\partial r}U\right)\frac{B}{A^2}\left(\frac{\partial}{\partial t} + U\frac{\partial}{\partial r}\right)\bar{v} = \left(\frac{\partial}{\partial t} + \frac{\partial}{\partial r}U\right)\frac{\partial\bar{h}}{\partial r}$$

$$= \frac{\partial}{\partial r}\left(\frac{\partial}{\partial t} + U\frac{\partial}{\partial r}\right)\bar{h} = \frac{\partial}{\partial r}\left[\left(\frac{\partial}{\partial t} + \frac{\partial}{\partial r}U\right)\bar{h} - U'\bar{h}\right]. \qquad (125)$$

Substituting (98b) in the first term on the r.h.s. of (125) yields:

$$\left(\frac{\partial}{\partial t} + \frac{\partial}{\partial r}U\right)\frac{B}{A^2}\left(\frac{\partial}{\partial t} + U\frac{\partial}{\partial r}\right)\bar{v} - \frac{\partial^2}{\partial r^2}(B\bar{v}) = -U'\frac{\partial\bar{h}}{\partial r} - U''\bar{h}, \qquad (126)$$

The convected Alfvén wave equation (126) simplifies to (104) in the case of uniform mean flow velocity $U' = 0$. In the case of a mean flow velocity proportional to the radius $U' \neq 0 = U''$, substitution of (98a) in (126) shows that the Alfvén wave equation for the (99a) modified velocity perturbation:

$$\left(\frac{\partial}{\partial t} + \frac{\partial}{\partial r}U\right)\frac{B}{A^2}\left(\frac{\partial}{\partial t} + U\frac{\partial}{\partial r}\right)\bar{v} - \frac{\partial^2}{\partial r^2}(B\bar{v})$$

$$= -\frac{U'B}{A}\left(\frac{\partial}{\partial t} + U\frac{\partial}{\partial r}\right)\bar{v}, \qquad (127)$$

has an extra term (on the r.h.s.) relative to the case (104) of non-uniform flow. It will be shown next that this extra term eliminates the cut-off frequency (116b).

The conservation of the mass flux (106) implies that a mean flow velocity proportional to the radius corresponds to a mass density decaying like the inverse cube of distance:

$$U(r) = u(r/r_0), \quad \rho(r) = \rho_0 (r_0/r)^3, \qquad (128a,b)$$

and hence (80b) to an Alfvén speed (81a) with $\nu = 1$. Substitution of this together with (80a, 128a) in the Alfvén wave equation (127) leads to the ordinary differential equation

$$\left(A^2 - U^2\right) \bar{V}'' - 2\left[i\omega U + \left(U^2 + 2A^2\right)/r\right] \bar{V}'$$
$$+ \left[\omega^2 - i\omega U/r + 6r^{-2}A^2\right] \bar{V} = 0, \quad (129)$$

for the modified velocity perturbation spectrum (102a). The change of independent (85a) and dependent (108) variable leads to the differential equation:

$$R^2 \left(1 - N^2 R^3\right) \Phi'' - 2R \left[2 + iNR^3 \left(\Omega - iN\right)\right] \Phi'$$
$$+ \left[6 + \Omega R^3 \left(\Omega - iN\right)\right] \Phi = 0, \quad (130)$$

involving two dimensionless parameters, namely the dimensionless frequency (85b) and the Alfvén number (61a).

The changes of independent (131a) and dependent (131b) variable:

$$\eta \equiv N^2 R^3 = u^2 r^3 / a\, r_0^3, \quad \Theta(\eta) \equiv R^{-2} \Phi(R) = (r_0/r) V(r; \omega) \qquad (131a,b)$$

transform the differential equation (130) to:

$$9 \left(1 - \eta\right) \eta \Theta'' + 6 \left[1 - \eta \left(4 + K_3\right)\right] \Theta' - 6 \left[1 - K_3 \left(5i - K_3\right)\right] \Theta = 0, \quad (132)$$

which involves only one dimensionless parameter

$$K_3 \equiv \Omega/N = r_0 \omega / u, \qquad (133)$$

combining the dimensionless frequency Ω and Alfvén number N. The differential equation (132) is of the Gaussian hypergeometric type (113), with variable η and parameters:

$$\gamma_2 = 2/3, \quad \alpha_2 = 1 + K_3/3, \quad \beta_2 = 2/3 + K_3/3, \qquad (134a,b,c)$$

The Alfvén wave equation (129) has a singularity where the mean flow velocity equals the Alfvén speed $U = A$, i.e. the location of the critical layer is specified by the equality of (128a) and (81a with $\nu = 1$), viz.:

$$U(r_3) = A(r_3): \ r_3 = r_0 R_3 = r_0 N^{-2/3} = r_0 \left(a/u\right)^{2/3}. \qquad (135)$$

Note that the critical layer $r = r_3$ is placed at the point unity $\eta_3 = 1$ by the change of independent variable (131a).

Among the six solutions of the Gaussian hypergeometric equation (132), with parameters (134a,b,c), the most convenient uses the variable $1 - 1/\eta > -1$ or $\eta > 1/2$, and thus applies to all radial distances larger than half the distance of the critical layer from the centre:

$$r > r_3/2 : \ V(r;\omega) = (r/r_0)\{C_3 V_3(r;\omega) + C_4 V_4(r;\omega)\} \qquad (136)$$

and in particular includes the asymptotic limit $r \to \infty$ and the critical layer $r = r_3$. One of the wave fields:

$$V_3(r;\omega) = \left(u^2 r^3 / a^2 r_0^3\right)^{-1-iK_3/3}$$
$$F\left(1 + i\,K_3/3, 4/3 + iK_3/3; 2 + 2iK_3/3; 1 - a^2 r_0^3/u^2 r^3\right), \qquad (137a)$$

is finite at the critical layer:

$$\bar{\bar{s}} \equiv 1 - (r/r_3)^3 : \ V_3(r;\omega) = 1 + O(\bar{\bar{s}}). \qquad (137b)$$

The other wave field:

$$V_4(r;\omega) = \left(u^2 r^3/a^2 r_0^3\right)^{iK_3/3}\left(1 - u^2 r^3/a^2 r_0^3\right)^{-1-2iK_3/3}$$
$$\times F\left(-iK_3/3, 1/3 - iK_3/3; -2iK_3/3; 1 - a^2 r_0^3/u^2 r^3\right), \qquad (138a)$$

is singular at the critical layer,

$$r \to r_3 : \ V_4(r;\omega) \sim \left[1 - (r/r_3)^3\right]^{-1-iK_3/3}, \qquad (138b)$$

i.e. the amplitude is singular as $\bar{\bar{s}}^{-1}$, and there is also an amplitude jump:

$$\left[1 - (r/r_3)^3\right]^{-1-2iK_3/3} \equiv \exp\left\{-(2iK_3/3)\log\left[1 - (r/r_3)^3\right]\right\}$$
$$= \exp\left\{-(2iK_3/3)\log\left|1 - (r/r_3)^3\right|\right\} \times \begin{cases} 1 & \text{if } r < r_3, \\ e^{-2\pi K_3/3} & \text{if } r > r_3. \end{cases} \qquad (139a,b)$$

The amplitude jump (139a,b) corresponds to a transmission factor and transition frequency:

$$S_3 = \exp(-2iK_3/3) = \exp(-\omega/\omega_3), \quad \omega_3 \equiv 3\omega/2\pi K_3 = 3u/2\pi r_0. \qquad (140a,b)$$

The transition frequency (140b) in the case of mean flow velocity proportional to the radius (128a), may be compared with (124b) for the case of uniform mean flow (107a); in the latter but not in the former case there is a cut-off frequency (116b).

The Alfvén wave equation for (§3.1) a uniform radial mean flow (104, 105, 109) and for (§3.2) a mean flow velocity proportional to the radius (127,

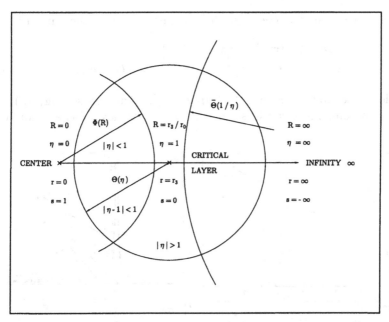

Fig. 11. Singularities of the wave equation at the center, infinity and the critical layer, and radius of convergence of the solutions around each singularity, showing how they cover the full range of radial distances with overlapping regions

129, 130) has three regular singularities. These are illustrated (Figure 11) in the latter case in terms of four variables, viz. the radial distance r, made dimensionless by dividing by the solar radius R in (85a), or η in (131a) or the dimensionless distance from the critical layer $\bar{\bar{s}}$ in (137b). The three singularities are regular : (i) the solution $\Phi(R)$ about the center $r = 0$, $R = 0$, $\xi = 0$ or $\bar{\bar{s}} = 1$, has radius of convergence $|\xi| \leq 1$ unity, limited by the critical layer; (ii) the solution $\Theta(\xi)$ about the critical layer $r = r_3$, $R = r_3/r_0$, $\xi = 1$ or $\bar{\bar{s}} = 0$, has radius of convergence unity $|\eta - 1| < 1$, limited by the center; (iii) the solution $\bar{\Theta}(1/\eta)$ about the point at infinity $r = \infty$, $R = \infty$, $\eta = \infty$, $\bar{\bar{s}} = -\infty$ converges outside the unit circle $|1/\eta| < 1$ or $|\eta| > 1$, as limited by the critical layer. The regions of overlap between the three pairs of solutions Φ, Θ, $\bar{\Theta}$ provide the analytic continuation over the entire range of radial distances $0 < r < \infty$, $0 < R < \infty$, $0 < \eta < \infty$ or $-\infty < \bar{\bar{s}} < 1$, of which only the part outside the solar radius is physically significant. The wave field component which is finite (137a) at the critical layer:

$$s \equiv r/r_3, \quad Q_+ \equiv V_3(r;\omega)$$
$$= s^{-1}F\left(3/2 + \delta, 1/2 + iK_2 + \delta; 2 + iK_2; (s-1)/(s+1)\right),$$
$$\text{(141a,b)}$$

and the wave field component which is singular (137a) at the critical layer:

$$Q_- \equiv V_4\left(r;\omega\right) = s(1-s^3)^{-1-i\phi_3}$$
$$\times F\left(-i\phi_3/2, 1/3 - i\phi_3; -i\phi_3; \left(1-s^3\right)\right), \qquad (142)$$

are plotted separating (Figure 12) amplitudes $|Q_\pm|$ and phases $\arg(Q_\pm)$. The wave field component which is finite at the critical layer has an amplitude

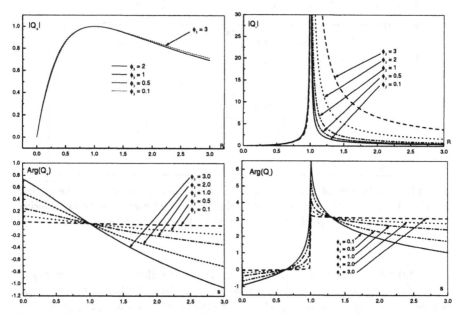

Fig. 12. Modulus (top) and phase (bottom) of velocity perturbation spectrum of Alfvén wave in a radial mean flow with velocity proportional to the radius; wave field which is finite (l.h.s.) or singular (r.h.s.) at critical layer, versus radial distance made dimensionless by dividing by the distance of the critical layer $R \equiv r/r_*$. The plots concern five values of the dimensionless frequency

$|Q_+|$ which (Figure 12, top left) rises rapidly to a maximum at the critical layer, and decays more slowly afterwards in a manner almost independent of frequency; the phase of the wave field component which is finite at the critical layer $\arg(Q_+)$ is a linear function of radial distance (Figure 12, bottom left), with slope proportional to the frequency, and unaffected by the critical layer. The phase of the wave field component which is singular at the critical layer $\arg(Q_-)$ has a phase jump of π at the critical layer (Figure 12, bottom right) and a larger variation on either side for higher frequency. The amplitude of the wave field component which is singular at the critical layer $|Q_-|$ is almost independent of frequency before the critical layer (Figure 12, top right) and is smaller for lower frequency afterwards, because of a faster decay after the critical layer. Thus the process of change of the wave spectrum with distance starts after the critical layer.

3.3   Effect of the transition layer on the properties of Alfvén waves

In the case of a uniform mean flow velocity (§3.1) and mean flow velocity proportional to the radius (§3.2), the Alfvén wave equation is solvable exactly in terms of Gaussian hypergeometric functions, because it has three regular singularities: the center $r = 0$, infinity $r = \infty$ and the critical layer $r = \bar{r}$. A change of variable of the type $r/\bar{r}$ places the three regular singularities at $0, 1, \infty$ as for the Gaussian hypergeometric equation. For other profiles of the mean flow velocity $U(r)$ as a function of the radius the reduction to a Gaussian hypergeometric equation is in general not possible, because besides the critical layer there is at least a fourth singularity, viz. a transition level. This will be shown by obtaining first the Alfvén wave equation for an arbitrary profile of the mean flow velocity $U(r)$, and then identifying the singularities. The linearized momentum (98a) and induction (98b) equations for the modified velocity (101a) and magnetic field (103b) perturbation spectra yield:

$$-i\omega \bar{V} + U\bar{V}' = \left(A^2/B\right) H', \tag{143a}$$

$$-i\omega \bar{H} + \left(U\bar{H}\right)' = \left(B\bar{V}\right)'. \tag{143b}$$

Elimination between (143a,b) leads to the Alfvén wave equation for the modified velocity perturbation spectrum:

$$\left(1 + iU'/\omega\right)\left(1 - U^2/A^2\right)\bar{V}''$$
$$+[2\left(B'/B\right)\left(1 + iU'/\omega\right) + 2i\omega\left(U/A^2\right)\left(1 + iU'/\omega\right)^2$$
$$-i\left(U''/\omega\right)\left(1 - U^2/A^2\right)]\bar{V}' + [\left(B''/B\right)\left(1 + iU'/\omega\right)$$
$$-\left(B'/B\right)\left(iU''/\omega\right) + U''U/A^2 + \left(\omega/A^2\right)\left(1 + iU'/\omega\right)]\bar{V} = 0, \tag{144}$$

The vanishing of the coefficient of $\bar{V}''$ shows that there at most two sets of singularities:

$$U(r_4) = A(r_4), \quad U'(r_5) = -i\omega, \tag{145a,b}$$

viz.: (i) a critical layer (145a) where the mean flow velocity equals the Alfvén speed; (ii) a transition level (145b) where the derivative of the mean flow velocity takes an imaginary value related to the wave frequency. An important distinction follows immediately from (145a,b): (i) the critical layer (145a) occurs for real r_4 and thus may correspond to a singularity of the wave field at real distance, besides limiting the radius of convergence of solutions around other points, e.g. the radius of convergence of the solution around

r_0 is $|r_4 - r_0|$; (ii) the transition level (145b) occurs for complex 'radius' $r_5 \neq \mathrm{Re}\,(r_5)$, and thus cannot cause a singular wave field at real distance, although it can still limit the radius of convergence of a solution about another point to $|r_5 - r_0|$.

The singularities of the Alfvén wave equation will be considered in more detail for a mean flow velocity varying like a power of radial distance:

$$U\,(r) = u\,(r/r_0)^{\nu}\,; \qquad (146)$$

the conservation of the mass flux (106) then implies the density stratification (80b) and Alfvén speed (81a). Substituting these together with the external magnetic field (80a) in the Alfvén wave equation (144), and using the notation (85a, 108), leads to the ordinary differential equation:

$$D\,(R)\,(1 - N^2 R^{\nu+2})\,\varPhi'' + \{2D\,(R)\,[-2 + i\Omega N R^3 D\,(R)]$$
$$-i\nu\,(\nu - 1)\,(N/\Omega)\,R^{\nu-1}\,(1 - N^2 R^{\nu+2})\}R\varPhi'$$
$$+\{D\,(R)\,[6 + \Omega^2 R^{4-\nu} D\,(R)] + 2i\,(N/\Omega)\,\nu\,(\nu - 1)\,R^{\nu-1}$$
$$+\nu\,(\nu - 1)\,N^2 R^{\nu+2}\}\varPhi = 0, \qquad (147)$$

where:

$$D\,(R) \equiv 1 + i\nu\,(N/\Omega)\,R^{\nu-1}, \qquad (148)$$

and two dimensionless parameters appear, namely the Alfvén number (61a) and dimensionless frequency (85b). The factor (148) specifies the transition level, and shows that it does not exist only in two cases, namely uniform mean flow velocity $\nu = 0$ (146), or mean flow velocity proportional to the radius $\nu = 1$, in which case the differential equation (147) simplifies respectively to (109) in §3.1 and to (130) in §3.2.

In all other cases $\nu \neq 0, 1$ it follows from (148) that the transition layer is located at the complex radius:

$$D\,(R_5) = 0 : \ \bar{r}_1 = r_0 R_1 = r_0\,(-i\Omega/N\nu)^{1/(\nu-1)}. \qquad (149)$$

The vanishing of other coefficient of $\varPhi''$ in (147) specifies the location of the critical layer at the distance r_4 which is the real root of:

$$r_4 = r_0 R_4 = r_0 N^{-1/(\nu+1)} = r_0\,(a/u)^{1/(1+\nu/2)}, \qquad (150)$$

e.g. (150) leads in the case (§3.1) of a uniform mean flow $\nu = 0$ to $r_4 = r_2$ in (117b), and for (§3.2) a mean flow velocity proportional to the radius $\nu = 1$ then $r_4 = r_3$ in (135). The only case in which a critical layer does not exist is $\nu = -2$, when the mean flow velocity (146) decays like the inverse square of distance (151a):

$$\nu = -2 : \ (r_0/r)^2 = U\,(r)\,/u = A\,(r)\,/a, \ \rho\,(r) = \rho_0, \qquad (151\mathrm{a,b,c})$$

as well as the Alfvén speed (81a) $\equiv$ (151b) implying also (80a) $\equiv$ (151c) a constant mass density. The Alfvén wave equation (147) was solved before in the cases (§3.1) of uniform mean flow $\nu = 0$, and (§3.2) mean velocity increasing with radial distance $\nu = 1$, in which there is a critical layer but no transition level; the next case to be considered (§3.3) is a mean flow velocity decreasing with distance $\nu = -2$, for which there is a transition level (149) at:

$$\nu = -2 : \ r_2 = r_0 \left(i\Omega/2N \right)^{-1/3}, \tag{152}$$

and the differential equation (147, 148) simplifies to:

$$\left(1 - N^2 \right) \left(1 + i\Omega R^3/2N \right) R^2 \Phi'' - \left(1 - N^2 - 4i\Omega R^3/N + \Omega^2 R^6 \right) R\Phi' + \left\{ i\Omega R^3 \left(N + 3/N \right) + 2\Omega^2 R^6 + i\Omega^3 R^9/2N \right\} \Phi = 0, \tag{153}$$

which is of the second-order for $N \neq 1$ and of first order for $N = 1$.

Note that in the case $\nu = -2$ the mean flow velocity and Alfvén speed (151a,b) are in a constant ratio:

$$U\left(r \right)/A\left(r \right) = u/a = N, \tag{154}$$

so that two cases arise: (i) if they are not equal at one point $N \neq 1$, they are not equal anywhere (hence no critical layer exists), and the differential equation (153) remains of second-order because Alfvén waves can propagate in opposite directions with phase speed $U\left(r \right) \pm A\left(r \right)$; (ii) if the mean flow velocity and Alfvén speed are equal at one point $N = 1$, then they are equal everywhere $U\left(r \right) = A\left(r \right)$, and the differential equation (153) reduces to first order:

$$N = 1 : \ \left(2 + i\Omega R^3 \right) R\Phi' + \left(4 - i2\Omega R^3 + \Omega^2 R^6/2 \right) \Phi = 0, \tag{155}$$

because Alfvén waves can propagate only in the direction of the flow with phase speed $U\left(r \right) + A\left(r \right) = 2A\left(r \right)$ (propagation opposite to the mean flow would give zero phase speed $U\left(r \right) - A\left(r \right) = 0$). In the case $N = 1$ of equal mean flow velocity and Alfvén speed everywhere, the solution of (155) leads to (108) the velocity perturbation spectrum:

$$V_5\left(r; \omega \right) = V\left(r_0; \omega \right) \left(r/r_0 \right)^{-3}$$
$$\times \exp\left[i\Omega \left(r/r_0 \right)^3/6 \right] \left\{ \left[2 + i\Omega \left(r/r_0 \right)^3 \right] / \left(2 + i\Omega \right) \right\}, \tag{156}$$

which scales for small radius like:

$$r \to 0 : \ V_5\left(r; \omega \right) \sim V\left(r_0; \omega \right) \left(r/r_0 \right)^{-3}, \tag{157}$$

and large radius like:

$$r \to \infty : \ V_5\left(r; \omega \right) \sim V\left(r_0; \omega \right) \exp\left[i\Omega \left(r/r_0 \right)^3/6 \right], \tag{158}$$

Since in the case $\nu = -2$ the mean flow velocity (151a) vanishes $U \to 0$ at infinity $r \to \infty$, one may compare with the solution (86a,b) for an outward propagating wave in a medium at rest:

$$V(r;\omega) = V(r_0;\omega)(r/r_0)^{3/2}$$
$$\times \left\{ H^{(1)}_{1/(4-\nu)}\left(2\Omega (r/r_0)^{2-\nu/2} / (2 - \nu/2) \right) / H^{(1)}_{1/(4-\nu)}\left(\Omega/(1 - \nu/4) \right) \right\} \tag{159}$$

for large radius:

$$r \to \infty: \quad V(r;\omega) \sim V(r_0;\omega)(r/r_0)^{1/2+\nu/4}$$
$$\times \exp\left[i\Omega (r/r_0)^{2-\nu/2} / (1 - \nu/4) \right]; \tag{160}$$

it is clear that (160) scales as (158) for $\nu = -2$, viz.:

$$\nu = -2: \quad V(r;\omega) \sim V(r_0;\omega) \exp\left[2i\Omega (r/r_0)^3 / 3 \right], \tag{161}$$

apart from the factor $2/3$ instead of $1/6$ in the exponential, which is due to the presence of mean flow in (158) and absence in (161) for smaller radii.

In the case $\nu = -2$ when the mean flow velocity and Alfvén speed are in a constant ratio but not equal $N \neq 1$, the second order differential equation has two singularities, at the origin $r = 0$ and infinity $r = \infty$. The transition layer (149) limits the radius of convergence of the solution around the centre:

$$r < |\bar{r}_1|: \quad V(r;\omega) = C_6 V_0(r;\omega) + C_6 V_{2/3}(r;\omega), \tag{162a}$$

which are a linear combination, with arbitrary constants, of:

$$\sigma = 0, 2/3: \quad V_\sigma(r;\omega) = (r_0/r) \sum_{n=0}^{\infty} a_n(\sigma) \left[i(\Omega/N)(r/r_0)^3 \right]^{n+\sigma}, \tag{162b}$$

with recurrence formula for the coefficients:

$$a_0(\sigma) = 1: \quad 6(1 - N^2)(n + \sigma)(3n + 3\sigma - 2)a_n(\sigma) +$$
$$[3(n + \sigma - 1)[3(n + \sigma - 2)(1 - N^2) - 2(1 - 3N^2) + 2(3 + N^2)]]a_{n-1}(\sigma)$$
$$= 2N^2(3n + 3\sigma - 8)a_{n-2}(\sigma) + N^2 a_{n-3}(\sigma). \tag{163}$$

For large radius the linear combination holds:

$$r \to \infty: \quad V(r;\omega) = C_+ V_+(r;\omega) + C_- V_-(r;\omega), \tag{164a}$$

with coefficients $C_\pm$, of the asymptotic expansions:

$$V_\pm(r;\omega) = (r_0/r) \exp\left\{ i[\Omega/3(N \mp 1)](r/r_0)^3 \right\}$$
$$\times \sum_{n=0}^{\infty} b_n^\pm(\sigma) \left[i(\Omega/N)(r/r_0)^3 \right]^{1/3-n}, \tag{164b}$$

where the coefficients satisfy the recurrence formula:

$$b_0^{\pm} = 1 : \pm 6nN b_n^{\pm}$$
$$= \left\{ (\pm 2N - 2 - 3n) \left[4 \left(2 - 2N^2 \pm N \right) + 3 \left(1 - N^2 \right) (n - 7/3) \right] \right\} b_{n-1}^{\pm}$$
$$- 6 \left(1 - N^2 \right) (n - 7/3) (3n - 5) b_{n-2}^{\pm}. \tag{165}$$

In the case $N = 1$ the leading term of $V_+ (r; \omega)$ in (164b) coincides with (158). For general $N \neq 1$, the leading exponential term of (164b) shows that: (i) the outward propagating wave V_+ has positive spatial phase for all $N > 0$, i.e. travels in the direction of increasing radius; (ii) the inward propagating wave V_- travels outward if the mean flow is superalfvénic $N > 1$, and travels inward if the mean flow is sub-Alfvénic.

In the case of a homogeneous medium (151c) when the mean flow velocity

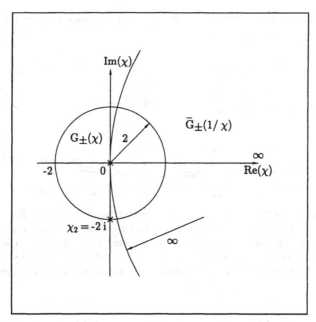

Fig. 13. Singularities of the Alfvén wave equation in a radial mean flow with velocity decaying like the inverse square of the radius. The transition layer on the imaginary axis limits the radius of convergence of the solution around the center, but does not limit the region of validity of the asymptotic expansion about infinity

(151a) decays like the Alfvén speed and hence their ratio is constant (154), the three singularities of the Alfvén wave equation (153) are (Figure 13), the origin $r = 0$, the point at infinity $r = \infty$ and besides, there is the transition layer at imaginary distance (152). The variable (166a) in the solutions (162b):

$$\chi \equiv i(\Omega/N)R^3 = i\omega r^3/ur_0^3, \; V_\sigma = R\sum_{n=0}^{\infty} a_n(\sigma)\chi^{n+\sigma}, \qquad (166\text{a,b})$$

places the transition layer (152) at $\chi_2 = -2i$, so that the radius of convergence of the solutions (166b) is two $|\chi| < 2$, in contrast to the radius of the solutions (164b) about the point at infinity which is infinite $|\chi| > 0$ or $|1/\chi| < \infty$, i.e. determined by the singularity at the center and unaffected by the transition layer. The two solutions (166b) are plotted (Figure 14) in the form:

$$G_+(|\chi|) \equiv R^{-1}V_0(|\chi|) = \sum_{n=0}^{\infty} a_n(0)\,|\chi|^n, \qquad (167\text{a})$$

$$G_-(|\chi|) \equiv R^{-1}V_{2/3}(|\chi|) = \sum_{n=0}^{\infty} a_n(2/3)\,|\chi|^{n+2/3}, \qquad (167\text{b})$$

so as to appear as real functions of $|\chi|$; substitution of $\chi = i\,|\chi|$ would

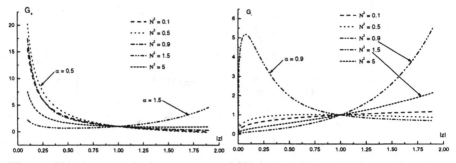

Fig. 14. Velocity perturbation spectrum of Alfvén wave in a radial flow with mean velocity decaying like the inverse square of the radius. Wave field component which is singular (top) and vanishes (bottom) at the center, normalized to value at reference radius $r = r_0$, versus radial distance made dimensionless by dividing by the reference radius $R \equiv r/r_0$. The plots concern five values of the dimensionless frequency

specify the real and imaginary parts of the wave fields. The first field (167a) is weakly dependent on the variable $|\chi|$ in (166a), except for an increase for small $|\chi|$ and larger Alfvén number. The second wave field (167b) again shows a weak dependence on $|\chi|$, except for Alfvén number closer to unity $N^2 = 0.9 - 1.5$ or $N = 0.95 - 1.22$. It follows that stronger interaction between Alfvén waves and the mean flow occurs for Alfvén number close to unity, suggesting that the transition level is a less important feature than the critical layer. Thus the latter is reconsidered in a final section (§3.4), using the mean flow velocity profile which approaches closely to that in the solar wind; the solution of the Alfvén wave equation can be obtained in this case with similar methods to §3.1-3.3, leading to series expansions that are more complicated than for the Gaussian hypergeometric case, i.e. the recurrence formulas for the coefficients are no longer simple, but rather multiple, viz. up to $n - 3$ in (163) and $n - 9$ in (190).

3.4 Wave transmission at the critical layer and modification of the wave fields

The last case to be considered is perhaps the most representative of the solar wind, with $\nu = 1/2$, the mean flow velocity (146) increasing like the square root of the radius:

$$U(r) = u\sqrt{r/r_0}, \; \rho(r)/\rho_0 = (r/r_0)^{-5/2}, \; A(r)/a = (r_0/r)^{5/4}, \quad (168a,b,c)$$

and hence the mass density (80b) ≡ (168b) and Alfvén speed (82a)≡(168c). This case, like all others except $\nu = 0, 1, -2$, combines the transition level (149)≡(169a) and critical layer (150)≡(169b):

$$r_3 = r_0 \left(2i\Omega/N\right)^{2/3}, \; r_6 = r_0 \left(a/u\right)^{4/5}. \quad (169a,b)$$

In order to specify the mean state in a self-consistent (106) manner: (i) the external magnetic field (80a) is specified by Maxwell's equation $0 = \nabla \cdot (B(r)\bar{e}_r) = r^{-2}\partial \left(r^2 B\right)/\partial r$; (ii) the mean flow velocity may be chosen at will, e.g. (146), but then it specifies the mass density through mass conservation (80b); (iii) the latter together with the external magnetic field specify through (8a) the Alfvén speed (81a); (iv) the gas pressure satisfies the momentum equation for the mean state, viz.:

$$dp = \rho U dU - \rho g dr, \quad (170)$$

bearing in mind that the external magnetic field is force free and the radial gravity field is given by (82b).

Substituting (80b, 146, 82b) in (170) specifies the gas pressure:

$$p(r) = p_\infty + [\rho_0 g_0 r_0/(\nu + 3)] (r/r_0)^{\nu+3} - [\rho_0 u^2 \nu/(\nu - 2)] (r/r_0)^{2-\nu},$$
$$(171)$$

where $p_\infty = p(\infty)$ is the gas pressure at infinity for $-3 < \nu < 2$. The case (§3.1) of uniform mean flow velocity:

$$\nu = 0: \; p(r) - p_\infty = (\rho_0 g_0 r_0/3)(r/r_0)^3 = (\rho_0 g_0 r_0/3)(\rho(r)/\rho_0)^{3/2}, \quad (172)$$

corresponds to a polytropic law with exponent 3/2. The case of mean flow velocity proportional to the radius leads to the gas pressure:

$$\nu = 1: \; p(r) - p_\infty = (\rho_0 g_0 r_0/4)(r/r_0)^4 + \rho_0 u^2 (r/r_0), \quad (173)$$

and the case when there is no critical layer to:

$$\nu = -2: \quad p(r) - p_\infty = \rho_0 g_0 r_0 \, (r_0/r) + \left(\rho_0 u^2/2\right) (r_0/r)^4 . \tag{174}$$

In the present case the gas pressure is given by:

$$\nu = 1/2: \quad p(r) - p_\infty = \left(2\rho_0 g_0 r_0/7\right) (r_0/r)^{7/2} + \left(\rho_0 u^2/3\right) (r_0/r)^{3/2} . \tag{175}$$

These representations concerning the background state are self-consistent but simple, so as to make possible the solution of the differential equation (147, 148), which is simplified by the changes of independent (176a) and dependent (176b) variable:

$$\vartheta \equiv \sqrt{R} = (r/r_0)^{1/2} , \quad G(\vartheta) \equiv \varPhi(R) = (r/r_0) \, V(r;\omega), \tag{176a,b}$$

leading to:

$$\begin{aligned}
(\vartheta + iN/2\varOmega) \left(1 - N^2\vartheta^5\right) \vartheta^2 G'' &+ [-4iN/\varOmega - 3\vartheta \left(3 + N^2\vartheta^5\right) \\
&- iN^3\vartheta^5/2\varOmega + 4i\varOmega N\vartheta^7]\vartheta G' + (4\varOmega^2\vartheta^8 + 4iN\varOmega\vartheta^7 \\
&- 2\,N^2\vartheta^6 + 24\vartheta + 10iN/\varOmega)G = 0
\end{aligned} \tag{177}$$

involving as dimensionless parameters the Alfvén number (61a) and dimensionless frequency (85b).

The vanishing of the coefficient of G'' shows that the differential equation (177) has eight regular singularities (Figure 15), viz.:

$$\vartheta_0 = 0, \quad \vartheta_3 = \infty, \quad \vartheta_2 = -iN/2\varOmega = -iu/ \left(2r_0\omega\right) = \sqrt{r_3/r_0}, \tag{178a,b,c}$$

the center (178a), the point-at-infinity (178b), the transition level (178c) $\equiv$ (169a), plus five other regular singularities

$$\vartheta_{1,4,5,6,7} = \sqrt[5]{1}N^{-2/5} = (a/u)^{2/5} \left\{1, e^{\pm i2\pi/5}, e^{\pm i4\pi/5}\right\}, \tag{179a-e}$$

of which only one is real, viz. $\vartheta_1 = (u/a)^{2/5} = \sqrt{r_6/r_0}$ the critical layer (169b) $\equiv$ (179a). The five singularities (179a-e) lie on the vertices of a regular pentagon of radius in a circle of radius $N^{-2/5}$; since the side of the pentagon is larger than the radius (which is equal to the side of a regular hexagon), the two singularities closer to the critical layer ϑ_4, ϑ_5 are farther than the origin $|\vartheta_4 - \vartheta_1| = |\vartheta_5 - \vartheta_1| > N^{-2/5} = \vartheta_1$. Also, the transition level is on the imaginary axis (178c), so the range of radial distances $0 < r < \infty$ is covered by three overlapping pairs of solutions, viz., below the critical layer $|\vartheta| < \vartheta_1$, above the critical layer $|\vartheta| > \vartheta_1$ and around the critical layer $|\vartheta - \vartheta_1| < \vartheta_1$.

Taking as example the solution around the critical layer suggests the change of variable:

$$s = 1 - \vartheta/\vartheta_1 = 1 - N^{-2/5}\vartheta = 1 - (a/u)^{2/5} (r/r_0)^{1/2} , \tag{180a}$$

$$(r/r_0) \, V(r;\omega) = \varPhi(r/r_0) = G\left(\sqrt{r/r_0}\right) = J\left(1 - (a/u)^{2/5} \sqrt{r/r_0}\right), \tag{180b}$$

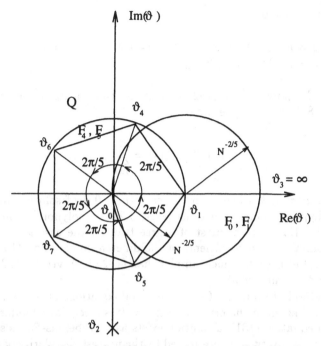

Fig. 15. Eight singularities of the Alfvén equation in a radial mean flow with velocity proportional to the square root of the radius, showing that the solutions around the center $\vartheta_0 = 0$, critical layer ϑ_1, and infinity $\vartheta_3 = \infty$, cover all radial distances, and their radius of convergence is not affected by the singularity at the transition layer ϑ_2

which leads to the differential equation:

$$\sum_{m=0}^{2} s^m j_m(s) \, d^m J/ds^m = 0, \tag{181}$$

where the critical layer $s = 0$ in (180a) is a regular singularity [66] , because the polynomial coefficients:

$$j_0(s) \equiv (1-s)^2 \left[1 - s + i/\left(2K_4\right)\right] \left(5 - 10s + 10s^2 - 5s^3 + s^4\right), \tag{182a}$$

$$j_1(s) \equiv (1-s) \left[4i/K_4 + 9(1-s) + (i/K_4)\left(1-s\right)^5 + 3(1-s)^6 - 4iK_4(1-s)^7\right], \tag{182b}$$

$$j_2(s) \equiv s \left[10i/K_4 + 24(1-s) - 2(1-s)^6 + 4iK_4(1-s)^7 + 4\left(K_4\right)^2\left(1-s\right)^8\right], \tag{182c}$$

which involve only one dimensionless parameter:

$$K_4 \equiv \Omega N^{-7/5} = \left(r_0\omega/a\right)\left(u/a\right)^{-7/5} = r_0\omega a^{2/5} u^{-7/5}, \tag{183}$$

are analytic at $s = 0$:

$$j_{0,0} \equiv j_0(0) = 0, \quad j_{1,0} \equiv j_1(0) = 12 + 5i/K_4 - 4iK_4, \qquad (184\text{a,b})$$
$$j_{2,0} \equiv j_2(0) = 5 + 5i/2\,K_4. \qquad (184\text{c})$$

Note that the polynomials (182a,b,c) in the coefficients of the differential equation (181) are of degree nine:

$$m = 0, 1, 2: \quad j_m(s) = \sum_{q=0}^{9} j_{m,q} s^q, \qquad (185)$$

in the case $\nu = 1/2$, whereas in the cases $\nu = 0, 1$ of solution in terms of Gaussian hypergeometric functions (§3.1, 3.2) the polynomials in the coefficients (111, 132) were at most of degree two; for the case $\nu = -2$ in §3.3 the polynomials in the coefficients of the differential equation (153) were of degree nine, but actually much simpler than (182a,b,c), viz. in (153) they are cubics of $R^3 \sim \chi$ in (166a).

The method of solution of the differential equations (111, 132, 153, 183) is the same, viz. since the critical layer $s = 0$ is a regular singularity of the differential equation (181), a solution exists as a Frobenius-Fuchs series [78], with radius of convergence determined by the nearest singularity of (178a,b,c; 179b,c,d,e), viz. the origin:

$$|s| < 1: \quad J(s) = \sum_{n=0}^{\infty} d_n(\sigma) s^{\sigma+n}, \qquad (186)$$

where the index σ and the recurrence formula for the coefficients $d_n(\sigma)$ are to be determined. Substitution of (185, 186) in (181) yields:

$$0 = \sum_{n=0}^{2} \sum_{q=0}^{9} j_{n,q} \sum_{n=0}^{\infty} d_n(\sigma) s^{n+\sigma+q} (n+\sigma)(n+\sigma-1)...(n+\sigma-m+1)$$

$$= \sum_{n=0}^{\infty} s^{n+\sigma} \sum_{n=0}^{\infty} d_{n-q}(\sigma) \sum_{m=0}^{2} j_{m,q}(n+\sigma-q)$$

$$\times (n+\sigma-1-q)...(n+\sigma-q-m+1); \quad (187)$$

equating to zero the coefficients of powers of s, leads to the recurrence formula for the coefficients:

$$0 = \sum_{q=0}^{9} d_{n-q}(\sigma) \sum_{m=0}^{2} j_{m,q}(n+\sigma-q)(n+\sigma-q-1)...(n+\sigma-q-m+1)$$

$$(188)$$

Setting $n = 0$, and noting that $0 = d_{-1}(\sigma) = d_{-2}(\sigma)$ yields the indicial equation:

$$n = 0: \quad 0 = d_0(\sigma)\left[j_{0,0} + (n+\sigma)\,j_{1,0} + (n+\sigma)(n+\sigma-1)\,j_{2,0}\right]. \tag{189}$$

Since $d_0(\sigma) \neq 0$, otherwise by (188) $d_n(\sigma) = 0$ for all n, resulting in a trivial solution $J(s) = 0$ in (186), the indices σ must be the roots of the binomial in square brackets in (189), viz.:

$$0 = \sigma\left\{5\sigma\left[1 + i/\left(2K_4\right)\right] + 13i/\left(2K_4\right) - 4iK_4\right\}, \tag{190}$$

where (184a,b,c) were used. The first root $\sigma = 0$ of (190) corresponds (186, 180b) to the wave field:

$$V_7(r;\omega) = (r_0/r)\sum_{n=0}^{\infty} d_n(0)\left(1 - \sqrt{r/r_1}\right)^n, \tag{191a}$$

which is finite at the critical layer

$$d_o(\sigma) = 1: \quad V_7(r_1;\omega) = r_0/r_1. \tag{191b}$$

The second root $\sigma = -1 - 4iK_4/5$ corresponds to the wave field:

$$V_8(r;\omega) = (r_0/r)\sum_{n=0}^{\infty} d_n\left(-1 - (4i/5)K_4\right)\left(1 - \sqrt{r/r_1}\right)^{n-1-4iK_4/5}, \tag{192a}$$

which is singular at the critical layer:

$$r \to r_1: \quad V_8(r;\omega) \sim \left(1 - \sqrt{r/r_1}\right)^{-1}\exp\left\{-(4iK_4/5)\log\left(1 - \sqrt{r/r_1}\right)\right\}, \tag{192b}$$

and has a phase jump:

$$\exp\left\{-(4iK_4/5)\log\left(1 - \sqrt{r/r_1}\right)\right\}$$
$$= \exp\left\{-(4iK_4/5)\log\left|1 - \sqrt{r/r_1}\right|\right\} \times \begin{cases} 1 & \text{if} \quad r < r_1, \\ e^{-4\pi K_4/5} & \text{if} \quad r > r_1, \end{cases} \tag{193a,b}$$

corresponding to a transmission factor and transition frequency:

$$S_4 = \exp\left\{-4iK_4/5\right\} = \exp\left(-\omega/\omega_5\right), \tag{194a}$$

$$\omega_4 \equiv s\omega/4\pi K_4 = (s/4\pi)\,r_0^{-1}a^{-2/5}u^{7/5}, \tag{194b}$$

which can be compared with (140a,b) for $\nu = 1$ in §3.2, and (122a,b) for $\nu = 0$ in §3.1, and (73a,b) in §2.3, which are other cases in which a critical layer exists.

The general wave field is a linear combination of (191a,192b) viz.:

$$V(r;\omega) = C_7 V_7(r;\omega) + C_8 V_8(r;\omega),\qquad(195)$$

where the arbitrary constants of integration correspond to $C_7 \equiv d_0(0)$ and $C_8 \equiv d_0(-1 - 4iK_4/5)$, that is why $d_0(\sigma)$ could be set to unity to make (191a, 192b) unique. In the literature it is usually assumed that the wave field is finite at the critical layer $C_8 = 0$, and the remaining constant of integration C_7 is determined from one boundary condition, e.g. the initial velocity perturbation spectrum at radius $r = r_0$, viz.:

$$C_7 = V(r_0;\omega) / V_7(r_0;\omega),\qquad(196a)$$

which would lead to resonances if frequencies $\bar{\omega}$ exist for which $V_7(r;\bar{\omega}) = 0$. The choice of $V(r;\omega) = C_7 V_7(r;\omega)$ may not satisfy a radiation condition at infinity, that waves have to propagate outward as $r \to \infty$. The imposition of a radiation condition at infinity $r \to \infty$, specifies a relation between C_7 and C_8 in (195), so that only one boundary condition of the type

$$V(r_0;\omega) = V_0(\omega),\qquad(197a)$$

is needed to determine the constants of integration. The constants of integration may be determined by two boundary conditions, e.g. (197a) and

$$V(r_1;\omega) = V_1(\omega),\qquad(197b)$$

in which case the wave field is specified for $r_0 < r < r_1$, e.g. between the sun $r = r_0$ and the earth $r = r_1$, and the radiation condition as $r \to \infty$ is no longer relevant. In the case (197a,b) the wave field would be generally singular at the critical layer; this is not surprising since the critical layer is a resonance of a linear, undamped system. The inclusion of either damping or non-linear effects would limit the wave amplitude at the critical layer.

The wave field components which are finite (191a) at the critical layer are:

$$s = 1 - \sqrt{r/r_1},\ J_0(s) \equiv V_7(r;\omega) = (1 - s)^{-2} \sum_{n=0}^{\infty} d_n(0)s^n,\qquad(198a,b)$$

and singular (192a) at the critical layer

$$J_2(s) \equiv sV_8(r;\omega) = s^{-4iK_4/5}(1 - s)^{-2} \sum_{n=0}^{\infty} d_n(-1 - 4iK_4/4)s^n,\qquad(199)$$

are plotted in Figure 16, separating amplitudes and phases for several values of the parameter $\alpha \equiv \Omega N^{-7/5}$. The wave field component which is finite at the critical layer (198b) has an amplitude $|J_0|$ which decreases smoothly through the critical layer (Figure 16, top left) and is almost independent of

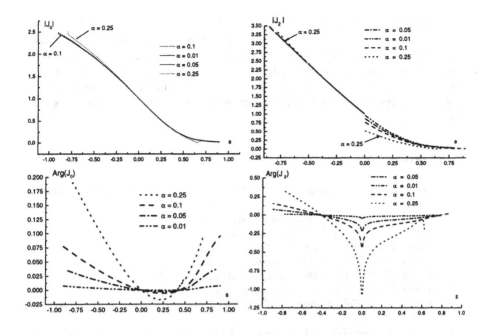

Fig. 16. Velocity perturbation spectrum of Alfvén wave in a radial mean flow with velocity proportional to the square root of the radius. Modulus (top) and phase (bottom) of the wave field component which is finite (l.h.s.) or singular (r.h.s.) at critical layer, the latter amplitude singularity removed. Both are plotted versus radial distance from critical layer made dimensionless by dividing by radius at critical layer. The plots concern five values of the dimensionless parameter $\alpha = \Omega N^{-7/5}$ combining the dimensionless frequency (84b) and Alfvén number (61a)

frequency and Alfvén number. The phase of the wave field component which is finite at the critical layer $\arg(J_0)$ varies more rapidly for larger frequency and smaller Alfvén number (Figure 16, bottom left) and has a minimum near the critical layer, implying that waves propagate inward before the critical layer and outward after the critical layer. The phase of the wave field component (199) which is singular at the critical layer $\arg(J_2)$ also corresponds (Figure 16, bottom right) to inward propagating waves below the critical layer and outward propagating waves above the critical layer, although the phase is now singular at the critical layer. Concerning the amplitude of the wave field component which is singular at the critical layer (192a), insertion of the factor s in (199) for J_2, eliminates the singularity s^{-1} in (192b) for the amplitude $|J_2|$ in (199), but still leaves the phase jump (193a,b), viz. (Figure 16, top right) the amplitude is almost independent of the dimensionless frequency and Alfvén number before the critical layer, and has a larger jump-type decay across it for larger frequencies. This shows that: (i) the process of change of the spectrum of Alfvénic perturbations with distance starts at the critical layer, as the mean flow velocity exceeds the Alfvén speed, and convects reflected waves outward; (ii) as the mean flow velocity becomes more superalfvénic, the amplitude is reduced more for the higher frequencies, leading to a steepening of the wave energy spectrum, as observed in the solar wind.

References

1. Abramowitz, M. & Stegun, I. (1965) Handbook of Mathematical Functions. Dover
2. Alkhaby, H. (1993) Geophys. Astrophys. Fluid Dyn. **72**, 197
3. Adam, J. A. (1977) Solar Phys. **52**, 293
4. Adam, J. A. (1982) Phys. Rep. **5**, 217
5. Alfvén, H. (1942) Ark. Mat. Astron. Fys. **B29**, 1
6. Alfvén, H. (1947) Month. Not. Roy. Astron. Soc. **107**,211
7. Alfvén, H. (1948) Cosmical Elecrodynamics. Oxford U. P.
8. Alfvén, H. & Falthammar, C. G. (1962) Cosmical Elecrodynamics. Oxford U. P.
9. An, C. H., Musielak, Z. E., Moore, R. L. & Suess, S. T. (1989) Astrophys. J. **345**, 597
10. An, C. H., Suess, S. T., Moore, R. L. & Musielak, Z. E. (1990) Astrophys. J. **350**, 309
11. Athay, R. G. (1976) The Quiet Sun, D. Reidel
12. Axford, W. I. (1985) Solar Phys. **100**, 575
13. Banos, A. (1955) Proc. Roy. Soc. Lond. **A233**, 350
14. Bavassano, B., Dobrowolny, M., Fanfoni, G., Mariani, F., & Ness, N. F. (1982) Solar Phys. **78**, 373
15. Bavassano, B., Dobrowolny, M., Mariani, F., & Ness, N. F., (1982) J. Geophys. Res. **87**, 3617
16. Bavassano, B. & Smith, E.J. (1986) J.Geophys. Res. **91**, 1706
17. Barkhudarov, M. R. (1991) Solar Phys. **135**, 131
18. Belcher, J. W. (1971) Astrophys. J. **168**, 509
19. Belcher, J. W. & Davies, L. (1971) J. Geophys. Res. **76**, 35
20. Biermann, L. (1948) Z. Naturwiss. **33**, 118
21. Belcher, J. W., Davies, L., & Smith, E.J. (1979) J. Geophys. Res. **74**, 2302
22. Bray, R. G. & Loughhead, R. E. (1974) The Solar Chromosphere. Chapman & Hall
23. Brillouin, L. (1926) C. R. Acad. Sci. **183**, 24
24. Bruno, R. & Dobrowolny, M. (1986) Ann. Geophys. **4**, 17
25. Bruzek, A. & Durrant, C. J. (1977) Glossary of Solar Physics. D. Reidel
26. Burlaga, L. F. & Turner, J. M. (1976) J. Geophys. Res. **81**, 73
27. Burlaga, L. F. (1984) Space Sci. Rev. **39**, 255
28. Cabannes, H. (1970) Theoretical Magnetofluid-dynamics. Academic P.
29. Cally, P. S. (1985) Solar Phys. **103**, 207
30. Campos, L. M. B. C. (1977) J. Fluid Mech. **81**, 529
31. Campos, L. M. B. C. (1983a) J. Phys. **A16**, 217

32. Campos, L. M. B. C. (1983b) Solar Phys. **82**, 355
33. Campos, L. M. B. C. (1983c) Wave Motion **5**, 1
34. Campos, L. M. B. C. (1983d) J. Mec. Theor. Appl. **2**, 861
35. Campos, L. M. B. C. (1984) Month. Not. Roy. Astron. Soc. **205**, 547
36. Campos, L. M. B. C. (1985) Geophys. Astrophys. Fluid Dyn. **32**, 217
37. Campos, L. M. B. C. (1987) Rev. Mod. Phys. **59**, 363
38. Campos, L. M. B. C. (1988a) Geophys. Astrophys. Fluid Dyn. **40**, 93
39. Campos, L. M. B. C. (1988b) J. Phys. **A21**, 2911
40. Campos, L. M. B. C. (1989) Month. Not. Roy. Astron. Soc. **241**, 215
41. Campos, L. M. B. C. (1990), Geophys. Astrophys. Fluid Dyn. **48**, 143
42. Campos, L. M. B. C. (1991) Wave Motion **15**, 103
43. Campos, L. M. B. C. (1992) Phys. Fluids **B4**, 2975
44. Campos, L. M. B. C. (1993a) Wave Motion **17**, 101
45. Campos, L. M. B. C. (1993b) Europ. J. Mechs. **B12**, 187
46. Campos, L. M. B. C. (1994) Eur. J. Mechs. **B13**, 613
47. Campos, L. M. B. C. (1998) Theor. Comp. Fluid Dyn. **10**, 37
48. Campos, L. M. B. C. (1999) Phys. Plasmas **6**, 57
49. Campos, L. M. B. C. & Gil, P. J. S. (1995) J. Fluid Mech. **301**, 153
50. Campos, L. M. B. C. & Gil, P. J. S. (1999) In preparation
51. Campos, L. M. B. C. & Isaeva, N. L. (1993) J. Plasma Phys. **48**, 415
52. Campos, L. M. B. C., Isaeva, N. L & Gil, P. J. S. (1999) Phys. Plasmas (to appear)
53. Campos, L. M. B. C. & Mendes, P. M. V. M. (1995) Month. Not. Roy. Astron. Soc. **276**, 1041
54. Campos, L. M. B. C. & Saldanha, R. S.(1991) Geophys. Astrophys. Fluid Dyn. **56**, 237
55. Campos, L. M. B. C., Saldanha, R. S. & Isaeva, N. L. (1999) in preparation
56. Coleman, P. J. (1968) Astrophys. J. **153**, 171
57. Cowling, T. G. (1960) Magnetohydrodynamics. Interscience
58. Dobrowolny, M., Mangeney, A. & Veltri, P. (1980) Astron. Astrophys. **83**, 26
59. Denskat, K. U. & Burlaga, L. F. (1977) J.Geophy. Res. **82**, 2693
60. Denskat, K.U. & Neubauer, F. M. (1982) J. Geophys. Res. **87**, 2215
61. Elsasser, M. (1956) Rev. Mod. Phys. **18**, 535
62. Erdelyi, A. (ed.) (1953) Higher transcendental functions. 3 vols. McGraw-Hill
63. Ferraro, C. & Plumpton, V. C. A. (1958) Astrophys. J. **127**, 459
64. Ferraro, C. & Plumpton, V. C. A. (1963) Magneto-fluid Dynamics. Oxford U. P.
65. Fichtner, H. & Fahr, H. J. (1991) Astron. Astrophys. **241**, 187
66. Forsyth, A. R. (1927) Treatise of differential equations. MacMillan
67. Foukal, P. V. (1990) Solar Astrophysics. Wiley
68. Goldstein, M. L., Roberts, D. A. & Matthaeus, W. H. (1995) Ann. Rev. Astron. Astrophys. **33**, 283
69. Heinemann, M. & Olbert, S. (1980) J. Geophys. Res. **85**, 1311
70. Herlofson, N. (1950) Nature **165**, 1020
71. Heyvaerts, J. & Priest, E. R. (1983) Astron. Astrophys. **117**, 120
72. Hollweg, J. V. (1972) Cosmic Electrodyn. **2**, 423
73. Hollweg, J. V. (1978) Solar Phys. **56**, 307
74. Hollweg, J. V. (1981) Solar Phys. **70**, 25
75. Hollweg, J. V. (1986) J. Geophys. Res. **96**, 411

76. Hollweg, J. V., (1990) J. Geophys. Res. **95**, 14873
77. Hu, Y. Q., Esser, R. & Habbal, S. R. (1997) J. Geophys. Res. **102**, 14461
78. Ince, E. L. (1954) Ordinary differential equations. Dover
79. Jeffreys, H. (1924) Proc. Lond. Math. Soc. **23**, 428
80. Kolgomorov, A. N. (1933) Foundations of the Theory of Probability. Springer
81. Kraichnan, R. H. (1965) Phys. Fluids **8**, 1385
82. Kramers, H. A., (1926), Zeits. Phys. **39**, 828
83. Krogulec, M., Musielack, Z. E., Suess, S. T., Nerney, S. F. & Moore, R. L. (1994) J.Geophys. Res. **99**, 23489
84. Kulsrud, R. M. (1955) Astrophys. J. **121**, 461
85. Landau, L. D. & Lifschitz, E. M. (1956) Electrodynamics of continuous media. Pergamon
86. Leroy, B. (1980) Astron. Astrophys. **81**, 136
87. Leroy, B. (1983) Astron. Astrophys. **125**, 371
88. Lightill, M. J. (1959) Phil. Trans. Roy. Soc. **A252**, 397
89. Lighthill, M. J. (1967) I. A. U. Symp. **28**, 429
90. Lighthill, M. J. (1978) Waves in Fluids. Cambridge U.P
91. Lou, Y.-Q. (1993a) J. Geophys. Res. **98**, 3563
92. Lou, Y.-Q. (1993b) J. Geophys. Res. **99**, 11483
93. Lou, Y.-Q. (1994) J. Geophys. Res. **99**, 4193
94. Lou, Y.-Q. & Rosner, R. (1994) Astrophys. J. **424**, 429
95. Lyons, P. & Yanowitch, M. (1974) J. Fluid Mech. **66**, 273
96. Marsch, E. (1991) Physics of Inner Heliosphere (ed. Schwenn, R. & Marsch, E.), Springer
97. Marsch, E., Muhlhauser, K. H., Schwenn, R., Rosenbauer, H., Pilipp, W. & Neubauer, F. M. (1982) J. Geophys. Res. **87**, 52
98. Marsch, E. & Mangeney, A. (1987) J. Geophys. Res. **92**, 7363
99. Massaglia, S. (1993) Astron. Astrophys. **267**, 595
100. McKenzie, J. F. (1973) J. Fluid Mech. **58**, 109
101. McKenzie, J. F. (1979) J. Plasma Phys. **22**, 361
102. McKenzie, J. F. (1994) J. Geophys. Res. **99**, 4193
103. McKenzie, J. F., Axford, W. I. & Banaszkiewicz, M. (1997) Geophys. Res. Let. **24**, 2877
104. McKenzie, J. F., Ip, W. H. & Axford, W. I. (1979) Astrophys. Space Sci. **64**, 183
105. McLellan, A. & Winterberg, F. (1968) Solar Phys. **4**, 401
106. Miles, J. F. (1961) J. Fluid Mech. **10**, 496
107. Moffat, H. K. (1978) Magnetic Field Generation in Electrically Conducting Fluids. Cambridge U. P.
108. Moore, D. W. & Spiegel, E. A. (1964) Astrophys. J. **139**, 48
109. Musielak, Z. E., Fontenla, J. M. & Moore, R. L. (1992) Phys. Fluids **B4**, 13
110. Musielak, Z. E. & Moore, R. L. (1995) Astrophys. J. **372**, C91
111. Nakariakov, Roberts, B. & Murawski, (1997) Solar Phys. **175**, 93
112. Nakariakov, Roberts, B. Murawski, (1998) Astron. Astrophys. **311**, 311
113. Nocera, L., Hollweg, J. V. & Priest, E. R. (1986) Geophys. Astrophys. Fluid Dyn. **35**, 111
114. Nocera , L., Leroy, O. & Priest, E. R. (1984) Astron. Astrophys. **133**, 387
115. Nye, A. H. & Thomas, J. H. (1976) Astrophys. J. **204**, 573
116. Oliver, R., Ballester, J. L. Hood, A. W. & Priest, E. R. (1993) Astron. Astrophys. **273**, 647

117. Orlando, S., Lou, Y.-Q., Rosner, R. & Peres, G. (1996) J. Geophys. Res. **101**, 24443
118. Osterbrock, D. E. (1961) Astrophys. J. **134**, 347
119. Parker, E. N. (1959) Interplanetary dynamical fields. Oxford U.P.
120. Parker, E. N. (1984) Geophys. Astrophys. Fluid Dyn. **29**, 1
121. Parker, E. N. (1989) Cosmical magnetic fields. Oxford U.P.
122. Richter, A. K., Hsieh, K. C., Rosenbauer, H. & Neubauer, F. M. (1986) Ann. Geophys. **4**, 3
123. Roberts, B. (1991) Geophys. Astrophys. Fluid Dyn. **62**, 83
124. Roberts, B. & Webb, A. R. (1978) Solar Phys. **56**, 5
125. Rosner, R, An, C. H. Musielak, Z. E., Moore, R. L. & Suess, S. T. (1991) Astrophys. J. **372**, L91
126. Scheuer, M. A. & Thomas, J. H. (1982) Solar Phys. **71**, 21
127. Schwartz, S. J., Bel, N. & Cally, P. S. (1984) Solar Phys. **92**, 133
128. Schwartz, S. J. & Leroy, B. (1982) Astron. Astrophys. **112**, 93
129. Schwarzschild (1948) Astrophys. J. **107**, 1
130. Similon, P. L. & Zargham, S. (1991) Astrophys. J. **388**, 644
131. Spruit, H. C. (1982) Solar Phys. **75**, 3
132. Stein, R. F. (1981) Astrophys. J. **246**, 966
133. Steinholfson, R. S. (1985) Astrophys. J. **295**, 213
134. Stix, M. (1989) The Sun. Springer
135. Summers, D. (1976) Quart. J. Mech. Appl. Math. **29**, 117
136. Thomas, J. H. (1983) Ann. Rev. Fluid Mech. **15**, 321
137. Thomas, J. H., Cram, L. E. & Nye, A. H. (1982) Nature **297**, 485
138. Tu, C. Y. (1988) J. Geophys. Res. **93**, 7
139. Tu, C. Y. & Marsch, E. (1990) J. Geophys. Res. **95**, 4337
140. Tu, C. Y. & Marsch, E. (1993) J. Geophys. Res. **98**, 1257
141. Tu, C. Y. & Marsch, E. (1997) Solar Phys. **171**, 363
142. Tu, C. Y., Marsch, E. & Rosenbauer, H. (1990) Geophys. Res. Lett. **17**, 283
143. Velli, M. (1993) Astron. Astrophys. **270**, 304
144. Velli, M., Grappin, R. & Mangeney, A. (1989) Phys. Rev. Lett. **63**, 1807
145. Verma, M. K. & Roberts, D. A. (1993) J.Geophys. Res. **98**, 5626
146. Wentzel, G. (1926) Zeits. Phys. **38**, 518
147. Whang, Y. C. (1973) J. Geophys. Res. **78**, 7221
148. Yanowitch, M. (1967) Can. J. Phys. **45**, 2003
149. Yu, C. P. (1965) Phys. Fluids **8**, 650
150. Zhugzhda, Y. D. (1971) Cosmic Electrodyn. **2**, 267
151. Zhugzhda, Y. D. & Dzhalilov, N. S. (1984a) Astron. Astrophys. **132**, 45
152. Zhugzhda, Y. D. & Dzhalilov, N. S. (1984b) Astron. Astrophys. **132**, 52
153. Zhugzhda, Y. D. & Dzhalilov, N. S. (1984c) Astron. Astrophys. **133**, 333
154. Zhugzhda, Y. D., Staude, J. & Locans, V. (1984) Solar Phys. **91**, 219
155. Zirin, H. (1988) Astrophysics of the sun. Cambridge U.P.

Relativistic Alfvén Solitons
and Acceleration of Cosmic Rays

G. Pelletier

Laboratoire d'Astrophysique de l'Observatoire de Grenoble and Institut Universitaire de France

Abstract. The environment of compact objects of astrophysics contains a relativistic plasma that is often confined by a magnetic field. Magnetic disturbances of those plasmas have specific properties that are presented in this paper. Nonlinear dynamics are different in a pair plasma and in proton dominated plasma. The nonlinear disturbances of Alfvén type are interesting to accelerate relativistic particles and the extension of Fermi acceleration to a relativistic plasma as an accelerating medium turns out to be far more efficient than in a non-relativistic plasma. If it happens in Nature, as suggested by some observations, a head on collision of relativistic Alfvén solitons is a very efficient event to accelerate cosmic rays.

1 Introduction

The environment of compact objects of astrophysics, such as galactic black holes, extragalactic black holes in Active Galactic Nuclei, neutron stars etc., contains a relativistic plasma revealed by the high energy radiation. They are often in relativistic expansion and magnetically confined, so that their magnetic disturbances manifest as waves and fronts propagating at relativistic velocity. Mostly incompressible because less damped than magnetosonic waves, they can be considered as a relativistic generalization of Alfvén waves; they are indeed fully electromagnetic because the displacement current is unavoidable.

The nonlinear behaviour of these relativistic fronts is an interesting and useful topic to understand the energetics of those environments, especially to understand how the magnetic field, concentrated by gravitation, allows to convert a fraction of the gravitational power into very high energy radiating particles. This investigation was opened recently (Pelletier and Marcowith 1998, hereafter PM) and the nonlinear dynamics turns out to be different for an electron-positron pair plasma and a proton dominated plasma. Particle acceleration in such relativistic plasmas is promising, for it is much more efficient than in a non-relativistic plasma as usually considered in the theory of Fermi processes.

The kinetic theory associated with these waves and solitary waves will be briefly presented. It is two-fold; on one hand high energy cosmic rays are accelerated by being resonantly scattered in the waveframe and it corresponds to a kind of synchrotron-Landau damping; on other hand, when the pressure of the relativistic plasma is close to the magnetic pressure, because of a degeneracy of the waves, intense compression effects are generated by the magnetic pressure of the Alfvén waves, then there is an important generation of a parallel electric field also contributing to particle energization.

2   The nonlinear relativistic wavefront

The general Alfvén velocity is defined by (see PM):

$$V_* = \frac{C}{\sqrt{1 + \frac{e+P}{2P_m}}} \, , \tag{1}$$

where e is the internal energy-mass density, P the plasma pressure and P_m the magnetic pressure. In non-relativistic plasmas, $e \simeq \rho C^2$ and is usually much larger than P_m and one has $V_* \simeq V_A = B_0/\sqrt{\mu_0 \rho}$. In an ultrarelativistic plasma, $e \simeq 3P$ and

$$V_* = \frac{C}{\sqrt{1 + \frac{2P}{P_m}}} \, , \tag{2}$$

so that the propagation velocity is relativistic for a confined plasma that has $P < P_m$, since V_* is larger than the relativistic sound velocity $C/\sqrt{3}$. A degeneracy occurs at pressure equipartition since the Alfvén and fast magnetosonic waves propagate at the same speed than the sound waves (slow magnetosonic waves) along the magnetic field.

Linear waves are of course well known, but the localized nonlinear relativistic waves, in particular relativistic solitons, are not well known. Such relativistic fronts has been recently studied (PM). It was argued that their dissipation by resonant interactions would lead to an efficient acceleration process, but the kinetics of suprathermal particles in these fronts was not addressed. In this section, the main properties are recalled and their derivation are indicated without entering in the technique of the so-called "reductive perturbation expansion"; some results are even indicated as non-perturbative results to stress their robustness.

Another important aspect of the kinetic corrections of these solitons is their modification as relativistic collisionless parallel shocks. The concept of collisionless parallel shock is not obvious, but unavoidable in astrophysics, especially in relativistic jets and gamma ray bursts. The following approach allows to make this concept precise.

2.1   Inertial effect in a cosmic ray plasma

Localized nonlinear wavefronts often result from a balance between nonlinear steepening (wave braking) and dispersion, which built solitons or solitary waves. In standard MHD, the waves are not dispersive and a dispersion effect is obtained at small scale by taking account of the inertial correction (often called "Hall effect"). The relative size of the correction for a mode of wavelength λ is of order of r_0/λ, where $r_0 \equiv V_A/\omega_{ci}$. Cosmic rays that have Larmor radii much larger than r_0 couple to the MHD of the thermal medium through resonant interactions with MHD waves. However, in a plasma containing proton cosmic rays, for large scale dynamics, more cosmic rays participate to the MHD and the radius r_0 is replaced by an energy dependent radius r_* that can be much larger than r_0 (PM). It could be thaught that r_* is simply r_0 multiplied by the averaged Lorentz factor of the cosmic ray population, $\bar{\gamma} \equiv < \epsilon > /mc^2$. This is the result that is found with a multi-fluid description. But the evaluation from kinetic theory (with Vlasov equation) gives a different estimate: $r_* = (< \gamma^2 > /\bar{\gamma})r_0$.

So for a given magnetic field, a relativistic plasma, because of its content of particles with high relativistic mass, has a much larger inertial effect. The inertial effect is described with the generalized Ohm's law which, in relativistic MHD, reads:

$$F_{\mu\nu}u^\nu = \eta J_\mu + \frac{1}{nq_p}F_{\mu\nu}J^\nu \ . \tag{3}$$

The general dispersion relation of Alfvén waves, including relativistic populations and inertial effects has been derived in (PM):

$$\omega^2(1 + \sum_a \frac{e_a + P_a}{2P_m}(1 \pm \chi_a(\omega))) - k_\parallel^2 C^2 = 0 \ ; \tag{4}$$

where e_a is the energy-mass density of the population labeled by "a" (thermal or relativistic electrons, thermal or relativistic protons) and P_a the pressure; the signe $\pm$ relates to right or left polarisation. For a non-relativistic population, the inertial correction reads:

$$\chi_a(\omega) = sgn(q_a)\frac{\omega}{\omega_{ca}} \ , \tag{5}$$

where ω_{ca} is the cyclotron pulsation; of course the dominant contribution comes from the proton population (the heaviest). For relativistic populations, the fluid theory leads to $\chi_a(\omega) = sgn(q_a)\bar{\gamma}\omega/\omega_{ca}$ whereas the kinetic theory with Vlasov equation leads to the correct result:

$$\chi_a(\omega) = sgn(q_a)\frac{1 + 5\beta_*^2}{4\beta_*^2}\frac{< \gamma^2 >}{\bar{\gamma}}\frac{\omega}{\omega_{ca}} \ . \tag{6}$$

Remark that the inertial effect vanishes at this order for a pure electron-positron plasma and the next order corrections need to be taken into account (as done in PM). Relativistic plasma dominated by the electron component (i.e. more massive than the protons) can also been envisaged in some situations. Note also that the cosmic rays can dominate the inertial effect even in a non-relativistic plasma like in the interstellar medium or in stellar winds, because even if the cosmic rays have a negligible contribution to the mass ($n_*\bar{\gamma} \ll n_{th}$, n_* being the numerical density of relativistic protons, and n_{th} the numerical density of thermal protons), they can bring the major contribution to the inertial effect if $n_* < \gamma^2 > /\bar{\gamma} \gg n_{th}$. At scales between r_0 and r_*, cosmic rays are not coupled to MHD, they participate at scales beyond r_* and a modification of the propagation through the dispersive effect stems from passing to this new dynamical regime. But in this paper, only a plasma dominated by the relativistic protons is considered.

The fact that the cosmic rays produce an inertial scale $r_* \sim (< \gamma^2 > /\bar{\gamma})r_0$ much larger than the usual one (r_0) is of great importance, because it is capable to maintain localized large perturbations at observable scales. Also shocks giving entropy to the cosmic ray population cannot have a front width smaller than this scale r_*, and they can also display oscillations due to the inertial effect. Thus a typical time scale (lower bound) for this MHD is $\tau_* \equiv r_*/V_* = T_g(< \gamma^2 > /\bar{\gamma})$.

One exemple of localized nonlinear perturbations is the well-known family of MHD solitons (Mjölhus 1976, Kaup and Newell 1978, Roberts1985). In those solitons, the nonlinear steepening of the wavefront is balanced by the inertial dispersion effect. Strictly speaking, it is difficult to assert that there are such MHD solitons in Nature, because it is an ideal concept based on fully integrable conservative PDEs. So when excitation and dissipation are taken into account, there are no more solitons but solitary waves, that do not have the same stability properties than ideal solitons. Anyway it is interesting to start with a soliton solution of a problem and then to introduce corrections in order to describe a more reallistic localized nonlinear wave, even possibly some kind of shock. In the purpose of cosmic rays acceleration, the emphazis is more on nonlinear relativistic fronts rather than on ideal solitons, especially when they collide. Ideal solitons are not affected by collisions, but of course real solitary waves are affected by collisions, mostly because of the dissipative effects, but often not completely destroyed when the dissipation remains smooth.

2.2 Relativistic MHD

Relativistic fronts are studied with the relativistic MHD; by "relativistic MHD", I understand a fluid description of fully electromagnetic disturbances such that the electromagnetic interaction with matter is mostly magnetic in the rest-frame of each disturbance (delocalized or localized, forward or backward wave). I consider only fronts that have a thickness smaller than their transverse size; thus I will analyse 1D relativistic dynamics only. The 1D relativistic MHD-system is written for the 4-specific momentum (u^0, u^1, u^2, u^3) (the velocity times the Lorentz factor of the flow, also called "unitary 4-velocity"), coupled with the transverse electromagnetic wave described by its reduced magnetic component $\mathbf{b}$: (b_1, b_2); the transverse flow is described by

$\mathbf{u}_\perp : (u^1, u^2)$ and the longitudinal flow due to compression is $u \equiv u^3$, and $u^0 = (1 + u^2 + \mathbf{u}_\perp^2)^{1/2}$.

In the front-frame, $u = u_* + \tilde{u}$ with $u_* = -\gamma_* \beta_*$ for a forward front, the compression is supposed either quasi static (off equipartition) or a wave (close to equipartition) calculated at second order in $\tilde{u}$, the system reads

- Parallel motion:

$$\frac{8}{3} u^0 P \partial_t u + (1 - 2u^2) \frac{\partial P}{\partial u} \partial_x u + P_m \partial_x |b|^2 = 0 . \tag{7}$$

- The transverse motion:

$$2u^0 P \partial_t \mathbf{u}_\perp + 2\partial_x (u P \mathbf{u}_\perp) = P_m \partial_x \mathbf{b} . \tag{8}$$

- Generalized transverse Ohm's law:

$$\mathbf{u}_\perp = u\mathbf{b} - \alpha \mathbf{e}_x \times \partial_x \mathbf{b} - \nu_m \partial_x \mathbf{b} . \tag{9}$$

The system is closed by inserting a barotropic law $P(\tilde{u})$. In this frame, because the magnetic pressure increases the velocity when $P < P_m$, and decreases it when $P > P_m$ (see later on), solitary waves exist only for $|b|^2 < 1$. It is therefore suitable to calculate the solitary wave in a perturbative theory, even if they can still exist for an amplitude $b_0 > 1$ but close to unity.

A localized wavefront undergoes an exponential decay, and some results can be derived from the asymptotic conditions at the linear approximation. The asymptotic solution is of the form, that involves four parameters a priori:

$$\mathbf{b} = b_0 e^{-a|y|} [\mathbf{e}_1 cos(\kappa y - \Omega t) \pm \mathbf{e}_2 sin(\kappa y - \Omega t)] . \tag{10}$$

with $y \equiv x - vt$. Two relations can be found between these parameters:

$$\gamma_* v = -2\alpha_0 \kappa , \tag{11}$$

that relates κ with the nonlinear modification of the soliton velocity; and the nonlinear oscillation is related to the width (a) and the nonlinear velocity (κ) such that

$$\gamma_* \Omega = \alpha_0 (a^2 + \kappa^2) , \tag{12}$$

where $\alpha_0 \equiv \alpha / \gamma_* \beta_*$.

However the relation between the width and the velocity with the amplitude is derived from the nonlinear theory. Off equipartition, the reductive perturbation expansion method allows to derive the relativistic DNLS-equation from the previous system. The properties of the DNLS-equation and its solitons are summarized in the next subsection.

2.3   Relativistic DNLS equations and properties

The simplest nonlinear equation is obtained by a perturbative method when off equipartition. The nonlinear compression $\tilde{u}$ is of order of $|b|^2$ in the front-frame. This is the relativistic version of the so-called DNLS-equation (Derivative NonLinear Schrödinger equation): (Mjölhus 1976, Mio et al. 1976):

$$\gamma_* \partial_t b - \frac{1}{2\delta} \partial_x |b|^2 b + i\alpha_0 \partial_x^2 b = 0 \;. \tag{13}$$

The coefficient $\alpha_0 = \frac{4}{3}\beta_*^2\gamma_*^2$ which equals to 2/3 at equipartition. But the most important coefficient is δ:

$$\delta \equiv \sigma \frac{\sqrt{2}}{3} \frac{P - P_m}{P_m} \;; \tag{14}$$

it measures the deviation to equipartition, and $\sigma = +1$ for forward propagation, -1 for bacward. This equation has been intensively studied (Kaup and Newell 1978, Mjölhus 1978, Spangler and Sheerin 1982, Kennel et al.1988), for it is one of the most famous exemple of soliton equation. The solitons have a width ξ inversely proportional to the square of their amplitude, an exponential decay of their envelope:

$$a = \zeta b_0^2 \;. \tag{15}$$

They do not propagate at the same speed; a nonlinear shift of their velocity proportional to the square of their amplitude, so that the larger solitons run faster than the weaker:

$$\kappa = \zeta' b_0^2 \;. \tag{16}$$

There exists a simple relation between the order one numbers ζ and ζ', ζ' being bounded beetween two values. They can cross each other and have head on collisions without destroying themselves as long as dissipation effects are neglected. These soliton properties are interesting for some astrophysical phenomena. For instance, the GRBs light curve and afterglow (Meszaros and Rees 1997) is explained if the relativistic expansion is organized such that the larger disturbances are faster and interact the first with the ambient medium and this organization of the flow must be realized without significant energy lost during the expansion; the concept of soliton help to account for such behaviour. The formation of wisps in pulsar nebulae (Scargle 1969, Klein et al. 1996) could be explained by relativistic soliton production by a kink instability. The wisps could then be the main energy carrier from the pulsar gyro-wave to relativistic particle.

2.4   Relativistic Hada's system

As can easily be seen, the DNLS-equation is not valid close to equipartition (δ small). It turns out that the compression effect becomes stronger and the scaling such that $\tilde{u}$ is of the order of b. Hada derived the correct system (Hada 93) with the appropriate reductive expansion method for the case of a non-relativistic plasma. This was done for relativistic plasma in PM. The relativistic system reads:

$$\frac{4}{3}\gamma_* \partial_t \tilde{u} + \partial_x(\frac{2}{3}\tilde{u}^2 + \delta\tilde{u} + \frac{1}{2}|b|^2) = 0 \qquad (17)$$

$$\gamma_* \partial_t b + \partial_x(\tilde{u}b) + i\alpha_0 \partial_x^2 b = 0 \qquad (18)$$

The interest of the system is to describe correctly the intensification of the compression effect of the Alfvén waves when approaching equipartition, which strengthens the absorption of energy by relativistic particles as claimed in PM. The kinetic understanding of that "anomalous" absorption will be explained in the next section and is related with the generation of a significant parallel component of the electric field. The enhancement of absorption in this regime has been checked numerically by Baciotti et al. (1998).

3   Generation of a parallel electric field

The nonlinear behaviour of Alfvén waves produce compression that tremendously intensifies when approaching equipartition as shown in PM. These compressions effects generate parallel components of the electric field that are more intense than by the usual weak turbulence theory situation. In the case of delocalized waves, this generation allows the Landau absorption to work and to accelerate low energy particle efficiently, so solving the injection problem. In the case of solitons, it modifies them into soliton-shocks.

3.1   Electrostatic potential in the wavefront

The main nonlinear effect of the parallel Alfvénic perturbation is the localized magnetic pressure that produces a local compression of the plasma. Since the inertia of the electrons is negligible, their pressure variation along the field line must be balanced by a parallel electric force.

$$\partial_\parallel P_e = n_e q_e E_\parallel \; . \qquad (19)$$

This effect can be described by the generalized Ohm's law in parallel direction. Since the electrons are in a quasi-static equilibrium, their pressure variation is such that $\delta P_e = T_e \delta n$. Now from the continuity equation, $\delta n/n_0 = -\tilde{u}/(u_* + \tilde{u})$. Since for a forward front $u_* < 0$, the solitary front exits only for $\tilde{u} < |u_*|$. For moderate perturbations, $\tilde{u}$ directly gives the electrostatic variation because

$$\frac{q_e V}{T_e} = -\frac{\tilde{u}}{|u_*|} \qquad (20)$$

which leads to the formation of a potential well for the electrons and a potential barrier for the protons in the front-frame.

The protons having a motion close to the front motion within an energy band fixed by the potential barrier are reflected by the solitary wave. They are more numerous to be reflected ahead. This leads to the formation of a collisionless parallel soliton-shock.

Electrons experience a static potential well in the front-frame. They cross the well as long as they do not suffer energy loss by radiation. The radiation loss can lead to trapping a part of electrons, those having a motion close to the front motion. Thus the soliton makes the plasma shining locally.

The previous features are typical of a localized BGK-mode (Bernstein, Green, Kruskal 1957). BGK-modes are exact nonlinear solutions of the 1D Vlasov-Poisson system, where the potential profile and the particle distributions are calculated self-consistently. It is particularly stimulating to envisage that such localized structures of phase-space can be generated in Nature by Alfvénic solitons. Similar structures were observed and analysed very recently in space (Goldman 1999).

Such structures can be viewed as a kind of coherent and localized realization of the nonlinear landau effect.

3.2   Nonlinear Landau damping of waves

The parallel electric field generated by magnetic compression is sensitive to that sort of Landau effect called "transit time magnetic damping", acting through the resonance $\omega - k_\| v_\| = 0$. In a wave-frame, from eq. 20 one has:

$$q_e E_\| = \frac{T_e}{\gamma_* \beta_*} \partial_x \tilde{u} \; . \tag{21}$$

For an ensemble of waves, this leads to a parallel momentum diffusion coefficient (in plasma frame) of the form:

$$\Gamma_\| = \frac{T_e^2}{\gamma_*^2 \beta_*^2} < (\partial_x \tilde{u})^2 > \tau_c \; , \tag{22}$$

where τ_c is the effective correlation time of the parallel electric field experienced by the particle in their motions. Off equipartition, in wave frame, $\tilde{u}$ is a response to the magnetic pressure $|b|^2$ and oscillates at Alfvén wave beating frequency. This is a possible generation of fast magnetosonic waves also, that Ragot and Schlickeiser (1998) have demonstrated their efficiency to inject electron in the cosmic ray population. Close to equipartition, there is no such well defined oscillation. The parallel electric field has an broadened frequency spectrum such that the resonance has a relaxation time by phase mixing. Since these intense compressions occurs at short scales (at scales larger than r_* but not very much), they are driven by magnetic perturbations that are dispersed by the inertial effect; therefore the characteristic time scale of the inertial dispersion effect surely measures the correlation time which should be of order τ_*. This issue would deserve a specific numerical simulation, postponed for fututre works.

Detailed calculation shows that the diffusion coefficient depends on the pitch angle of the particle, not on its energy! Moreover the interaction does not require a high energy threshold like the Landau-synchrotron resonnance necessary in the Fermi processes. This means that this process could probably solve the long standing issue of the injection problem of cosmic rays...

4   Electron-positron plasma

Pair plasma has been considered in astrophysics in pulsar magnetosphere, and also in the vicinity of black holes of stellar mass or in active galactic nuclei where black holes of 10^6 to 10^9 solar masses could reside. Gamma ray emission can reveal the existence of such relativistic pair plasmas (see Henri et al. 1999) and the fast dynamics of such a plamsa is interesting to account for both the spectrum shape and the variability. In a Pair-plasma, the dispersion effect is of higher order and does not appears with the former scalings as seen in the previous section. However the dispersion effect can compete with the magnetic pressure effect at the appropriate scale. It can be incorporated by using the modified scaling that does not change the derivation of the nonlinear terms as done in the previous section. The scaling is $\delta = o(\varepsilon)$, $b = o(\varepsilon)$ and $u = o(\varepsilon)$, but $\partial_x = o(\varepsilon^{1/2})$ and $\partial_t = o(\varepsilon^{3/2})$ (thus $\varepsilon = \bar{r}_*^2/l^2$). The system is now

$$\frac{4}{3}\gamma_* \partial_t u + \partial_x \left(\frac{2}{3}u^2 + \delta u + \frac{1}{2}|b|^2\right) = 0 \tag{23}$$

$$\gamma_* \partial_t b + \partial_x(ub) + \alpha \partial_x^3 b = 0 \tag{24}$$

(the coefficient α changes its sign with the direction of propagation and $\alpha = \frac{4}{3}\beta_* \gamma_* P/P_m$ which equals $\sigma 4/3\sqrt{2}$ at equipartition).

Like in the previous case when $\delta = o(1)$, $u = o(\varepsilon^2)$, and the system reduces to the mKdV equation:

$$\gamma_* \partial_t b - \frac{1}{2\delta}\partial_x |b|^2 b + \alpha \partial_x^3 b = 0 . \tag{25}$$

The cosmic ray and Pair-systems have common constants of motion. The equations themselves are conservation equations for the integrals $\int u dx$ and $\int b dx$. The systems are invariant under the gauge transformation $b \rightarrow e^{i\theta_0}b$ which implies that $\int(|b|^2 + u^2)dx$ is a constant of the motion. They keep the magnetic helicity constant: $\frac{1}{2}\int(a^*b + ab^*)dx$. However only their asymptotic approximations for $\delta_P = o(1)$ (DNLS and mKdV) have an infinite sequence of constants of motion. Those invariants are adiabatic invariants of the systems when the plasma is significantly off equipartition. They loose their adiabatic invariants in the vicinity of equipartition because the sonic waves generated by the generalized Alfvén waves intensify close to the degeneracy (equipartition) and goes from second order to first order in the perturbation theory. Beyond equipartition, the wave packets does not steepen and a pair-plasma having a pressure larger than the magnetic pressure is no more efficiently heated by the relativistic fronts. Therefore it suffer radiation cooling and tends to come back to a condition of rough equipartition.

Like for the cosmic ray system, following Hada (1993), one can take into account oblique propagation contributions at order ε, which modify the system by introducing $\nabla_\perp = \partial_1 + i\partial_2$ according to:

$$\frac{4}{3}\gamma_* \partial_t u + \partial_x(\frac{1}{2}u^2 + \delta u + \frac{1}{2}|b|^2) - \frac{1}{2}(\nabla_\perp^* b + \nabla_\perp b^*) = 0 \qquad (26)$$

$$\gamma_* \partial_t b + \partial_x(ub) + \alpha\partial_x^3 b - \nabla_\perp u = 0 \qquad (27)$$

The transverse or obliquity effects are not addressed in this paper. However some parameter regimes might have interesting consequences in the prospect of particle heating (Passot, Sulem, Sulem 1994).

5   Relativistic Fermi acceleration

By "relativistic Fermi acceleration" I understand acceleration of suprathermal particles by relativistic MHD disturbances that propagate at a velocity close to the velocity of light. Those disturbances are very likely in the form of delocalized Alfvén waves or localized nonlinear Alfvén wavefronts, because the magnetosonic perturbations are more damped in astrophysical plasmas that have a pressure comparable with the magnetic pressure. The detailed theory is presented in a forthcoming paper (Pelletier 1999).

5.1   The acceleration scheme

The interaction with forward waves can be presented in the following way:

$$(p_1, \mu_1) \xrightarrow{L_{\beta_*}} (p_1', \mu_1') \xrightarrow{S} (p_2', \mu_2') \xrightarrow{L_{\beta_*}^{-1}} (p_2, \mu_2) \qquad (28)$$

The Lorentz transform L_{β_*} is such that

$$p_1' = \gamma_*(1 - \beta_* \mu_1)p_1 \qquad (29)$$

$$\mu_1' = \frac{\mu_1 - \beta_*}{1 - \beta_* \mu_1} \qquad (30)$$

The scattering S does not change the energy, the pitch angle is changed randomly: $p_2' = p_1'$ and $\mu_1' \mapsto \mu_2'$, with a conditional probability density

$K^+_{\Delta t'}(\mu'_2|\mu'_1)$ during a time $\Delta t'$ measured in the wave-frame. This kernel is normalized such that

$$\int_{-1}^{+1} K^+_{\Delta t'}(\mu'_2|\mu'_1)d\mu'_2 = 1 \ , \ \forall \mu'_1 \ . \tag{31}$$

Moreover

$$\lim_{\Delta t' \to 0} K^+_{\Delta t'}(\mu'_2|\mu'_1) = \delta(\mu'_2 - \mu'_1) \ . \tag{32}$$

Then the reversed Lorentz transform gives the momentum after the interaction in the plasma frame:

$$p_2 = \gamma_*(1 + \beta_*\mu'_2)p'_1 \tag{33}$$

$$\mu_2 = \frac{\mu'_2 + \beta_*}{1 + \beta_*\mu'_2} \tag{34}$$

Because μ_1 and μ'_2 can take any value between -1 and $+1$, some particles can undergo a large energy gain by a factor of order γ_*^2:

$$p_2 = \gamma_*^2(1 - \beta_*\mu_1)(1 + \beta_*\mu'_2)p_1 \tag{35}$$

The energy jump in a progressive wave is then

$$\Delta p^+ = \beta_*\frac{\mu_2 - \mu_1}{1 - \beta_*\mu_2}p_1 \tag{36}$$

The jump is large for most particles because the Lorentz transform concentrates the pitch-angle α_2 in a narrow cone of half-angle $\alpha_* = arcsin(1/\gamma_*)$. This is similar to the inverse Compton effect in Thomson regime.

If only forward waves (linear or nonlinear) would be considered, then they would tend to isotropize the tail of the suprathermal distribution with respect to the wave-frame. Fermi acceleration works only if both forward and backward waves come into play. So now consider a mixture of forward and backward waves (in proportion a^+ and a^- respectively, with $a^+ + a^- = 1$); then the distribution function is changed during Δt according to the following law:

$$2\pi p^2 f(p,\mu,t) = \int_{-1}^{+1} d\mu_1 \int_0^\infty dp_1[K^+_{\Delta t}(\mu|\mu_1)a^+\delta(p - p_1 - \Delta p^+) +$$
$$K^-_{\Delta t}(\mu|\mu_1)a^-\delta(p - p_1 - \Delta p^-)]2\pi p_1^2 f(p_1,\mu_1,t - \Delta t) \tag{37}$$

The probability density $K'^-_{\Delta t'}$ in the backward frame is not significantly different from the forward probability density. However the probability density $K^-_{\Delta t}$ differs from $K^+_{\Delta t}$ because of the different composition with the Lorentz transforms.

$$K^\pm_{\Delta t}(\mu|\mu_1) = \frac{1}{\gamma_*^2(1 \mp \beta_*\mu)^2}K'^\pm_{\Delta t'}(\mu'|\mu'_1) \tag{38}$$

This expression 37 holds if the interaction of a particle with a forward wave is independent of its interaction with a backward wave. This is true because the interaction occurs under a resonance condition that differs for forward and backward waves (see section 6 for details).

Several types of interaction can occur.

- i) Mirror reflection:

$$K'^{\pm}_{\Delta t'}(\mu'|\mu'_1) = \delta(\mu' - \mu'_1)(1 - \frac{c\Delta t'}{l'}) + \frac{c\Delta t'}{l'}\delta(\mu' + \mu'_1) , \qquad (39)$$

where l' is the mean free path of the particle colliding with magnetic perturbations. This was the assumption used by Fermi in his historical paper (1949). However he assumed that the perturbations (clouds) were propagating in every directions, whereas Alfvén perturbations propagates along the field lines.

- ii) Fast random scattering:

$$K'^{\pm}_{\Delta t'}(\mu'|\mu'_1) = \delta(\mu' - \mu'_1)(1 - \frac{c\Delta t'}{\xi}) + \frac{c\Delta t'}{\xi}\frac{1}{2} \qquad (40)$$

That kind of scattering is relevent to describe the interaction with large localized wavefronts of width ξ (see section 6).

- iii) Angular diffusion:

$$K'^{\pm}_{\Delta t'}(\mu'|\mu'_1) = \frac{1}{\sqrt{2\pi}\sigma_{\mu'_1}}exp[-\frac{(\mu' - \mu'_1)^2}{2\sigma^2_{\mu'_1}}] \qquad (41)$$

When a diffusion process occurs, it is entirely governed by the pitch angle frequency in the wave-frame:

$$\nu'_s \equiv \frac{<\Delta\alpha'^2>}{\Delta t'} . \qquad (42)$$

and $\sigma^2_{\mu'_1} = \nu'_s(1-\mu'^2_1)\Delta t'$. This diffusive approximation is used in the usual 1st and 2nd Fermi processes since the seventies (Jokipii 1966, Melrose 1968). This is based on the so-called quasi-linear theory leading to a Fokker-Planck equation. This cannot be used in the relativistic regime.

- iv) Anomalous diffusion:

$$K'^{\pm}_{\Delta t'}(\mu'|\mu'_1) = \alpha\frac{|\mu' - \mu'_1|^{\alpha-1}}{(\nu'_a\Delta t')^{\beta}}g(\frac{|\mu' - \mu'_1|^{\alpha}}{(\nu'_a\Delta t')^{\beta}}) ; \qquad (43)$$

where the function g is normalized to unity:

$$\int g(y)dy = 1 . \qquad (44)$$

If g has a momentum: $\int |y|^{2/\alpha} g(y) dy = s < \infty$, then the anomalous diffusion is such that:

$$< (\Delta\mu')^2 > = s(\nu'_a \Delta t')^{2m} ; \tag{45}$$

with $m = \beta/\alpha$ and the process is said to be subdiffusive if $m < 1/2$, diffusive for $m = 1/2$ and superdiffusive for $m > 1/2$. To describe pitch angle scattering of particle in magnetic disturbances, the usual diffusion description does not work at all pitch angles, because the quasi-linear pitch angle frequency vanishes at 90^0, indicating that a more non-linear theory must be developed. The problem was pointed out by R. Schlickeiser (1994) who proposed a linear remedy by introducing a damping broadening of the resonance. Another linear possibility to cure this problem is to take into account the inertial dispersion effect. Indeed the passing through 90^0 is essential for Fermi acceleration to work. This issue will be revisited in section 6.

5.2   The non-relativistic second and first order processes

The concept of first and second order Fermi processes hold only in the case of non-relativistic MHD disturbances, when the Alfvén velocity is much smaller than the velocity of light $V_A \ll C$. Indeed these processes are derived by expanding the "collision" operator at the second order in powers of β_*. When the cosmic rays are not carried by a decelerating flow, like in shock, the acceleration process is at the second order and a pitch angle diffusion in waveframe leads to an energy diffusion in plasma frame. Indeed

$$\Delta p_0 = \gamma_* (\Delta p'_0 + \beta_* \Delta p'_\parallel) \tag{46}$$

$$\Delta p_\parallel = \gamma_* (\Delta p'_\parallel + \beta_* \Delta p'_0) \tag{47}$$

with $\Delta p'_0 \simeq 0$ and $\Delta p'_\parallel \simeq p'_0 \Delta\mu'$, and the following diffusion coefficients are then derived, knowing that $\Delta t = \gamma_* \Delta t'$:

$$\Gamma_\parallel \equiv \frac{< \Delta p^2_\parallel >}{2\Delta t} = \frac{1}{2} \gamma_* p_\perp^2 (1 + \beta_*^2 \mu^2 - 2\beta_* (a^+ - a^-)\mu)\nu'_s ; \tag{48}$$

$$\Gamma_0 \equiv \frac{< \Delta p^2_0 >}{2\Delta t} = \beta_*^2 \Gamma_\parallel ; \tag{49}$$

The last result clearly shows the superiority of relativistic waves having $\beta_* \simeq 1$, compared to nonrelativistic ones that have $\beta_* \ll 1$, in the purpose of particle acceleration. But the expansion in power of β_* cannot be done; in particular there is no more a scale separation between a fast isotropization and a longer energy diffusion, the energy diffusion being as fast as the pitch angle diffusion.

In a decelerating flow, there is a well known first order contribution to the acceleration. It can be described by the inertial force in the waveframe $F'_j = -p'_i \nabla_i u_j$, whose angular average power is

$$P_a = -\frac{1}{3} p v \, div \mathbf{u} \, . \tag{50}$$

It is often said that the first order process is more efficient than the second order process because at each shock crossing the relative gain is such that $\Delta p/p \sim (u_1 - u_2)/c$ whereas the second order process leads to V_A^2/c^2; this is not plainly true. Schlickeiser et al. (1993) realized that the second order process is never negligible behind a shock. Jones (1994) argued that first and second Fermi processes are not very different in terms of efficiency, the interest of the first order acceleration at adiabatic shocks is to give universal power law distribution downstream. This can be seen as follows (see henri et al. 1999 for details). The rate of change of the energy of a particle crossing a non-relativistic shock many times is given by:

$$\frac{< \Delta p >}{\Delta t} = \frac{r - 1}{3 t_r} p \, , \tag{51}$$

where r is the shock compression ratio and t_r the residence time of a relativistic particle behind the shock, that depends on the diffusion coefficient D through: $t_r = 2D/u_2^2$, u_2 being the downstream flow speed. For $D \simeq D_\parallel \simeq \frac{1}{3}\frac{v^2}{\nu_s}$, the first order acceleration time is thus:

$$t_1 = t_r \sim \frac{c^2}{u_2^2} \nu_s^{-1} \, ; \tag{52}$$

whereas the second order acceleration time is:

$$t_2 \sim \frac{c^2}{V_A^2} \nu_s^{-1} \, . \tag{53}$$

These two time scales are not significantly different because u_2 is often of the order of V_A. The result has a simple meaning: at each scattering there is a small energy diffusion; the energy jump is larger when the particle crosses the shock front, but these crossings are not frequent because their frequency is $\nu_c \sim (u_2/c)\nu_s$. When it can work at quasi perpendicular shocks, the first order acceleration becomes more efficient because t_1 is changed into $t_{1\perp} = t_1(\nu_s/\omega_s)^2$ with transverse diffusion.

The results for t_1 and t_2 indicate of course that first and second order Fermi acceleration tend towards the same maximum efficiency when V_A and u_2 are close to the velocity of light; which emphasizes once more the interest to work with relativistic waves.

5.3   The efficiency of the relativistic Fermi acceleration

The efficiency of relativistic acceleration has two origins: first, as seen in subsection 5.1, the momentum variation during scattering is much larger than in nonrelativistic waves; second, as seen in subsection 5.2, the acceleration rate reaches its highest possible value which is given by the pitch angle frequency. Therefore relativistic waves leads to an acceleration time close to the pitch angle scattering time and this time becomes closer (but still longer) to the gyro-period T_g of the considered particle for large wave amplitude. Large amplitude relativistic waves are the best and only candidates to achieve the expected efficiency of the acceleration process to explain high energy cosmic rays with an acceleration time scale $t_{acc} = AT_g(\epsilon)$, with $A \sim 10$. If the very high energy cosmic rays would not be accounted by these physical assumptions, then exotic phenomena such as topological defects would be invoked.

However it was recently pointed out by Gallant and Achterberg (1999) that the particle energy gain in a relativistic shock is by a factor γ_s^2 (γ_s being the shock Lorentz factor); this increment is obtained at the first crossing, further crossings being inefficient to give a further significant acceleration. Since a particle cannot travel accross several strong relativistic shocks, its repeated interactions with relativistic waves and relativistic solitary waves become a very competitive process.

6   Astrophysics consequences and head on collision of solitons

The detailed theory, not completed in this paper, must develop both the case of delocalized waves and the case of localized fronts. Some preliminary results will be presented for both cases. However the general statements presented in section 5 allow to anticipate some astrophysical consequences already that will motivate further theoretical developments.

The efficiency gain with delocalized relativistic waves is interesting per se, however spectacular acceleration could be achieved with head on collisions of relativistic fronts. This is somehow a rather primitive concept of particle accelerator to make heavy particles (here the relativistic solitons)to collide each other to produce new energetic particles. Indeed the head on collision of two solitons of bulk Lorentz factor γ_* produces the following energy amplification of protons

$$p_2 \sim \gamma_*^4 p_1 \tag{54}$$

within a time of few gyro-periods.

Is it reasonable to expect such head on collisions in Nature? Yes, if we believe in the existence of relativisitic expansion flows. Any relativistic expansion of a relativistic plasma emating from the environment of a compact object (single or twin in coalescence) will produce both forward fronts because of perturbations at the source) and backward fronts because of the interaction with the ambient medium.

A forward front observed in a relativistic flow of bulk reduced velocity β_J has a reduced velocity:

$$\beta_+ = \frac{\beta_J + \beta_*}{1 + \beta_J \beta_*} \tag{55}$$

whereas for a backward:

$$\beta_- = \frac{\beta_J - \beta_*}{1 - \beta_J \beta_*} \tag{56}$$

When the expansion is superAlfvénic, they are both seen to advance, but with different speed, and of course, they collide. Such events could be expected in relativistic extragalactic and galactic jets, and very likely in Gamma Ray Bursts and in pulsar nebulae. VLBI observations (Guirado 1995) revealed the possibility of such collision of relativistic knots in relativistic extragalactic jets and those events are being currently analysed.

Now relativistic motions of such fronts could occur in hot spots as well, even if there is no more relativisitic bulk motions in them because they results from the terminal shocks of the jets. Disturbances with relativistic motions will develop if they are dominated by the cosmic ray pressure. That is what could make hot spot interesting site of very high energy cosmic ray generation. Extended lobes could also have the same properties if they are still magnetically confined.

To get these cosmic ray generation events, ideal solitons are not necessary, relativitic fronts are enough, even if they are destroyed after collision. The only advantage of the soliton is that it can survive after collision and the ground conditions, in particular the magnetic topology, are restaured in the plasma after it passed away.

Assume that the acceleration process is efficient enough to reach locally a maximum energy which is a significant fraction of the accelerator size limit such that $\epsilon_m = aZeBR$ (with $a \sim 0.1$, say) and assume that the integrated energy distribution results from the distribution of the product BR. Then if the local energy distribution function, $f(\epsilon, \epsilon_m(z))$, is not a power law, but any self-similar function of the form:

$$f(\epsilon, \epsilon_m(z)) = \frac{n_*}{\epsilon_m(z)} g(\frac{\epsilon}{\epsilon_m(z)}) \tag{57}$$

The probablity density to find a particle of energy ϵ in the global flow of cross section $S(z)$ is given by

$$\rho(\epsilon) = \int dz n_* S \frac{1}{\epsilon_m(z)} g(\frac{\epsilon}{\epsilon_m(z)}) . \tag{58}$$

Assume a field distribution such that $B \propto R^{-m}$, ($1 < m < 2$ is expected in a jet), assume $dR/dz \propto R^{-n}$ for the flow (for a conical jet $n = 0$, a confined jet $n > 0$, a widening flow $n < 0$) and assuming the same probability of escape along the flow, the energy distribution of the cosmic ray leaving the source is

$$p(\epsilon) \propto \epsilon^{-\frac{m+n}{m-1}} . \tag{59}$$

For instance, a confined jet, such that $m = 2, n = 0.5$, gives $\rho(\epsilon) \propto \epsilon^{-2.5}$; a conical jet with $m = 1.5$ gives ϵ^{-3}.

7   Cosmic rays scattering off magnetic disturbances

High energy particles undergo pitch angle scattering in the waveframe (forward or backward) and this leads to important energy variation in the plasma frame. A preliminary investigation of this scattering is presented in this section. Since the plasma electromagnetic waves give energy to the particles through this Fermi process, in turn, a kind of collisionles damping occur, which is nothing but a generalization of the Landau-synchrotron absorption. This require some resonance between the gyro-motion of the particle and the wave oscillations, as will be recall further on.

The motion of a particle that interacts with quasi parallel propagating magnetic modes, characterized by a vector potential $\mathbf{A}(z) \equiv B_0 \xi \mathbf{a}(z)$, ξ being a typical variation length along the average magnetic field B_0, is governed by a simple non-linear system that reads :

$$\dot{\alpha} = \frac{qB_0\xi}{m\gamma} \mathbf{e}_\perp(t).\frac{\partial}{\partial z}\mathbf{a}(z(t)) \tag{60}$$

$$\dot{z} = v\cos\alpha \tag{61}$$

This is an Hamiltonian system of one degree of freedom, but depending on time through the gyro-motion described by the transverse unit vector $\mathbf{e}_\perp(t)$ that rotates at the gyro-pulsation ω_s almost. The Hamilton function can be written in terms of the two conjugate variables (x, α):

$$H(\alpha, x) = \sin\alpha - \bar{\omega}\mathbf{e}_\perp(t).\mathbf{a}(x) , \tag{62}$$

where $x \equiv z/\xi$, the time unit is the travel time over the characteristic length ξ, namely $\tau_c \equiv \xi/v$, $\bar{\omega} \equiv \omega_g\tau_c = 2\pi\tau_c/T_g$. The dynamics is different for an ensemble of plane waves and for a localized soliton.

7.1   Pitch angle scattering by delocalized waves

For an ensemble of Fourier modes in the waveframe (forward propagation say):

$$\mathbf{a}(x) = \sum_n a_n(\mathbf{e}_1\cos(k_nx) + \varepsilon_c\mathbf{e}_2\sin(k_nx)) , \tag{63}$$

where $\varepsilon_c = +1$ for right handed polarized modes and -1 for left handed polarized modes. In fact these Fourier components are not necessarly eigen modes of the plasma; they are eigen modes for quasi parallel propagation only; but for oblique propagation the Alfvén waves have a linear polarization. However as long as $k_\perp r_g \ll 1$, r_g being the Larmor gyro-radius, the effect of the transverse wave length is unimportant in the particle dynamics. So, even in case of linearly polarized Alfvén waves, the previous expansion 63 is assumed. The Hamilton function is then

$$H(\alpha, x) = sin\alpha + \bar{\omega} \sum_n a_n cos(k_n x - \varepsilon \bar{\omega} t + \phi_0) \qquad (64)$$

A resonance occurs for different values α_n of α such that $k_n \dot{x} = \varepsilon \bar{\omega}$, or $k_n \mu_n = \varepsilon \bar{\omega}$ with $\mu_n = cos\alpha_n$. A negative charge moving forward resonates with a right mode ($\epsilon = 1$) whereas it resonates with a left mode if it moves backward and vice versa for a positive charge. The opposite conclusions held for backward wave propagation. These are the synchrotron resonances.

When the resonances are isolated, the hamiltonian can be approximate by a pendulum resonance Hamiltonian in the vicinity of each resonance: setting the canonical transform $\theta = k_n x - \varepsilon \bar{\omega} t + \phi_0$, $J = (\alpha - \alpha_n)/k_n$, $H' = H - \varepsilon \bar{\omega} \alpha / k_n + Cst$, the approximate Hamiltonian is

$$H'(J, \theta) = -k_n^2 sin\alpha_n (\frac{J^2}{2} - \Omega_n^2 cos\theta) , \qquad (65)$$

where the nonlinear pulsation Ω_n is such that

$$\Omega_n^2 = \frac{\bar{\omega} a_n}{k_n^2 sin\alpha_n} . \qquad (66)$$

The pendulum approximation differs from the exact Hamiltonian by oscillating terms.

The half-width of the nonlinear resonance in (J, θ) phase-space is $\Delta J = 2\Omega_n$ and resonances overlap when this half-width is larger the half-spacement between resonances $\Delta \alpha_n / k_n$; which leads to the Chirikov criterium (Chirikov 1972) for stochasticity:

$$\bar{\omega} a_n sin\alpha_n > (\Delta \mu_n)^2 / 4 . \qquad (67)$$

As is well known chaos occurs even at a lower threshold. The dynamics described by H' differ from the exact one by oscillating contributions, among those there are the contribution of the backward waves. The particle cannot resonate simultaneously with a forward and backward wave.

The smallest value of μ_n control the jump around the pitch angle of 90^0. The particle can jump from the resonance with the right mode ($\varepsilon = 1$) to the resonance with the left mode ($\varepsilon = -1$) if $k_n^2 a_n > \bar{\omega}/4$. This is the nonlinear solution to the momentum turn over problem.

When the modes are sufficient above the stochasticity threshold, the chaotic jumps of the pitch angle behave like a diffusion process. Only the momentum turn over might be slown down by sticky regimes, leading to subdiffusion.

The acceleration process is suitably described by the response to an initial monoenergetic distribution of the form $\delta(p - p_0)/4\pi p_0^2$. The solution at time t is given by eq. 37:

$$f_0(p, \mu, t) = \frac{1}{4\pi p^2 p_0 \beta_*} \sum_{\pm} a^{\pm}(1 \mp \beta_* \mu) P^{\pm}(p, \mu, t) , \tag{68}$$

where

$$P^{\pm}(p, \mu, t) = \frac{1}{\sqrt{2\pi}\sigma(t)} exp[-\frac{(\Delta\mu^{\pm})^2}{2\sigma(t)^2}] , \tag{69}$$

with

$$\Delta\mu^{\pm} \equiv \frac{p - p_0}{\beta_* p_0}(1 \mp \beta_* \mu) . \tag{70}$$

The standard deviation is such that $\sigma(t)^2 = (1 - \mu^2)\nu_s^{\pm}t$ for μ sufficiently different from ± 1; whereas for μ close to ± 1, $\sigma(t)^2 \sim (\nu_s^{\pm}t)^2$ (because of the superdiffusion of μ). This response explicitly showes that in the cone $\alpha < \alpha_*$ the particle gain an energy of order $\gamma_*^2 p_0$ in a scattering time whereas outside the cone the gain factor is only β_*^2. Indeed the isocountours (defined by constants C_{ic}) of the distribution are such that, for $\alpha > \alpha_*$,

$$p_{ic}(t) \simeq p_0(1 + C_{ic}\beta_*^2\sigma(t)) , \tag{71}$$

whereas for $\alpha < \alpha_*$,

$$p_{ic}(t) \simeq p_0(1 + C_{ic}2\gamma_*^2\sigma(t)) , \tag{72}$$

Let end this subsection with some remarks about anomalous diffusion. Obviously, if the pitch angle scattering is an anomalous diffusion process, then the energy variations undergo an anomalous diffusion also with the same secular behaviour. In that case the stationary energy spectrum would significantly be modified compared to the ordinary Fermi processes. Spatial diffusion derived from anomalous pitch angle scattering is also anomalous. Recently Kirk et al. (1996) analyzed the consequences of assumed braided magnetic field lines behind a shock. In particular, the index of the cosmic ray power law could be changed because of the change of the effective compression ratio due to anomalous diffusion (Ragot and Kirk 1997). Anomalous pitch angle scattering would produce also these deviations to the normal laws in shock acceleration theory. On the subset of anomalous diffusion points, the distribution function is not differentiable.

7.2 Scattering by a localized nonlinear wavefront

In the frame of a soliton, the cosmic rays undergo pitch angle variations that are governed by a similar system.

$$\dot{\alpha} = \bar{\omega}\phi(x)cos[\theta(x) - \varepsilon\bar{\omega}t] \tag{73}$$

$$\dot{x} = cos\ \alpha \tag{74}$$

The profile decays exponentially on the scale ξ and the phase has a non-linear variation: $\theta'(x) = \kappa + c_0\phi^2(x)$, c_0 being a coefficient of order unity. The dynamics is then controlled by three parameters $\bar{\omega}$, b_0 and κ. The particle interacts with the soliton during few time units and numerical computations exhibit few typical behaviours. First, even for large amplitude perturbations (say $b_0 \sim 1$), particles having a large Larmor radius compared to the soliton width are characterized by a small parameter $\bar{\omega}$. Therefore they clearly suffer a small deterministic variation of their pitch angle. Second, in the opposite extreme, particles that have a small Larmor radius ($\bar{\omega}$ large) undergo tremendous pitch angle variations, unless the amplitude b_0 is very small. They behave for a while like in a broad bank spectrum of plane waves and the pitch angle undergoes a diffusion process because of the resonances. The third interesting case is that of particles having a Larmor radius comparable to the width of a soliton ($b_0 \sim 1$). Numerical solutions show irregular motions with a possible reflection, the momentum turn over occuring even for a single polarization. The pitch angle jump is practically unpredictable.

In summary, a forward soliton transforms the distribution function of cosmic rays having a Larmor radius r_g larger than r_* upto few ξ, cosmic rays having a Larmor radius much larger than ξ are not scattered ($\bar{\omega} \ll 1$); that part of the distribution function is fastly isotropized in the soliton frame and thus these suprathermal tail has the bulk motion of the solitons $\simeq \beta_*$. The scattering is suitably described by the fast random scattering of subsection 2.1. In the plasma frame, this tail is concentrated in a narrow cone of half angle $\alpha_* \simeq 1/\gamma_*$ and the energy of the particle in this cone has been amplified by a factor γ_*^2. No further acceleration is possible with other forward solitons. Further acceleration is possible with an incoming backward soliton that produces a new amplification of the energy by a factor γ_*^2, changing thus the initial energy of the cosmic ray by a factor γ_*^4. If several head on collisions could occur, the energy gain would be by a factor γ_*^4 each time within few Larmor periods.

8 Conclusion

These investigations are still in a preliminary stage. For instance quasi parallel propagation of the fronts only has been considered and their stability to transverse perturbations are not yet known. It would be very important in the purpose of cosmic ray acceleration that these fronts would keep an important cross section for collisions with cosmic rays.

Also from the viewpoint of mathematical physics many studies are to be done to understand the lost of integrability of the PDEs and its consequences. The chaotic dynamics of cosmic rays in these fronts need also more developments; it was stressed that it controls the momentum turn over of the particles which is crucial for the efficiency of Fermi acceleration.

Anyway I think that the preliminary investigations revealed interesting consequences for astrophysics that seems robust and would contribute to master the long standing problem of the origin of the high energy cosmic rays. Head on collisions of relativistic fronts would probably be the most efficient occurence of generalized Fermi acceleration. Anyway, even with delocalized relativistic waves, the generalized Fermi acceleration process is quite efficient. Relativistic waves and solitary waves are very competitive for cosmic ray acceleration even compared to a strong relativistic shock, because particle energy is amplified by a factor γ_*^2 at each scattering, whereas it is amplified mostly in a single step by a factor γ_s^2 for ever in a relativistic shock.

References

1. Aharonian F. A., & Cronin J. W., 1994, Phys. Rev. D, 50, 1892.
2. Baciotti F., Chiuderi P., & Pouquet A., 1997, ApJ, 478, 594.
3. Bernstein I.B., Green J.M., & Kruskal M.D., 1957, Phys. Rev., 108, 546.
4. Bonazzola S., & Peter P., 1997, Astropart. Phys., 7, 161.
5. Fermi E., 1949, Phys. Rev., 75, 1169.
6. Gallant Y., & Achterberg A., 1999, MNRAS, 305, L6.
7. Gibbs K. G., 1997, proceedings of the 32nd rencontres de Moriond "Very high energy phenomena in the Universe" p. 203.
8. Goldman M., 1999, in "Nonlinear MHD waves and turbulence", Springer- Verlag, proceeding Nice workshop 1998, Passot & Sulem ed..
9. Guirado J.C., et al., 1995, the Astronomical Journal, 110, 2586.
10. Henri G., Pelletier G., Petrucci P.O., & Renaud N., 1999, Astroparticle Physics, (in press).
11. Gardner C.S., Greene J.M., Kruskal M.D., & Miura R.M., 1967, Phys. Rev. L., 14, 1095.
12. Hada T., 1993, GRL, 20, 2415.
13. Hansen P.J., & Nicholson D.R, 1979, Am. J. Phys., 47, 769.
14. Jokipii J.R., 1966, ApJ, 146, 480; 1987, ApJ, 313, 842.
15. Jones F., 1994, ApJS, 90, 561.
16. Kaup D.J., & Newell A.C., 1978, J. Math. Phys., 11, 798.
17. Kennel C.F., Buti B., Hada T., & Pellat R., 1988, Phys. Fluids, 31, 1949.
18. Kirk J.G., Duffy P., & Gallant Y.A., 1996, A&A, 314, 1010.
19. Klein, R.I. et al., 1996, ApJ, 469L, 119.
20. Lichti et al., 1995, A&A, 298, 711.
21. Mannheim K., & Biermann P., 1992, A&A, 253, L21
22. Melrose D.B., 1968, "Plasma Astrophysics", vol. 2, Gordon and Breach.

23. Melrose D.B., 1986, "Instabilities in space and laboratory plasmas", Cambridge Univ. Press.
24. Meszaros P., & Rees M.J., 1997, ApJ, 482, L29.
25. Von Montigny et al., 1995, ApJ, 440, 525.
26. Mio K., Ogino T., Minami K., & Takeda S., 1976, J. Phys. Soc. Jpn., 41, 265.
27. Mjölhus E., 1976, J. Plasma Phys., 16, 321.
28. Mjölhus E., 1978, J. Plasma Phys., 19, 437.
29. Passot T., Sulem C., & Sulem P-L, 1994, Phys. Rev. E, 50, 1427.
30. Pelletier G., 1999, A&A (in press).
31. Pelletier G., & Marcowith A., 1998, ApJ, 502, 598.
32. Protheroe R.J., & Stanev T., 1992, in High Energy Neutrino Astronomy, eds. V.J. Stenger et al. (World Scientific, Singapore), p. 40.
33. Ragot B.R., & Kirk J.G., 1997, A&A, 327, 432.
34. Ragot B.R., & Schlickheiser R., 1998, APh, 9, 79.
35. Roberts B., 1985, Phys. Fluids, 28, 3280.
36. Sarazin C., et al., 1999, ApJ, 510, 90.
37. Scargle, J.D., 1969, ApJ, 156, 401.
38. Schlickeiser R., Campeanu A., & Lerche L., 1993, A&A, 276, 614.
39. Schlickeiser R., 1994, ApJS, 90, 929.
40. Spangler S.R., & Sheerin J.P., 1982, J. Plasma Phys., 27, 193.
41. Taniuti T., & Wei C.C., 1969, J. Phys. Soc. Jpn., 26, 1305.
42. Weiss J., Tabor M., & Carnevale G., 1983, J. Math. Phys., 24, 522.
43. Zaslavsky G.M., & Chirikov B.V., 1972, Sov. Phys. Uspekhi, 14, 549; Chirikov B.V., 1979, Phys. Reports, 52, 265.

Reduced Models of Magnetohydrodynamic Turbulence in the Interstellar Medium and the Solar Wind

A. Bhattacharjee and C.S. Ng

Department of Physics and Astronomy, The University of Iowa, Iowa City, Iowa 52242, USA

Abstract. Recent developments in the derivation of reduced models for weakly compressible magnetohydrodynamic (MHD) turbulence are discussed. A four-field system of equations has been derived from the compressible magnetohydrodynamic (MHD) equations to describe turbulence in the interstellar medium and the solar wind. These equations apply to a plasma permeated by a spatially varying mean magnetic field when the plasma beta is of the order unity or less. In the presence of spatial inhomogeneities, the four-field equations predict pressure fluctuations of the order of the Mach number of the turbulence, as observed by Helios 1 and 2. In the presence of a uniform background field and a spatially homogeneous plasma, the four-field system reduces to the so-called nearly incompressible system. In the weak-turbulence limit, dominated by three-wave interactions, the anisotropic energy spectrum is deduced by a combination of exact analytical results and numerical simulations.

1 Introduction

Magnetohydrodynamics (MHD), despite its significant limitations as a model, provides the principal framework for the theoretical description of turbulence in the interstellar medium (ISM) and the solar wind. There is strong observational evidence that the effect of plasma compressibility cannot be neglected in theoretical studies of magnetohydrodynamic (MHD) turbulence in these systems. This should not be all that surprising, for the plasma beta in the solar wind and the ISM is often of the order unity or greater. Consequently, the sound speed is of the same order of magnitude or greater than the Alfvén speed. (In a truly incompressible plasma, the sound speed is infinitely large.) Hence, the assumption of constant plasma density (ρ), which is sufficient to ensure that the plasma flow velocity v obeys the incompressibility condition $\nabla \cdot v = 0$ by the continuity equation $\partial \rho / \partial t + \nabla \cdot (\rho v) = 0$, cannot be sustained.

Since the observed fluctuations in the solar wind and the ISM involve density variations, it is widely appreciated that the effects of plasma compressibility should be incorporated in a viable theory of MHD turbulence for such systems. However, theoretical attempts to grapple with compressible turbulence from first principles are hampered by the formidable analytical (as well as numerical) complexities of the fully compressible MHD equations which involve eight scalar variables (i.e., three components each for the fluid velocity and magnetic fields, the plasma density, and pressure) and several nonlinearities. Even in simple magnetic geometry, it is difficult to obtain numerically a reliable inertial range spectrum (spanning a few decades in wavenumber) for compressible MHD turbulence.

In view of these difficulties, attempts have been made (primarily in the theoretical solar wind literature) to derive reduced MHD models that can enable the computation of compressible turbulence as perturbative corrections to a leading-order incompressible description. Perhaps the most well-known of such attempts is the nearly incompressible MHD (NI-MHD) model of Zank and Matthaeus [1,2] (hereafter, ZM). The NI-MHD model formulated by ZM has its antecedents in the pseudosound theory of Lighthill [3] for compressible hydrodynamic (HD) turbulence. Lighthill's work provided the stimulus for analogous MHD applications by Montgomery et al. [4], Shebalin and Montgomery [5], and Matthaeus and Brown [6] as well as the development of nearly incompressible HD [7] which preceded the development of NI-MHD by ZM. Independently, Grappin et al. [8] have pointed out that the MHD variant of pseudosound theory can also be viewed as an extension of the HD theory of Kliatskin [9].

In its basic form, the equations of the NI-MHD model are similar to those of the reduced MHD (RMHD) model [10] which has proved to be extremely useful in describing nonlinear MHD dynamics in tokamak or solar coronal plasmas. (The derivation of NI-MHD given by ZM suggests that RMHD is a subset of NI-MHD.) Tokamak as well as solar coronal plasmas are typically permeated by a strong magnetic field and characterized by low values of the plasma beta (β) (that is, $\beta \ll 1$). Exploiting the fact that solar wind turbulence in the plasma frame is characterized by low values of the turbulent Mach number, ZM show, by means of an asymptotic expansion in powers of the Mach number, that the equations of compressible MHD reduce to the NI-MHD equations.

One of the interesting but surprising conclusions of ZM is that the NI-MHD equations are valid for solar wind plasmas even when $\beta \sim 1$. Since the pressure fluctuations in the NI-MHD equations are decoupled, at leading order, from the dynamical equation for the velocity field fluctuations, ZM claim that compressible fluctuations are enslaved to incompressible dynamics. If true, this result is a remarkable simplification of the problem of compressible MHD turbulence in the solar wind (as well as in parts of the ISM) because one can then rely on analytical or numerical results derived from a leading-order incompressible MHD calculation and calculate the higher-order density fluctuations convected passively by an incompressible flow field.

The NI-MHD theory predicts that the root-mean-square pressure fluctuation (normalized by the background plasma pressure) in the solar wind should be $O(M^2)$, where M is the Mach number of the turbulence. This prediction has been tested by detailed comparison with observations from Voyager as well as Helios 1 and 2 [11–13]. From Fig. 3 of Tu and Marsch [12], who considered a large dataset from Helios 1 and 2 between 1 and 7 AU, we see that although there some data points are $O(M^2)$ scaling, most of the data points are distributed in the range between $O(M^2)$ and $O(M)$.

Motivated in part by these discrepancies between observations and the NI-MHD model, Bhattacharjee, Ng and Spangler [14] (hereafter, BNS) have formulated a new system of reduced equations for MHD turbulence, referred to hereafter as the four-field equations, valid for $\beta \sim 1$ plasmas. They too assume, following ZM, that the plasma is permeated by a background magnetic field and that the Mach number of the turbulence is small. However, unlike ZM who assume that the background magnetic field is spatially uniform, BNS allow the background field to be spatially nonuniform and capable of sustaining a nonvanishing plasma current density and pressure gradient. BNS then show that the compressible, three-dimensional (3D) MHD equations can be simplified to a system involving four scalar variables: the magnetic flux, the parallel vorticity (i.e., the component of vorticity parallel to the mean magnetic field), the perturbed pressure, and the parallel flow. This four-field system represents a generalization of NI-MHD (a two-field system at leading order) to plasmas with $\beta \leq 1$, and includes NI-MHD as a special case. The four-field system reduces to NI-MHD when (i) the background magnetic field is constant in space and all spatial inhomogeneities in the background quantities are neglected, and/or (ii) when $\beta \ll 1$. This reduction delineates clearly the restricted domain of applicability of NI-MHD, and suggests that although NI-MHD may apply to specific numerical experiments or to a selected subset of observations, it cannot be expected to account for as broad a dataset as shown in [12].

An important attribute of the four-field equations is that in a spatially inhomogeneous $\beta \sim 1$ plasma, the four field variables, that is, magnetic flux, parallel vorticity, pressure and parallel flow, are coupled to each other. This coupling has significant implications for the scaling of root-mean-square pressure fluctuations. In the presence of spatial inhomogeneities, the first-order pressure or density fluctuations evolve to non-zero values even if they are chosen to be zero initially, and modest inhomogeneities are enough to raise the level of these fluctuations to $O(M)$ values. Thus, we suggest that the level of fluctuations seen in [12] can be attributed to the effect of spatial inhomogeneities in the background plasma pressure and magnetic fields, neglected in NI-MHD.

The four-field model also offers an answer to the fundamental question: Do the effects of plasma compressibility enter the fluctuation dynamics at leading order for $\beta \sim 1$ plasmas, or are the effects of compressibility a higher-order phenomenon enslaved to a leading-order incompressible description, as seen in the NI-MHD model? It is shown that when there are spatial inhomogeneities in the background magnetic field and plasma pressure, the effects of plasma compressibility enter the dynamical equation for the pressure fluctuation at leading order.

Another outstanding question confronting MHD turbulence theory is the nature and scaling of the turbulent energy spectrum. Evidence for anisotropic, probably magnetic field-aligned turbulence is strongly suggested by radio wave scintillation studies of heliospheric and interstellar turbulence. Radio wave scattering in the solar wind close to the Sun is highly anisotropic, with the "long axis" of the irregularities being in the radial, magnetic field-oriented direction and with axial ratios in excess of 10:1 [15,16]. Anisotropic scattering, indicative of anisotropic irregularities, is also a general characteristic of interstellar scattering [17–19]. The observed ratios are smaller than in the heliospheric case, but this may be due to averaging by integration along the line of sight. It seems likely that the axial ratio of the density irregularities exceed the maximum observed factors of 4:1 [18].

The existence of anisotropy in both heliospheric and interstellar plasmas and its implications for spectral scaling laws is a significant challenge for theory. Even when the background consists of a spatially uniform plasma embedded in a constant magnetic field, the precise nature of wave-wave interactions and the form of the energy spectrum has been a subject of some debate [20–25]. All of the studies cited concur, however, that the energy spectrum is strongly anisotropic in the presence of a uniform magnetic field. In the weak-turbulence limit, we present here some new analytical and numerical results on the anisotropic energy spectrum.

The following is a layout of this paper. In Sect. 2, we review the main assumptions and equations underlying the four-field model. In Sect. 3, we derive the NI-MHD model as a special case of the four-field model. In Sect. 4, we present numerical simulation results based on the four-field equations that illustrate the connection between spatial inhomogeneities and enhanced pressure fluctuations. In Sect. 5, we review some exact analytical results on the dominance of three-wave interactions in weak turbulence theory and present the results of a numerical experiment that relies on the analytical results to obtain the anisotropic energy spectrum.

2   The four-field model

The compressible resistive MHD equations are

$$\rho \left(\frac{\partial}{\partial t} + \boldsymbol{v} \cdot \nabla \right) \boldsymbol{v} = -\nabla p + \frac{1}{4\pi} \left(\nabla \times \boldsymbol{B} \right) \times \boldsymbol{B} \,, \tag{1}$$

$$\frac{\partial \boldsymbol{B}}{\partial t} + \nabla \times \left(\boldsymbol{B} \times \boldsymbol{v} + \eta c \boldsymbol{J} \right) = 0 \,, \tag{2}$$

$$\frac{\partial \rho}{\partial t} + \nabla \cdot \left(\rho \boldsymbol{v} \right) = 0 \,, \tag{3}$$

$$\frac{d}{dt} \left(\frac{p}{\rho^{\gamma}} \right) = \frac{\gamma - 1}{\rho^{\gamma}} \eta \left| \boldsymbol{J} \right|^2 \,, \tag{4}$$

where ρ is the plasma density, v is the fluid velocity, p is the plasma pressure, B is the magnetic field, $J = c\nabla \times B/4\pi$ is the current density, c is the speed of light, η is the plasma resistivity and γ is the ratio of specific heats. We cast (1)–(4) in dimensionless form by scaling every dependent variable by its characteristic (constant) value, designated by a subscript c (which should be distinguished from the speed of light). We define the sound speed, Alfvén speed, Mach number, Alfvén Mach number, and the plasma beta, respectively, by the relations

$$C_s^2 \equiv (\partial p/\partial \rho)_c \, , \tag{5a}$$
$$V_A^2 \equiv B_c^2/4\pi\rho_c \, , \tag{5b}$$
$$M^2 \equiv v_c^2/C_s^2 \, , \tag{5c}$$
$$M_A^2 \equiv v_c^2/V_A^2 \, , \tag{5d}$$
$$\beta \equiv 4\pi p_c/B_c^2 = M_A^2/\gamma M^2 \, . \tag{5e}$$

In terms of scaled (dimensionless) variables, (1), (2), and (4) can be written

$$\rho\left(\frac{\partial}{\partial t} + v \cdot \nabla\right)v = \frac{1}{\varepsilon^2}\left[-\nabla p + \frac{1}{\beta}(\nabla \times B) \times B\right] \, , \tag{6a}$$

$$\frac{\partial B}{\partial t} + \nabla \times (B \times v + \eta J) = 0 \, , \tag{6b}$$

$$\frac{d}{dt}\left(\frac{p}{\rho^\gamma}\right) = \frac{\gamma - 1}{\beta\rho^\gamma}\eta |J|^2 \, , \tag{6c}$$

where $\varepsilon \equiv \sqrt{\gamma}M$, with $J = \nabla \times B$, and (3) retains its present form. (Distance and time are scaled by the system size and the Alfvén time, respectively.) Since (1)–(4) are invariant under Galilean transformations, we choose to work in a reference frame that is stationary with respect to the plasma. The parameter ε, which is essentially the Mach number of the turbulence, is assumed to be much smaller than one and provides the basis for an asymptotic expansion of the dependent variables. We write

$$B = B_0 + B_1 + \cdots \, ,$$
$$v = v_1 + \cdots \, ,$$
$$\rho = \rho_0 + \rho_1 + \cdots \, ,$$
$$p = p_0 + p_1 + \cdots \, , \tag{7}$$

where all $O(1)$ quantities are denoted by subscript zero, and $O(\varepsilon)$ quantities by the subscript one. Note that any large-scale, uniform flow of the plasma with respect to the laboratory frame has been transformed away by moving to the plasma frame (i.e., $v_0 = 0$).

In the four-field model, the background quantities obey the magnetostatic equilibrium condition

$$\nabla p_0 = \frac{1}{\beta}(\nabla \times B_0) \times B_0 \, , \tag{8}$$

which continues to hold even as the (first-order) fluctuations evolve in time. In the NI-MHD model, the background quantities are assumed to be

$$p_0 = 1, \boldsymbol{B}_0 = \hat{\boldsymbol{z}}, \tag{9}$$

which is a special solution of (8). We shall consider the special case (9) in Sect. 3, but propose to develop our model with more general solutions of (8) in mind, allowing for the presence of spatial inhomogeneities in the background plasma variables.

The form of (6a) and (8) motivate the transformations $\boldsymbol{B} \to \sqrt{\beta}\boldsymbol{B}$, $\boldsymbol{J} \to \varepsilon\sqrt{\beta}\boldsymbol{J}$, $\eta \to \eta/\varepsilon^2$, $\boldsymbol{v} \to \boldsymbol{v}/\varepsilon$, $\nabla \to \varepsilon\nabla$, so that $\boldsymbol{B} \sim O(1/\sqrt{\beta})$, $\boldsymbol{v} \sim O(\varepsilon)$, and $\nabla \sim O(1/\varepsilon)$. Under these transformations, (6a) and (8) become

$$\rho\left(\frac{\partial}{\partial t} + \boldsymbol{v} \cdot \nabla\right)\boldsymbol{v} = -\nabla p + (\nabla \times \boldsymbol{B}) \times \boldsymbol{B}, \tag{10}$$

and

$$\nabla p_0 = (\nabla \times \boldsymbol{B}_0) \times \boldsymbol{B}_0, \tag{11}$$

respectively. Also, (3) and (4) may be combined to yield

$$\frac{\partial p}{\partial t} + \boldsymbol{v} \cdot \nabla p + \gamma p \nabla \cdot \boldsymbol{v} = (\gamma - 1)\eta |\boldsymbol{J}|^2. \tag{12}$$

Under these transformations, (6b) remains unchanged while (6c) returns to the form (4), with $\boldsymbol{J} = \nabla \times \boldsymbol{B}$.

As shown by BNS, the momentum equation (10) yields the condition

$$p_1 + \boldsymbol{B}_0 \cdot \boldsymbol{B}_1 = 0. \tag{13}$$

Hence, the magnetic field fluctuation can be represented as

$$\boldsymbol{B}_1 = \nabla_\perp A \times \boldsymbol{b} - p_1 \boldsymbol{b}, \tag{14}$$

where $\boldsymbol{b} \equiv \boldsymbol{B}_0/B_0^2$ and A is the perturbed flux function. To $O(1)$, (12) yields the condition

$$\nabla_\perp \cdot \boldsymbol{v}_1 = 0, \tag{15}$$

which implies that the perturbed flow is incompressible in the (local) two-dimensional plane perpendicular to $\boldsymbol{B}_0$. By (15), the velocity fluctuation can be represented as

$$\boldsymbol{v}_1 = \nabla_\perp \phi \times \boldsymbol{b} - v_1 \boldsymbol{b}, \tag{16}$$

where ϕ is a stream function and v_1 is the parallel flow. Equations (14) and (16) contain four field variables: A, p_1, ϕ, and v_1. BNS demonstrate that the compressible resistive MHD equations (1)–(4) can be reduced to the four-field system:

$$\frac{dA}{dt} = \boldsymbol{B}_0 \cdot \nabla \phi + \eta \nabla_\perp^2 A, \tag{17}$$

$$\rho_0 \frac{d\Omega}{dt} = DJ + 2\boldsymbol{b} \times \nabla P \cdot \nabla_\perp p_1 - \left(\boldsymbol{b} \cdot \nabla B_0^2\right) J, \tag{18}$$

$$\frac{dp_1}{dt} = -\boldsymbol{v}_1 \cdot \nabla p_0 + \frac{\gamma p_0}{\gamma p_0 + B_0^2} \left[2\boldsymbol{v}_1 \cdot \nabla P + Dv_1 + \eta \nabla_\perp^2 p_1\right], \tag{19}$$

$$\rho_0 \frac{dv_1}{dt} = Dp_1 + \boldsymbol{B}_1 \cdot \nabla p_0, \tag{20}$$

Here

$$\Omega \equiv -\nabla_\perp^2 \phi, \tag{21a}$$

$$J \equiv -\nabla_\perp^2 A, \tag{21b}$$

$$P \equiv p_0 + B_0^2/2, \tag{21c}$$

$$\frac{d}{dt} \equiv \frac{\partial}{\partial t} + \boldsymbol{v} \cdot \nabla_\perp, \tag{21d}$$

$$D \equiv (\boldsymbol{B}_0 + \boldsymbol{B}_{1\perp}) \cdot \nabla, \tag{21e}$$

$$\nabla_\perp \equiv \nabla - \hat{\boldsymbol{B}}_0 \hat{\boldsymbol{B}}_0 \cdot \nabla, \quad \hat{\boldsymbol{B}}_0 \equiv \boldsymbol{B}_0/B_0, \tag{21f}$$

$$\rho_0 \equiv p_0^{1/\gamma}. \tag{21g}$$

The derivation of the four-field equations by BNS does not restrict the background inhomogeneous magnetic field and pressure in any way except for the requirement that they satisfy (8) which permits an infinity of solutions with spatial dependencies in the magnetic field as well as the pressure. For very low or zero values of β, (8) permits spatially dependent force-free magnetic fields of which the vacuum field is a special case. The solution (9) which is the starting point of NI-MHD is, in fact, the simplest non-trivial example of a vacuum field. In Sect. 3, we derive the equations for the NI-MHD model as a special case of the four-field equations.

3   The NI-MHD model

In order to derive the NI-MHD equations from the more general four-field equations, it is convenient to write

$$\boldsymbol{B}_0 = \frac{\hat{\boldsymbol{z}}}{\sqrt{\beta}} + \boldsymbol{B}_s, \tag{22}$$

where $\boldsymbol{B}_s$ represents the spatially inhomogeneous part of the background magnetic field which has been separated from a constant part. The separation assumed in writing (22) is not required for the validity of the four-field equations, but allows us to establish the conditions under which the NI-MHD equations hold. We now observe:

(i) If $\boldsymbol{B}_s \to 0$, then (17) and (18) simplify to the NI-MHD system with two field variables A and ϕ. Equations (19) and (20) decouple from (17) and (18) in this limit, and we recover the results of ZM. The equations for A and ϕ in the NI-MHD model are

$$\frac{dA}{dt} = \boldsymbol{B}_0 \cdot \nabla\phi + \eta\nabla_\perp^2 A \,, \tag{23}$$

$$\rho_0 \frac{d\Omega}{dt} = DJ \,. \tag{24}$$

(ii) The case $\beta \ll 1$ is effectively the same as $\boldsymbol{B}_s \to 0$, and conclusion (i) again holds.

In general, if $\beta \sim 1$, the effect of compressibility enters the dynamics at leading order and cannot be enslaved to an incompressible flow field. In this general case, $p_1 \neq 0$ and $\rho_1 \neq 0$ which differs from the predictions of the NI-MHD model. Due to the dynamical coupling at leading order between the pressure fluctuations and the flow field, brought about by the presence of spatial inhomogeneities, the first-order pressure and density fluctuations, even if they are small initially, can increase to the level of the Mach number.

Before we conclude this section, we remark on the role of a pressure-balanced structure (PBS) in the four-field model. An interesting simplification of the four-field equations occurs if we specialize to a PBS which obeys the relation $P \equiv p_0 + B_0^2/2 = $ constant. Then (17) and (18) decouple from (19) and (20), but (18) contains the additional term $(\boldsymbol{b} \cdot \nabla B_0^2)\,J$, absent in NI-MHD. This implies that pressure-balanced structures obey dynamical equations slightly more general than NI-MHD. However, we repeat for emphasis that the equation governing pressure fluctuations associated with a PBS in the four-field model remains decoupled from (18) and (19) (as in NI-MHD). This suggests that the mere presence of spatial inhomogeneities in a PBS is not enough to raise the level of pressure (or density) fluctuations to order M if they are small initially. In other words, if we are to understand the data on pressure fluctuations presented in Fig. 3 of [12] within the framework of the four-field model, we cannot do so by merely generalizing (8) to include only pressure-balanced structures. The background magnetic field must also include inhomogeneities associated with the curvature of magnetic field lines, absent in a PBS. The initial condition discussed in Sect. 4 does not obey the condition $P \equiv p_0 + B_0^2/2 = $ constant which, as we show below, leads to the generation of density fluctuations of order M.

4 Four-field simulations: implications for pressure fluctuations

We represent the initial magnetic field as

$$\boldsymbol{B}_0 = \nabla_{\perp 0} A_0(x, y) \times \hat{\boldsymbol{z}} + B_0 \hat{\boldsymbol{z}} , \tag{25}$$

where

$$\nabla_{\perp 0} = \hat{\boldsymbol{x}} \frac{\partial}{\partial x} + \hat{\boldsymbol{y}} \frac{\partial}{\partial y} , \tag{26}$$

and $A_0(x, y)$ is the mean flux representing the spatially inhomogeneous component of the initial magnetic field in the x-y plane. Assuming periodic boundary conditions in x and y, an exact solution of (11) is

$$A_0 = a \left(\cos 2\pi x - \sin 2\pi y \right) , \tag{27}$$

with

$$p_0 = 1 + 2\pi^2 A_0^2 . \tag{28}$$

In the initial state, we take the plasma β to be equal to one, held fixed at this value in all the four-field runs. Since the four-field equations have been derived with the ordering $p_0 \leq O(1)$, and p_0 is assumed to have the functional form (28), we must constrain the function A_0 by the inequality $2\pi^2 A_0^2 \leq 1$. Since the maximum magnitude of A_0 is $2a$, the parameter a is constrained by the inequality $a < 0.1125$.

In order to follow the time-evolution of this initial state, we have developed a 2 1/2-D pseudo-spectral code for the four-field equations. Periodic boundary conditions are applied in x and y, and all dependent field variables are represented in Fourier series, such as

$$\phi(x, y) = \sum_{mn} \phi_{mn} e^{2\pi i (mx + ny)} , \tag{29}$$

where m and n are integers. We refer the reader to BNS for more details of the numerical method, and limit ourselves here to a discussion of the link between spatial inhomogeneities and enhanced pressure fluctuations.

Figure 1 shows a contour plot of level surfaces of the mean flux A_0 on the x-y plane. This is a 2D projection of four flux tubes carrying currents, two parallel and two anti-parallel to $\hat{\boldsymbol{z}}$. The tubes are unstable to the coalescence instability which arises from the tendency of two opposite current-carrying tubes to attract each other. The tubes coalesce, squeezing magnetic flux and generating thin current sheets in-between them. Due to the presence of finite resistivity, magnetic reconnection occurs at the separatrices, facilitating the process of coalescence.

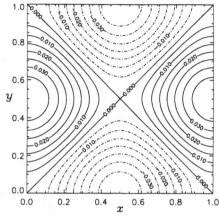

Fig. 1. Contour plot of the background inhomogeneous field A_0 [given by (25)]

The results of NI-MHD are recovered as a smooth limit of four-field MHD when the limit $a \to 0$ is taken. When a is very small ($\leq 10^{-6}$), numerical results demonstrate that the overall dynamics of the four-field system is very similar to NI-MHD dynamics ($a = 0$). However, even with moderate values a ($= 10^{-4}, 10^{-3}$), there are significant dynamical differences between four-field and NI-MHD dynamics.

Figure 2 shows the level of p_{rms} in the quasi-saturated state (indicated by diamonds) as a function of M for different values of a, with $M(t = 0) \approx 0.01$. In order to quantify the approximate saturation level of the fluctuations, two lines are drawn: a solid line corresponding to the level $p_{\mathrm{rms}} = M$, and a dashed line corresponding to the level $p_{\mathrm{rms}} = M^2$. We note that saturated values of p_{rms} approach the line $p_{\mathrm{rms}} = M$ as the parameter a increases. Even for values of a as small as 0.04, the condition $p_{\mathrm{rms}} = M$ is attained. This trend persists for higher values of a. Similar qualitative features also appear for higher initial values of M, as shown by data from runs with $M(t = 0) \approx$ 0.1 (indicated by stars).

Note that when a is small enough ($\leq 10^{-4}$) the quasi-saturated level of the pressure fluctuation lies below the $p_{\mathrm{rms}} = M^2$ line. This implies that the contribution from p_1 to the total pressure fluctuation may actually become subdominant to the second order contribution p_2 which is not zero in general. In the small a limit, we thus recover ZM's result that the pressure fluctuations are second order.

We have focused above on the issue of the scaling of the pressure fluctuations with the Mach number of the turbulence, identified in [12] as an outstanding feature of observations. The numerical results show that in the presence of spatial inhomogeneities, the four-field model produces density fluctuations that increase from $O(M^2)$ to $O(M)$ as the inhomogeneity parameter is increased. We thus conclude that if heliospheric and interstellar turbulence exists in a plasma with large-scale, non-turbulent spatial gradi-

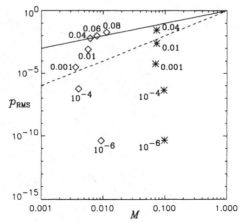

Fig. 2. The level of the root-mean-square value of the first order pressure p_{rms} in the quasi-saturated state as a function of the Mach number M for different values of a, for $M(t = 0) \approx 0.01$ (diamonds) and $M(t = 0) \approx 0.1$ (stars). The solid line corresponds to $p_{rms} = M$, and the dashed line to $p_{rms} = M^2$

ents, one expects the pressure fluctuations to be of significantly larger magnitude than suggested in nearly incompressible models such as pseudosound.

5   Anisotropic spectra in weak turbulence theory

In the presence of a directed magnetic field, MHD turbulence tends to exhibit a pronounced anisotropy. The main goal of this section is to examine, in the limit of weak turbulence, the nature and scaling of the anisotropic spectrum in a plasma permeated by a spatially uniform magnetic field $B = B_0\hat{z}$. As discussed above, the four-field equations reduce in this simple case to the NI-MHD (or RMHD) equations.

Weak magnetohydrodynamic turbulence in the presence of a uniform magnetic field is dominated by three-wave interactions that mediate the collisions of shear-Alfvén wave packets propagating in opposite directions parallel to the magnetic field. This dominance of three-wave interactions has been known since the advent of the Iroshnikov–Kraichnan theory [26,27], and has been the subject of a few recent papers [21,22,25]. Using the ideal NI-MHD equations, Ng and Bhattacharjee (NB) [22] calculate in closed form the three-wave and four-wave interaction terms, and show the former to be asymptotically dominant if the wave packets have non-zero $k_{\parallel} = 0$ components. To keep this discussion self-contained, we begin with a summary of relevant results by NB.

The ideal NI-MHD equations (23) and (24) can be rewritten as the system of equations,

$$\frac{\partial \Omega}{\partial t} - \frac{\partial J}{\partial z} = [A, J] - [\phi, \Omega],$$

$$\frac{\partial A}{\partial t} - \frac{\partial \phi}{\partial z} = -[\phi, A], \tag{30}$$

where the magnetic field is given by $\boldsymbol{B} = \hat{\boldsymbol{z}} + \nabla_\perp A \times \hat{\boldsymbol{z}}$ with A as the magnetic flux function, the flow velocity is given by $\boldsymbol{v} = \nabla_\perp \phi \times \hat{\boldsymbol{z}}$ with ϕ as the stream function, and $[\phi, A] \equiv \phi_y A_x - \phi_x A_y$. The parallel vorticity is then $\Omega = -\nabla_\perp^2 \phi$, and the parallel current density is $J = -\nabla_\perp^2 A$. Note that we have normalized the background uniform magnetic field in the $\hat{\boldsymbol{z}}$-direction to have unit magnitude, and the density has been chosen so that the Alfvén speed $V_A = 1$.

For weak interactions between two colliding shear-Alfvén wave packets $f^\pm$ traveling in the $\pm\hat{\boldsymbol{z}}$ directions, we write perturbative solutions of the form

$$\phi = f^-(\boldsymbol{x}_\perp, z^-) + f^+(\boldsymbol{x}_\perp, z^+) + \phi_1 + \phi_2 + \cdots,$$

$$A = f^-(\boldsymbol{x}_\perp, z^-) - f^+(\boldsymbol{x}_\perp, z^+) + A_1 + A_2 + \cdots, \tag{31}$$

where $\boldsymbol{x}_\perp = (x, y)$ is perpendicular to $\hat{\boldsymbol{z}}$ and $z^\pm = z \mp t$. Here $f^\pm(\boldsymbol{x}_\perp, z^\pm)$ represents Alfvén wave packets that propagate non-dispersively with the Alfvén speed $V_A = 1$. These exact solutions propagate inward from $z = \mp\infty$ at $t \to -\infty$, retaining their form, until they collide. Because of the intrinsic nonlinearity of (30), the interaction of two colliding wave packets cannot be simply described by linear combinations of $f^\pm(\boldsymbol{x}_\perp, z^\pm)$, for the linear combinations are not exact solutions of (30). For given zero-order fields $f^\pm$, we can then calculate the first-order fields in (31) from the equations

$$\frac{\partial \Omega_1}{\partial t} - \frac{\partial J_1}{\partial z} = 2\left\{[f^+, \nabla_\perp^2 f^-] + [f^-, \nabla_\perp^2 f^+]\right\} \equiv F, \tag{32a}$$

$$\frac{\partial A_1}{\partial t} - \frac{\partial \phi_1}{\partial z} = 2[f^-, f^+] \equiv G. \tag{32b}$$

This is a radiation equation for the first-order fields, with the source term determined by the overlap of the given zero-order fields f^+ and f^-. The source term is localized both in space and in time, assuming that the functional forms of $f^\pm$ are chosen so that the wave packets are localized in z. The asymptotic expression of ϕ_1, A_1 can be written,

$$\phi_1(\boldsymbol{x}_\perp, t \to \infty) \to f_1^-(\boldsymbol{x}_\perp, z^-) + f_1^+(\boldsymbol{x}_\perp, z^+), \tag{33a}$$

$$A_1(\boldsymbol{x}_\perp, t \to \infty) \to f_1^-(\boldsymbol{x}_\perp, z^-) - f_1^+(\boldsymbol{x}_\perp, z^+), \tag{33b}$$

where

$$f_1^\pm(\boldsymbol{x}_\perp, z) = \pi \int \left[\tilde{F}'(\boldsymbol{k}, \pm k_z) \mp \tilde{G}(\boldsymbol{k}, \pm k_z)\right] e^{i\boldsymbol{k}\cdot\boldsymbol{x}} d\boldsymbol{k}, \tag{34}$$

with $\tilde{F}'(\boldsymbol{k}, \omega) \equiv \tilde{F}(\boldsymbol{k}, \omega)/k_\perp^2$ and $\tilde{F}(\boldsymbol{k}, \omega)$ is the Fourier transform of $F(\boldsymbol{x}, t)$, defined by

$$F(\boldsymbol{x}, t) = \int \tilde{F}(\boldsymbol{k}, \omega) e^{i(\boldsymbol{k} \cdot \boldsymbol{x} - \omega t)} d\boldsymbol{k} d\omega .$$

The Fourier transform $\tilde{G}(\boldsymbol{k}, \omega)$ is similarly defined.

To simplify the calculation that follows, we consider the case in which the functions $f^\pm(\boldsymbol{x}_\perp, z)$ are separable, i.e., $f^\pm(\boldsymbol{x}_\perp, z) = f_\perp^\pm(\boldsymbol{x}_\perp) f^\pm(z)$. (The calculation can also be carried through in the more general case when $f^\pm(\boldsymbol{x}_\perp, z)$ can be written as a sum of such separable terms, which is always possible as long as the boundary conditions in $\boldsymbol{x}_\perp$ are periodic.) Then we can write

$$F(\boldsymbol{x}, t) = F_\perp(\boldsymbol{x}_\perp) f^+(z^+) f^-(z^-) , G(\boldsymbol{x}, t) = G_\perp(\boldsymbol{x}_\perp) f^+(z^+) f^-(z^-) ,$$

$$\tilde{F}(\boldsymbol{k}, \omega) = \frac{1}{2} \tilde{F}_\perp(\boldsymbol{k}_\perp) \tilde{f}^+(\kappa^+) \tilde{f}^-(\kappa^-) , \tilde{G}(\boldsymbol{k}, \omega) = \frac{1}{2} \tilde{G}_\perp(\boldsymbol{k}_\perp) \tilde{f}^+(\kappa^+) \tilde{f}^-(\kappa^-) ,$$

where $\kappa^\pm \equiv (k_z \pm \omega)/2$, $\tilde{f}(\kappa^\pm)$ is the one dimensional Fourier transforms of $f^\pm(z^\pm)$, $\tilde{F}_\perp$ and $\tilde{G}_\perp$ are the two dimensional Fourier transforms of $F_\perp$ and $G_\perp$. We obtain

$$f_1^\pm(\boldsymbol{x}_\perp, z^\pm) = \pi u_\perp^\pm(\boldsymbol{x}_\perp) \tilde{f}^\mp(0) f^\pm(z^\pm)/2 , \tag{35}$$

where

$$u_\perp^\pm(\boldsymbol{x}_\perp) = \int \left[\tilde{F}_\perp' \mp \tilde{G}_\perp \right] e^{i\boldsymbol{k}_\perp \cdot \boldsymbol{x}_\perp} d\boldsymbol{k}_\perp , \tag{36}$$

and $\tilde{f}^\mp(0)$ is the $k_z = 0$ Fourier component of $f^\pm(z)$, with $\tilde{F}_\perp'(\boldsymbol{k}_\perp) \equiv \tilde{F}_\perp(\boldsymbol{k}_\perp)/k_\perp^2$.

We note that the expression (36) for three-wave interactions preserves the z-dependence of the zero-order fields. This implies that there is no energy transfer parallel to the magnetic field. However, as pointed out by NB, four-wave interactions do not generally preserve the z-dependence of the zero-order fields and exhibit harmonic generation, and can, in principle, contribute to parallel energy transfer. Since three-wave couplings are much larger than four-wave couplings for weak turbulence, we neglect the effect of parallel energy transfer. Using (36), we can then calculate explicitly the scaling of three-wave interactions. Specifically, our objective is to calculate the spectral indices of the three-wave fields as functions of the spectral indices of the zero-order fields.

Imposing periodic boundary condition in $\boldsymbol{x}_\perp$, we can write

$$f_\perp^\pm(\boldsymbol{x}_\perp) = \sum_{mn} f_{mn}^\pm e^{2\pi i(mx + ny)} , \tag{37}$$

where $f_{mn}^\pm$ are constants. Let the energy spectra have the form

$$E_\pm(k_\perp) \propto k_\perp^{-\mu_\pm} \text{ or } |f_{mn}^\pm| \propto (m^2 + n^2)^{-(3+\mu_\pm)/4} , \tag{38}$$

where $\mu_{\pm}$ are the spectral indices. The profiles of the three-wave interaction fields can then be calculated by

$$u_{\perp}^{\pm}(\boldsymbol{x}_{\perp}) = \sum_{mn} [F'_{mn} \mp G_{mn}] e^{2\pi i(mx+ny)} = \sum_{mn} u_{mn}^{\pm} e^{2\pi i(mx+ny)} , \qquad (39)$$

having spectral indices $\nu_{\pm}$, i.e., $|u_{mn}^{\pm}| \propto (m^2 + n^2)^{-(3+\nu_{\pm})/4}$, where $F'_{mn} = F_{mn}/(2\pi)^2(m^2 + n^2)$ with

$$\sum_{mn} F_{mn} e^{2\pi i(mx+ny)} = 2\left\{[f_{\perp}^{+}, \nabla_{\perp}^2 f_{\perp}^{-}] + [f_{\perp}^{-}, \nabla_{\perp}^2 f_{\perp}^{+}]\right\} ,$$

$$\sum_{mn} G_{mn} e^{2\pi i(mx+ny)} = 2[f_{\perp}^{-}, f_{\perp}^{+}] . \qquad (40)$$

It is found both analytically and numerically that [28]

$$\nu_{+} \approx \mu_{-} , \quad \nu_{-} \approx \mu_{-} - 2 \quad \text{for} \quad \mu_{+} \gg \mu_{-} , \qquad (41a)$$

$$\nu_{+} \approx \mu_{+} - 2 , \quad \nu_{-} \approx \mu_{+} \quad \text{for} \quad \mu_{+} \ll \mu_{-} . \qquad (41b)$$

Our main objective now is to determine how the spectrum of an Alfvén wave packet changes in time after many collisions with wave packets coming from the opposite direction. To be specific, let us consider the evolution of a f^{+} field interacting with a sequence of random f^{-} fields. From (39) and (40), we deduce that

$$\frac{\partial \Psi^{+}}{\partial t} = -[f^{-}, \Psi^{+}] + [f_{x}^{-}, f_{x}^{+}] + [f_{y}^{-}, f_{y}^{+}] , \qquad (42)$$

where $\Psi^{+} = -\nabla_{\perp}^2 f^{+}$. Numerically, the Fourier amplitudes f_{mn}^{-} are randomly chosen for a given spectral index μ_{-} in every time step τ_{A}. The time step has to be chosen small enough to satisfy the weak turbulence assumption and to represent the fact that each wave packet in the sequence of f^{-} is uncorrelated with each other. Also, in order to have a better resolved inertial range of the f^{+} spectrum, a hyper-dissipation term of the form $\eta \nabla_{\perp}^6 \Psi^{+}$ is added to the right hand side of (42) with a suitably chosen η so that the inertial range is resolved with an index that is insensitive to the value of η. Equation (42) is then solved by a pseudo-spectral method for different values of μ_{-} and for different levels of resolution, up to 1024^2, until the f^{+} spectrum reaches a quasi-steady state when inertial range index is roughly a constant with only small temporal fluctuations.

Figure 3 shows the f^{+} spectra for the case with $\mu_{-} = 2$ for different resolution levels. We see that the inertial range for all runs roughly have the same index, $\mu_{+} \approx 2$. In this case, we obtain the anisotropic energy spectrum $k_{\perp}^{-2}$. This result has been derived earlier by dimensional analysis [20,22,24] and more recently, by careful analytical and numerical work based on a weak turbulence formalism [25].

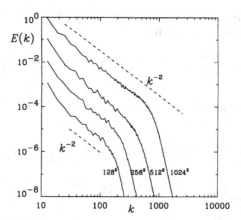

Fig. 3. The spectra of the f^+ field with $\mu_- = 2$ for different resolution level indication beside each curve. A vertical separation is added in between each pair of spectra for clarity

Acknowledgments

This research is supported by the National Science Foundation Grant No. ATM-9801709.

References

1. Zank, G. P., Matthaeus, W. H. (1992) The equations of reduced magnetohydrodynamics. J. Plasma Phys. **48**, 85
2. Zank, G. P., Matthaeus, W. H. (1993) Nearly incompressible fluids. II: Magnetohydrodynamics, turbulence, and waves. Phys. Fluids A **5**, 257
3. Lighthill, M. J. (1952) On sound generated aerodynamically, I. General Theory. Proc. R. Soc. London Ser. A **211**, 564
4. Montgomery, D., Brown, M., Matthaeus, W. H. (1987) Density fluctuation spectra in magnetohydrodynamic turbulence. J. Geophys. Res. **92**, 282
5. Shebalin, J. V., Montgomery, D. (1988) Turbulent magnetohydrodynamic density fluctuations. J. Plasma Phys. **39**, 339
6. Matthaeus, W., Brown, M. R. (1988) Nearly incompressible magnetohydrodynamics at low Mach-number. Phys. Fluids **31**, 3634
7. Zank, G. P., Matthaeus, W. H. (1991) The equations of nearly incompressible fluids. I. Hydrodynamics, turbulence and waves. Phys. Fluids A **3**, 69
8. Grappin, Velli, R. M., Mangeney, A. (1991) "Alfvénic" versus "standard" turbulence in the solar wind. Ann. Geophys. **9**, 416
9. Kliatskin, V. I. (1966) Homogeneous isotropic turbulence in weakly compressible media. Izv. Atmos. Oceanic Phys. **2**, 474
10. Strauss, H. R. (1976) Nonlinear three-dimensional magnetohydrodynamics of noncircular tokamaks. Phys. Fluids **19**, 134

11. Matthaeus, W. H., Klein, L. W., Ghosh, S., Brown, M. R. (1991) Nearly incompressible magnetohydrodynamics and solar wind fluctuations. J. Geophys. Res. **96**, 5421

12. Tu, C.-Y., Marsch, E. (1994) On the nature of compressive fluctuations in the solar wind. J. Geophys. Res. **99**, 21481

13. Bavassano, B., Bruno, R., Klein, L. (1995) Density-temperature correlation in Solar-wind magnetohydrodynamic fluctuations — a test for nearly incompressible models. J. Geophys. Res. **100**, 5871

14. Bhattacharjee, A., Ng, C. S., Spangler, S. R. (1998) Weakly compressible magnetohydrodynamic turbulence in the solar wind and the interstellar medium. Astrophys. J. **494**, 409

15. Armstrong, J. W., Coles, W. A., Kojima, K., Rickett, B. J. (1990) Observations of field-aligned density fluctuations in the inner solar wind. Astrophys. J. **358**, 685

16. Grall, R. R., Coles, W. A., Spangler, S. R., Sakurai, T., Harmon, J. K. (1997) Observations of field-aligned density microstructure near the Sun. J. Geophys. Res. **102**, 263

17. Wilkinson, P. N., Narayan, R., Spencer, R. E. (1994) MNRAS **269**, 67

18. Frail, D. A., Diamond, P. J., Cordes, J. M., Van Langevelde, H. J. (1994) Anisotropic scattering of OH/IR stars toward the galactic center. Astrophys. J. **427**, L43

19. Molnar, L. A., Mutel, R. L., Reid, M. J., Johnston, K. J. (1995) Interstellar scattering toward cygnus X-3: measurements of anisotropy and of the inner scale. Astrophys. J. **438**, 708

20. Sridhar, S., Goldreich, P. (1994) Toward a theory of interstellar turbulence. I. weak Alfvénic turbulence. Astrophys. J. **432**, 612

21. Montgomery, D., Matthaeus, W. H. (1995) Anisotropic modal energy transfer in interstellar turbulence. Astrophys. J. **447**, 706

22. Ng, C. S., Bhattacharjee, A. (1996) Interaction of shear-Alfvén wave packets: implication for weak magnetohydrodynamic turbulence in astrophysical plasmas. Astrophys. J. **465**, 845

23. Chen, S, Kraichnan, R. H. (1997) Inhibition of turbulence cascade by sweep. J. Plasma Phys. **57**, 187

24. Goldreich, P., Sridhar, S. (1997) Magnetohydrodynamic turbulence revisited. Astrophys. J. **485**, 680

25. Galtier, S., Nazarenko, S., Newell, A. C., Pouquet, A. (1998) A weak turbulence theory for incompressible magnetohydrodynamics. These proceedings

26. Iroshnikov, P. S. (1963) The turbulence of a conducting fluid in a strong magnetic field. Astron. Zh. **40**, 742

27. Kraichnan, R. H. (1965) Inertial-range spectrum of hydromagnetic turbulence. Phys. Fluids **8** 1385

28. Ng, C. S., Bhattacharjee, A. (1997) Scaling of anisotropic spectra due to the weak interaction of shear-Alfvén wave packets. Phys. Plasma **4**, 605

Alfvénic Turbulence and Wave Propagation in the Corona and Heliosphere

M. Velli[1,2]

[1] Dipartimento di Astronomia e Scienza dello Spazio,
  Università di Firenze, I-50125 Firenze, Italy
[2] JPL, 4800 Oak Grove Drive, 91109 Pasadena, CA

Abstract. The properties of Alfvén wave propagation through the solar corona and heliosphere are reviewed, with particular emphasis on the role of large scale radial gradients in determining the variation of amplitude with distance. Some comments on the role of photospheric vorticity as a source for Alfvén waves are presented, 1-D and 2-D MHD couplings of the waves as they climb out of the coronal density gradient are described (parametric decay, sound wave generation), while the final section is devoted to the 1-D propagation of Alfvén type solitons in the solar wind. Throughout, the focus is on the relevance of theory to the turbulent fluctuations measured in the wind and remote sensing observations of the corona: many puzzles and problems are highlighted.

1 Alfvén waves observed in the solar wind

Fluctuations in the high-speed solar wind streams with periods below a few hours, and down to periods of minutes and less are found to be dominated by what is known as Alfvénic turbulence, that is a well developed turbulence spectrum which has all the properties of a flux of large amplitude, constant magnetic field magnitude Alfvén waves propagating away from the sun. The properties of such fluctuations have been summarized in Grappin et al. (1993) [1] as far as Helios observations are concerned, while the observations within the high speed flow at polar latitudes by the Ulysses spacecraft are described in Horbury et al. (1996) [2]. Denoting the magnetic fluctuations and velocity fluctuations by **b** and **v** respectively, and defining $z^\pm = \mathbf{v} \mp \text{sign}(\mathbf{B})\mathbf{b}/\sqrt{4\pi\rho}$, (we have incorporated changes in the sign of the average field in the definition of Alfvén waves), we may characterize Alfvénic turbulence by the relations $\delta|\mathbf{B}|^2 \ll |\mathbf{b}|^2$, i.e. small total magnetic intensity fluctuations; $|z^+| >> |z^-|$, i.e. outward propagating waves dominate ; $|\delta\rho/\rho|^2 << |\mathbf{v}/C_s|^2 = M_T^2$, where C_s is the sound speed and M_T the turbulent Mach number. In standard MHD turbulence on the other hand, all the $\ll, \gg$ above become $\simeq$. With little exceptions, at at least at solar minimum, solar wind turbulence varies continuously between the Alfvénic state (in the polar wind and in trailing edges of high speed streams in the ecliptic plane) and the standard state (slow wind at magnetic sector crossings). Incompressible MHD turbulence predicts Alfvénic turbulence as the asymptotic outcome when initial conditions have $u \simeq \delta b / \sqrt{(4\pi\rho)}$ (u being the absolute value of velocity fluctuations). There is some indication that this result is also valid in compressible MHD, while the observed evolution with heliocentric distance is such that Alfvénic turbulence decays towards "standard": the power index of the transverse magnetic field spectrum is typically $\alpha \simeq -1$ for lower frequencies close to the sun, decreasing to the Kolmogorov value $\alpha \simeq -1.6$ at higher frequencies. The bend in the spectrum moves to lower frequencies with increasing distance from the sun, the evolution being somewhat faster within high-speed streams in the ecliptic plane and slower in the polar wind. Together with the evolution in the shape of the spectrum, the specific energy in the fluctuations also varies with distance from the sun, in a way which is roughly consistent, $e \sim r^{-1}$ (r being heliocentric distance, normalized to the solar radius), with the conservation of wave action at the lowest frequencies ($\simeq 10^{-3}$Hz). Suggestions to solve this paradox have included nonlinear evolution due to the in situ generation of inward modes in the solar wind (such modes are necessary, in incompressible MHD, to have nonlinear interactions) and the interaction of the waves with the large scale magnetic field and velocity shears in the current sheet and between fast and slow streams (Veltri, 1999 [3], these proceedings). In this paper we discuss some aspects of the formation and evolution of Alfvénic turbulence. First we will summarize the properties arising from linear propagation in the coronal and solar wind gradients. Then we will discuss possible generation mechanisms close to the sun and some non-linear couplings during the outward propagation. Finally, we will discuss the 1D evolution of Alfvénic solitons, which have been suggested as a possible basic ingredient of the turbulence observed in the solar wind.

2   Basic features of linear propagation

2.1   Propagation through a static atmosphere

The basic equations for transverse magnetic field (**b**) and incompressible velocity (**v**) fluctuations may be written in terms of the Elsasser variables (defined above) which in a homogeneous medium describe Alfvén waves propagating in opposite directions along the average magnetic field $\mathbf{B}_0$:

$$\frac{\partial \mathbf{z}^{\pm}}{\partial t} \pm \mathbf{V}_a \cdot \nabla \mathbf{z}^{\pm} \mp \mathbf{z}^{\mp} \cdot \nabla \mathbf{V}_a \pm \frac{1}{2}(\mathbf{z}^{\mp} - \mathbf{z}^{\pm})\nabla \cdot \mathbf{V}_a = 0, \qquad (1)$$

where $\mathbf{V}_a$ is the mean (large-scale) Alfvén velocity. The first two terms in eq. (1) describe wave propagation; the third term describes the reflection of waves by the gradient of the Alfvén speed along the fluctuations (which vanishes for a vertical field in a planar atmosphere, but is different from zero in the more realistic case of a spherically or supraspherically diverging flux tube); the fourth term describes the WKB amplitude variation (which occurs because energy flux must be conserved in the medium with variable wave speed) and the isotropic part of the reflection. In eq. (1) gravity and terms involving the gradients of the average density along the fluctuation polarisation are absent: this is because the average magnetic field and gravity are assumed to be collinear. Equation (1) then describes the parallel propagation of fluctuations in the plane perpendicular to **B**, or in the case of spherical or cylindrical symmetry, the propagation of toroidal fluctuations in the equatorial plane. In more general cases the magnetic, velocity and density fluctuations are coupled together via magnetoacoustic modes, a process we neglect here but will come back to in subsequent sections. Conservation of net upward energy flux may be written as

$$S^+ - S^- = S_\infty, \qquad S^\pm = FV_a|\mathbf{z}^\pm|^2/8, \qquad (2)$$

where S_∞ is the constant flux and $F = \rho r^\sigma$ (r is the normalized radial distance from the base of the atmosphere and σ is the infinitesimal flux tube expansion factor: $\sigma=0$, 2 in a plane and spherical atmosphere respectively). One may then define the transmission coefficient across an atmospheric layer bounded by regions of constant Alfvén speed by applying the boundary condition that only an outward propagating wave should exist above the layer in question: T is then given by $T = S_\infty/S_0^+$, where S_∞ coincides with the energy flux carried by the outwardly propagating wave, while S_0^+ is the outward propagating energy flux at the atmospheric base. Note that for waves of frequency ω and wavevector $k = \omega/V_a$, eq. (1) becomes, after elimination of the systematic amplitude variation of $z^\pm$ through the renormalization $z^\pm = \rho^{1/4}z^\pm$,

$$z^{\pm\prime} \mp ikz^\pm - \frac{1}{2}\frac{k'}{k}z^\mp = 0, \qquad (3)$$

(a prime denotes differentiation with respect to r). With the propagation equation written in this form, it becomes obvious that the relative importance of reflection (the term coupling of z^+ and z^-) and propagation are determined by the non-dimensional ratio $\epsilon_a = |k'/2k^2| = |V_a'/2\omega|$. Velli (1993) [4] discusses the properties of eq. (3) in detail, and develops a general formalism for obtaining approximate analytical solutions by dividing the region of propagation into intervals where $\epsilon_a < 1$ (propagation dominates over reflection), and regions where $\epsilon_a > 1$ (reflection dominates). An important point to recall is that for very long wave-length waves propagating over a region with varying Alfvén speed the transmission coefficient may be simply written as

$$T = \frac{4V_{al}V_{ar}}{(V_{al} + V_{ar})^2}, \qquad (4)$$

where $V_{al,r}$ indicate the Alfvén speed on either side of the layer.

For an isothermal, static, spherical corona and a radial magnetic field the Alfvén speed depends on radius as

$$V_a = \frac{V_{a0}}{r^2}\exp((\frac{\alpha}{2}(1 - \frac{1}{r})), \qquad \alpha = \frac{GM_0}{R_0C_s^2}, \qquad 1 < r < \infty, \qquad (5)$$

where C_s is the isothermal sound speed, R_0 the coronal base radius. The parameter α, for the sun, typically lies in the range $4 \leq \alpha \leq 15$ for coronal temperatures between $8.0\ 10^5 - 3.0\ 10^6$ °K. For this family of profiles, the Alfvén speed first increases exponentially, has a maximum in $r = \alpha/4$ and then decreases asymptotically as $V_a \sim r^{-2}$. The general behaviour of T may be gleaned from the low frequency approximation eq. (4) if one is careful to remember that the thickness of the reflection-dominated layer depends on the frequency and extends from $r = 1$ (for frequencies such that $\epsilon(r = 1) \ll 1$ to the distance r_ω where $\epsilon = 1$. Writing $\Omega = \omega R_0/V_{a0}$ one has

$$\epsilon = 1 \rightarrow \Omega \simeq 1/r_\omega^3 \exp(\alpha/2). \tag{6}$$

In other words the corona becomes transparent to Alfvén waves at a distance $r_\omega \simeq \exp(\alpha/6)\Omega^{-1/3}$, where the value of the Alfvén speed is

$$V_a = V_{a0}\Omega^{2/3}\exp\left(\frac{\alpha}{6}\right). \tag{7}$$

Substitution into eq. (4) then shows that transmission for very low frequencies should increase like $\Omega^{2/3}$, reach a maximum where $V_a = V_{a0}$ and then decrease again as $\Omega^{-2/3}$. This was shown in [4], where the value of the transmission at the maximum was computed both numerically and analytically (its value is not 1, but about 0.6). The above result holds true provided α is large enough ($\alpha \geq 8$) so that the region where the Alfvén speed reaches a maximum (and where propagation always dominates) is small enough (in fact, it is the presence of the maximum and hence of two distinct non-propagating regions on either side of the maximum to yield the factor 0.6 mentioned above). For smaller values of α one only sees a monotonic increase of the transmission with frequency going as $\Omega^{2/3}$. For the range of temperatures compatible with the solar corona wave periods below 15 minutes are completely transmitted, and even for periods of a few hours transmission is above 50%.

2.2 Propagation in the solar wind

In the above discussion the presence of the solar wind, which becomes fundamental above a few solar radii. In this case the wave propagation equation becomes ([5]),

$$\frac{\partial \mathbf{z}^\pm}{\partial t} + (\mathbf{U} \pm \mathbf{V}_a) \cdot \nabla \mathbf{z}^\pm + \mathbf{z}^\mp \cdot \nabla(\mathbf{U} \mp \mathbf{V}_a) + \frac{1}{2}(\mathbf{z}^- - \mathbf{z}^+)\nabla \cdot \left(\mathbf{V}_a \mp \frac{1}{2}\mathbf{U}\right) = 0, \tag{8}$$

where $\mathbf{U}$ is the average wind velocity. The wind and Alfvén speed profiles corresponding to the previous isothermal atmosphere are now (introducing the mach number $M = U/C_s$)

$$\left(M - \frac{1}{M}\right)M' = \frac{2}{r} - \frac{\alpha}{r^2}, \qquad V_a = V_{a0}/r(U/U_0)^{1/2}. \tag{9}$$

For waves propagating in the wind, the energy flux is no longer conserved since the wave pressure does work in the expansion. The wave-action however is still conserved,

$$S^+ - S^- = S_\infty, \qquad S^\pm = F\frac{(U \pm V_a)^2}{V_a}|z^\pm|^2/8, \qquad (10)$$

where F is the geometrical factor defined previously. With a conserved flux (in this case the wave action) one can still associate a transmission coefficient as long as there is a position where there is no "inward propagating" wave. From eq. (10) it follows that the inward flux vanishes at the Alfvénic critical point where the solar wind speed equals the Alfvén speed, so that $T = S_\infty/S_0^+$, where now however the wave action flux is determined by the amplitude of the outwardly propagating wave at the critical point as $S_\infty = S_c^+ = F|z^+|_c^2/2$. Remarkably, the transmission T for moderate to high frequencies in this case parallels the static computation exactly, and the conclusions of the previous paragraph remain valid. At the lowest frequencies instead one finds a transmission coefficient which is significantly enhanced. This may be understood by rewriting eq. (8) in the low frequency case in terms of $y^\pm = (U \pm V_a)z^\pm$:

$$y^{\pm\prime} - \frac{1}{2}\frac{V_a'}{V_a}y^\mp - \frac{1}{2}\Big(\frac{U'}{U} + \frac{V_a'}{V_a}\Big)y^\pm = 0 \qquad (11)$$

which has the two solutions (as in the static case) $y^+ = \pm y^-$. Imposing that y^- vanish at the critical point then gives ([5],[6], [7])

$$z^\pm(r) \sim \frac{1}{U \pm V_a}\Big(\frac{U}{U_c}\Big)^{1/2}\Big(\frac{V_a}{V_{ac}} \pm 1\Big),$$

and the subscript denotes quantities calculated at the Alfvénic critical point. For the low frequency limit of the transmission we obtain

$$T = \frac{4U_0 V_{a0}}{(U_0 + V_{a0})^2}\frac{z_c^{+2}}{z_0^{+2}} = \frac{4V_{a0}V_{ac}}{(V_{a0} + V_{ac})^2}.$$

Thus we see that it is possible to have perfect transmission at low frequencies, if the Alfvén speed at the coronal base and the critical point are 'tuned' close to the same value. We remark that the above result is independent of the position of the coronal base, provided the geometry allows for the propagation of a pure Alfvén-type wave.

The meaning of the transmission coefficient into the wind requires some discussion however since we have seen that it is calculated exactly at the Alfvén critical point. Beyond this point both "inward" and "outward" modes are carried together outwards because the wind speed is greater than the mode propagation speed, so that in this sense it is a good definition. However it is not true that at greater distance the amplitude of the inward mode vanishes, on the contrary, the normalized cross-helicity

$$\sigma_c = \frac{z^{+2} - z^{-2}}{z^{+2} + z^{-2}} \tag{12}$$

which in the static case is by definition equal to one when there is only an outward propagating wave (in which case the specific energy in velocity and magnetic field fluctuations is the same) continues to evolve. In the spherically expanding case, the behaviour of σ_c beyond the critical point depends on the frequency. This may be seen using a particularly simple model for wind velocity and Alfvén speed at large distances: $U = U_\infty$, a constant, so the radial Alfvén speed goes as $V_a = V_{a\infty}/r$. Equation (8) may now be rewritten as

$$\left(U_\infty \pm \frac{V_{a\infty}}{r}\right)z^{\pm\prime} - i\omega R_0 z^\pm + \frac{1}{2r}(z^+ + z^-)\left(U_\infty \mp \frac{V_{a\infty}}{r}\right) = 0. \tag{13}$$

An eikonal expansion which treats the boundary condition at the critical point correctly (see e.g. Barkhudarov (1991) [8], Velli et al. (1991) [9]) then shows that at large distances

$$z^+ \simeq -\left(\left(1 - 4\omega^2 V_{a\infty}^2 R_0^2/U_\infty^4\right)^{\frac{1}{2}} - 2i\omega V_{a\infty} R_0/U_\infty\right)^{-1} z^- + O(1/r)z^-. \tag{14}$$

For all frequencies greater than $\omega_0 = U_\infty^2/2V_{a\infty}R_0$ the normalized cross helicity σ_c increases with distance beyond the Alfvén critical point to a frequency-dependent limiting value which tends to one at high frequencies as $1 - (\omega_0/\omega)^2$. At frequencies below ω_0 however σ_c decreases with distance and tends asymptotically to 0, i.e., we have total reflection at infinity. This critical frequency has a straightforward physical interpretation, in terms of the relative strength of the wave coupling and gradient, or rather expansion effects. An Alfvén wave is a coupling of transverse magnetic and velocity fluctuations in which the underlying field-line tension provides the restoring force. In the presence of a wind, the equations are modified by the outward flow, but the angular momentum and magnetic flux must be conserved. This translates into the appearance of a decaying term in u/r for the transverse velocity fluctuation u in the momentum equation, and a decaying term b/r for the transverse magnetic field b in then induction equation, which disappears if b is renormalized with the square root of the density (i.e. one writes it in terms of the transverse alfven velocity $b = b/\sqrt{(4\pi\rho)}$ as was done earlier in defining the Elsasser variables). The equations for the fluctuations then become

$$\frac{\partial u}{\partial t} + U_\infty u' - \frac{V_{a\infty}}{r}b' + \frac{U_\infty}{r}u = 0, \tag{15}$$

$$\frac{\partial b}{\partial t} + U_\infty b' - \frac{V_{a\infty}}{r} u' = 0, \tag{16}$$

which are the same as eq. (13) where we have neglected the gradients of the Alfvén speed with respect to the divergence of the bulk velocity field in the last term in parentheses of eq. (13).

At large distances and to lowest order oscillations with frequency ω have a wave-number given by $kR_0 = \omega/U_\infty$ (first two terms in eqs.(15,16). If the Alfvén speed is vanishingly small (i.e. low-frequency oscillations) the magnetic and velocity fields are decoupled entirely, the transverse velocity decays as $u \sim 1/r$ while the magnetic field is constant. This translates into a cross-helicity which tends to zero at great distances. If the Alfvén speed is not negligible, one may substitute for the terms of type $V_{a\infty}b'/r$ the value obtained with the wave-number kR_0 to get $V_{a\infty}R_0\omega b/U_\infty r$. This term depends on r in the same way as the angular momentum conservation term $U_\infty u/r$ in eq. (15), the relative magnitude of the two being given dimensionally by $\omega V_{a\infty}R_0/U_\infty^2 = \omega/2\omega_0$. Therefore, for frequencies much larger than the critical one, the Alfvénic coupling is important, and in this regime u, b are constrained to evolve together (i.e. reflection may be neglected), both fields decaying asymptotically $u, b \sim 1/\sqrt{(r)}$. For frequencies below the critical one, the fields of course decouple as shown more rigorously by the expansion eq. (14).

The critical frequency is a number of some importance: in the high-speed solar wind streams, where $U \simeq 800$ km/sec, and assuming a typical value for the Alfvén speed $V_a \simeq 50$ km/sec at $R = 1$ AU, we obtain $\omega_0 \simeq 4.27\ 10^{-5}$sec, corresponding to a period of about 41 hours. This is quite a long period, while Alfvénic turbulence is seen at periods substantially lower, from several hours to a few minutes, and indeed the specific energy in this range appears to fall as r^{-1}, which is consistent with $u, b \sim 1/\sqrt{r}$.

We have seen that some aspects of the turbulence observed in the wind are consistent with a flux of Alfvén waves of coronal origin. On the other hand, the origin of the spectrum, and its evolution, remain mysterious. In the next section we briefly investigate the possible origin of the waves below the corona, where the stratification becomes stronger and geometry more complex. Correspondingly, our discussion will be more qualitative.

3   Observations of Alfvén wave generation

In searching for the solar source of the outward propagating wave flux one must rely mostly on observational diagnostics based on velocity rather than magnetic field fluctuations, given the difficulty of measuring fields above the photosphere. The solar atmosphere abounds with dynamic phenomena that could either be a source of waves or the direct consequence of wave generation below, for example spicules and macrospicules which are ubiquitous outside active regions.

Spicules are jets of material seen in H_α outlining the chromospheric network, with diameters of about 10^3 km, that extend up to $6\ 10^3 - 10^4$ km in the corona. Their lifetime is of order $5 - 10$ min, vertical velocities are typically $v_z \simeq 20 - 30$ km/sec and the jets are observed to fade out, diffuse or fall back down. Characteristic electron temperatures and densities are of order $T_e \simeq 1.5\ 10^4$K, $n_e \simeq 1 - 2\ 10^{11}$ cm^{-3}. Observations by Pasachoff et al.(1968) [10] and subsequently Cook (1991) [11] seemed to show torsional motions in spicules, possibly up to speeds of order 50 km/sec.

Macrospicules are larger versions of $H\alpha$ spicules with diameters up to 10^4 km, that extend higher up, from $2 - 4\ 10^4$ km into the corona, have longer lifetimes, up to about 45 min, and higher vertical velocities, up to about 150 km/sec. They are seen in lines formed between $6\ 10^4$ K and $3\ 10^5$K and in radio observations (6 cm) where they appear to tilt and expand at their tops. Macrospicules seen in the EUV are associated with $H\alpha$ macrospicules, and may be associated with coronal X-ray bright points. Pike and Harrison (1997) reported observations of a macrospicule using the CDS instrument on SOHO and found relative Doppler velocities of up to 150 km/sec in OV emission, which they interpreted as an outward flow possibly accelerating into the solar wind.

Pike and Mason (1998) [13] found that CDS macrospicules observed on the limb show red and blue shifts on opposing sides of the macrospicule axis. The estimated velocities are just about symmetric with respect to the axis and increase from $0 - 10$ km/sec at the surface to about $20 - 50$ km/sec at heights of $1.5\ 10^4$ km. They interpreted such structures as rotating features, with rotational velocities increasing with height, possibly superimposed on an overall axial plasma acceleration. As to the structure of the rotating jet, observation of a dark lane in the OV emission lead them to conclude that it is made of a cooler material core (presumably emitting in $H\alpha$) surrounded by a transition region sheath of hotter material. The authors suggested the reconnection of emerging twisted flux tubes with pre-existing fields as the cause for the formation of such "tornadoes", along the lines of work reviewed in Shibata (1997) [14]. A twisted magnetic loop reconnecting with an open field line and unwinding does so principally by emitting torsional Alfvén waves. However, such waves are also generated directly by photospheric turbulent velocity fields and the waves may only differ in the overall duration of individual packets as well as intrinsic frequencies excited.

The propagation of waves generated in the photosphere/chromosphere up through the transition region and corona has been the subject of extensive investigations in both linear and non-linear regimes (see, e.g. [15]). A linear wave propagating upwards from the photosphere suffers significant reflection in the higher chromosphere and transition region, if one assumes horizontal homogeneity and the radial density stratification described by standard one dimensional models of the solar atmosphere: in this respect, generation via magnetic reconnection at transition region heights would circumvent the transmission problem by shifting the generation height of the waves to the corona, where we have seen that transmission is not a problem, as well as creating higher frequency waves as will be shown below. On the other hand, the stratification in magnetic regions of the solar atmosphere probably deviates significantly from the one described above, so that the Alfvén velocity gradient as observed while moving upwards within a thin expanding photospheric flux tube could be significantly weaker than the corresponding gradient in the weakly magnetized atmosphere outside the flux tubes, allowing a significant energy flux to enter the corona.

Shibata and Uchida (1986) [16] first proposed the reconnected, untwisting jet mechanism for the solar atmosphere and carried out 2D cylindrically symmetric simulations of a two-temperature atmosphere (representing the chromosphere and corona) with gravitational stratification. They started with a field twisted in its inner core extending from the chromospheric layer into the lower corona. As the twist is released it creates a jet propagating upwards and fanning out with height, attaining speeds that are a significant fraction of the ambient Alfvén speed, when the twist of the initial tube is high (i.e., they always considered azimuthal field strengths of the same order of magnitude or larger than the axial field strengths).

Hollweg et al. (1982) [17] and subsequently Hollweg (1992) [15], carried out detailed 1D simulations of the propagation of a torsional pulse (of 90 second duration) within a fixed axially symmetric magnetic field geometry, and observed the formation of a fast and slow shock as the pulse propagated upward, as well as reflection at the transition region and upwelling of the latter, which was tentatively interpreted as formation of a spicule. More recently, Kudoh and Shibata (1997) [18] carried out a 1D simulation similar in spirit, but with a continuously varying random azimuthal velocity at the base of the flux tube.

The given flux tube geometry is such that the flux tube expands from a 280 km width, 1600 G field intensity in the photosphere, to a diameter of $4 \ 10^3$ km, with a 7.8 G field intensity in the corona, while the photospheric density is $3 \ 10^{-7}$ g/cm^3 and falls to 10^{-15} g/cm^3 in the corona. As a result, the Alfvén speed varies from 8.25 km/sec in the photosphere to 695 km/sec in the corona. The initial rms photospheric azimuthal velocity of $\simeq 1$ km/sec becomes an rms velocity of $\simeq 20$ km/sec in the corona, a factor of 5 less than expected from a WKB (0 reflection, conserved energy flux) estimate of the velocity. The variations in amplitude of the driver, combined with the background stratification, create a wave associated force which drives a slow (sound) wave upwards as well as a jet. The rms axial coronal velocity is found to be 9.62 km/sec. The azimuthal forcing drives the transition region upwards from $2 \ 10^3$ km to $5 \ 10^3$ km and higher. Concerning the time-scale of the driver, the variations in azimuthal velocity were composed of a rapid (less than 1 minute period) variation superimposed on a more regular 5 minute oscillation of 2 km/sec amplitude.

Although the calculated coronal velocities are of the correct order of magnitude for observed spicules, the flux tube model used may underestimate the transmitted power. The reason is that a 1600 G flux tube is close to equipartition with the photospheric pressure, and must therefore be significantly evacuated. If this is the case, the Alfvén speed in the tube is much higher than the estimate given above, and therefore the overall Alfvén speed gradient is greatly reduced, allowing for improved wave transmission. This implies that 20 - 40 km/sec speeds in the corona might result from photospheric velocity fields with an rms amplitude well below 1 km/sec.

Consider now the characteristics of an Alfvén wave generated by reconnection: if a twisted flux tube of radius r and length L emerges with a winding angle $N = LB_\phi/2\pi r B_z \geq 1$, (where B_ϕ, B_z are the poloidal and axial magnetic fields respectively), and reconnects, a wave-packet with typical period $\tau = L/NV_a$ and overall duration $T = L/V_a$ is emitted. One may estimate dimensionally the amplitude of the wave to be the greater of $B_\phi, R_c B_z/L$ where R_c was the original curvature of the emerging flux tube. For reconnection to be considered as a trigger type process, it must occur on a timescale not appreciably greater than the times mentioned above, which for a length of order $10^3 - -10^4$ km and Alfvén speeds in the range $V_a \sim 10^2 - -10^3$ km/sec translate into $T \sim 10 - 100$ secs. This is close to the wave periods considered in the papers described above, and is within the time frame of spicules and chromospheric jets. Macrospicules however last appreciably longer, and one must invoke either much longer flux tubes (unlikely) or a very slow reconnection rate to increase T. An alternative may be given by the dynamics of photospheric vorticity.

Photospheric supergranulation flows concentrate magnetic flux at the network boundaries, where the resulting flux tubes exceed dynamic pressure equipartition and are close to evacuation pressure balance. Vorticity obeys the same equation as magnetic induction, and it is therefore plausible that vorticity and magnetic flux concentrations coincide. Vorticity filaments are the natural dissipative structures of 3D hydrodynamic turbulence and are observed to form in simulations of the solar convection zone [19]. Simon and Weiss (1997) [20] have given several examples of vorticity sinks, associated with photospheric downdrafts, at mesogranular scales. They fit the observed vorticity with a profile

$$\omega(r) = (V/R)\,\exp(-r^2/R^2),$$

where V is the characteristic rotational velocity associated with the vortes, and R their characteristic radius. A typical photospheric vortex lasts several hours and takes about two hours to develop. The strongest vortex they observed had $|\omega| \simeq 1.4\ 10^{-3}$ sec^{-1}, a size $R \simeq 2.5\ 10^3$ km and a maximum azimuthal velocity of order 0.5 km/sec. Such a vortex would produce an Alfvén wave packet, as the magnetic field lines are entrained by the rotational motion, whose frequency is given by the vorticity itself, and duration coincides with the life of the vortex. It is an intriguing coincidence that the lower bound of the so called Alfvénic range of turbulence in the solar wind resides at frequencies comparable to the intensity, i.e. vorticity or frequency, of the vortex filament found above.

We have compared the typical characteristics of reconnection generated Alfvén waves with those which should arise naturally above photospheric downdrafts with associated vorticity sinks, and have shown that the latter typically have longer durations and lower frequencies than waves emitted during reconnection processes. The macrospicule jet would in our scenario be associated with the establishment of a vorticity sink, which could however last much longer than the associated macrospicule. The macrospicule jet might decay, depending on the details of the magnetic geometry and thermodynamics, even though torsional Alfvén waves are being emitted until the vortex filament disrupts. This is because the spicule is associated with the initial piston provided by the development of the Alfvén wave packet (as will be described in the next section) as the vortex filament develops. Depending on the overall geometry (i.e. the nozzle shape) and temperature profile, such a jet might never reach a stationary state or provide the source of a "funnel wind flow" as envisaged by Tu & Marsch (1997) [21]. From a numerical point of view, simulations of Alfvén waves generated by a discrete set of vortex filaments with a somewhat simplified description of the photosphere, chromosphere and transition region in 2 and 3 dimensions are also being planned to provide a quantitative evaluation of the ideas described here. Initial results are presented in the next section. From an observational point of view, it would clearly be of interest to determine the photospheric velocity fields at the same time and place where a macrospicule is occurring, to verify whether or not an association with vortex filaments exists.

4   Non-linear propagation in the lower atmosphere: Parametric decay and filamentation

We have seen how large amplitude Alfvén waves might play an important role also in the lower solar atmosphere, contributing to the formation of spicules and field aligned jets. In a homogeneous medium, such waves represent an exact solution of the MHD equations, provided the total magnetic field intensity is a constant. However, such waves may be unstable to decay processes in which the energy is gradually transferred to compressible fluctuations and daughter Alfvén waves, so that the initial state tends to be destroyed, provided some initial random noise is present in the system. Compressible fluctuations are typically subject to stronger damping rates than the incompressible Alfvén wave, and this may be an important channel for the heating of an extended atmosphere, if indeed the decay instability, which is a resonant process, can be shown to survive in the much less idealized situations of incoherent wave-trains especially for low values of $\beta = C_s^2/V_a^2$ (as was shown in an overall periodic geometry by Malara & Velli, 1996 [22]) propagating in a stratified medium. In the following paragraphs we will show how stratification modifies the decay process both for coherent and incoherent wave-trains in an open geometry, i.e. in an atmospheric section with non-reflecting boundary conditions on both sides. The instability is still present both in the homogeneous and stratified case, and, in two dimensions, leads to a strong density filamentation in directions perpendicular to the direction of wave propagation.

Consider a flux of Alfvén waves propagating upwards through a section of the solar atmosphere, which for simplicity we shall consider isothermal. Let z denote the vertical axis (to be identified with the radial direction), along which there is an average uniform magnetic field B_0. If the propagating Alfvén wave is of sufficiently high frequency, the WKB approximation holds and in this limit the equations for the average atmospheric structure read

$$\frac{d}{dz}\left(C_s^2 \rho_0 + \frac{\mathbf{B}_\perp^{0\,2}}{8\pi}\right) = -\rho_0 g \qquad (17)$$

$$\frac{d}{dz}\left(V_a \frac{\mathbf{B}_\perp^{0\,2}}{8\pi}\right) = 0; \qquad (18)$$

here g is the gravitational acceleration while $\mathbf{B}_\perp^0$ is the wave magnetic field. The static density profile is modified by the contribution of Alfvén wave pressure, which is determined self-consistently via the requirement of conservation of net energy flux.

Eqs. (17, 18) are valid only if the wave is spherically polarized, i.e. if locally, on the scale of the wave, the magnetic field fluctuation is constant in absolute value, otherwise the perturbed magnetic pressure arising from the wave would drive compressible motions at twice the wavenumber of the Alfvén wave (the initial perturbation of the atmosphere created by the generation of such an Alfvén wave could be related to the formation of spicules , but a systematic parametric study of such a process has yet to be carried out). In other words, we write

$$\mathbf{B}_\perp^0(z,t) = B_\perp^0\left\{\cos[k_0(z - V_a t)]\mathbf{e}_x + \sin[k_0(z - V_a t)]\mathbf{e}_y\right\}. \qquad (19)$$

The density, pressure and field-aligned velocity fluctuations thus vanish to lowest order while the perpendicular velocity field satisfies

$$\mathbf{v}_\perp^0 = -\frac{\mathbf{B}_\perp^0}{\sqrt{4\pi\rho_0}}. \qquad (20)$$

However, at first order in the parameter $\epsilon = V_a g/(8\omega_0 C_s^2)$, $(\omega_0 = k_0 V_a)$, which is assumed to be vanishing in WKB, one must include the reflected

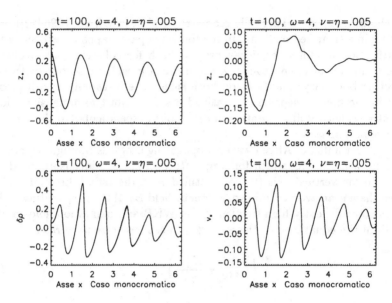

Fig. 1. Profiles of the y component of the fluctuations clockwise from top left, mother Alfvén wave, backscattered Alfvén wave, compressive velocity fluctuation and density fluctuation, at saturation for the monochromatic, homogeneous, case.

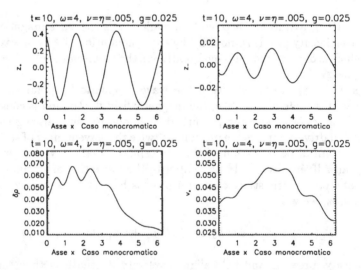

Fig. 2. Profiles of the y component of the fluctuations clockwise from top left, mother Alfvén wave, backscattered Alfvén wave, compressive velocity fluctuation and density fluctuation, just before the development of decay, in the monochromatic stratified case.

Alfvén wave (Velli et al. 1989) [23] which, though obeying a relation of type (4) with opposite sign, propagates in the same direction as the mother wave, with a $\pi/2$ phase shift, and amplitude ϵ times smaller. Since the growth rate of the parametric decay instability is proportional, in the small β limit, to the mother wave amplitude, and generates a backscattered Alfvén with normal phase characteristics, it is unclear a-priori how a moderate stratification will affect the mode coupling. One may carry out a linearization of the MHD equations around the basic state "stratified atmosphere + upwardly propagating Alfvén wave" to obtain a local dispersion relation which is a sixth order polynomial in ω for given k and not especially transparent . Another way to attack the problem is to consider numerical simulations of an upward propagating Alfvén wave in a noisy medium, where because of the stratification the usual periodic boundary conditions are not permitted. As a result, we may also follow the instability into its nonlinear phase and saturation.

We have developed a one- and two-dimensional compressible MHD simulation code (periodic, pseudo-spectral in the direction perpendicular to the radial, compact finite differences in the radial direction, third order Runge Kutta time advance) which uses a non-singular formulation of projected characteristics along the radial direction to impose transparent (non-reflecting) boundary conditions at the outer boundary and an influx of Alfvén waves at the lower boundary. Because of the vertical stratification, it is in general impossible to have perfect transmission of sound waves at the upper and lower boundaries of the numerical domain, because of the contemporary requirement of maintaining an on average constant density at one of these boundaries. If one did not constrain the density in any way, the perturbation introduced by the upcoming Alfvén wave would cause the density to slowly drift away from its initial value leading either to a secular increase or decrease of mass in the box. On the other hand, imposing the density leads to strong reflection of sound waves. A practical way to avoid this problem is to introduce a force in the continuity equation at the lower boundary, proportional to the deviation of the density from its average value and with a time constant much greater than the wave-periods of interest. This guarantees that over long time scales the density will tend to return to its mean value while at the same time allowing waves with sufficient frequency an almost perfect transmission.

Consider first 1D simulations, with no stratification. Here one expects the decay instability to occur, though with slighlty different conditions because of the non-reflecting boundary conditions imposed. In Fig. (1) we show the profiles of the mother Alfvén wave, the backscattered Alfvén wave, and the fluctuations ρ, v_z just before saturation. The amplitude of the mother waves is 1/5 that of the background magnetic field B_0 (B_0=410 km/sec in velocity units, while the plasma $\beta = 0.1$). Notice how the backscattered Alfvén wave has vanishing amplitude at the outer boundary and how this increases towards the base, while the compressible fluctuations have steepened into a train of shocks. Notice also how the backscattered wave has a frequency close to half the mother wave frequency (a characteristic feature of parametric decay). The length of the box is $l = 9. \, 10^4$ km and the the wave period is $\tau_0 = 40$ secs.

Let us now introduce the gravitational stratification, which in adimensional units ($g = 8\epsilon$, with ϵ defined above) is $g = 0.025$. In Fig. (2) we show a snapshot of the same field quantities though at a much shorter time, just as the decay instability is about to develop. One may observe how the wavelength increases with height, and the phase-shift of the reflected daughter wave. At longer times, parametric decay takes over, the behaviour of the

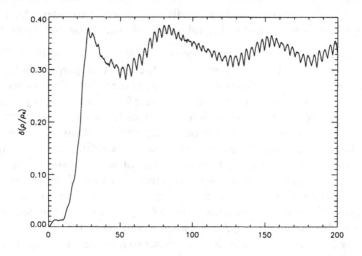

Fig. 3. Rms density fluctuations as a function of time, showing the evolution of parametric decay with moderate stratification.

rms density fluctuation being shown in Fig. (3). A clear, exponential increase in density fluctuations occurs between $t = 10.$ and $t = 30$ after which nonlinear saturations occurs, leaving an rms density fluctuation of around 35%. The field profiles after saturation are shown in Fig. (4); notice the slightly different profile, with respect to the homogeneous case, of the steepened compressible fluctuations, and the usual behaviour of the backscattered wave.

Numerical simulations with the same non-monochromatic Alfvén waves considered in [22] yield very similar results to those presented above. In order to get a better, more realistic understanding of the effect, a limited number of simulations in two dimensions have also been carried out, with both monochromatic and non-monochromatic spectra in stratified and non stratified atmospheres. Though the linear growth rates are comparable, the field profiles change significantly in that the wave profiles, initially homogeneous

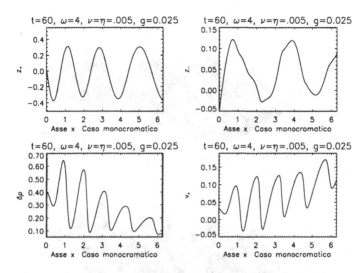

Fig. 4. Profiles of the y component of the fluctuations clockwise from top left, mother Alfvén wave, backscattered Alfvén wave, compressive velocity fluctuation and density fluctuation, at saturation, in the monochromatic stratified case.

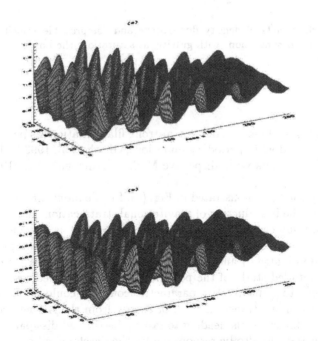

Fig. 5. Surface plots of density fluctuations and compressible velocity v_z for the two dimensional simulation, with no gravity, at saturation, the box width is equal to box length.

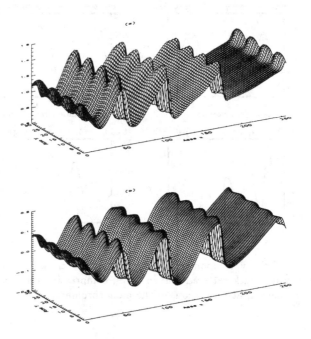

Fig. 6. Surface plots of density fluctuations and compressible velocity v_z for the two dimensional simulation, with gravity, at saturation, the box width is equal to box length.

in the transverse direction, develop a strong filamentation (as partly expected on the basis of doubly periodic simulations, Ghosh et al. 1993 [24], as well as studies of wave collapse in dispersive MHD, Champeaux et al. 1997 [25]).

An example is demonstrated in Fig. (5), for the monochromatic, homogeneous case. The introduction of gravitational stratification changes the shape of the fields somewhat, Fig. (6), but the main effect is to cause reflections of the fast magnetoacoustic part of the daughter waves, leaving a transverse structure dominated by anti-correlation of magnetic and kinetic pressures. Although a detailed study of the parameter space is required, our simulations clearly seem to indicate that parametric coupling could be a very important process in the transformation of energy from the Alfvén wave channel to compressible channels, leading to faster, distributed dissipation. A flux of Alfvén waves could also be responsible for the development of very fine scale structures in density orthogonal to the magnetic field, structures whose presence has been identified via radio sounding measurements (Woo and Habbal, 1997 [26]).

5 Alfvén solitons in the solar wind

It has often been suggested that solar wind Alfvénic turbulence might be interpreted as a collection of solitary waves propagating away from the sun. Recently, Buti et al. (1998) [27] have studied how such modes, exact solutions of the DNLS equation, evolve within the framework of the dispersive Hall-MHD system of equations. The DNLS is derived from Hall-MHD in the limits of ratio of kinetic to magnetic pressure $\beta > 1$ or $\beta < 1$, neglecting higher order terms above cubic nonlinearities. In an effort to apply a similar kind of analysis to the evolution of Alfvén waves in the solar wind, Buti et al. (1998) [28] have incorporated the effects of a spherical background wind into the weakly nonlinear DNLS type development, deriving a modified DNLS equation which does not appear to have soliton solutions. The numerical simulations presented in [28] showed that soliton type initial conditions evolve into fairly well developed spectra for the transverse magnetic fields as well as the densities.

In a more realistic situation, soliton initial conditions must evolve both because of higher order nonlinearities as well as any background inhomogeneity introduced by the combined effects of gravitational stratification and the solar wind. To understand such processes simulations of soliton evolution both in the Hall-MHD system and the Expanding Box Hall-MHD system were carried out([30], [29]), to ascertain the relative importance of neglected nonlinarities and background inhomogeneities. The expanding box framework consists in modeling the overall solar wind expansion in the spirit of the final part of section 2, in the sense that a constant solar wind flow speed is considered and one follows a plasma parcel advected by the wind. Locally, the plasma is considered homogeneous, and the overall large scale spatial inhomogeneity is recovered via a time dependence of the average densities, magnetic field etc. in a way consistent with conservation of mass flux, magnetic flux, and angular momentum. The scaling laws presented earlier for linear waves are indeed recovered in simulations with small amplitude waves. Such WKB effects are fundamental for small amplitude solitons as well, while for larger amplitude solitons the solar wind expansion effects on the wave-forms are negligible compared to what occurs naturally in the "homogeneous" Hall-MHD system.

To understand the effect of solar wind expansion on a soliton propagating away from the sun, one must first compare the time-scales over which such a soliton evolves due to the higher order nonlinearities in the Hall-MHD system with the typical expansion time (denoted τ_{NL}, τ_E respectively). Clearly, the first time-scale must depend on the soliton amplitude, while the second is given essentially by the transport time in the solar wind. To obtain a quantitative measure of the evolution time τ_{NL} we have carried out a series of simulations using a spectral code to solve the homogeneous Hall MHD system of equations in one spatial dimension x, the direction of propagation, which we take to coincide with the mean field direction (times are normalized to the cyclotron time and the soliton amplitude to that of the mean

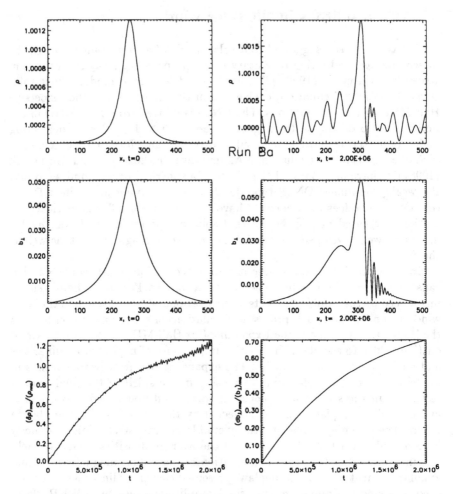

Fig. 7. Initial and final profiles of density (top panels), magnetic fields (middle panels), and time variation of the measure eq. (22) for the density (bottom left) and the magnetic field (bottom right); x is the abscissa in units of grid point number.

field). In the subsequent runs in the expanding solar wind this also coincides with the radial direction from the sun.

The soliton initial conditions ($b_y + ib_z = B_\perp$) are given by

$$B_\perp(x, t_0) = \frac{(\sqrt{2} - 1)^{\frac{1}{2}} B_s e^{i\theta(x)}}{(\sqrt{2}\cosh(2V_s x) - 1)^{\frac{1}{2}}}, \tag{21}$$

where B_s is the overall envelope amplitude,

$$\theta(x) = -V_s x + \frac{3}{8(1 - \beta)} \int_{-\infty}^{2x} |B_\perp|^2 \, dx'$$

is the phase, and

$$V_s = \frac{(\sqrt{2}-1)B_s^2}{8(1-\beta)}$$

is the soliton speed in a frame of reference moving at the Alfvén speed based on the background magnetic field (which in the units presented is $B_0 = 1$).

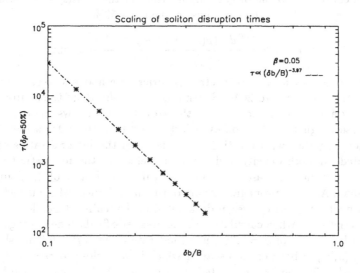

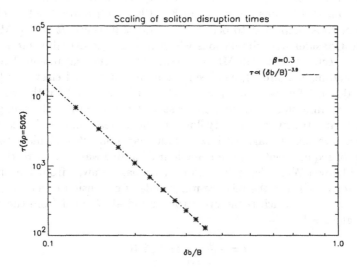

Fig. 8. Soliton disruption times as a function of soliton amplitude ($\delta b/B = B_s$) for values of plasma β (a) $\beta = 0.05$, and (b) $\beta = 0.3$

The evolution is best followed in a frame of reference moving with the soliton speed V_s, so that in the absence of evolution all fields in the simulation would remain constant. To determine how τ_{NL} scales with the envelope amplitude B_s one must define the time-scale for the soliton initial condition to break up unambiguously: the adopted choice here is to define τ_{NL} as the time after which the L_2 (i.e. square-integral) measure of the change in the density, compared to the original L_2 measure of the density fluctuation, has become significant, i.e. the ratio has reached say 50%:

$$\tau_{NL} : \delta\rho^2 = \frac{\int dx' \left(\rho(x', \tau_{NL}) - \rho(x', 0) \right)^2}{\int dx' \rho(x', 0)^2} = 50\%. \tag{22}$$

The code uses a pseudo-spectral (Fourier) method and resolution typically in the range 512 to 2048 (depending on amplitude of the soliton), with resistivity and viscosity adapted to the grid. Fig. (7) shows the density and transverse magnetic field profiles at initial and final times for a low β value $\beta = 0.05$, together with the time dependence of the integral ratios eq. (22) calculated for both density and magnetic field. For the magnetic field, our quadratic measure increases more gradually in time because of the dominance of the main Alfvénic propagating part of the disturbance, which is lacking for density fluctuations. However, it does not alter the scaling dependence of τ_{NL}, which is the same when calculated on the magnetic fields. The scalings of τ_{NL} both for low values of $\beta = 0.05$ and a moderate value of $\beta = 0.3$ are shown in Fig. (8) (a,b); in both cases a scaling which is very close to $\tau_{NL} \sim B_s^{-4}$, the time it takes for the first neglected term in the DNLS expansion to become $O(1)$, is recovered.

The spectra of magnetic field and density at different times in the homogeneous Hall-MHD runs are shown in Fig. (9a,b), for normalized amplitudes typical of those expected to occur at an initial distance $R = 0.1$ AU from the sun. Consider now simulations which include the overall solar wind expansion effects on the Hall-MHD system. The numerical model, known as the Expanding Box Model, has been extensively described elsewhere [29]. In this model one follows a plasma parcel as it is advected by the wind. It's transverse dimensions therefore increase in time while the threading magnetic field and transverse velocity fluctuations decrease in time in accordance with conservation of magnetic flux and angular momentum. This is the minimal set of requirements for the behaviour of simple waves to correctly follow the well known WKB decay laws as they propagate away from the sun. The parameter which sets the relative magnitude of expansion effects, expressed in our time unit, which is the cyclotron period at $R = 0.1$ from the sun, is, for a high speed stream,

$$\epsilon = \frac{U_0}{R_0} \frac{1}{\Omega_{ci}} \simeq 6.25 \ 10^{-7},$$

an extremely small number (which is one of the reasons an Eulerian simulation of the expanding solar wind including kinetic effects exceeds present

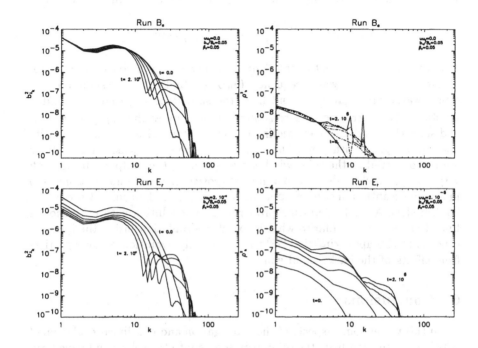

Fig. 9. Magnetic field and density spectra for the low β run and no expansion (a,b) top panels left and right, and for the same initial conditions but taking the solar wind expansion into account (c,d), bottom panels left and right.

capabilities). In our simulations the importance of the expansion effects are increased by a factor of about 3, to $2\ 10^{-6}$, to reduce the duration of the simulations carried out.

The evolution of magnetic field and density spectra is shown in Fig. (9c,d), covering a distance from 0.1 to 0.5 AU. The main effect of the expansion is to cause the overall wave amplitudes to decrease, in the standard "WKB" way, (i.e., the magnetic fluctuation amplitude in Alfvén speed units drops as R^{-1}). At the smallest scales, nonlinear effects have time to evolve the fluctuations, and changes in shape, and the magnetic field evolution is very similar to what is observed in the homogeneous case. There is however an increase in relative density fluctuations as compared to the homogeneous case, a feature which is typical of simulations in the expanding box model, and is due to the different effects of the wind expansion on the evolution of average and fluctuating densities with distance from the sun.

For sound waves, the conserved wave-action flux (the invariant analogous to the Alfvén wave-action considered in previous sections), may be written

$$S_c = F \frac{(U \pm C_s)^2}{2\omega C_s} |(u \pm C_s \delta\rho/\rho)|^2.$$

F is the geometrical factor ρr^σ defined earlier which is constant in a stationary wind, ω the wave frequency (also constant). The $\pm$ signs refer to sound waves propagating to and from the sun. Recalling that the wind is substantially supersonic, the constancy of S_c implies that $\delta\rho/\rho \sim C_s^{-1/2}$, and since the wind in the expanding box cools adiabatically, relative density fluctuations grow with time. The reason such enhanced density fluctuations are not observed in the solar wind is twofold: first, the temperature falls off very slowly, which leads only to a modest growth in density fluctuations; and more importantly Landau damping must occur, a mechanism not present in fluid models. A final interesting property of the evolution with expansion is that it is also very similar to what is found in simulations using the inhomogeneous DNLS approximation [28] where the only evolution occurring arises from effects of the background wind.

6   Conclusions

In this paper some aspects of the propagation and non-linear evolution of Alfvénic fluctuation from the solar atmosphere into the solar wind have been considered. The topics covered are by no means complete, and their selection has been largely due to personal interests. The most important couplings only touched upon here are those arising from the true 3D nature of the wave generation and turbulence development problem. On the other hand, the richness and relevance of the linear theory is often neglected by those interested in studying turbulence proper, and this has lead to some confusion over the importance of the different effects that have been, it is hoped, at least partly clarified here.

The question of the origin of Alfvénic turbulence clearly remains an open one, together with the associated question of how much energy is drained from such modes through compressional couplings as they propagate outward through the solar atmosphere. Such fascinating questions should come within the scope of a quantitative theory both thanks to full 3D simulations as well as the analysis of simplified problems such as those discussed in this paper. In particular the question of propagation through the 3D chromosphere and transition region is one that we hope to address in the near future.

Part of this work was carried out while the author held a National Research Council-JPL Research Associateship. I would like to thank Bruce Goldstein, Paulett Liewer, Bimla Buti, Marsha Neugebauer, Roland Grappin for the many and long discussions on the topics covered in this paper as well as the collaborations which have led to the original results presented here.

References

1. Grappin, R., Velli, M. & Mangeney, A., (1991): Ann. Geophys. **9**,416
2. Horbury, T.S., Balogh, A., Forsyth, R.J. and Smith, E.J., (1996): JGR **101**,405
3. Veltri, P.L., (1999): these proceedings
4. Velli, M., (1993): A&A **270**, 304
5. Heinemann, M., and Olbert, S., (1980): JGR **85**,1311
6. Leer E., Holzer, T.E., and Fla,T., (1982): Space Sci. Rev. **33**, 161
7. Hollweg, J.V., & Lee, M.A., (1989): GRL **16**,919
8. Barkhudarov M.N. (1991): Solar Phys. **135**,131
9. Velli, M., Grappin, R. & Mangeney, A. (1991): GAFD **62**,101
10. Pasachoff, J.M., Noyes, R.W. & Beckers, J.M., (1969): Sol. Phys.**5**, 131
11. Cook, J.W., (1991): in Mechanisms of Chromospheric and Coronal Heating, P.Ulmschneider, E.R. Priest & R. Rosner Eds., Springer Berlin, 93
12. Pike, C.D. & Harrison, R., (1997): Solar Phys. **175**,457
13. Pike, C.D. & Mason, H.E., (1999): Solar Phys., in press
14. Shibata, K., (1997): in Fifth SOHO Workshop: The Corona and Solar Wind near Minimum Activity, ESA SP-404, 103
15. Hollweg, J., (1992): ApJ **389**, 731
16. Shibata, K. & Uchida, Y., (1986): Solar Phys. **103**,299
17. Hollweg, J., Jackson, S. and Galloway, (1982): Solar Phys. **75**,35
18. Kudoh, T. and Shibata, K., (1997): in Fifth SOHO Workshop: The Corona and Solar Wind near Minimum Activity, ESA SP-404, 477
19. Stein, R.F. & Tuominen, I., (1996): J Fluid Mech. **306**
20. Simon, G.W. and Weiss,N.O., (1997): ApJ **489**, 960
21. Tu, C.Y., & Marsch, E., (1997): Solar Phys. **176**, 87
22. Malara, F. and Velli, M. (1996): Phys. Plasmas **3**,4427
23. Velli, M., Grappin, R. and Mangeney, A., (1989): PRL **63**, 1807
24. Ghosh, S., Vinas, A.F. and Goldstein, M.L., (1993): JGR **98**,15561
25. Champeaux, S., Gazol, A., Passot, T. & Sulem, P.L. (1997): ApJ **486**, 477
26. Woo,R. and Habbal, S.R., (1997): ApJ **474**, L139
27. B.Buti, Jayanti, V., Vinas, A.F., Ghosh, S., Goldstein, M.L. , Roberts, D.A., Lakhina, G.S. and Tsurutani, B.S., (1998): GRL **25**, 2377
28. B.Buti, Galinski, V.L., Shevchenko, V.I., Lakhina, G.S., Tsurutani, B.T., Goldstein, B.E., Diamond, P. and Medvedev, M.V., (1999): "Evolution of nonlinear Alfvén wave trains in streaming inhomogeneous plasmas", ApJ , in press
29. R. Grappin, and M. Velli, (1996): JGR **101**, 425
30. M.Velli, B.Buti, Goldstein, B.E. and Grappin, R. (1999): "Propagation and Disruption of Alfvénic Solitons in the Expanding Solar Wind" in Solar Wind 9, S.R. Habbal et al. Eds., AIP, in press

Nonlinear Alfvén Wave Interaction with Large-Scale Heliospheric Current Sheet

P. Veltri[1,2], F. Malara[1,2], and L. Primavera[1,2]

[1] Dipartimento di Fisica, Universitá della Calabria
  I-87036 Arcavacata di Rende (Cosenza), Italy
[2] Istituto Nazionale Fisica della Materia, Unitá di Cosenza,
  I-87036 Arcavacata di Rende (Cosenza), Italy

Abstract. The in-situ measurements of velocity, magnetic field, density and temperature fluctuations performed in the solar wind have greatly improved our knowledge of MHD turbulence not only from the point of view of space physics but also from the more general point of view of plasma physics.

These fluctuations which extend over a wide range of frequencies (about 5 decades), a fact which seems to be the signature of turbulent non-linear energy cascade, display, mainly in the trailing edge of high speed streams, a number of features characteristic of a self-organized situation: (i) a high degree of correlation between magnetic and velocity field fluctuations, (ii) a very low level of fluctuations in mass density and magnetic field intensity. These features are locally lost, in the presence of large scale inhomogeneities of the background medium, like velocity shears due to the stream structure and current sheets at magnetic sector boundaries. Such nonuniformities generate a coupling among different modes and tend to destroy the equilibrium solution represented by outward propagating Alfvénic fluctuations. The Alfvénicity is also reduced with increasing distance from the Sun, and this could be in part due to the effects of the large-scale inhomogeneity related to the solar wind expansion.

The effects of inhomogeneities of the background medium on a MHD turbulence have been studied, from a theoretical point of view, by a number of numerical models. In this paper we briefly review such models, discussing the main results and their limitations, and comparing with observed features of the solar wind fluctuations.

1 Introduction

Solar wind offers us a unique opportunity to study MHD turbulence. In fact solar wind represents a supersonic and superalfvénic flow, inside which space experiments have furnished to the scientific community a wealth of data (velocity, magnetic field, plasma density, temperature etc. or also particle distribution functions) at a resolution which is not available in any earth laboratory.

At high latitudes, the wind consists of a remarkably homogeneous (at least at solar minimum) flow with wind speed in the range 750-800 km/sec, while at low latitudes the wind flows with a speed of 300-400 km/sec. The high speed wind originates from the open magnetic field lines in the coronal holes, while the low speed streams must originate from field lines adjacent to if not within the coronal activity belt (coronal streamers and active regions).

The heliospheric current sheet separating the global solar magnetic polarities is embedded within the low speed wind (Fig. 1). Due to the bending

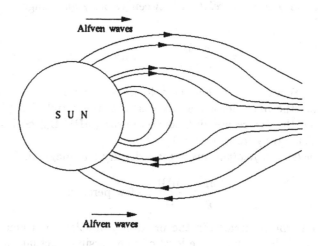

Fig. 1. A schematic view of solar wind magnetic field

of the heliospheric current sheet, in the ecliptic plane, which represents the region of the interplanetary medium most widely studied by space experiments, the plasma flow is structured in high and low speed streams, and is extremely variable.

Inside this flow, which can be considered as a sort of *wind tunnel*, large amplitude fluctuations, with frequencies lower than the ion-cyclotron frequency, extend over a very wide range

$$10^{-6} \, Hz < f < 1 \, Hz \tag{1}$$

In analogy with wind tunnel, the time variations of these fluctuations, observed in the rest frame of the satellite are assumed as being due to spatial variations convected by solar wind velocity (Taylor's hypothesis [1]). Assuming a solar wind velocity $v_{SW} \sim 400 \, km/sec$ it can be seen that the wavelengths, $l = v_{SW}/f$, corresponding to (1), range from $400 \, km$ to about $1 \, AU$.

All over this frequency range these fluctuations display a power law spectrum

$$E(k) \propto k^{-\alpha} \tag{2}$$

with spectral indices α comprised between 1 and 2 [2,3]. This fact seems to be the signature of a fully developed MHD turbulence resulting from a non-linear energy cascade.

In spite of this, the fluctuations display, mainly in the trailing edges of high speed streams and at small scales ($1 min < T < 1 day$), some striking features, which seem to show that these fluctuations are in some sense *organized* [4]:
(i) *a high degree of correlation* between velocity and magnetic field fluctuations [3]

$$\delta \mathbf{v} \simeq \frac{\sigma\, \delta \mathbf{B}}{\sqrt{4\pi\rho}} \qquad \text{with} \qquad \sigma = \pm 1 \qquad (3)$$

($\delta\mathbf{v}$ and $\delta\mathbf{B}$ are respectively the velocity and magnetic field fluctuations, ρ represents the mass density and the sign of the correlation ($\sigma = \pm 1$) turns out to be that corresponding to non-linear ($|\delta\mathbf{B}|/|\mathbf{B}_0| \simeq 1$) Alfvén waves propagating away from the Sun)
(ii) *a low level of fluctuations* in mass density and magnetic field intensity [3]

$$\frac{\delta\rho}{\rho} \simeq \frac{\delta|\mathbf{B}|}{|\mathbf{B}|} \simeq \text{few percents} \qquad (4)$$

In the low speed streams in the proximity of solar wind neutral sheet, the correlation is lower, while the level of compressible fluctuations is higher.

The discovery of the correlation (3),[3], has represented the motivation of a lot of theoretical work. Dobrowolny et al. [5], initially suggested that this high level of correlation should be due to a sort of self-organization produced by the natural dynamical evolution of incompressible MHD turbulence. This conjecture about the MHD turbulence evolution, has subsequently been confirmed by rather different mathemathical techniques: numerical integration of statistical equations, obtained via closure hypothesis [6,7]; simplified models for the nonlinear energy cascade [8,9]; direct numerical simulation of MHD equations in both 2D [10,11] and 3D [12,13].

The picture of the evolution of *incompressible* MHD turbulence, which comes out from these theoretical models is rather nice, but the solar wind turbulence, which stimulated all this work, displays a more complicated behavior. Data analysis by Roberts *et al.* [15,16], Bavassano and Bruno [17], Grappin *et al.* [18] shows that solar wind turbulence evolves in the reverse way: the correlation is high near the Sun. At larger radial distances from 1 AU to 10 AU the correlation is progressively lower, while the level of fluctuations in mass density and magnetic field intensity increases. The spectra initially flatter than a Kolmogorov's ($\alpha = 5/3$) or Kraichnan's ($\alpha = 3/2$) spectrum, increase their indices up to $\alpha = 5/3$ at 10 AU.

What is more difficult to understand is the reason why correlation is progressively destroyed in the solar wind if the natural evolution of MHD turbulence is towards a state of maximal normalized cross-helicity. A possible solution to such paradox can be found in the fact that solar wind is neither *incompressible* nor *statistically homogeneous*. Some attempts to take into account compressibility and/or inhomogeneity of the solar wind have thereinafter been performed.

Roberts *et al.* [16] suggested that stream shear velocity gradients should be directly responsible for the decrease of correlation. Moreover Roberts *et al.* [19,20] simulated the evolution of Alfvénicity (the correlation mentioned in (3)) near a magnetic neutral sheet, showing that in this case the decay of the correlation is accelerated.

Veltri *et al.* [21] performed numerical simulations which show that, in a compressible medium, the interaction between small scale waves and large scale magnetic field gradients on the one hand, the parametric instability on the other hand, reduce the correlation between the velocity and magnetic field fluctuations and let develop a compressive component of the turbulence characterized by $\delta\rho \neq 0$ and $\delta|\mathbf{B}| \neq 0$.

Grappin *et al.* [22] observed that the overall solar wind expansion increases the lengths normal to the radial direction, thus producing a sort of inverse energy cascade which competes with the direct non-linear energy cascade. As a result non-linear interactions are slowed down, at least at large scales. To describe the effect of the solar wind expansion they have built up a numerical simulation where MHD equations are solved in an expanding box comoving with the solar wind (Expanding Box Model). The results of their simulations show that, after a first stage of evolution, non-linear interactions are effectively stopped.

In conclusion it is now clear that in a *compressible* and *inhomogeneous* medium there are lot of processes which may be responsible for the decorrelation of the turbulence and for the development of a compressive component of the fluctuations. To explain the fact that, in fast streams, the correlation lives longer (up to 1 AU), Veltri *et al.* [21] proposed that Landau damping could play a role in keeping the density and magnetic field intensity fluctuations at their observed low level.

2   A model for Alfvén wave interaction with heliospheric current sheet

The scenario to explain the high degree of correlation between velocity and magnetic field fluctuations near the Sun, is now based on the idea that the main source of low frequency fluctuations is the Sun. In fact, Alfvén waves propagating in opposite directions are both convected by solar wind only beyond the Alfvénic point, where the flow speed becomes greater than the Alfvén speed. One should then expect that only those Alfvén waves, which propagate outward can pass through the critical point and leave the Sun carried by the solar wind [3]. Also if the situation is somewhat more complicated [23], this explanation is widely accepted.

In this scenario, large amplitude Alfvén waves, generated at the footpoints of open magnetic field lines, converge towards the heliospheric current sheet (see Fig. 1) where they interact with large scale inhomogeneities associated to the current. Since the superposition of Alfvénic fluctuations and current sheet gives rise to a non-equilibrium state, it is worth to study what is the dynamical evolution of this state and to try to compare the results with solar wind measurements, in order to test the suitability of the overall model.

A two dimensional numerical model of MHD turbulence in presence of a large scale current sheet has been built up by Roberts et al. [20]. Stribling et al. [24] performed the three dimensional extension of this model. The main results of these simulations can be summarized as follows: (i) the correlation of velocity and magnetic fields fluctuations is progressively reduced; (ii) energy spectra anisotropic, with wave vectors perpendicular prevailing on wave vectors parallel to the background magnetic field, are finally obtained.

At variance with these studies Malara, Primavera and Veltri [25,26,47] considered some physical aspects of the problem of the interaction between large amplitude Alfvén wave and heliospheric current sheet which are very peculiar. First of all, the nonlinear Alfvénic solution in a compressible medium is characterized not only by the correlation (3) but also by B^2 and ρ both uniform so it is worth to study the modification in this kind of solution due to the inhomogeneity in the background medium. The condition $B^2 = const.$ ensures that dynamical effects induced by the ponderomotive force associated with $|\mathbf{B}|$ spatial variations are suppressed at the initial time. This kind of perturbation propagates without distortion in a uniform background magnetic field, but is subject to a dynamical evolution when propagating in a current sheet.

Secondly, in the simulations by Roberts et al. [20] and Stribling et al. [24], the authors assumed the same sign of the correlation between magnetic and velocity fluctuations on both sides of the current sheet. On the contrary, the Alfvénic fluctuations which, after starting from the Sun, follow magnetic field lines which converge on the two sides of the heliospheric current sheet, should propagate in the same (outward) direction on both sides of the current sheet. This means that they have opposite Alfvénic correlation, i.e. σ must change sign. When we try to model this situation we are led to assume that there is a region inside the current sheet, where the velocity field is not solenoidal, i.e. a region where a source of compressions is present.

2.1 Numerical model

The basic equations of our model are the compressible, dissipative, MHD equations in dimensionless form

$$\frac{\partial \rho}{\partial \tau} + \nabla \cdot (\rho \mathbf{v}) = 0 \tag{5}$$

$$\frac{\partial \mathbf{v}}{\partial \tau} + (\mathbf{v} \cdot \nabla)\mathbf{v} = -\frac{1}{\rho}\nabla(\rho T) + \frac{1}{\rho}(\nabla \times \mathbf{b}) \times \mathbf{b} + \frac{1}{\rho S_\nu}\nabla^2 \mathbf{v} \qquad (6)$$

$$\frac{\partial \mathbf{b}}{\partial \tau} = \nabla \times (\mathbf{v} \times \mathbf{b}) + \frac{1}{S_\eta}\nabla^2 \mathbf{b} \qquad (7)$$

$$\frac{\partial T}{\partial \tau} + (\mathbf{v} \cdot \nabla)T + (\gamma - 1)T(\nabla \cdot \mathbf{v}) = \frac{\gamma - 1}{\rho}\left[\frac{1}{S_\kappa}\nabla^2 T + \frac{1}{S_\nu}\left(\frac{\partial v_i}{\partial x_j}\frac{\partial v_i}{\partial x_j}\right) + \right.$$
$$\left. + \frac{1}{S_\eta}(\nabla \times \mathbf{b})^2\right] (8)$$

where space coordinates are normalised to the typical shear length a of the large scale current sheet, magnetic field components $\mathbf{b}$ to a characteristic value B_0, mass density ρ to a characteristic value ρ_0, velocities $\mathbf{v}$ to the corresponding Alfvén velocity $c_A = B_0/(4\pi\rho_0)^{1/2}$, times τ to the ideal Alfvén time $\tau_A = a/c_A$, and temperature T to $\mu m_p c_A^2/k_B$ (μ is the mean molecular weight, m_p the proton mass k_B the Boltzmann constant and $\gamma = 5/3$ the adiabatic index). The quantities S_ν and S_η represent respectively the kinetic and magnetic Reynolds numbers, while S_κ is the inverse of the normalised heat conduction coefficient. Since dissipative coefficients are very low in solar wind we have used for the above quantities the largest values allowed by computer limitations: $S_\nu = S_\eta = S_\kappa/(\gamma - 1) = 1400$.

The considered geometry is $2\frac{1}{2}$-D: all quantities depend on two space variables (x and y), but vector quantities have three nonvanishing components. The equations (5) - (8) have been numerically solved in a rectangular spatial domain $D = [-\ell, \ell] \times [0, \pi R\ell]$, with free-slip and periodic boundary conditions, along x and y, respectively. The x axis represents the cross-current sheet direction, while the y axis represents the propagation direction of the initial perturbation. The domain width ($2\ell = 8$) has been chosen sufficiently larger than the shear length ($\simeq 1$) while the domain length is equal to the largest wavelength λ_{max} of the perturbation in the periodicity (y) direction. The aspect ratio has been chosen $R = 0.15$ corresponding to $\lambda_{max}/a = 3\pi/5$. Using the Taylor's hypothesis [1], the corresponding frequency, in the spacecraft reference frame, is

$$f_{min} \sim \frac{v_{sw}}{\lambda_{max}} = 2\frac{a}{\lambda_{max}}\frac{v_{sw}}{2a}. \qquad (9)$$

Assuming $2a/v_{sw} \leq 12$ h we find $f_{min} \geq 2.4 \times 10^{-5}$ Hz. Then, the lower limit of the spectrum in our model roughly corresponds to the lower limit of the Alfvénic range of the solar wind turbulence.

2.2 Numerical technique

The equations (5)-(8) have been numerically solved using a $2\frac{1}{2}$-D pseudospectral code. The periodic boundary conditions in the y-direction allow to perform a Fourier expansion with a finite number of harmonics in this direction. Any physical quantity $f(x, y, \tau)$ can then be expressed as

$$f(x, y, \tau) = \sum_{m=-M/2}^{M/2} f_m(x, \tau) \exp[2imy/(Rl)] \tag{10}$$

where $f_m(x, \tau)$ is the complex m-th Fourier harmonic of $f(x, y, \tau)$, and M is the number of Fourier harmonics.

In consequence of the interaction between the perturbation and the inhomogeneity of the equilibrium, small scale structures are expected to be generated near the center ($x = 0$) of the domain D, where the inhomogeneity is localized. Then, an enhanced spatial resolution is required close to $x = 0$. For this reason we used a multi-domain technique: the domain D has been divided in two subdomains $D = D_1 \cup D_2$, where $D_1 = [-l, 0] \times [0, \pi Rl]$ and $D_2 = [0, l] \times [0, \pi Rl]$.

A Chebyshev expansion along the x-direction is performed in each subdomain. This furnishes the highest spatial resolution at $x = 0$ and $x = \pm l$, where the density of the Chebyshev meshpoints is the highest. At each time step the equations (5)-(8) are separately solved in each subdomain, and the continuity of any quantity, as well as of first-order space derivatives, is imposed at the internal boundary $x = 0$, in order to fulfill regularity conditions in the whole domain D.

The time dependence in the equations (5)-(8) is treated by a semi-implicit method [27], which is numerically stable also for relatively large time steps. This allows to get rid of the limitation imposed by the Courant-Frederichs-Levy stability condition, which in the present case would be very severe, in consequence of the high density of Chebyshev meshpoints at the border of both subdomains. A more detailed description of the numerical method can be found elsewhere [28–30].

2.3   Initial condition

The initial condition is set up in order to represent an Alfvénic perturbation superposed on a background medium which reproduces the main properties of the heliospheric current sheet. Direct measurement in the solar wind have shown that, crossing the heliospheric current sheet, the magnetic field rotates (changing, for instance, from an inward to an outward orientation) but the intensity remains roughly constant. In a similar way, the density, the velocity and the proton temperature remain on the average constant.

On the basis of these considerations we have assumed that at the initial time the total magnetic field is given by

$$\mathbf{b}(x, y, 0) = A\Big\{\varepsilon \cos[\phi(y)]\ \mathbf{e}_x + \sin(\alpha)F(x)\ \mathbf{e}_y$$

$$+\sqrt{1 - \sin^2(\alpha)F^2(x) + \varepsilon^2 \sin^2[\phi(y)]}\ \mathbf{e}_z\Big\}. \tag{11}$$

The function $F(x)$ is defined by

$$F(x) = \frac{\tanh x - \dfrac{x}{\cosh^2 \ell}}{\tanh \ell - \dfrac{\ell}{\cosh^2 \ell}}. \tag{12}$$

It monotonically increases with increasing x, and it is consistent with the free-slip boundary conditions at $x = \pm \ell$.

The equilibrium magnetic field $\mathbf{b}_{eq}(x)$ is obtained setting $\varepsilon = 0$ in the expression (11), and it models the current sheet associated to a sector boundary of the solar wind. The magnetic field rotates by an angle 2α (we used $\alpha = \pi/4$ and $A = \sqrt{2}$) and its y component changes sign across the current sheet. The associated current $\mathbf{j}_{eq}$ is in the yz plane and its maximum is on the line $x = 0$, where $\mathbf{j}_{eq}$ is in the z direction. The current sheet width is ~ 1; we will refer to the remaining part of the spatial domain as "homogeneous region".

The perturbation amplitude has been chosen $\varepsilon = 0.5$, and its spectrum is determined by the choice of $\phi(y)$. We used a power-law spectrum function:

$$\phi(y) = 2 \sum_{m=1}^{m_{max}} (mk_0)^{-5/3}(\cos mk_0) \tag{13}$$

where $k_0 = 2\pi/\lambda_{max}$, and we have chosen $m_{max} = 32$.

The fluctuating part δf of any quantity f is defined as

$$\delta f(x, y, \tau) = f(x, y, \tau) - \frac{1}{\pi R \ell} \int_0^{\pi R \ell} f(x, y, \tau) \, dy, \tag{14}$$

the solutions being periodic along the y direction. Using this definition to calculate the fluctuating part of the initial magnetic field (11), the initial velocity field is given by

$$\mathbf{v}(x, y, 0) = \sigma(x) \frac{\delta \mathbf{b}(x, y, 0)}{\sqrt{\rho(x, y, 0)}} \tag{15}$$

with $\sigma(x) = \tanh(x/\delta)$ and $\delta = 0.05$. The expression (15) satisfy the condition (a) everywhere, except in a thin layer around $x = 0$, where $\sigma(x)$ changes sign. This choice of $\sigma(x)$ corresponds to a continuous change of the Alfvénic correlation sign from -1 to $+1$. This is necessary in order to have the same propagation direction of the initial Alfvénic perturbation on both sides of the current sheet, the y component of $\mathbf{b}_{eq}$ changing sign across the current sheet. The divergence of the initial velocity field is nonvanishing in the region where the correlation $\delta \mathbf{v}$-$\delta \mathbf{b}$ changes sign.

The initial density and temperature are

$$\rho(x, y, 0) = 1 \quad \text{and} \quad T(x, y, 0) = T_0 \tag{16}$$

where T_0 is a free parameter of the model, which determines the sound velocity $c_s = (\gamma T_0)^{1/2}$ and the plasma β, defined by $\beta = c_s^2/c_A^2 = \gamma T_0/A^2$. The plasma β represents a critical parameter with respect to the compressible fluctuation behavior. Then we have performed runs with different values of β (from $\beta = 0.2$ up to $\beta = 1.5$).

3   Numerical results

As expected, the time evolution of our system consists in the generation of compressive fluctuations, mainly inside the current sheet region. These fluctuations then propagate in our simulation box. It is worth noting that, since in our model the wavevectors necessarily lie on the xy plane, all the perturbations propagate obliquely with respect to the equilibrium magnetic field, the minumum propagation angle being equal to $\pi/4$. In particular, at the center of the current sheet, where $\mathbf{b}_{eq}$ is parallel to the z axis, the wavevectors are perpendicular to the equilibrium magnetic field. Let us consider in details some features of the time evolution for which a fine comparison with solar wind observations is allowed.

3.1   Time evolution of the Alfvénic correlation

The study of the time evolution of the Alfvénic correlation is best performed by introducing the Elsässer variables $\mathbf{z}^{\pm}$. In nondimensional units they are defined by

$$\mathbf{z}^{\pm}(x,y,\tau) = \mathbf{v}(x,y,\tau) \pm \frac{\mathbf{b}(x,y,\tau)}{\sqrt{\rho(x,y,\tau)}}.$$

From these variables the pseudo-energies $e^{\pm}$ associated with the fluctuations of $\mathbf{z}^{\pm}$ can be calculated:

$$e^{\pm}(x,y,\tau) = \frac{|\mathbf{z}^{\pm}(x,y,\tau)|^2}{2} - \frac{|\mathbf{z}_0^{\pm}(x,\tau)|^2}{2} \qquad (17)$$

where the index 0 indicates a space average along the periodicity y direction (see the equation (14)).

In Fig. 2, the y-integrated normalized cross-helicity obtained for $\beta = 0.2$

$$h(x,\tau) = \frac{\epsilon^+(x,\tau) - \epsilon^-(x,\tau)}{\epsilon^+(x,\tau) + \epsilon^-(x,\tau)} \qquad (18)$$

is shown at different times, $\epsilon^{\pm}(x,\tau)$ being

$$\epsilon^{\pm}(x,\tau) = \int_0^{\pi R\ell} e^{\pm}(x,y,\tau) \, dy. \qquad (19)$$

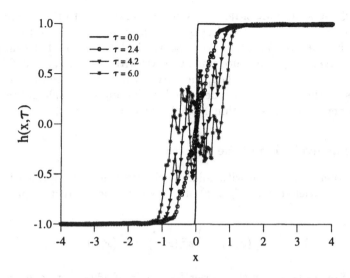

Fig. 2. $h(x, \tau)$ is represented at various times, for $\beta = 0.2$

At the initial time the correlation $\delta\mathbf{v}$-$\delta\mathbf{b}$ is either +1 or -1 in each subdomain, the width δ of the region where the correlation changes its sign being very small. After few Alfvén times the correlation is lost all over the current sheet region, i.e. $-1 < x < 1$, where the pseudo-energies become of the same order of magnitude. Moreover, in the region $-0.5 < x < 0.5$ after a certain time $(\tau = 4.2)$ an inversion in the initially dominant sign of correlation can be observed. However, the decorrelation region, where the initial Alfvénic character of the perturbation is destroyed, remains confined within the current sheet region; in the region where $\mathbf{b}_{eq}$ is homogeneous the velocity and magnetic field perturbations remain correlated up to the end of the simulation, i.e. during several Alfvén times. More or less the same evolution is observed for different values of β, showing that the value of β does not affect the evolution of Alfvénic fluctuations, at least at large scales.

In order to compare with the observed loss of Alfvénic correlation going away from the Sun in the ecliptic plane, we must recall that in the simulations the time has been normalized to the Alfvén time

$$\tau_A = \frac{a}{c_A} = \frac{1}{2}\frac{2a}{v_{sw}}\frac{v_{sw}}{c_A}, \tag{20}$$

v_{sw} being the solar wind bulk velocity. In order to estimate τ_A we assume for the Alfvén velocity the value $c_A \sim 70$ km s^{-1} and for the characteristic solar wind velocity in a slow speed stream (where the current sheet is localized) the value $v_{sw} \sim 350$ km s^{-1}. The ratio $2a/v_{sw}$ represents a lower limit for the time which a spacecraft takes to cross the current sheet (in our model the current sheet width is equal to $2a$). In particular this time would be equal to

$2a/v_{sw}$ if the spacecraft trajectory were perpendicular to the current sheet. We assume $2a/v_{sw} \leq 12$ h. Using the above values we get the estimate $\tau_A \leq 30$ h. The time $T_{1AU} = 1$ AU$/v_{sw}$ which the slow speed stream plasma takes to travel along a distance equal to 1 AU with a speed $v_{sw} \sim 350$ km s^{-1}, is given by $T_{1AU} \sim 120$ h. Then, the Alfvén time τ_A is at most 1/4 of T_{1AU} and most of the time evolution described in the simulations should take place in the inner heliosphere (within 1-1.5 AU from the Sun).

3.2   Time evolution of the spectra

A further comparison with solar wind data can be performed by calculating the Fourier spectra $e_m^{\pm}(x, \tau)$ of the pseudo-energy fluctuations, defined by

$$e_m^{\pm}(x, \tau) = \frac{1}{2} z_m^{\pm}(x, \tau) \cdot z_m^{\pm *}(x, \tau) \tag{21}$$

where $z_m^{\pm}(x, \tau)$ is the m-th Fourier harmonic of $z^{\pm}(x, y, \tau)$ defined by the equation (10), and the asterisk indicates complex conjugate. We will also study the power spectra of density and magnetic field intensity fluctuations

$$e_m^{\rho}(x, \tau) = \rho_m(x, \tau)\rho_m^*(x, \tau) \tag{22}$$

$$e_m^{b}(x, \tau) = b_m(x, \tau)b_m^*(x, \tau) \tag{23}$$

with $b(x, y, \tau) = |\mathbf{b}(x, y, \tau)|$. The x-dependence in the quantities defined by the equations (21)-(23) has been displayed because these spectra calculated inside or outside the current sheet show different behaviors. At the initial time e_m^- (inward propagating Alfvén waves) is vanishing while e_m^+ (outward propagating Alfvén waves) roughly follows a power law. At the time $\tau = 1.2$ (Fig. 3) the spectrum of e^- is completely formed and it is superposed to e_m^+; for $m \leq 10 - 15$ both spectra follow a power-law which is less steep than the initial e^+ spectrum and is close to a $k^{-5/3}$ power-law (which is shown for comparison). For later times the two spectra remain very close to each other and they are gradually dissipated. A behavior similar to that of the e^- spectrum is observed in the density and magnetic field intensity fluctuation spectra, i.e. both spectra completely form during the first stage of the time evolution, and presents a power-law in the wavenumber range $m \leq 10 - 15$. During all the time evolution the e^{ρ} and e^b spectra remain very close to each other.

The situation in the homogeneous region is rather different (Fig. 3). In the large scale range, i.e. for $m \leq 10 - 15$ the slope of e_m^+ is larger than the corresponding slope in the current sheet region and the reverse situation holds for e_m^-. Moreover, the energy in the e_m^- spectrum remains much smaller than that in the e^+ spectrum, while the two spectra are superposed for higher

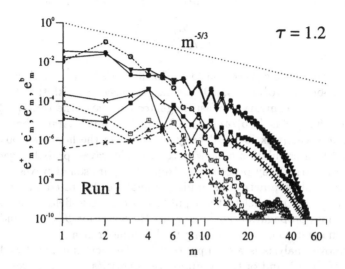

Fig. 3. The Fourier spectra of pseudoenergies fluctuation $e_m^+(x, \tau)$ (circles), $e_m^-(x, \tau)$ (triangles), density fluctuation e_m^ρ (squares) and magnetic field intensity fluctuation e_m^b (crosses) are represented as functions of the wavenumbers at $x = 0.2$ (full lines) and at $x = 3.01$ (dashed lines) at the time $\tau = 1.2$, for $\beta = 0.2$

wavevectors. The behavior of the density and magnetic field intensity spectra in this range follows quite well that of e_m^-, also in this case. At an intermediate wavenumber range ($20 < m < 50$) both the $e^\pm$ spectra present a *plateau*, followed by another steep decrease of both pseudo-energies.

These structures are probably due to a nonlocal mechanism like a parametric decay [31,32]. This hypothesis is supported by the fact that, in the same frequency range, at the time $\tau = 1.2$, the density spectrum presents a local maximum; this phenomenon is totally absent in the e^b spectrum, which is neatly steeper than e^ρ. Compressive fluctuations in this frequency range are characterized by a ratio $e_m^b/e_m^\rho \lesssim 10^{-2}$. This can be explained by the presence of slow magnetosonic fluctuations in oblique propagation. Actually, considering a small amplitude magnetosonic wave, the magnetic field intensity to density fluctuation ratio is approximated by

$$\frac{\delta|\mathbf{b}|}{\delta\rho} \simeq \frac{\mathbf{b}_{eq} \cdot \delta\mathbf{b}_f}{\delta\rho} = \frac{b_{eq}}{\rho_{eq}} \frac{(\delta b_f/b_{eq}) \sin\theta_p}{(\delta\rho/\rho_{eq})},$$

$\delta\mathbf{b}_f$ representing the magnetic field fluctuation and θ_p the propagation angle, i.e. the angle between $\mathbf{b}_{eq}$ and the propagation direction. In our geometry, the minimum propagation angle is $\theta_p = \pi/4$, corresponding to fluctuations propagating in the periodicity y-direction. In that case the ratio

$$\frac{(\delta b_f / b_{eq})}{(\delta \rho / \rho_{eq})} \gtrsim 0.1 \qquad (24)$$

for slow magnetosonic waves, at $\beta = 0.2$ [33]. Using $b_{eq} = A = \sqrt{2}$ and $\rho_{eq} = 1$ we obtain $(\delta|\mathbf{b}|/\delta\rho)^2 \gtrsim 10^{-2}$, which is consistent with the above value of e_m^b/e_m^ρ. On the contrary, for fast magnetosonic waves the ratio (24) would be $\gtrsim 0.9$. This yields $(\delta|\mathbf{b}|/\delta\rho)^2 \gtrsim 0.8$, which is much larger than e_m^b/e_m^ρ. In conclusion, the peak in the density spectrum in the homogeneous region correspond to a population of slow magnetosonic waves, probably generated by the parametric decay process [32]. Note that pure sound waves, as in the one-dimensional parametric instability, cannot be generated in our model, since propagation parallel to the equilibrium magnetic field is not allowed (the minimum propagation angle is $\pi/4$). However, for $\beta < 1$ sound waves and slow magnetosonic waves belong to the same branch.

The above analysis is also supported by the fact that for $\beta = 1.5$, the behavior of the $e^\pm$ and of the compressive quantities spectra is very similar to that observed for $\beta = 0.2$, when calculated at locations inside the current

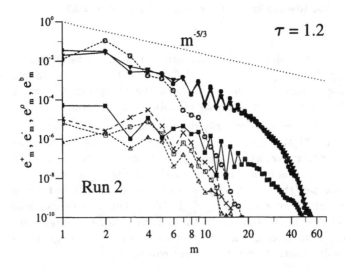

Fig. 4. The Fourier spectra of pseudoenergies fluctuation $e_m^+(x, \tau)$ (circles), $e_m^-(x, \tau)$ (triangles), density fluctuation e_m^ρ (squares) and magnetic field intensity fluctuation e_m^b (crosses) are represented as functions of the wavenumbers at $x = 0.2$ (full lines) and at $x = 3.01$ (dashed lines) at the time $\tau = 1.2$, for $\beta = 1.5$

sheet region (Fig. 4). On the contrary, in the homogeneous region it is seen that the *plateau* observed in the e^+ and e^- spectra which is found at intermediate wavenumbers for $\beta = 0.2$, does not form for $\beta = 1.5$. Moreover, the e^ρ and e^b spectra are essentially superposed. This confirms the hypothesis that the *plateau* was due to the occurrence of a parametric instability, which is suppressed for $\beta > 1$ [31].

Let us compare these results with the corresponding spectra typically found for the solar wind fluctuations. We recall that in our model, the longest wavelength in the fluctuation spectrum in the y direction roughly corresponds to the lower limit of the Alfvénic range of the solar wind turbulence, i.e. $f_{min} \geq 2.4 \times 10^{-5}$ Hz. Spectra of e^+ and e^- calculated inside solar wind slow speed streams reveal a behavior typical of a fully developed turbulence [34]: the two spectra are close to each other (with e^+ slightly more intense than e^-), and both have a slope near to that of the Kolmogorov spectrum ($k^{-5/3}$). A similar slope has been found also for the proton density fluctuation spectrum.

These features are qualitatively reproduced by our model. Inside the current sheet, due to the form chosen for the initial condition $\mathbf{B}^2 = const$, the main dynamical effect is represented by the interaction between the inhomogeneous structure and the propagating waves. This dynamical effect, within few Alfvén times ($\tau \lesssim 1 - 2$), is able to form spectra of e^+ and e^- which are essentially superposed. At the same time, also the e^ρ and e^b spectra form, with a slope close to that of the $e^\pm$ spectra (Fig. 3).

Far from the magnetic current sheet we have seen that the dynamical evolution we find should be due to the development and growth of an instability which transfers energy non locally in the wavevector space (parametric decay). A similar behavior has been observed also in the solar wind fluctuations spectra, where also during Alfvénic periods, the high frequency part (2×10^{-4} Hz $\leq f \leq 10^{-2}$ Hz) of the e^{in} and density fluctuation spectrum, becomes rather flat and superpose to the e^{out} spectrum [34,35]. Moreover, in the same range the magnetic field intensity spectrum is much steeper than the density spectrum.

Concerning the comparison between the spectral indices measured in the solar wind and those derived from the simulation, it is worth to note that the latter are strongly affected by the low Reynolds number used; no clear distinction between an inertial and a dissipative range can be made so the values obtained in the simulation can be sistematically higher than those one could expect using more realistic Reynolds numbers. Nevertheless the comparison with solar wind observations diplays a clear qualitative agreement. Let us note however that the lower frequency Alfvènic range (10^{-5} Hz $\leq f \leq 10^{-2}$ Hz) where e^{out} is very flat is not reproduced by our model.

3.3 Density magnetic field intensity correlation

Compressive fluctuations with $\delta\rho$ and δb both correlated and anticorrelated have been detected in our simulation [25]. Among the latter, there are magnetic flux tubes (i.e., structures where magnetic field is higher and density lower than in the surrounding medium) and tangential discontinuities, both pressure-balanced and quasi-static in the plasma reference frame, mostly concentrated in the current sheet region, where $\mathbf{b}_{eq}$ is quasi-perpendicular to the propagation direction. Positively correlated $\delta\rho$-δb fluctuations, which belong to the fast magnetosonic mode, have been found propagating in oblique directions in the xy plane. The most intense among them form fast magnetosonic shocks, which gradually dissipate.

Vellante and Lazarus [36], in their studies of the $\delta\rho\ \delta b$ correlation on solar wind data, have shown that for fluctuations at time scales larger (smaller) than $t_c \sim 10$ h, positive (negative) correlations prevail. Let us focus now on the correlation between the density and the magnetic field intensity fluctuations found in the simulation results, in order to study in particular the distribution of this correlation at different spatial scales. Since the correlation appears to change going from the current sheet to the homogeneous region, we studied its dependence on the transverse x direction. Finally, as the propagation properties of the compressive modes change considerably when the value of β crosses 1, we have also studied the behavior of the correlation for different values of β.

We define the density-magnetic field intensity correlation as

$$\sigma_{\rho b} = \frac{\langle \Delta\rho\ \Delta b \rangle_{\Delta x}}{\sqrt{\langle (\Delta\rho)^2 \rangle_{\Delta x}\ \langle (\Delta b)^2 \rangle_{\Delta x}}} \tag{25}$$

where angular parentheses indicate a running average taken over a length Δx:

$$\langle f \rangle_{\Delta x} = \frac{1}{\Delta x} \int_{-\Delta x/2}^{\Delta x/2} f(x + \xi, y, \tau)\, d\xi, \tag{26}$$

$\Delta f = f - \langle f \rangle_{\Delta x}$ represents the contribution of scales smaller than Δx (f being either ρ or b). Due to the symmetry of the problem, quantities have been periodically extended for $|x| > \ell$, to calculate the running average (26) also in points closer to the boundaries $x = \pm\ell$ than $\Delta x/2$.

To compare the results of the simulations with measures in slow speed streams of the solar wind, we have performed the same kind of analysis on data taken by the Helios spacecraft in the inner heliosphere. So, we have selected some periods of data, each one within a slow speed stream and containing crossings of the interplanetary current sheet. The detailed analysis for all the periods is reported elsewhere [26]. Here we limit to present two typical behaviors: the first one for $\beta \sim 1$, the second one for $\beta > 1$.

Let us consider first the results of the low-β run ($\beta = 0.2$). In the upper part of Fig. 5 the correlation $\sigma_{\rho b}$ is plotted at the time $\tau = 4.8$, i.e. several Alfvén times after the initial time. This quantity is represented as a function of the x coordinate, transverse to the current sheet, for a given value of y

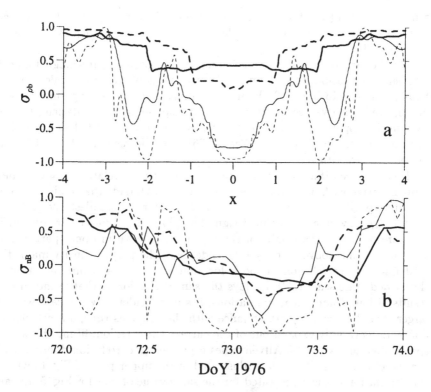

Fig. 5. (a) Numerical simulation: σ_{pb} for $\Delta x = 0.5$ (thin dashed line), $\Delta x = 1$ (thin line), $\Delta x = 2$ (thick dashed line), and $\Delta x = 4$ (thick line), are shown as functions of x, for $y = 1.6$, $\tau = 4.8$ and $\beta = 0.2$ and (b) Helios data: correlation σ_{nB} for $\Delta t = 3$ hours (thin dashed line), $\Delta t = 7$ hours (thin line), $\Delta t = 13$ hours (thick dashed line), $\Delta t = 25$ hours (thick line), as functions of time (time units are day of the year 1976), with a value of $\beta \sim 1$ (the neutral sheet crossing is located at $DoY = 73.3$)

($y = 1.6$). We can see that, on average, positive (negative) correlations prevail in the whole domain at large (small) spatial scales Δx.

A closer examination reveals that, when going from the current sheet region to the homogeneous region, the correlation σ_{pb} shows a different behavior at different scales. In particular, in the region close to the current sheet σ_{pb} is slightly positive at scales $\Delta x \gtrsim 1$ while it is clearly negative for $\Delta x \lesssim 1$. In the homogeneous region σ_{pb} is positive at large scales, while it does not show a definite sign at small scales. This behavior indicates that anticorrelated density-magnetic field fluctuations, like pressure balanced flux tubes and tangential discontinuities prevail in the current sheet region. On the contrary, correlation corresponding to fast magnetosonic fluctuations dominate far from the current sheet at scales of the order of, or larger than the current sheet width (which has been used as normalization length in the present similation). When performing simulations up to $\beta = 1$, $\sigma_{\rho b}$ displays the same behavior [26].

On the Helios data we have calculated the proton density-magnetic field intensity correlation σ_{nB} at four different values of the time scale: $\Delta t = 3$ hours, 7 hours, 13 hours, 25 hours. Assuming a shear crossing time $t_a \sim 6$ hours, the above values of Δt roughly correspond to the scale lengths Δx used in the upper part of Fig. 5. The correlation σ_{nB} is plotted as a function of time (in units of Day of the Year 1976) in the lower part of Fig. 5, for a period of data with $\beta \sim 1$.

It can be seen that close to the current sheet the correlation σ_{nB} becomes more negative with decreasing the time scale Δt. Outside that region, σ_{nB} is definitely positive at all the time scales, except for the smallest one ($\Delta t = 3$ hours) σ_{nB} having a less defined sign. The plots in the lower part of Fig. 5 compare well with the results of the $\beta \leq 1$ runs shown in upper part of the same figure: the above described dependence of the proton density-magnetic field intensity correlation on both location (close or far from the current sheet) and time scale (Δt) displays the same behavior as that found in the results of low β numerical simulations. Les us consider now the case with β larger than 1 ($\beta = 1.5$). At variance with the previous runs, anticorrelated fluctuations rapidly expand out of the current sheet region, filling the whole spatial domain. After 4-5 Alfvén times negative $\sigma_{\rho b}$ correlation dominate all the structure. This situation is illustrated in the upper part of Fig. 6, where the correlation $\sigma_{\rho b}$ is represented for the same value of τ as for Fig. 5. In can be seen that the correlation is strongly negative at any position and for all the considered spatial scales Δx, except for the smallest scale where localized spikes of positive correlation are present.

In the lower part of Fig. 6 we have represented the analysis of the correlation performed on Helios data during a period where the value of β was larger than 1. In this case the correlation σ_{nB} appears to be definitely negative at all the considered time scales. At the smallest time scale the correlation sporadically rises to large positive values. The location of current sheet crossings, which occurs at $t = 101.6$ DoY, seems not to affect the correlation σ_{nB}. Comparing with the results of numerical simulation reported in the upper part of Fig. 6, it can be seen that both the dominant negative correlation and the spikes of positive correlation at the smallest scale are reproduced by the numerical model.

4 The effects of an inhomogeneous entropy distribution

Compressive perturbations in slow speed streams of solar wind can also be characterized through the study of the correlation between proton density (δn) and temperature (δT_p) fluctuations. This correlation has been studied by a number of authors [37–40]. In particular, Bavassano et al. [40] carried

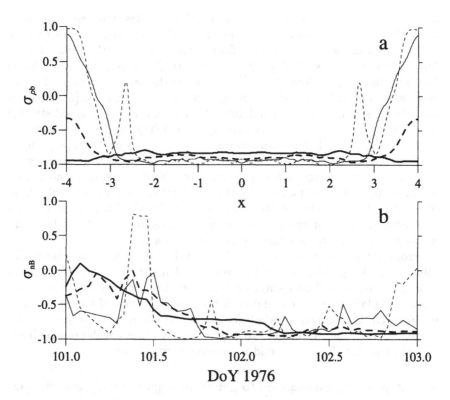

Fig. 6. (a) Numerical simulation: σ_{pb} for $\Delta x = 0.5$ (thin dashed line), $\Delta x = 1$ (thin line), $\Delta x = 2$ (thick dashed line), and $\Delta x = 4$ (thick line), are shown as functions of x, for $y = 1.6$, $\tau = 4.8$ and $\beta = 1.5$ and (b) Helios data: correlation σ_{nB} for $\Delta t = 3$ hours (thin dashed line), $\Delta t = 7$ hours (thin line), $\Delta t = 13$ hours (thick dashed line), $\Delta t = 25$ hours (thick line), as functions of time (time units are day of the year 1976), with a value of β larger than 1 (the neutral sheet crossing is located at $DoY = 101.6$)

out a detailed analysis of the density-temperature correlation as a function of the solar wind speed and the radial distance from the Sun, in the inner heliosphere, using Helios data. They found that in general, cases with a well defined sign of the density-temperature correlation are seldom observed in solar wind, where very few cases reach a value larger than 0.8 (as absolute value) for the correlation coefficient. Moreover, they found that on smaller scales, the sign of the correlation is mainly positive in fast streams, while both signs are present in slow streams.

From a theoretical point of view, correlation between density ρ and temperature T in MHD turbulence has been considered within the so-called "Nearly-incompressible magnetohydrodynamics" (NI-MHD) theory [41–45], in which the limit of small sonic Mach number M is studied, i.e., small departures from incompressibility are considered. If the fluid is considered as a poly-trope (heat-fluctuations-modified fluid (HFMF)) this theory predicts positive ρ-T correlations, as well as density fluctuations scaling as the squared sonic Mach number ($\delta\rho/\rho \propto O(M^2)$). On the contrary, if heat conduction is allowed for (heat-fluctuations-dominated fluid (HFDF)) the density-temperature correlation is expected as negative, and $\delta\rho/\rho \propto O(M)$. Klein et al. [37] have analyzed such scalings in solar wind data. They found that slow speed streams fit well the predictions of NI-MHD theory for a HFDF, whilst the situation is rather ambiguous in fast streams. More recently, Bavassano et al. [40] considered the scalings of the density fluctuations with the sonic Mach number for the cases where the sign of the correlation was better defined, without finding the expected trend foreseen by the NI-MHD theory. These results have to be related to the ones of Matthaeus et al. (1991) [43] who performed an analogous analysis for the aforementioned scalings in the outer heliosphere without finding a clear evidence of them in the data.

From the above results, we have been pushed to look for an alternative solution to the problem posed by presence of negative n-T_p correlation in slow speed streams. This solution is based on the idea that the main ingredient to explain negative correlation is the inhomogeneity of the background medium. On the contrary, turbulence models which describes a statistically homogeneous situation (like the NI-MHD theory) neglect such an ingredient, and assume that the background is spatially homogeneous.

4.1   A physical mechanism to generate negative n-T_p correlation

In order to illustrate the physical mechanism which we propose, let us consider the evolution equation of the entropy per-mass-unit s, within the MHD framework [33]:

$$\rho T(\frac{\partial}{\partial t} + \mathbf{v} \cdot \nabla)s = Q + \nabla \cdot \mathbf{q}. \tag{27}$$

In this equation Q represents heat sources, $\mathbf{q}$ is the heat flux, while $\mathbf{v}$, ρ, and T are the velocity, the density and the temperature, respectively. The right-hand side contains the time derivative of s along the flow lines. In an ideal case, when energy dissipation and heat conduction are both neglected ($Q = 0$, $\mathbf{q} = 0$), the entropy s is simply convected by the fluid motion. So, if s is spatially homogeneous at the initial time, it will remain uniform all over the time. Such a configuration is referred to as *isoentropic*. In an isoentropic situation, compressive perturbations necessarily have density and temperature *positively correlated*. In fact, since in a perfect gas

$$s \propto \ln(\frac{T}{\rho^{\gamma-1}}) \tag{28}$$

(γ being the adiabatic index), $s =$ uniform during all the time implies

$$T \propto \rho^{\gamma-1} \tag{29}$$

i.e., positive (negative) variations of ρ correspond to positive (negative) variations of T. For instance, this situation is typically recovered when (as usual in many MHD turbulence models) the polytropic equation $p/\rho^\gamma =$ constant is assumed, which is equivalent to the condition (29).

Let us note also that when the above polytropic equation is assumed, only magnetosonic waves are found as small amplitude compressive perturbation. In such waves density and temperature are always positively correlated. In a more general case, when the polytropic equation is relaxed, also entropy waves can be found among the small amplitude compressive perturbations [33]. In entropy waves density and temperature fluctuations are anticorrelated, and the isoentropic condition ($s =$ uniform) is clearly violated.

From the above discussion it is clear that density-temperature negative correlations require a non-uniform entropy distribution. In the statistically homogeneous situation considered by the NI-MHD heat-fluctuations-dominated fluid model [42], this is achieved by including the effects of the heat conduction. Heat conduction represents a source for entropy modulation (see equation (27)), and it can generate density-temperature *negative correlations* also in an initially isoentropic configuration. On the contrary, in the model which we propose the entropy modulation is not due to non-ideal effects like heat conduction, but it is assumed to be present from the outset, in the large scale inhomogeneity of the background structure.

Slow speed streams are colder and denser than the surrounding fast speed streams. So, when moving from a slow to a fast stream, the density and the temperature variations at large scale are anticorrelated. In other words, the entropy per-mass-unit s changes, being smaller in a slow stream than in the surrounding fast streams. We will now explicitly took into account this large scale variation of s in our model, by including it in the background structure.

We expect that, due to the Alfvénic perturbations initially superposed on the background, a spectrum of velocity fluctuations forms. As a result, the coupling between the large scale entropy inhomogeneity and the Alfvénic perturbation (represented by the second term in the left-hand side of equation (27)), will move the entropy modulation to increasingly smaller scales. This mechanism could produce density-temperature negative correlations at all the scales. On the other hand, as we have just seen, the coupling between the Alfvénic perturbation and the large scale current sheet produces also magnetosonic-like fluctuations, in which density and temperature are positively correlated. As a result, the actual sign of the ρ-T correlation, at differents scales and locations, will be determined by the competition between these two mechanisms (entropy cascade and production of magnetosonic-like fluctuations).

4.2   A numerical model with inhomogeneous entropy distribution

To describe the entropy modulation at large scale, associated with mass density and temperature variations around the heliospheric current sheet, we must modify the initial conditions of our model described in the previous sections.

The slow speed solar wind streams, in which the current sheet is embedded, are colder and denser than the surrounding fast streams. In particular, the ion density and temperature are anticorrelated on a time scale ~ 1 day, the density displaying in many cases a rough maximum close to the current sheet location. In our model the background density and temperature vary along the cross-current-sheet (x) direction. The density is maximum and the temperature is minimum at the center $x = 0$ of the current sheet, while the associated variation length $a_e = 2$ is larger than the current sheet width a. The product ρT is initially uniform to ensure gas pressure equilibrium in the background structure. Correspondingly, the specific entropy s varies on the same scale a_e. As discussed above we expect that the nonlinear evolution of the perturbation induces an entropy cascade to smaller scales.

In order to single out the effects due to entropy modulation on the dynamical evolution, we neglect the velocity large scale variation associated to the stream structure. We assume a uniform background velocity, which is vanishing in a reference frame moving with the plasma. solar wind observations [46] show that in many cases the magnetic sector boundary is totally embedded into slow-speed streams and that the typical length scale for the associated current sheet (some hours) is generally much less than the typical width of slow-speed streams (some days). In such cases neglecting background velocity inhomogeneities related to the stream structure is a reasonable approximation.

Taking into account the above considerations, we have only to modify the initial conditions for density and temperature, described in (16) in the following way:

$$\rho(x, y, t = 0) = \rho_0 \left\{ 1 + \Delta \left[\frac{1}{\cosh^2(x/a_e)} + p \left(\frac{x}{a_e} \right)^2 \right] \right\} \qquad (30)$$

$$T(x, y, t = 0) = \rho_0 T_0 / \rho(x, y, t = 0). \qquad (31)$$

In the equation above

$$p = \frac{\tanh(l/a_e)}{\left(\frac{l}{a_e} \right) \cosh^2(x/a_e)}$$

is a parameter that ensures the fulfillment of the boundary conditions for the density and temperature, $\Delta = 1/4$ is the amplitude of the entropy inhomogeneity, and the product $\rho_0 T_0$ is the total, constant, kinetic pressure.

We point out that in this initial condition, due to the density and temperature modulation (equations (30) and (31)) both the Alfvén speed and the sound speed $c_s = (\gamma T_0(x, y, t = 0))^{1/2}$ (where γ is the adiabatic index) are not uniform, both becoming larger with increasing the distance from the current sheet. However, the plasma $\beta = c_s^2/c_A^2$ is still uniform in the entire domain. Also in this model we carried out numerical Runs using different values of β [47]. In the following we will refer to two Runs with respectively $\beta = 0.2$ and $\beta = 1.5$ In these Runs to give a general information about the behaviour of the correlation coefficient across the inhomogeneity, we have also averaged the correlation in the y direction, always considering the quantity $<\sigma_{fg}>_y$.

The initial condition is different from the one used in the previous Sections, in that both density and temperature in the background structure are spatially modulated. This further inhomogeneity contributes to the dynamical evolution of the initial Alfvénic perturbation, and the resulting ρ-b correlation could in principle be different from that found in the previous Sections. For this reason, we have performed a study of the density–magnetic-field-intensity correlation and we have compared it, with that resulting from the previous model. We have found [47] that the behaviour of the ρ-b correlation in the two models is very similar. Then the presence of the entropy modulation in the background structure does not modify the behaviour of the ρ-b correlation, at the considered scales.

4.3   Density–temperature correlations

In the upper part of Fig. 7 we show the plots of the quantity $<\sigma_{\rho T}>_y$ at different length scales $\Delta x = 0.5, 1.0, 2.0$ and 4.0 for $\beta = 1.0$. We remember that the length unit corresponds to the half-width of the current sheet. The quantities are plotted at a time $\tau = 4.8$. This time corresponds to few eddy turnover times, so nonlinear effects have had enough time to built up the spectrum. One can see that, when averaging on the whole domain, a negative density-temperature correlation prevails at large scale, while $<\sigma_{\rho T}>_y$ tends to become increasingly positive with decreasing the scale Δx.

The detailed behaviour of $<\sigma_{\rho T}>_y$ is different at different scales Δx: at large scale the correlation is negative in the whole domain; decreasing the scale $<\sigma_{\rho T}>_y$ remains slightly negative at the center of the domain, while it becomes more and more positive close to the boundaries, i.e. far from the inhomogeneity region.

The negative correlation at larger scales essentially reflects the entropy modulation of the background structure, which was present in the initial condition. The behaviour of $<\sigma_{\rho T}>_y$ at smaller scales indicates that in the region where the background structure is more inhomogeneous small scale entropy fluctuations prevail; such fluctuations should result from an entropy cascade from large to small scales. In the region where the background structure is more homogeneous, magnetosonic-like fluctuations dominate. In this same region, the density–magnetic-field intensity correlation is positive too (see Fig. 5). This indicates that small scale compressive fluctuations in the homogeneous region essentially have properties similar to those of the fast

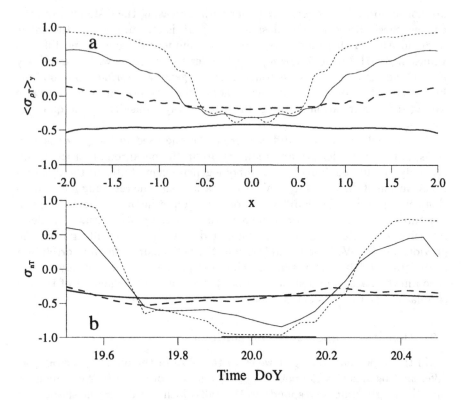

Fig. 7. (a) Numerical simulation: $< \sigma_{pb} >_y$ for $\Delta x = 0.5$ (thin dashed line), $\Delta x = 1$ (thin line), $\Delta x = 2$ (thick dashed line), and $\Delta x = 4$ (thick line), are shown as functions of x, for $\tau = 4.8$ and $\beta = 1.0$ and **(b)** Helios data: correlation σ_{nT} for $\Delta t = 2$ hours (thin dashed line), $\Delta t = 6$ hours (thin line), $\Delta t = 12$ hours (thick dashed line), $\Delta t = 24$ hours (thick line), as functions of time (time units are day of the year 1976), with a value of $\beta \simeq 1$ (the neutral sheet crossing is located at $DoY = 20.1$)

magnetosonic mode. Runs carried out with bigger values of β up to $\beta = 1.5$ give rise to results which are very similar to those obtained in the case $\beta = 0.2$ [47].

Let us now consider the analogous analysis for the observations. For this we used data from the Helios 2 mission. We studied the correlation between proton density n and temperature T_p, assuming the behaviour of the proton temperature in slow speed streams as representative of that of the total temperature $T_p + T_e$ [38]. The correlation σ_{nT} has been calculated for several periods, each containing a sector boundary. We selected, in particular periods in which the large scale structure is more similar to the one used in our model: namely, the maximum of proton density is near to the location of the heliospheric current sheet.

In the lower part of Fig. 7 we show the correlation coefficient $\sigma_{\rho T}$ calculated on various time scales $\Delta t = 2, 6, 12, 24$ hours. Assuming a shear crossing time $t_a \sim 6$ h, those times roughly correspond to thescale lengths Δx used in the simulations. The data have been hourly averaged before working out the correlations, in order to filter out the oscillations uninteresting for our comparison and that degenerate the clearness of the plots. For this period, $\beta \sim 1$. The magnetic field changes sign in proximity of $t \sim 20.1$ DoY 1976, while the position of the heliospheric current sheet is denoted by a thick segment on horizontal axis. At that location the density has a bump and the temperature a hole. These conditions correspond to those we used in our simulations. It is apparent from the plot that, corresponding to the density peak, σ_{nT} has a negative sign at all time scales. Far from both the current sheet and the density maximum the sign of correlation becomes increasingly positive by going towards smaller scales. The qualitative behaviour is in fairly good agreement with the trend observed in the simulations.

In summary, the behaviour displayed by σ_{nT} as a function of both position (close or far from the current sheet and density maximum) and time scale, in the considered periods of the Helios data set, is essentially reproduced by our model. Moreover, the above described features of σ_{nT} appears to be essentially independent of the value of β; this is also verified for the ρ-T correlation, as it results from our model.

5   Discussion and conclusions

The results of this model and comparisons with solar wind data indicate that the large scale inhomogeneity associated to slow speed streams and to the heliospheric current sheet plays an important role in determining major features of compressive fluctuations.

In particular, the numerical simulations described in the previous section have allowed us to study in detail the time evolution of an initial condition modelling the physical situation produced by the convergence of large amplitude Alfvénic fluctuations on the two sides of the heliospheric current sheet. Since these fluctuations propagate in the same (outward) direction, they have opposite sign of the $\delta \mathbf{v}$-$\delta \mathbf{B}$ correlation. Moreover, we have tried to set up an initial condition which reproduces an important feature of large amplitude Alfvénic fluctuations in a compressible medium; namely, we have assumed that the total magnetic field (structure+perturbation) is initially uniform. This implies that in the homogeneous region, the initial perturbation locally represents an exact solution of MHD equations.

We have found that in a short time ($t \leq \tau_A$) the initial $\delta \mathbf{v}$-$\delta \mathbf{b}$ correlation is greatly reduced in the current sheet region. With increasing time the decorrelation effect remains strictly limited to the current sheet region; in the region where $\mathbf{b}_{eq}$ is homogeneous the velocity and magnetic field perturbations remain correlated also after several Alfvén times.

Along with a $\delta\mathbf{v}$-$\delta\mathbf{b}$ decorrelation, density and magnetic field intensity fluctuations are generated. These compressive fluctuations do not remain confined within the current sheet but, with increasing time, they expand in the direction perpendicular to the current sheet, also in the region where $\mathbf{b}_{eq}$ is homogeneous. In the solar wind, the level of density fluctuations around the heliospheric current sheet is higher than the average value; however, the increase in the level of those fluctuations is not limited to the current sheet, but it is observed in the whole slow speed stream. Then, the presence of compressive fluctuations in a wide layer around the heliospheric current sheet seems to be in accordance with the results of our simulations.

We have found that the dependence of σ_{nB} on location (close or far from the current sheet), on the fluctuation scale and on the value of β are qualitatively reproduced by the numerical model. The similarity between simulation results and solar wind data shows that the main physical mechanisms which determine the density-magnetic field intensity correlation observed in slow speed streams have been included in our numerical model. This allows us to try to give a physical interpretation for the $n - B$ correlation observed in the vicinity of the heliospheric current sheet. Density and magnetic field fluctuations, both positive- and negative-correlated, are produced around the current sheet by the above-described mechanisms, but, in order to escape from that region they must propagate at large angles with respect to the background magnetic field. Actually, close to the current sheet the background magnetic field is mainly perpendicular to the cross-current sheet direction. This fact represents a limitation for the propagation of negative-correlated fluctuations; indeed, slow magnetosonic perturbations (which are characterized by anticorrelated ρ and $|\mathbf{B}|$ fluctuations) have a vanishing phase velocity when the wave vector is perpendicular to the background magnetic field, so they tend to remain confined close to the current sheet. On the contrary, this limitation does not affect positive-correlated fluctuations, which are free to propagate away from the current sheet. As a consequence, the region around the current sheet will be dominated by negative-correlated compressive structures with wave vectors nearly perpendicular to the large scale magnetic field (flux tubes and tangential discontinuities).

Since the above mechanisms produce compressive fluctuations inside the current sheet, in such a region the level of compressive fluctuations is higher and nonlinear effects are more intense than outside. Then, a nonlinear cascade is produced [25] which transfer negative-correlated fluctuations toward small scales. This could explain why σ_{nB} becomes more negative with decreasing the scale Δt in the current sheet region. Outside such a region, positive-correlated compressive fluctuations (fast mode) propagating away from the current sheet are essentially present. For this reason, moving away from the current sheet σ_{nB} becomes positive, at least at large scales and in situations with $\beta \lesssim 1$.

When $\beta > 1$, both in solar wind data and in our simulation negative-correlated fluctuations are observed also outside the current sheet; moreover, negative correlation prevails at all time scales. This could be related to the tendency of a magnetofluid with $\beta > 1$ to develop compressive structures in which magnetic and gas pressure fluctuations are anticorrelated. Actually, in the current sheet region the interaction between the initial Alfvénic perturbation and the large scale inhomogeneity generates both magnetic pressure and density fluctuations. If the sound velocity is larger than the Alfvén velocity, such density perturbations can propagate sufficiently fast to balance the magnetic pressure fluctuations by opposite gas pressure fluctuations. This argument could explain why anticorrelated $n - B$ fluctuations are more frequent than positively correlated fluctuations, for β sufficiently larger than 1.

The simulations performed in presence of an initial inhomogeneity in the entropy distribution seem also to show that the observed density-temperature correlation can be due to the presence of an entropy cascade and to the generation of a spectrum of magnetosonic fluctuations. Both phenomena are driven by the dynamical interaction between Alfvén waves propagating away from the Sun and nonuniformities intrinsic to the large scale configurations of the background medium. Then, models treating the MHD turbulence in slow speed streams should include inhomogeneity effects.

It is clear that our model for the current sheet is oversimplified and does not take into account other phenomena, both physical (velocity shear) and geometrical (solar wind expansion). However we think that in studying the behavior of a complex system, like solar wind, it is important to be able to single out physical effects in order to understand what is the respective importance in determine the observed characteristics. In this respect it is worth noting that the choice we have made in the initial condition, namely to start with the nonlinear Alfvénic solution in a compressible homogeneous medium has allowed us to outline important physical effects which should be at work in the solar wind current sheet and far from it.

Acknowledgements
This work is part of a research program which is financially supported by the Ministero dell'Università e della Ricerca Scientifica e Tecnologica (MURST), the Consiglio Nazionale delle Ricerche (CNR), contract 98.00148 CT02 and the Agenzia Spaziale Italiana (ASI), contract ARS 98-82.

References

1. Taylor, G. I. (1938) The spectrum of turbulence. Proc. R. Soc. London Ser. A **164**, 476

2. Coleman, P. J. (1968) Turbulence, viscosity and dissipation in the solar wind plasma. Astrophys. J. **153**, 371

3. Belcher, J. W., Davis, L. (1971) Large-amplitude Alfvén waves in the interplanetary medium. J. Geophys. Res. **76**, 3534–3563

4. Veltri, P. (1980) An observational picture of Solar Wind MHD Turbulence. Il Nuovo Cimento **C3**, 45–55

5. Dobrowolny, M., Mangeney, A., Veltri, P. (1980) Fully developed anisotropic hydromagnetic turbulence in interplanetary space. Phys. Rev. Lett. **45**, 144–147

6. Grappin, R. Frisch, U., Leorat J., Pouquet, A. (1982) Alfvén fluctuations as asymptotic states of mhd turbulence. Astron. Astrophys. **105**, 6–14

7. Carbone, V., Veltri, P. (1990) A shell model for anisotropic magnetohydrodynamic turbulence Geophys. Astrophys. Fluid Dyn. **52**, 153–181

8. Gloaguen, C., Leorat, J., Pouquet, A., Grappin R. (1985) A scalar model of MHD Turbulence. Physica **17D**, 154–182

9. Carbone, V., Veltri, P. (1987) A simplified cascade model for m.h.d. turbulence Astron. Astrophys. **188**, 239-250

10. Grappin, R. (1986) Onset and decay of two-dimensional magnetohydrodynamic turbulence with velocity magnetic field correlation. Phys. Fluids **29**, 2433–2443

11. Ting, A., Matthaeus, W. H., Montgomery D. (1986) Turbulent relaxation processes in magnetohydrodynamics. Phys. Fluids **29**, 3261–3274

12. Meneguzzi, M., Frisch, U., Pouquet, A., (1981) Helical and nonhelical turbulent dynamos. Phys. Rev. Lett. **47**, 1060–1064

13. Pouquet, A., Meneguzzi, M., Frisch, U. (1986) The growth of correlations in mhd turbulence. Phys. Rev. **A33**, 4266–4276

14. Grappin, R. (1985) Quelques aspects de la turbulence MHD développée. These de Doctorat d'Etat en Sciences Physiques, Université de Paris VII, Paris

15. Roberts, D. A., Goldstein, M. L., Klein, L. W., Matthaeus, W. H. (1987) Origin and evolution of fluctuations in the solar wind: Helios observations and Helios-Voyager comparisons. J. Geophys. Res. **92**, 12023–12035

16. Roberts, D. A., Goldstein, M. L., Klein, L. W. (1990) The amplitudes of interplanetary fluctuations: stream structure, heliocentric distance, and frequency dependence. J. Geophys. Res. **95**, 4203–4216

17. Bavassano, B., Bruno, R. (1989) Evidence of local generation of Alfvénic turbulence in the solar wind. J. Geophys. Res. **94**, 11,977–11,982

18. Grappin, R., Mangeney, A., Marsch, E. (1990) On the origin of solar wind turbulence: Helios data revisited. J. Geophys. Res. **95**, 8,197–8,210

19. Roberts, D. A., Ghosh, S., Goldstein, M. L., Matthaeus, W. H. (1991) MHD simulation of the radial evolution and stream structure of the solar wind turbulence. Phys. Rev. Lett. **67**, 3741–3744

20. Roberts, D. A., Goldstein, M. L., Matthaeus, W. H. Ghosh, S. (1992) Velocity shear generation of solar wind turbulence. J. Geophys. Res. **97**, 17115–17130

21. Veltri, P., Malara, F., Primavera L. (1992) Correlation, anisotropy and compressibility of low frequency fluctuations in solar wind, in *Solar Wind Seven*, E. Marsch and R. Schwenn *ed.s*, Pergamon Press, Oxford, Cospar Colloquia Series, Vol. 3, pp. 559–564

22. Grappin, R., Mangeney, A., Velli, M. (1992) MHD turbulence: theory and simulations, in *Solar Wind Seven*, E. Marsch and R. Schwenn *ed.s*, Pergamon Press, Oxford, Cospar Colloquia Series, Vol. 3, pp. 451-456

23. Velli, M. (1992) On the propagation of ideal, linear Alfvén waves in radially stratified stellar atmospheres and winds. Astron. Astrophys. **270**, 304-314

24. Stribling, T., Roberts, D. A., Goldstein, M. L. (1996) A tridimensional MHD model of the inner heliosphere. J. Geophys. Res. **101**, 27,603-27,623

25. Malara, F., Primavera L., Veltri, P. (1996) Compressive fluctuations generated by time evolution of Alfvénic perturbations in the solar wind current sheet. J. Geophys. Res. **101**, 21,597-21,617

26. Malara, F., Primavera L., Veltri, P. (1997) Nature of the density–magnetic-field–intensity correlation in the solar wind. Phys. Rev. E **56**, 3508-3514

27. Harned, D. S., Kerner W. (1985) Semi–implicit method for three dimensional compressible magnetohydrodynamic simulation. J. Comput. Phys. **60**, 62-75

28. Malara, F., Veltri, P., Carbone, V. (1992) Competition among nonlinear effects in tearing instability saturation. Phys. Fluids **B**, **4**, 3070-3086

29. Malara, F. (1996) A Chebyshev pseudospectral multi-domain method for a boundary-layer problem. J. Comput. Phys. **124**, 254-270

30. Malara, F., Primavera, L., Veltri, P. (1996) Formation of small scales formation via Alfvén wave propagation in compressible nonuniform media. Astrophys. J., **459**, 347-364

31. Goldstein, M. L. (1978) An instability of a finite amplitude circularly polarized Alfvén wave. Astrophys. J. **219**, 700-704

32. Viñas, A. F., Goldstein, M. L. (1991) Parametric instabilities of circularly polarized large-amplitude dispersive Alfvén waves: excitation of obliquely-propagating daughter and side-band waves. J. Plasma Phys. **46**, 129-152

33. Akhiezer, A. I., Akhiezer, I. A., Polovin, R. V., Sitenko, A. G., Stepanov K. N. (1975) Plasma Electrodynamics, vol.1, *Nonlinear Theory and Fluctuations*, Pergamon Press, New York

34. Marsch, E., Tu, C.-Y. (1990) Spectral and spatial evolution of compressible turbulence in the inner solar wind. J. Geophys. Res. **95**, 11,945-11,956

35. Grappin, R., Velli, M., Mangeney, A. (1991) "Alfvénic" versus "standard" turbulence in the solar wind. Ann. Geophysicae **9**, 416-426

36. Vellante, M., Lazarus., A. J. (1987) An analysis of solar wind fluctuations between 1 and 10 AU. J. Geophys. Res.**92**, 9893-9900

37. Klein L., Bruno, R., Bavassano, B., Rosenbauer H. (1993) Scaling of density fluctuations with Mach number and density-temperature anti-correlations in the inner heliosphere. J. Geophys. Res. **98**, 7837-7841

38. Marsch E., Tu, C. Y., (1993) Correlations between the fluctuations of pressure, density, temperature and magnetic field in the solar wind. Ann. Geophys. **11**, 659-677

39. Tu, C. Y., Marsch, E. (1994) On the nature of compressive fluctuations in the solar wind. J. Geophys. Res. **99**, 21,481-21,509

40. Bavassano, B., Bruno, R., Klein, L. W. (1995) Density temperature correlation in solar wind MHD fluctuations: a test for nearly incompressible models. J. Geophys. Res. **100**, 5871-5875

41. Matthaeus, W. H., Brown, M. R. (1988) Nearly incompressible magnetohydrodynamics at low Mach-number. Phys. Fluids **31**, 3634-3644

42. Zank, G. P., Matthaeus, W. H. (1990) Nearly incompressible hydrodynamics and heat conduction. Phys. Rev. Lett. **64**, 1243-1246

43. Matthaeus, W. H., Klein, L. W., Ghosh, S., Brown, M. R. (1991) Nearly in-compressible magnetohydrodynamics, pseudosound and solar wind fluctuations. J. Geophys. Res. **96**, 5421–5435

44. Zank, G. P., Matthaeus, W. H. (1991) The equation of nearly incompressible fluids, I, Hydrodynamics, turbulence, and waves. Phys. Fluids **A3**, 69–82

45. Zank, G. P., Matthaeus, W. H. (1993) Nearly incompressible fluids. II: Magne-tohydrodynamics, turbulence, and waves. Phys. Fluids **A5**, 257–273

46. Pilipp, W. G., Miggenrieder, H., Mühlhäuser, K.-H., Rosenbauer H., Schwenn R. (1990) Large–Scale Variations of Thermal Electron Parameters in the Solar Wind Between 0.3 and 1 AU. J. Geophys. Res. **95**, 6305–6329

47. Malara, F., Primavera L., Veltri, P. (1999) Nature of the density–temperature correlation in the solar wind Phys. Rev. E, in press

Coherent Electrostatic Nonlinear Waves in Collisionless Space Plasmas

C. Salem, A. Mangeney, and J.-L. Bougeret

DESPA, Observatoire de Paris-Meudon, F-92195 Meudon Cedex, France

Abstract. This paper presents an overview of plasma wave observations in the collisionless plasmas of the solar wind and of the Earth's environment (auroral regions, magnetotail, etc) from different space experiments. One of the striking results obtained in the recent years is that these waves are basically electrostatic and coherent, in the form of solitary structures (weak double layers and solitary waves having respectively a net and no net potential drop) or modulated wavepackets. This electrostatic "activity" occurs quite frequently but is not a permanent feature of collisionless space plasmas. The details of these waveforms appear to depend on the region of observation which determines the plasma regime. We shall first present new available observational evidence of such waves in the solar wind and in many other regions of the Earth's environment. Then we shall compare their respective properties and discuss the analogies and differences between the different observations. We shall finally discuss the relevant theoretical interpretations and numerical simulations that have been put forward to explain these observations.

1 Introduction

The solar wind and the planetary magnetospheres provide very good laboratories for the study of the physics of collisionless plasmas, covering a very wide range of physical parameters, from the strongly magnetized $((\frac{\omega_p}{\omega_c})_{e,i} < 1)$, low β auroral plasma to the very weakly magnetized $((\frac{\omega_p}{\omega_c})_{e,i} > 1)$, high β solar wind plasma (we use here the usual definitions of the electron and proton plasma frequencies, ω_{pe}, ω_{pi} and cyclotron frequencies ω_{ce}, ω_{ci}). They have been extensively explored and have yielded a wealth of results about fundamental physical processes such as the physics of collisionless shocks, MHD turbulence, wave-particle interactions etc...... However there remains a number of basic processes which are not properly understood, such as the transformation of large scale "macroscopic" kinetic or magnetic energy into microscopic "thermal" energy. A key point in this respect concerns the role of electromagnetic or electrostatic waves.

Most of the available (considerable !) observational information accumulated on the properties of these waves concerns their spectral properties ; it is natural enough, then, to assume that the effective dissipative processes are of the quasilinear type, involving a resonant interaction between charged particles and waves with random phases. However, the development of observations with very high temporal resolution has allowed to get an idea about their waveforms (i.e. to combine informations about their phase as well as on their energy distribution in physical and spectral spaces). The striking result of these investigations is that the observed wavefields have often a higher degree of coherence than their spectral appearance would have suggested : localized wave packets, isolated solitary-like structures and weak double layers have been observed in many different regions of space.

Does that mean that the effective dissipative processes occuring in space plasmas imply a significant amount of particle trapping and coherent nonlinear waves ?

In this paper, we present some recent observations made in the solar wind and review some of the available observations of coherent waveforms in other regions of space.

2   Coherent ion acoustic waves in the solar wind

A good example of the contribution of high temporal resolution to the understanding of the strucure of wavefields in natural plasmas is given by the so called "Ion acoustic waves" observed in the solar wind (see for example Gurnett and Anderson, 1977) in the range of frequency f between the proton plasma frequency f_{pi} and the electron plasma frequency f_{pe}. In this frequency range, a very bursty electric activity occurs more or less continuously in the inner heliosphere. While the first observations were limited to the ecliptic plane, recent measurements by the URAP instrument onboard Ulysses show that this activity is also present at high heliographic latitudes, with, perhaps a lower intensity (MacDowall et al., 1996). Some properties of the waves observed in this frequency range, the so-called "ion acoustic range" (referred to in what follows as the IAC range) are now well established (see for example the review by Gurnett, 1991, and the recent work by Mangeney et al., 1999).

They are basically electrostatic and propagate along the ambient magnetic field **B** : their electric field **E** is parallel to their wavevector **k**, which is itself parallel to **B** ; typical values for the field amplitude are $E \sim 0.1$ mV/m while the wavelengths λ are of the order of $10 - 50$ electron Debye length, λ_D. In the solar wind plasma frame, they have a relatively low frequency $f_0 \leq f_{pi}$ and a phase velocity v_ϕ which is much smaller than the solar wind velocity V_{sw}, $v_\phi \ll V_{sw}$. Therefore they are strongly affected by the Doppler shift due to the solar wind velocity with respect to the spacecraft and, in the spacecraft frame, their frequency is shifted upwards into the range $f_{pi} \leq f < f_{pe}$.the corresponding electric energy density is small compared to the thermal energy density of the plasma : indeed, $\epsilon_0 E^2/(2 N k_B T_e)$ is only about 10^{-7} to 10^{-5} during intense bursts, and much smaller on the average (Gurnett, 1991). (ϵ_0: vacuum dielectric constant, N: particle number density, k_B: Boltzmann constant, T_e : electron temperature).

These properties led Gurnett and Frank (1978), soon after the first observations, to propose an interpretation in terms of ion acoustic waves propagating along the ambient magnetic field. However, it is hard to understand how such ion accoustic waves could survive the strong Landau damping which is expected in the solar wind where, most of the time, the ratio between the electron and proton temperatures T_e and T_p lies in the range 0.5–2. This has remained a puzzling problem for space plasma physics and one may say that neither the wave mode nor the source of these waves have yet been unambiguously identified (see Gurnett, 1991). Several instabilities have been suggested to be responsible for their occurrence. For example, the electron heat flux (Forslund, 1970 ; Dum *et al.*, 1981) may be a source of free energy for the electrostatic IAC mode. But Gary (1978) has shown that other, electromagnetic, heat flux instabilities have much larger growth rates in conditions typical of the solar wind. Another attractive possibility would be ion beams (Gary, 1978 ; Lemons *et al.*, 1979 ; Marsch, 1991) such as those observed more or less continuously in the fast solar wind and drifting with roughly the local Alfvén speed with respect to the core ion distribution. These ion beams could excite IAC waves. However, Gary (1993) concludes that for the range of temperature ratios $T_e/T_p \leq 5$ usually observed in the solar wind the instability threshold is never reached.

Recent observations with high time resolution have shed a new light on the nature of the "ion acoustic" electrostatic turbulence. Indeed when observed with instruments having relatively poor time and frequency resolutions, the IAC electrostatic turbulence appeared as made of very bursty broadband emissions lasting from a few hours to a few days. At higher time resolution (Kurth *et al.*, 1979) the emission begun to show some structure in the form of brief narrowband bursts with rapidly drifting center frequency. More recently, Mangeney *et al.* (1999) analyzed waveform data obtained with the Time Domain Sampler (TDS) experiment on the WIND spacecraft (Bougeret *et al.*, 1995) with a time resolution reaching 120,000 points per second. At this temporal resolution, the wave activity in the ion acoustic range appears to be highly coherent, made of a mixture of electrostatic *wave packets* (W.P.) and more or less *isolated electrostatic structures* (I.E.S.), lasting less than 1ms and similar in many respects to those observed in different regions of the Earth's environment (see section 3 ; Temerin *et al.*, 1982 ; Matsumoto *et al.*, 1994 ; Mottez *et al.*, 1997). Note that these observations were obtained in the vicinity of the ecliptic plane when WIND was at a distance of the Earth greater than 200 R_E, near the Lagrange point L_1, therefore minimising the perturbing influence of the earth.

It is also interesting to note that when observed with a spectral receiver having a limited bandwidth, there are periods of hours or days when no IAC wave activity is detected. This is probably due to the fact that the Doppler effect is not sufficient, during these periods, to bring the waves into the receiver bandwidth, either because the magnetic field is perpendicular to the wind velocity or because this velocity is too small. On the other hand, when observed with the TDS experiment, the "*electrostatic activity*", in the range ($f_{pi} \leq f \leq f_{pe}$) apparently occurs continuously in the solar wind, i.e. without gaps of the order of one hour or larger. At least, this is the case outside the regions of high density ($N > 10 - 15$ cm^{-3}) where the TDS observations are plagued by artefacts which hide the natural waves (see Mangeney et al., 1999).

We reproduce here as Figure 1 and Figure 2, two of the figures of the paper by Mangeney et al. (1999) which display typical IAC waveforms as seen by the TDS at the highest time resolution. The panels 1b and 1c of Figure 1 display two examples of low frequency wave packets. The first one (Figure 1b) is a narrow band signal with a centre frequency $f \simeq 3.4$ kHz ($f_{pe} \simeq 17$ kHz). The second one (Figure 1c) is also a narrow band sinusoidal wave at a frequency $f \simeq 2.4$ kHz, ($f_{pe} \simeq 18$ kHz) but with a much shorter envelope of temporal width $\simeq 5$ ms. In both cases the maximum electric field is of the order of 0.1 mV/m, close to the values observed for the IAC waves on Helios 1 by Gurnett and Anderson (1977). The waveform of Figure 1d is nearly periodical at 2.4 kHz ($f_{pe} \simeq 25$ kHz) but it has a strongly non-sinusoidal appearance, indicating the presence of significant nonlinear effects. In all three cases the wavelength lies in the range $10 < \lambda/\lambda_D < 50$.

Finally, the bottom panels 1e and 1f display two examples of isolated spikes of duration 0.3 to 1 millisecond with amplitudes similar to those of the wave packets described just above ; since their velocity in the plasma frame is small with respect to the wind velocity, they behave as frozen structures convected past the spacecraft so that durations can be transformed to spatial widths Δx in the range $10 < \Delta x/\lambda_D < 40$. In Figures 1e and 1f the IES are clearly isolated, whereas in Figure 1d the spikes are so closely packed that the resulting waveform does not appear very different from the modulated wave packet of Figure 1b.

The highest intensity and level of wave activity in the range of frequency $f_{pi} \leq f < f_{pe}$ is found for relatively low proton temperatures, at low heliomagnetic latitudes. Towards higher latitudes, the frequency of occurrence of wave packets decreases, and one observes mainly weak double layers deep into the fast solar wind. No clear relation with important plasma parameters like the electron to proton temperature ratio was found.

Assuming that the IES are one-dimensional structures varying only along the magnetic field, and knowing the solar wind velocity with respect to the spacecraft we can easily determine the spatial profiles of the electric field **E** and of the corresponding electric potential ϕ, from the signal shown in Figure 1. An example is given in Figure 2. The left top panel shows the electric field (in mV/m, and smoothed over 10 points in order to eliminate the high frequency noise) during a time interval of about ≈ 4 ms around the central spike of Figure 1f while the left bottom panel shows the short WP of Figure 1c. The corresponding electric potentials, normalized to the local electron temperature T_e, are plotted in the right panels; both structures (IES and WP) exhibit a finite potential drop $e\Delta\phi/k_B T_e \approx 2\,10^{-4}$. This example

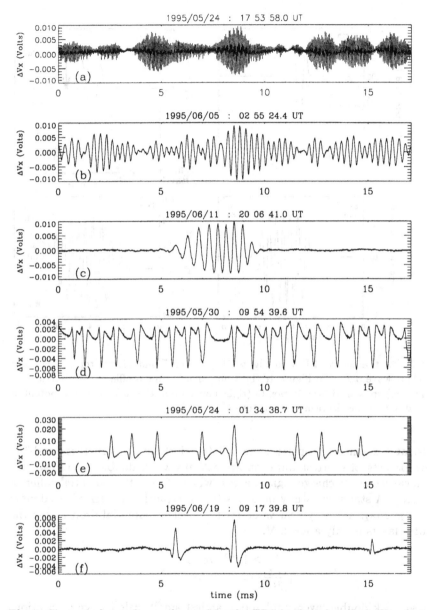

Fig. 1. Six typical waveforms observed by the TDS in the solar wind. The electric potential ΔV_x in Volts is measured at the terminals of the x-antenna, $L_x = 50m$: (a) Langmuir waves. (b) and (c) Low frequency quasi-sinusoidal wave packets (WP). (d), (e) and (f) Non sinusoidal wave packets (WP) and isolated electrostatic structures (IES) [figure from Mangeney *et al.*, 1999].

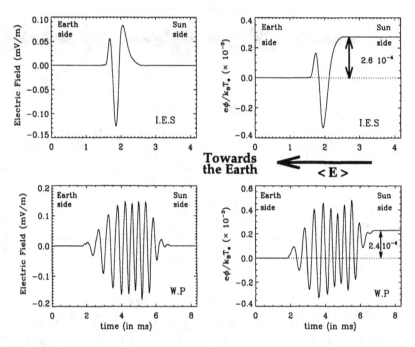

Fig. 2. For the central I.E.S. in Figure 1f, between 7 and 11 ms (top panels) and the short wave packet in Figure 1c (bottom panels), time profile of the electric field E (mV/m) smoothed over 10 points (right panels) and the corresponding potential normalized to the electron temperature $e\Delta\phi/k_BT_e$.

illustrates a general property : the isolated electrostatic structures and the wave packets of shortest duration are actually weak double layers (WDL), i.e. localized space charge regions across which the electric potential suffers a net drop. A statistical study indicates that a typical thickness of the double layers is $25\lambda_D$ and a typical value for the electric potential drop across the double layers is only a few mV:

$$\begin{cases} \Delta\phi \geq 10^{-3} \text{ Volts} \\ \frac{e\Delta\phi}{k_BT_e} \approx 10^{-4} \text{ to } 10^{-3} \end{cases}$$

These weak double layers probably manifest small scale charge separations due to a partial decoupling between electrons and protons on distances comparable to a Debye length scale. The potential usually decreases towards the Earth : it varies in the same sense as the interplanetary electric potential Φ_{SE} which tends to decelerate the solar wind outward propagating electrons, and to accelerate the protons ; it is therefore tempting to speculate that this potential difference $\Phi_{SE} \sim 1000$ Volts (see for example Meyer-Venet, 1999 and references therein) between the solar corona and the earth orbit is actually established through a succession of small potential jumps across a multitude of WDLs, say $\sim 10^6$ WDLs if one assumes a roughly constant amplitude for the individual WDL potential drops ; thus one should find about one WDL every $150-200$ km along a magnetic field line. This is much larger than what may be inferred from the WIND observations, which suggest something like one event every 10 millisecond, i.e. one WDL every 4 km along the sun-earth direction, assuming a typical solar wind velocity of 400 km/s. It should be stressed that this estimate of the rate of occurrence of WDLs in the solar wind is at most a reasonnable guess since only small selected intervals (17, 70 or 270 milliseconds of data, depending on the bit rate available to the experiment) are transmitted roughly every 5 minutes ! Whatever the actual value may be, it is clear from the observations that there are much more WDLs in the quiet solar wind than one every $150-200$ km. If one combines this with the fact that the sign of the potential jumps across WDLs is predominantly but not always the same as Φ_{SE}, one is led to propose that, locally, there exists in the solar wind, random electric potential differences, along the magnetic field, which are significantly greater than the large scale one, Φ_{SE}.

A look at Ohm's law in a plasma with a non uniform electron pressure p_e (neglecting electron inertia):

$$\mathbf{E} + \mathbf{v} \times \mathbf{B} = \frac{1}{Ne} \nabla p_e$$

shows that this is likely to occur in the solar wind. Assume for example that the fluctuations in electronic pressure are due essentially to density variations δN ; a typical value for the associated field aligned potential fluctuations is then:

$$\frac{e \delta \Phi}{k_B T_e} \sim \frac{\delta N}{N}$$

The spectrum of density fluctuations in the solar wind has been studied extensively ; Celnikier et al. (1987) give an average value of $\frac{\delta N}{N} \sim 0.05$ at temporal scales (in the spacecraft frame) of 0.1 seconds. If there is p WDLs over the corresponding distance with each a potential drop of $\Delta \phi$, then $\delta \Phi = p \Delta \phi$ so that there should be $p \sim 500$ WDLs in a time interval of 0.1 second, using the estimates for $\Delta \phi$ given above. This is larger than but not incompatible with the estimate one WDL every 10 millisecond.

It is also interesting to note that bursts of Langmuir waves (at f_{pe}), with levels far above the noise level and not related to type III solar bursts, have also been observed in the solar wind (Gurnett and Anderson, 1977 ; Gurnett et al., 1979 ; MacDowall et al., 1996 ; Mangeney et al., 1999). The top panel (1a) of Figure 1 is a good example of such modulated high frequency Langmuir wave packets, oscillating at the local electron plasma frequency, $f_{pe} \simeq 18$ kHz. Its appearance and its amplitude ($\Delta V_x \simeq 0.01$ V at the antenna terminals i.e. an electric field of $\simeq 0.2$ mV/m) are similar to the weakest events observed in the Earth foreshock (Bale *et al.*, 1997 ; Kellogg *et al.*, 1996) and in the upstream solar wind (Bale *et al.*, 1996). These bursts of Langmuir waves are frequently associated with magnetic holes (Lin *et al.*, 1995) and their intensity appears to be significantly higher at high heliographic latitudes (*i.e.* in the fast speed wind, MacDowall *et al.*, 1996). A question which remains to be investigated is wether this high frequency noise is related in some way with the IAC wave activity.

3   Review of the observational evidence on coherent waves in the Earth's environment

Solitary waves (SW) and weak double layers (WDL) seem to be a common feature of space plasmas, since they have been observed in other regions of the Earth's environment, *i.e.* at different altitudes in the auroral regions, in the magnetotail and also in the shock transition region. These regions and the corresponding observations are indicated schematically in Figure 3 and 4. In this section, we first briefly review these different observations, and then make a comparative analysis of the properties of the waves.

3.1   S3-3, VIKING, POLAR and FAST observations in the Earth's auroral regions

The S3-3 satellite provided in 1976 (Temerin *et al.*, 1982 ; Mozer and Temerin, 1983) the first evidence on the existence of Solitary Waves and Weak Double Layers in the mid-altitude auroral magnetosphere (between altitudes of 6000 to 8000 km). These observations were confirmed by the Viking spacecraft in 1986 (Boström *et al.*, 1988; Boström *et al.*, 1989; Koskinen *et al.*, 1989) and more recently by the POLAR satellite (Mozer *et al.*, 1997), so that more detailed information on these small-scale plasma structures is now available. This region of the magnetosphere is strongly out of equilibrium : one finds a cool, strongly magnetized background plasma : $(\frac{\omega_p}{\omega_c})_{e,i} \sim 0.1-0.2$, with typical values of $n_e = 1-10$ cm^{-3} and $T_e = 1-10$ eV, submitted to an electric potential drop of a few kilovolts as well as upward propagating ion beams (with energies of typically 0.5 to 1 keV or velocities of about 100 to 400 km/s for protons) and downward accelerated electron beams.

Fig. 3a shows an example, lasting 300 msec, of simultaneous density (measured with a Langmuir probe) and electrostatic potential variations measured by VIKING at an altitude of 10450 km. It is seen that the electric signal is made of isolated structures, of amplitude $\Delta\phi \sim$ few volts ($e\Delta\phi \lesssim k_B T_e$, the corresponding electric field amplitude being about 100 mV/m) lasting from 1 to 10 millisecond, corresponding to sizes along the magnetic field of the order of 50 to 100 m, *i.e.* 5–50 λ_D ; some are negative pulses (WDL) while others have a more symmetric structure (SW) ; they are associated with localized density depletions in the background plasma of up to 50% . The WDL show

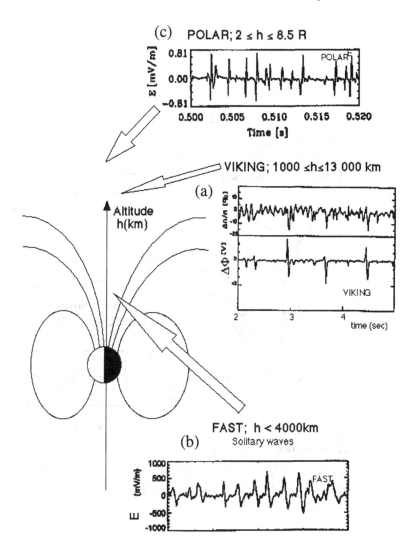

Fig. 3. Schematical view of the Earth's auroral regions : (a) example of mid-altitude auroral magnetosphere observations by VIKING of solitary waves and weak double layers (Bostrom *et al.*, 1989); (b) example of lower altitude observations in the upward current region by FAST (Ergun *et al.*, 1998a); (c) example of high altitude observations by POLAR (Franz *et al.*, 1998).

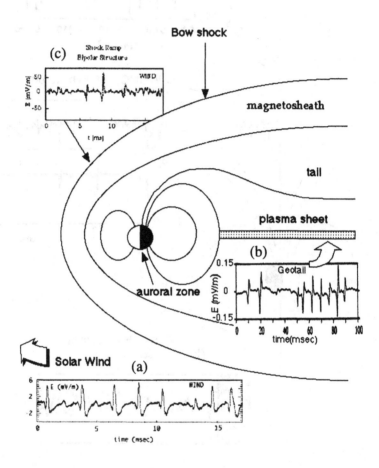

Fig. 4. Schematical large-scale view of the Earth's magnetosphere with some typical observations of solitary structures. The 3 panels displays the electric field in mV/m as a function of time in millisecond : (a) example of the WDL observed in the solar wind by WIND (Mangeney *et al.*, 1999), (b) example of the ESW observed in the Plasma Sheet Boundary Layer (PSBL) by GEOTAIL (Matsumoto *et al.*, 1994), and (c) example of bipolar structures observed by WIND in the ramp of the bow shock (Bale *et al.*, 1998).

a net potential drop $\Delta\phi$ of up to a few volts which is generally directed upwards. These structures propagate upwards with a relatively high velocity of more than 100 km/s (there is some discrepancy between the S3-3 and POLAR velocity estimations and those obtained from VIKING ; see McFadden (1998) for a possible explanation based on the operation of Langmuir probes in a low density plasma), corresponding to what may be expected for the ion acoustic velocity c_s, indicating a significant role for the ions in the formation of these structures. The SW and WDL are often observed in association with periodic Electrostatic Ion Cyclotron (EIC) waves with frequencies (of about 140 Hz) just above the local proton gyrofrequency, propagating perpendicular to the ambient magnetic field (Ergun et $al.$, 1998a).

In addition, FAST observations have revealed a new type of solitary waves, called "Fast Solitary Waves" (FSW) in the regions of the mid-altitude auroral zone where downwards directed current are found (Ergun et $al.$, 1998b) and in association with energetic, up-going electron beams. These FSW have a very short duration of $\sim 50\mu s$ to $\sim 200\mu s$, propagate upwards with velocities between 500 to 5000 km/s, so that their spatial size is of $\sim 2\lambda_D$. Similar structures have also been observed in the low altitude (~ 6000 km) POLAR data (Mozer et $al.$, 1997).They have an electric signature which is that of a positive charge (or electron hole) moving with the beam. An example of such solitary waves is given in Fig. 3b (Ergun et $al.$, 1998b).

Preliminary results from POLAR, at higher altitudes ($\sim 6\,R_E$), have also provided evidence for fast moving solitary waves with much lower amplitude (~ 1 mV/m) than those observed at lower altitudes (Franz et $al.$, 1998 ; Cattell et $al.$, 1998). These structures have no magnetic signature. An example is displayed in Fig. 3c. They look very similar to those observed by GEOTAIL (see below) in the PSBL.

3.2   GEOTAIL and WIND recent observations in other parts of the magnetosphere

The plasma wave instrument on the GEOTAIL satellite has observed similar electrostatic solitary waves (ESW) in the deep tail region of the Earth's magnetosphere to radial distances beyond 70 R_E (Matsumoto et $al.$, 1994). In the Plasma Sheet Boundary Layer (PSBL), the plasma environment is not well known but a density of about 0.5 cm^{-3} and an electron temperature of about 200 eV can be taken as reasonable estimates. In this region, one observes an electrostatic noise with a brodband spectrum (known under the name BEN, Scarf et $al.$, 1974 ; Gurnett et $al.$, 1976) associated with regions of high ion temperatures as well as a narrowband electrostatic noise (known as NEN, Coroniti et $al.$, 1990), and Langmuir waves. Once again, high time resolution has renewed our understanding of this wave activity, by replacing a structureless BEN noise by a succession of electrostatic solitary waves as is apparent in the example is given in Fig. 4b, which displays 100 msec of data taken in the PBSL (Matsumoto et $al.$, 1994). The polarization of the ESW is almost parallel to the ambient magnetic fields ; their duration is about 2 to 5 ms with an amplitude of about 0.1 mV/m and their waveform is that of bipolar pulses, propagating at velocities of about 1000 km/s, much faster than the solitary structures observed by S3-3 or VIKING, implying spatial sizes for these ESW of a few tens of λ_D. Typical time scales are of the order of 10~100 ms (Kojima *et al.*, 1997), implying that these waves are related to the dynamics of electrons rather than ions. The same structures have been observed by GALILEO, (Mottez *et al.*, 1997), deeper in the magetotail, in correlation with magnetic flux ropes, where the proton distribution presents a double beam structure.

The same kind of bipolar structures, though more intense, have also been detected at bow shock crossings by GEOTAIL and WIND (respectively, Matsumoto *et al.*, 1997 and Bale *et al.*, 1998). They have amplitudes greater than 100 mV/m and durations of the order of a tenth of a millisecond corresponding to spatial sizes of about 2 to 7 λ_D. Fig. 4c displays an example of a series of such bipolar structures detected by WIND in the ramp of the bow shock.

3.3   Comparative analysis of the wave properties

In tables 1 and 2, we give an overview of the observations described above: Table 1 contains a summary of the general wave and plasma properties while Table 2 contains the characteristics of the observed structures and their interpretation.

The solitary waves (with no net potential drop) observed in the different regions have similar shapes though their properties (amplitude, size and velocity) are different. They appear as bipolar pulses, *i.e.* two successive electric pulses of roughly equal amplitude but of opposite sign, moving along the ambient magnetic field. No systematic relation have been found between velocity, amplitude, or dimension of the structures as expected for bona-fide solitary waves and the usual interpretation is that they are electron holes, *i.e.* a depletion in electron density.

The double layers (with a net potential drop) have particular interest since they have been suggested as a field-aligned, particle acceleration mechanism for the auroral plasma (Hudson *et al.*, 1983 ; Mozer and Temerin, 1983 ; Eriksson and Boström, 1993). The similarity of the WDL observed in the solar wind with those observed in the auroral region is striking although the peak amplitude of the latter reach much higher values, *i.e.* $e\Delta\phi/k_B T_e \approx 1$ (Mälkki *et al.*, 1993) while remaining very weak compared to those observed in numerical simulations (see Borovsky, 1984) or in laboratory plasmas (see Fälthammar, 1993). They have essentially negative potential pulses so that they may be interpreted as ion holes.

Table 1. Waveforms observations : general wave and plama properties

Spacecraft	Region of Observations	Waveform	Plasma β	Direction of Propagation	T_e/T_p	$e\Delta\phi/k_B T_e$
WIND	Solar Wind	WDL (IES)	~ 1	$\parallel \mathbf{B}$	0.2 to 5	10^{-4} to 10^{-3}
WIND	Ramp of Bow Shock	B.S.	~ 1	$\parallel \mathbf{B}$		No potential drop
GEOTAIL	PSBL and MS	ESW	$\gg 1$	$\parallel \mathbf{B}$		No potential drop
FAST	L.A. Auroral Zone	FSW	$\ll 1$	$\parallel \mathbf{B}$		No potential drop
POLAR	H.A. Auroral Zone (cusp)	FSW	< 1	$\parallel \mathbf{B}$		No potential drop
VIKING	Mid-altitude auroral zone	SW and WDL	$\ll 1$	$\parallel \mathbf{B}$	$\simeq 1$	for WDL ≤ 1

4 Theoretical interpretations

In this section, we present a brief discussion about the relevant theoretical interpretations and numerical simulations that have been put forward to explain these observations.

Basically, they all imply the deformation of the charged particle distribution functions in a localized region of phase space, called ion or electron hole if it corresponds to a depletion with respect to the surrounding phase space density. These holes are associated self consistently with a non-linear electrostatic wave, which forms the hole by trapping or reflecting part of the particle population. Stationary solutions of the Vlasov-Poisson equations corresponding to these structures are well known (BGK modes, Bernstein *et al.*, 1957, see also Schamel, 1986); on the other hand such holes are very often observed in the nonlinear stages of the development of kinetic instabilities where the linear growth rate often presents a well defined peak, so that a large part of the theoretical work has been devoted to identify an instability, where the source of free energy was indicated by the observations, and leading to the formation of holes similar to those observed.

The first attempts in this direction were based on ion acoustic waves, in the nonlinear regime driven by an electron current or by an ion beam (see for

Table 2. Waveforms observations : wave properties and interpretations

Spacecraft	Region of Observations	Waveform	Size	Amplitude	Velocity	Interpre- tation
WIND	Solar Wind	WDL (IES)	$\simeq 25\lambda_D$	0.2 mV/m	$v_\phi \ll V_{sw}$	Ion holes ?
WIND	Ramp of Bow Shock	B.S.	2 to $7\lambda_D$	100 mV/m	$\sim$ 300 km/s ?	Electron holes
GEOTAIL	PSBL and MS	ESW	few 10s λ_D	0.1 mV/m	1000 km/s	Electron holes
FAST	L.A. Auroral Zone	FSW	$2\lambda_D$	200 mV/m	500 to 5000 km/s	Electron holes
POLAR	H.A. Auroral zone (cusp). Mid-altitude auroral zone	FSW	few λ_D	1 mV/m ($7\ R_E$) 200 mV/m ($1\ R_E$)	1000 km/s	Electron holes
VIKING	Mid-altitude auroral zone	SW and WDL	5 to 50 λ_D	100 mV/m	5 to 50 km/s	Ion holes

example, Lokto and Kennel, 1983; Lokto, 1983); the study of the nonlinear development of ion acoustic turbulence has generated a number of numerical works (Sato and Okuda, 1981 ; Hudson *et al.*, 1983 ; Chanteur *et al.*, 1983 ; Chanteur, 1984 ; Barnes *et al.*, 1985).

However, in most cases, the WDL form only for electron drift exceeding some threshold for both electrostatic ion cyclotron and ion acoustic waves, or for high values of the electron to proton temperature ratio, what is not supported by the observations (Koskinen and Mälkki, 1993). An alternative theory, the nonlinear phase-space ion hole theory, originally developped by Dupree (1982), was applied to VIKING observations by Tetreault (1988, 1991). It predicts that a nonlinear interaction between background electrons and ions results in the growth of the holes for any finite drift between these two populations, even below the thresholds for linear ion acoustic or electrostatic ion cyclotron instabilities. Though there exists only a limited number of one-dimensional simulations of phase-space ion holes which have been carried out (Berman *et al.*, 1986), giving results quite similar to those obtained on ion acoustic turbulence, this theory still remains among the most serious alternatives (Mälkki *et al.*, 1989).

It has been known in laboratory experiments and computer simulations that electron beam instabilities can lead to formation of electron holes in phase space which are some kind of BGK equilibrium (Bernstein *et al.*, 1957 ; Krasovsky *et al.*, 1997). In the case of GEOTAIL observations, a close similarity has been found between the observed solitary waves and the electron holes formed in simple simulations of two-stream instabilities (Omura *et al.*, 1994 ; Mottez *et al.*, 1997). In these simulations, counter-streaming electron beams with comparable densities interact strongly to form large electrostatic potential fluctuations which trap a significant portion of the distribution function, leadind to the formation of holes. In the presence of "hot" ions, the ion acoustic modes which may tap the free energy are severely Landau damped, thereby allowing the ESW to develop (Omura *et al.*, 1996). This mechanism of selection does not work in the solar wind, where the IAC waves appear to be insensitive to the proton to electron temperature ratio (Mangeney *et al.*, 1999). Besides the two-stream instability, a variety of electron beam instabilities (for example bump on tail instability, see Omura *et al.*, 1996) can lead also to the formation of the solitary waves.

5   Conclusion

Although considerable progresses have been made, some important parameters remain poorly known such as the frequency of occurrence of these structures. On the theoretical side, it is not completely clear that some kinetic instabilities are involved in the formation of the WDL or SW ; almost identical structures were indeed produced in numerical simulations of a stable plasma but with time varying boundary conditions (Mangeney, 1999). At least in the solar wind, where no source of free energy appears clearly in the observed particle distribution functions, it is highly probable that these structures are the result of the changing electric potential due to larger scale MHD-like fluctuations.

References

1. Bale S.D., Burgess D., Kellogg P.J., Goetz K., Howard R.L., Monson S.J. (1996), Phase coupling in Langmuir wave packets: possible evidence of three-wave interactions in the upstream solar wind, Geophys. Res. Lett. **23**, 109–112
2. Bale S.D., Burgess D., Kellogg P.J., Goetz K., Monson S.J. (1997), On the amplitude of intense Langmuir waves in the terrestrial electron foreshock, J. Geophys. Res. **102**, 11281–11286
3. Bale S.D., Kellogg P.J., Larson D.E., Lin R.P., Lepping R.P. (1998), Bipolar electrostatic structures in the shock transition region : evidence of electron phase space holes, Geophys. Res. Lett. **25**, 2929–2932

4. Barnes C., Hudson M.K., Lokto W. (1985), Weak double layers in ion acoustic turbulence, Phys. Fluids **28**, 155

5. Berman R.H., Dupree T.H., Tetreault D.J. (1986), Growth of nonlinear intermittent fluctuations in linearly stable and unstable plasma, Phys. Fluids **29**, 2860

6. Bernstein I.B., Greene J.M., Kruskal M.D. (1957), Exact nonlinear plasma oscillations, Phys. Rev. **108**, 546

7. Borovsky J.E. (1984), A review of plasma double-layer simulations, in *Second Symposium on Plasma Double Layers and Related Topics*, R. Schrittwieser and G. Eder editors, Innsbruck, 33–54

8. Boström R., Gustafsson G., Holback B., Holmgren G., Koskinen H.E.J., Kintner P. (1988), Characteristics of solitary waves and weak double layers in the magnetospheric plasma, Phys. Rev. Lett. **61**, 82–85

9. Boström R., Holback B., Holmgren G., Koskinen H.E.J. (1989), Solitary structures in the magnetospheric plasma observed by VIKING, Phys. Scr. **39**, 782–786

10. Bougeret J.-L., Kaiser M.L., Kellogg P.J., Manning R., Goetz K., Monson S.J., Monge N., Friel L., Meetre C.A., Perche C., Sitruk L., S. Hoang (1995), WAVES: the radio and plasma wave investigation on the WIND spacecraft, Space Science Reviews **71**, 231–263

11. Cattell C.A., Wygant J., Dombeck J., Mozer F.S., Temerin M., Russell C.T. (1998), Observations of large amplitude parallel electric field wave packets at the plasma sheet boundary, Geophys. Res. Lett. **25**, 857–860

12. Celnikier L.M., Muschietti L., Goldman M.V. (1987), Aspects of interplanetary plasma turbulence, Astron. Astrophys. **181**, 138–154

13. Chanteur G., Adam J.C., Pellat R., Volokhitin A.S. (1983), On the formation of ion acoustic double layers, Phys. Fluids **26**, 1584

14. Chanteur G. (1984), Vlasov simulations of ion acoustic double layers, in *Computer Simulation of Space Plasmas*, Eds. H. Matsumoto and T. Sato, Terra Scientific Publishing Company, 279–301

15. Coroniti F.V., Greenstadt E.W., Tsurutani B.T., Smith E.J., Zwickl R.D., Gosling J.T. (1990), Plasma waves in the distant geomagnetic tail : ISEE 3, J. Geophys. Res. **95**, 20977–20995

16. Dum C.T., Marsch E., Pilipp W.G., Gurnett D.A. (1981), Ion sound turbulence in the solar wind, in *Solar Wind Four*, 299–304

17. Dupree T.H. (1982), Theory of phase-space density holes, Phys. Fluids **25**, 277

18. Ergun R.E., Carlson C.W., MacFadden J.P., Mozer F.S., Delory G.T., Peria W., Chaston C.C., Temerin M., Elphic R., Strangeway R., Pfaff R., Cattell C.A., Klumpar D., Shelley E., Peterson W., Moebius E., Kistler L. (1998a), FAST satellite observations of electric field structures in the auroral zone, Geophys. Res. Lett. **25**, 2025–2028

19. Ergun R.E., Carlson C.W., MacFadden J.P., Mozer F.S., Delory G.T., Peria W., Chaston C.C., Temerin M., Roth I., Muschietti L., Elphic R., Strangeway R., Pfaff R., Cattell C.A., Klumpar D., Shelley E., Peterson W., Moebius E., Kistler L. (1998b), FAST satellite observations of large amplitude solitary structures, Geophys. Res. Lett. **25**, 2041–2044

20. Eriksson A.I. and Boström R. (1993), Are weak double layers important for auroral particle acceleration ?, in *Auroral Dynamics*, Geophys. Monogr. Ser. **80**, edited by R.L. Lysak, AGU, Washington D.C., 105–112

21. Fälthammar C.G. (1993), Laboratory and space experiments as a key to the plasma universe, in *Symposium on Plasma-93*, Allahabad, India

22. Forslund D.W. (1970), Instabilities associated with heat conduction in the solar wind and their consequences, J. Geophys. Res. **75**, 17–28

23. Franz J.R., Kintner P.M., Pickett J.S. (1998), POLAR observations of coherent electric field structures, Geophys. Res. Lett. **25**, 1277–1280

24. Gary S.P. (1978), Ion-acoustic-like instabilities in the solar wind, J. Geophys. Res. **83**, 2504–2510

25. Gary S.P. (1993), Theory of space plasma microinstabilities, Cambridge University press

26. Gurnett D.A. (1991), Waves and instabilities, in *Physics of the inner heliosphere* II, Eds. R. Schwenn and E. Marsch, Springer-Verlag, 135–157

27. Gurnett D.A., Frank L.A., Lepping R.P. (1976), Plasma waves in the distant magnetotail, J. Geophys. Res. **81**, 6059–6071

28. Gurnett D.A. and Anderson R.R. (1977), Plasma wave electric fields in the solar wind: initial results from Helios 1, J. Geophys. Res. **82**, 632–650, 1977

29. Gurnett D.A., Frank L.A. (1978), Ion acoustic waves in the solar wind, J. Geophys. Res. **83**, 58–74

30. Gurnett D.A., Marsch E. , Pilipp W., Schwenn R. , Rosenbauer H. (1979), Ion acoustic waves and related plasma observations in the solar wind, J. Geophys. Res. **84**, 2029–2038

31. Hudson M.K., Lokto W., Roth I., Witt E. (1983), Solitary waves and double layers on auroral field lines, J. Geophys. Res. **88**, 916

32. Kellogg P.J., Monson S.J., Goetz K., Howard R.L., Bougeret J.L., Kaiser M.L. (1996), Early Wind observations of bow shock and foreshock waves, Geophys. Res. Lett. **23**, 1243–1246

33. Kojima H., Matsumoto H., Chikuba S., Horiyama S., Ashour-Abdalla M., Anderson R.R. (1997), GEOTAIL waveform observations of broadband/narrowband electrostatic noise in the distant tail, J. Geophys. Res. **102**, 14439

34. Koskinen H.E.J., Lundin R., Holback B. (1990), On the plasma environment of solitary waves and weak double layers, J. Geophys. Res. **95**, 5921–5929

35. Koskinen H.E.J. and Mälkki A. (1993), Auroral weak double layers : a critical assessment, in *Auroral Dynamics*, Geophys. Monogr. Ser. **80**, edited by R.L. Lysak, AGU, Washington D.C., 97–104

36. Kurth W.S., Gurnett D.A., Scarf F.L. (1979), High-resolution spectrograms of ion acoustic waves in the solar wind, J. Geophys. Res. **84**, 3413–3419

37. Lemons D.S., Asbridge J.R., Bame S.J., Feldman W.C., Gary S.P. and Gosling J.T. (1979), The source of electrostatic fluctuation in the solar wind, J. Geophys. Res. **84**, 2135

38. Lin N., Kellogg P.J., MacDowall R.J., Balogh A., Forsyth R.J., Phillips J.L., Buttighoffer A., Pick M. (1995), Observations of plasma waves in magnetic holes, Geophys. Res. Lett. **22**, 3417–3420

39. Lokto W. (1983), Reflection dissipation of an ion-acoustic soliton, Phys. Fluids **26**, 1771–1779

40. Lokto W. and Kennel C.F. (1983), Spiky ion acoustic waves in collisionless auroral plasma, J. Geophys. Res. **88**, 381–394

41. MacDowall R.J., Hess R.A., Lin N., Thejappa G., Balogh A., Phillips J.L. (1996), Ulysses spacecraft observations of radio and plasma waves: 1991-1995, Astron. Astrophys. **316**, 396–405

42. MacFadden J. (1998), What is known and unknown about particle acceleration on auroral field lines, EOS Trans. AGU **79**, S316

43. Mälkki A., Koskinen H.E.J., Boström R., Holback B. (1989), On theories attempting to explain observations of solitary waves and weak double layers in the auroral magnetosphere, Phys. Scr. **39**, 787–793

44. Mälkki A., Eriksson A.I., Dovner P., Boström R., Holback B., Holmgren G., Koskinen H.E.J. (1993), A statistical survey of auroral solitary waves and weak double layers, 1.- Occurrence and net voltage, J. Geophys. Res. **98**, 15521–15530

45. Mangeney A., Salem C., Lacombe C., Bougeret J. L., Perche C., Manning R., Kellogg P. J., Goetz K., Monson S. J., Bosqued J.-M. (1999), WIND observations of coherent electrostatic waves in the solar wind, Ann. Geophysicae **17**, 307–320

46. Mangeney A. (1999), in preparation

47. Marsch E. (1991), Kinetic physics of the solar wind plasma, in *Physics of the inner heliosphere* II, Eds. R. Schwenn and E. Marsch, Springer-Verlag, 45–133

48. Matsumoto H., Kojima H., Miyatake T., Omura Y., Okada M., Nagano I., Tsutsui M. (1994), Electrostatic solitary waves (ESW) in the magnetotail: BEN wave forms observed by GEOTAIL, Geophys. Res. Lett. **21**, 2915–2918

49. Matsumoto H., Kojima H., Kasaba Y., Miyatake T., Anderson R.R., Mukai T. (1997), Plasma waves in the upstream and bow shock regions observed by GEOTAIL, Adv. Space Res. **20**, 683–693

50. Meyer-Vernet N. (1999), How does the solar wind blow ? A simple kinetic model, Eur. J. Phys. **20**, 167–176

51. Mottez F., Perraut S., Roux A., Louarn P. (1997), Coherent structures in the magnetotail triggered by counterstreaming electron beams, J. Geophys. Res. **102**, 11399–11408

52. Mozer F.S. and Temerin M. (1983), Solitary waves and double layers as the source of parallel electric fields in the auroral acceleration region, in *High Latitude Space Plasma Physics*, B. Hultqvist and T. Hagfors eds., Plenum Publ. Corp., London, England, 453–467

53. Mozer F.S., Ergun R.E., Temerin M., Cattell C.A., Dombeck J., Wygant J. (1997), New features of time domain electric-field structures in the auroral acceleration region, Phys. Rev. Lett. **79**, 1281–1284

54. Omura Y., Matsumoto H., Miyake T., Kojima H. (1996), Electron beam instabilities as generation mechanism of electrostatic solitary waves in the magnetotail, J. Geophys. Res. **101**, 2685–2697

55. Scarf F.L., Frank L.A., Ackerson K.L., Lepping R.P. (1974), Plasma wave turbulence at distant crossings of the plasma sheet boundaries and the neutral sheet, Geophys. Res. Lett. **1**, 189–192

56. Schamel H. (1986), Electron holes, ion holes and double layers : Electrostatic phase space structures in theory and experiment, Physics Reports **140**, 161–191

57. Sato T. and Okuda H. (1981), Numerical simulations on ion acoustic double layers, J. Geophys. Res. **86**, 3357

58. Temerin M., Cerny K., Lotko W. and Mozer F.S. (1982), Observations of double layers and solitary waves in the auroral plasma, Phys. Rev. Lett. **48**, 1175–1179

59. Tetreault D. (1988), Growing ion holes as the cause of auroral double layers, Geophys. Res. Lett. **15**, 164

60. Tetreault D. (1991), Theory of electric fields in the auroral acceleration region, J. Geophys. Res. **96**, 3549–3563

Modeling the Dissipation Range
of Magnetofluid Turbulence

M.L. Goldstein[1], S. Ghosh[1,2], E. Siregar[1,2], and V. Jayanti[1,3]

[1] NASA Goddard Space Flight Center, Greenbelt MD, 20771 USA
[2] Space Applications Corporation, Largo MD, 20774 USA
[3] Universities Space Research Association, 7501 Forbes Blvd, Suite 206, Seabrook, MD 20706 USA

Abstract. Fluid descriptions of plasma phenomena are valuable tools for simulating large-scale phenomena. Fluid simulations, however, generally use idealized models to describe the dissipation of energy at small scales. The physical dissipation scale is usually considerably smaller than what can be evaluated numerically when using fluid descriptions of macroscopic phenomena. Here we review several approaches for describing dissipation in magnetofluid turbulence. These simulations divide into two general classes: those that employ mathematical models of the dissipation to concentrate dissipation to regions of large gradients in the fluid parameters, and those that use generalizations of Ohm's law to model kinetic effects more completely described by the Vlasov-Maxwell equations. The former class includes using either hyperresistivity and hyperviscosity or nonlinear dissipation operators to locate dissipation in regions of strong gradients. These terms replace the standard Navier-Stokes dissipation term in the magnetohydrodynamic (MHD) equations. The second class of models includes one that modifies the magnetofluid equations by including the Hall and Finite Larmor radius corrections to Ohm's Law and another approach that uses a coarse-grained fluid description to describe the effect of the ion cyclotron instability on the elements of the pressure tensor.

1 Observational background

Plasmas are both notoriously difficult to control in the laboratory and to describe theoretically. Laboratory plasmas resist attempts at confinement and are often subject to rapid instabilities that further complicate efforts to understand their behavior. Because laboratory devices tend to be rather small, it is a challenge to devise situations in which the boundary conditions do not dominate the evolution. Plasmas are also ubiquitous in space, both in the solar system and in the interstellar and galactic media. In the large volumes of space that plasmas fill, boundary conditions are not dominant in controlling the flow and the plasma is free to evolve as dictated by the initial conditions.

In the solar system, two plasma regimes that have been studied intensively *in situ* since the beginning of the space age are the magnetosphere and solar wind. Because the magnetosphere is relatively small and is confined by the external solar wind, boundaries (and kinetic effects) are significant in controlling its evolution. In addition, both the ionosphere and the solar wind are significant sources of magnetospheric plasma, further increasing the complexity of the system. In contrast, the solar wind appears, at least superficially, to be simpler. Boundaries are relatively unimportant except in the vicinity of planets and comets, and, except in the outer heliosphere where interstellar pickup ions become energetically important, the sole source of solar wind plasma is the solar corona. Furthermore, because of its large scale relative to scales at which kinetic effects become important, many aspects of the solar wind can be described using fluid approximations to the kinetic equations.

Even before spacecraft measured the properties of the solar wind, Parker [43] used fluid equations to predict its existence along with many of its most salient properties. Of particular importance, was the prediction that the solar wind beyond a few solar radii would be supersonic and super-Alfvénic. Analyses of the earliest measurements of solar wind velocity and magnetic fields [11–13] suggested strongly that the solar wind was turbulent. The evidence was two-fold: first, the power spectra of the velocity and magnetic field fluctuations had a power law shape reminiscent of fluid turbulence, *i.e.*, there appeared to be an inertial range where the spectral index was close to the Kolmogorov [36] value of $-5/3$ which was known to characterize fluid turbulence. Second, there was sufficient energy in the interactions between the fast and slow solar wind streams to drive a turbulent cascade [13].

Subsequent research has confirmed this picture of the solar wind as a turbulent magnetofluid. Interested readers are referred to a review of magnetohydrodynamic (MHD) turbulence theory in general by Pouquet [46] and to reviews of the application of MHD turbulence to solar wind observations in particular by Marsch [39] and Goldstein [29,30]. The great wealth of fields and particle data have enabled us to characterize the spatial scales of the solar wind from tens of astronomical units (AU) down to a few kilometers, thus providing an excellent laboratory for the study of magnetofluid turbulence. Observations of the solar wind from 0.3 AU and beyond show that the expansion of the wind into the heliosphere, together with the stirring of the medium by the interaction of fast and slow streams, leads to a dynamical mixture of MHD fluctuations, convected structures, microstreams, and propagating compressive structures. As evidenced by the Kolmogorov inertial range spectrum, these dynamical processes are all interacting nonlinearly to drive a flow of energy in wave number space that is predominantly from large to small scales where the dissipation of the turbulent energy occurs.

One of the more curious features of solar wind fluctuations, is that in spite of its close resemblance to fluid turbulence, one can very often identify what appear to be almost pure Alfvén waves in the flow [3]. Alfvénic fluctuations are characterized as ones for which $\mathbf{b} = \pm \mathbf{v}/\sqrt{4\pi\rho}$, where $\mathbf{b}$ and $\mathbf{v}$ are the fluctuating components of the magnetic and velocity fields, respectively, and ρ is the mass density. (Fluctuations with positive (negative) correlation between b and v are often referred to as being "outward" ("inward") propagating with positive (negative) cross helicity.)

Once generated, Alfvénic fluctuations are difficult to damp, so it is not surprising that they are a common feature of the solar wind, at least in the inertial range of the spectrum [40,49,48]. The observed sign of the b − v correlation indicates that most Alfvénic fluctuations are propagating outward from the Sun. (Here, we will adopt the convention that outward propagating fluctuations have positive cross helicity regardless of whether the background magnetic field is directed toward or away from the Sun.) The most obvious explanation for this is that nearly all solar wind Alfvénic fluctuations are generated in the lower (subAlfvénic) solar corona. At the "critical" point where the wind becomes superAlfvénic, only outward propagating fluctuations are convected into the heliosphere. Any inward propagating Alfvén waves observed in the solar wind must have originated beyond the critical point. This implies that until the wind has had a chance to generate waves *in situ*, the predominant direction of propagation will be outward. This simple idea has been a powerful diagnostic for the analysis of the solar wind dynamics.

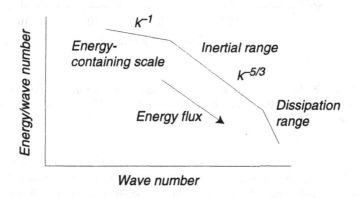

Fig. 1. A schematic representation of the power spectrum of solar wind fields. (After [29].)

Although studies of turbulence in the solar wind cover temporal scales ranging from years to milliseconds, it is impractical to construct spectra covering the entire range of observations. The general shape of the spectrum can, however, be illustrated schematically as in Fig. 1 [29] which divides the interplanetary spectrum into three ranges: the largest, "energy containing" scales provide the reservoir that is tapped by the turbulent cascade; the intermediate "inertial range" [36], characterized by power-law spectra, in which the nonlinear inertial term in the equations of motion dominates over the dissipation. Here energy "cascades" from the energy-containing portion of the spectrum to the small-scale, "dissipation range," where the fluctuations are finally converted to thermal energy.

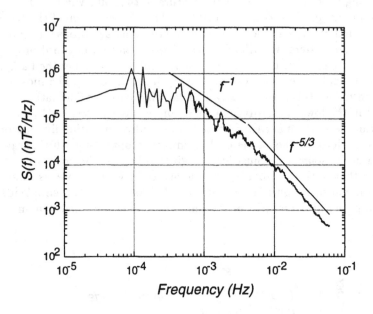

Fig. 2. Trace if the power spectral matrix constructed from 8-s averaged Helios B data obtained near 0.3 AU. Lines with slopes of $-5/3$ and -1 are shown for comparison.

An example showing the transition from the energy-containing scale to the inertial range is shown in Fig. 2. The spectrum is computed from 8 s averaged magnetometer data from Helios B when the spacecraft was at 0.3 AU. The spectrum is the trace of the power spectral matrix of the vector components of the fluctuating magnetic field. For comparison with Fig. 1, reference lines indicating $-5/3$ and -1 spectral slopes are also included.

Fig. 3 shows an example of an inertial range spectrum constructed from Voyager 2 magnetometer data obtained at 10 AU. The trace of the power spectral tensor of the magnetic fluctuations is plotted, along with a reference line indicating an $f^{-5/3}$ spectrum. By 10 AU the energy containing, f^{-1}, portion of the spectrum has almost disappeared.

Fig. 4 shows an example of the dissipation range of interplanetary turbulence from Mariner 0 data obtained near 0.5 AU [28]. The magnetometer data used in this spectrum was averaged to 0.12 s resolution. In addition to the trace of the power spectral matrix of the magnetic field, the plot includes the spectrum of the positive and negative values of the magnetic helicity [40].

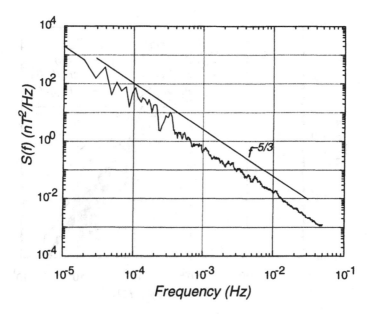

Fig. 3. A power spectrum of Voyager 2 data from 10 AU. A $-5/3$ line is shown for comparison.

In the inertial range, the magnetic helicity spectrum, which is a measure of the sense of twist of the magnetic field [41], is randomly signed [40,27]. Near the dissipation range, the magnetic helicity is often single-signed, and, if one assumes that the fluctuations are propagating outward from the Sun as they are at lower frequencies, then the sign of the magnetic helicity is often consistent with the waves being right-hand polarized [28]. More detailed analyses of the symmetry and polarization properties of fluctuations in the dissipation range using high time resolution data from the Wind spacecraft at 1 AU can be found in Leamon *et al.* [38].

Near the ion-cyclotron scale one expects kinetic processes to control the dissipation. Fig. 5 shows a dynamic spectrum of the normalized magnetic helicity, again from the magnetometer on Mariner10, but from day 79 of 1974 (March 20). During this time, the spacecraft crossed a sector boundary, going from an inward to an outward and back again to an inward directed sector. The sector crossings appear to be associated with strong circular polarization (with positive magnetic helicity) in the magnetic fluctuations (whether the polarization is right- or left-handed cannot be determined). Phenomena such as this cannot be described using a fluid description.

One expects that the dissipation of magnetic fluctuations near the proton Larmor radius will be determined in large measure by ion cyclotron damping.

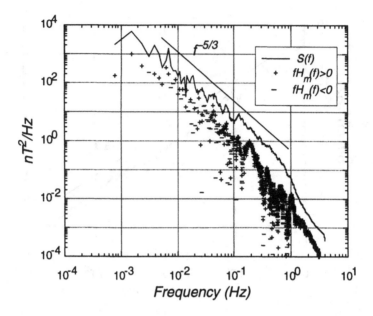

Fig. 4. A power spectrum of Mariner 10 data from 0.5 AU showing the dissipation range of magnetic fluctuations. Also shown are positive and negative values of the (reduced) magnetic helicity spectrum multiplied by frequency f.

As we shall discuss below, for parallel propagating waves there is a complex interaction between ion cyclotron damping and the temperature anisotropy. For obliquely propagating fluctuations, $e.g.$ kinetic Alfvén waves or other compressive modes, the situation is even less well understood (see, $e.g.$, [38,61]. The spectrum shown in Fig. 4 indicates that damping does occur. In contrast, a power spectrum (not shown) computed from data taken close to the sector crossing contained in the interval used in Fig. 5, shows a small, but noticeable, enhancement in power near the cyclotron frequency ($cf.$ Fig. 8, [28]). Leamon et $al.$ [38] found a similar variety of spectra in the dissipation range.

The observed change in solar wind temperature with distance indicates that the solar wind is heated significantly [15,16,20–22,50,60], however, the source of heating is unclear. Certainly the dissipation of shock waves and the pickup of interstellar ions in the outer heliosphere contribute. However, shock waves generated by the interaction of fast and slow wind generally do not form inside $1-2$ AU, and there have been periods in the outer heliosphere nearly devoid of shocks [8], so that shocks alone are inadequate to explain the observations. Thus, it is quite plausible that heating by turbulent dissipation is important, especially in the inner heliosphere.

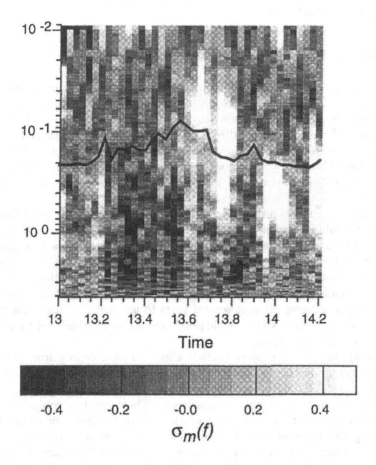

Fig. 5. A dynamic spectrum of the normalized magnetic helicity constructed from the Mariner10 data used in Fig. 4. The panel consists of 40 individual spectra each with 22 degrees of freedom and 467 frequencies ranging from 1.78×10^{-2} to 4 Hz. The solid line indicates the proton cyclotron frequency. For details of how the magnetic helicity was obtained, see [28].

The role of turbulent heating in Alfvénic and nonAlfvénic flows in the solar wind has been estimated (see, [57,60]). Verma *et al.* assumed that the turbulence was either Kolmogorov-like with power indices of $-\frac{5}{3}$, or was Kraichnan-like, with spectral indices of $-\frac{3}{2}$. They concluded that turbulent heating was likely to be an important contributor to the evolution of solar wind temperature and that the Kolmogorov phenomenology provided qualitatively reasonable agreement with Voyager observations, at least for non-Alfvénic flows. A complete description of turbulent dissipation of solar wind fluctuations thus appears necessary if we are to understand the temperature evolution of the solar wind.

2   More realistic dissipation operators

The MHD equations themselves cannot describe the physics of the dissipation range. Even when dissipation terms are added to the ideal equations, the form they take is analogous to the viscous term in the Navier-Stokes equation and are not derived directly from physical arguments. Indeed, the dissipation terms are often included more to control numerical instabilities in the solutions than for physical reasons. Furthermore, the resistive and viscous terms, although weighted somewhat to damp preferentially small scales and large gradients, affect a broad range of the spectrum.

Ideally, to study realistically the dissipation range of solar wind fluctuations one would like to use the full set of Vlasov-Maxwell equations. Present computational capabilities, however, make it impractical to solve large-scale problems with complicated initial and boundary conditions. This leaves the theoretical description of the three-dimensional evolution of the solar wind largely dependent on solutions of magnetofluid equations. While there have been attempts during the past several years to use new algorithms for particle simulations to attack large-scale problems, the results have yet to achieve adquate resolution [6,7,42].

One approach to including particle effects in fluid descriptions is to use "test particles" moving in the electric and magnetic fields computed from solutions to the MHD equations. While this technique can indicate important physical effects [2], it is not self-consistent. There are two complementary motivations for refining dissipation coefficients in the fluid equations: first, many algorithms, most notably spectral methods, require some form of dissipation to simulate turbulent evolution. Some finite difference algorithms contain numerical dissipation that provides the necessary stabilization. This dissipation may be present throughout the solution, or, as is the case in the flux-corrected-transport (FCT) algorithm [62,14] only where sharp gradients appear. See [26] for a recent application to solar wind turbulence. Dissipation within finite difference codes will not be discussed further here. The second rationale for including dissipation in fluid algorithms is to modify the fluid equations in ways that mimic salient physical properties of the plasma. We shall discuss several approaches for doing this in the remainder of this paper, concentrating on numerical techniques that use spectral or pseudospectral algorithms.

2.1   Hyperresistivity and hyperviscosity

The standard viscous dissipation operator in the Navier Stokes and MHD equations is $\nu \nabla^2$, which, when transformed into Fourier space has the form

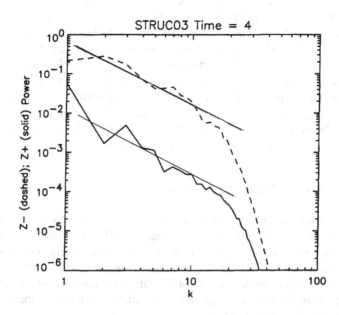

Fig. 6. Spectra of "outward" (z^+; dashed) and "inward" (z^-; solid) fluctuations at $T = 4$ in a 2-1/2-dimensional simulation with structures and waves. The straight lines have slopes of $-5/3$. (From [47].)

νk^2. Unless the simulation is very large, dissipation will affect the intermediate as well as the small scales. One technique for further concentrating the dissipation at the smallest scales, is to add higher order terms [1,4,5,45]. In Fourier space, the dissipation terms in the equations are modified so that the viscosity and resistivity become

$$\nu(k) = \nu_0 \left[1 + (k/k_{eq})^2\right]$$
$$\mu(k) = \mu_0 \left[1 + (k/k_{eq})^2\right] \tag{1}$$

The parameter k_{eq}^2 controls the strength of the bi-Laplacian contribution to the dissipation. With careful adjustment of the values of ν_0, μ_0, and k_{eq}, one can obtain very low dissipation for most wave vectors and approximate a nearly dissipation-free inertial range power spectrum.

As an example, in Fig. 6 we show a spectra computed from a 2-1/2-dimensional simulation [47] with 256×256 Fourier modes. The two curves are spectra of outward (positive cross helicity) and inward (negative cross helicity) fluctuations. The simulation was designed to explore the interaction between Alfvénic fluctuations and structures. A well-defined inertial inertial range is evident. The bi-Laplacian is not the only generalization of the standard Navier Stokes formulation; other artificial dissipation coefficients have been developed and used successfully (see, *e.g.*, Umeki and Terasawa [59]).

Hyperviscosity and hyperresistivity, because they dramatically reduce damping of intermediate scale fluctuations, are very useful for exploring conditions under which inertial range spectra might form and for determining what the slope of the inertial range will be. In addition, they are not very computationally expensive to implement. Nonetheless, such artificial modifications of the magnetofluid equations cannot be used to provide physically meaningful information about the dissipation range.

2.2   Generalized dissipation operators

A more physically motivated approach to generalizing dissipation in fluid models was developed by Passot and Pouquet [45] and Passot *et al.* [44]. Transport operators are constructed satisfying certain general conditions including a requirement that dissipation lead to an increase of entropy independent of reference frame. Additionally, one can require that dissipation be greatest in regions of large gradients in velocity, density, or current, and minimal in homogeneous regions.

The technique was implemented for MHD by Siregar *et al.* [51]. In the MHD formalism, two viscosity functions are defined that adjust to local gradients for both the compressional and solenoidal components of velocity. A nonlinear conductivity is also included which decreases quadratically as the current increases so that the resistivity becomes large in current sheets while remaining asymptotically small where the magnetic field is relatively unstructured. This form for the resistivity has the property that it produces an anomalous increase for currents exceeding a particular "critical" value, thus mimicking the behavior of microinstabilities which limit current growth.

The dissipation operators have been used to study various physical situations, including reconnection and the evolution of a von Kárman vortex street (see [53]) and the results were compared with those obtained using standard Navier-Stokes and bi-Laplacian dissipation. The Kárman vortex street has the property that there are readily identifiable regions of large spatial gradients. These can be used to compare linear and nonlinear dissipation operators. Fig. 7 shows profiles of an arbitrary cut of the viscous dissipation functions at $T = 3$ from pseudospectral simulations using the standard (dotted line) and nonlinear (solid line) dissipation terms. Although the maxima in the dissipation occur at the same locations, the nonlinear dissipation is better confined to regions of large gradients.

The method has also been applied to the study of turbulent dissipation. In those simulations, two neutral sheets were embedded in regions of oppositely directed uniform magnetic fields. Fig. 8 compares the omnidirectional (modal) internal energy, kinetic energy, and magnetic energy at $T = 3$ from

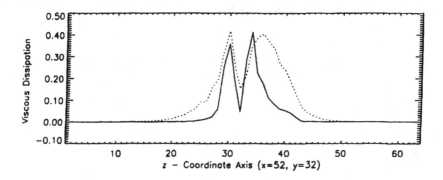

Fig. 7. Spatial profiles of the viscous dissipation function associated with a velocity shear flow which produces a vortex street. (From [51].)

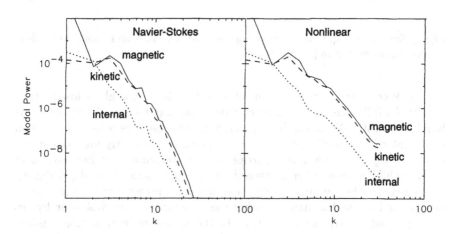

Fig. 8. The omnidirectional (modal) internal energy (dotted line), kinetic energy (dashed line), and magnetic energy (solid line) at $T = 3$ from simulations that compare the Navier-Stokes and nonlinear dissipation operators in a reconnecting current sheet. (From [51].)

the standard and nonlinear simulations of magnetic reconnection. The omni-directional modal spectrum, $P_G(k)$, is defined by $G = \sum P_G(k)n(k)$, where G is the average value of a physical quantity, $P_G(k)$ is the average value of the Fourier component in each wave number bin, and $n(k)$ is the number of Fourier modes in each isotropic k^{th} integer bin, $(k - 1/2) \leq |\mathbf{k}| < (k + 1/2)$. At low wave numbers the spectra are nearly identical; however, for $k > 5$, the standard dissipation coefficient results in lower overall power, suggesting that significant dissipation is occurring even at low values of k. In contrast, the nonlinear operators damp only the highest wave numbers and preserve an inertial range out to the Nyquist frequency.

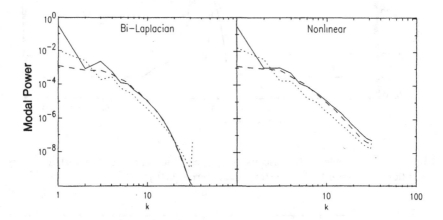

Fig. 9. Similar to Fig. 8, but comparing the "bi-Laplacian" and "nonlinear" dissipation operators. (From [24].)

Fig. 9 compares a similar run, also at $T = 3$, at a slightly lower Mach number (≈ 0.2) using the bi-Laplacian dissipation (left) (eq. 1) and nonlinear function (right). While the bi-Laplacian dissipation does a better job than the standard Navier-Stokes coefficient at preserving a nearly dissipation-free inertial range, the power law portion of the spectrum does not extend to nearly as high wave number as when the nonlinear function is used (*cf.* Fig. 8).

In general, the nonlinear dissipation operators preserve more information at intermediate scales than do either the standard Navier-Stokes or hyperviscosity and hyperresistivity terms. In the magnetic reconnection studies, quadrupoles in vorticity and current are particularly well defined. In addition, the nonlinear resistivity operators allow for the formation of a self-consistent resistive length scale which in turn supports the formation of multiple $X-$points in reconnection layers.

2.3   Hall MHD and finite Larmor radius corrections

The methods described above make no attempt to incorporate kinetic aspects of the damping into the fluid formalism. Even the approach using entropy production near strong gradients cannot account for changes in, *e.g.*, the temperature anisotropy that accompanies ion cyclotron damping. To include such kinetic processes within the fluid formalism requires that the fluid equations themselves be modified. The simplest modification is to use a more generalized Ohm's Law that incorporates some of the kinetic modifications which occur near the proton cyclotron frequency and Larmor radius. Two terms, both of equal order, must be addressed, *viz.*, the Hall and Finite Larmor Radius correction terms.

Motivated by the question as to whether the break in the power spectrum near the (Doppler shifted) proton cyclotron frequency (*cf.* Fig. 4) might arise naturally from cyclotron interactions, Ghosh *et al.* [24] added the Hall term to the MHD equations and solved the equations. The Hall term adds a dispersive term to the usual Alfvén wave dispersion relation. Thus polarization becomes a parameter one can study in Hall MHD (*cf.* Fig. 5). Resonant couplings of MHD modes will also be affected, which could lead to changes in the overall energy cascade at the proton cyclotron scales.

Another property of the Hall MHD system is that the cross helicity (global) invariant H_c [18,32,56] is replaced by a hybrid helicity invariant $\tilde{H}_h$ [24,58]. Because the Hall term itself is not dissipative, spectral method solutions of the Hall-MHD equations must include dissipation explicitly. In [24] a bi-Laplacian was used and the initial results were limited to a relatively high plasma beta ($\beta = 4$) in an effort to model solar wind parameter regimes. In more recent work, Jayanti *et al.* [35], included the Finite Larmor Radius (FLR) correction. Although formally of the same order as the Hall term, the FLR term contains additional dependence on plasma β. When both terms are included, the dimensionless compressible Hall-MHD system of equations has the form:

$$\frac{\partial}{\partial t}\rho = -\nabla \cdot (\rho \mathbf{u}) \tag{2a}$$

$$\frac{\partial}{\partial t}\mathbf{u} = -\mathbf{u} \cdot \nabla \mathbf{u} - \frac{1}{\rho M_{s0}^2}\nabla \cdot [P\mathbf{I} + \varepsilon \Pi]$$
$$+ \frac{1}{\rho}\mathbf{J} \times \mathbf{B} + \frac{1}{\rho}\nu\nabla^2\mathbf{u} + \frac{1}{\rho}\left(\zeta + \frac{1}{3}\nu\right)\nabla\left(\nabla \cdot \mathbf{u}\right) \tag{2b}$$

$$\frac{\partial}{\partial t}\mathbf{A} = \left(\mathbf{u} - \varepsilon\frac{\mathbf{J}}{\rho}\right) \times \mathbf{B} - \mu\mathbf{J} + \nabla F \tag{2c}$$

where the isotropic pressure $P = \rho^\gamma/(\gamma M_{s0}^2)$ with $\gamma = 5/3$ and $M_{s0} = 0.25$ is the sonic Mach number (the Alfvén Mach number is assumed to be unity). In addition, $\mathbf{B}=B_0\hat{\mathbf{x}} + \nabla \times \mathbf{A}$, $\varepsilon = \omega_A/\Omega_i = \omega(k_{\min})/\Omega_i$ (ω_A is the Alfvén frequency of the lowest wave number), Ω_i is the proton cyclotron frequency, and ν and μ are given by eq. 1. In constructing the solutions, the Coulomb Gauge is assumed, so that $F \to F_k = ik \cdot [\mathbf{u} \times \mathbf{B}]_k/k^2$. The FLR term, $\varepsilon\Pi$, is given by

$$\Pi_{xx} = 0,$$

$$\Pi_{yy} = -\Pi_{zz} = -\frac{P}{2}\left[\frac{\partial u_z}{\partial y} + \frac{\partial u_y}{\partial z}\right],$$

$$\Pi_{yz} = \Pi_{zy} = \frac{P}{2}\left[\frac{\partial u_y}{\partial y} - \frac{\partial u_z}{\partial z}\right],$$

$$\Pi_{zx} = \Pi_{xz} = P\left[\frac{\partial u_y}{\partial x} + \frac{\partial u_x}{\partial y}\right],$$

$$\Pi_{xy} = \Pi_{yx} = P\left[\frac{\partial u_z}{\partial x} - \frac{\partial u_x}{\partial z}\right],$$

and the Hall term is $\varepsilon \mathbf{J}/\rho$ with $\mathbf{J} = \nabla \times \nabla \times \mathbf{A}$.

The initial studies included only the Hall term and the behavior of the cross helicity and magnetic helicity were investigated for various values of β and ε[24,23]. Here we illustrate how inclusion of the Hall term changes both the polarization and power spectra. The results from a 256×256 simulation are summarized in Fig. 10. The run was initialized with a spectrum of left-circularly polarized waves propagating parallel to the mean magnetic field and $\beta = 4$. The solution with nonzero Hall term ($\varepsilon = 1/20$) has a steeper power spectrum at high wave number and the polarization above wave number 20 is nearly -1, indicating that, at least for large β, left-hand polarized waves disappear near the cyclotron cutoff. Large anisotropies were found in the power spectra for small and large plasma β, but the anisotropies were significantly smaller for $\beta \approx 1$ [23]. If the steeper slope is caused by the suppression of nonlinear cascades due to high cross helicity, it is plausible that the spectral steepening adjacent to $k = 1/\varepsilon = 20$ is due to additional suppression of nonlinear cascades caused by the large value of the generalized hybrid helicity $\hat{H}_h$ [24]. If the nonlinear cascade of power is suppressed by H_c in the inertial range, and further suppressed by the hybrid invariant $\hat{H}_h$ near $k = 1/\varepsilon$, then the total energy should decrease when $\varepsilon = 1/20$ and somewhat more energy should be lost for the $\varepsilon = 0$ runs, which is what was found.

The Hall term alone, however, does not account for all of the observations. One issue is that $-5/3$ inertial-range slopes are often seen in conjunction with the spectral turnover at f_p (e.g., Fig. 4). Because velocity measurements are unavailable, it is not possible to make a detailed correspondence between the simulations, which are sensitive to the polarization and cross helicity, and the observations, for which we only have the magnetic helicity. However, the simulations suggest that a spectral break at $k = 1/\varepsilon$ occurs only when the inertial-range spectrum is already steeper than $k^{-5/3}$ and the cross helicity is large. Therefore, if the cross helicity of the interplanetary fluctuations is indeed large, then the spectral break near f_p may be due to factors outside the scope of Hall MHD.

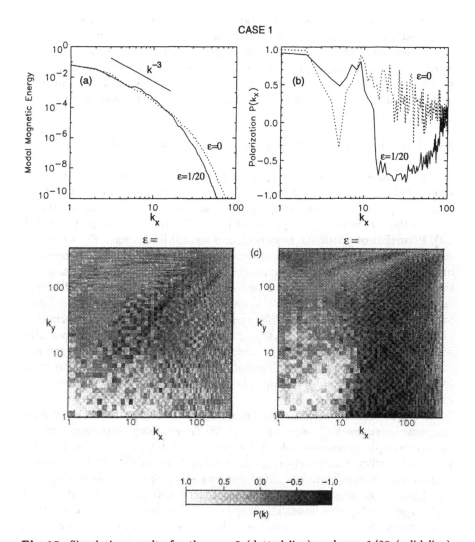

Fig. 10. Simulation results for the $\varepsilon = 0$ (dotted line) and $\varepsilon = 1/20$ (solid line) comparing (a) the modal omnidirectional magnetic energy power $S_m(k)$ spectrum at $T = 4$ (a k^{-3} slope is included between $3 \le k \le 15$ for reference); (b) the time-averaged polarization $P(k_x)$ along $\mathbf{B}_0$; and (c) in gray scale, the time-averaged polarization $P(\mathbf{k})$ in the first quadrant of the wave number space $\mathbf{k}$. (From [24].)

Preliminary results from [35], suggest that the FLR term produces less steep spectra than does the Hall term. These results, taken together with the earlier ones using the Hall term alone, suggest that the polarization enhancements observed in the simulations are consistent with the observations [28], supporting the expectation that cyclotron resonance plays an important role in determining the nature of the spectrum near the dissipation range.

2.4   Kinetic effects: Quasi-invariants and coarse-graining

One could go further in incorporating physically accurate descriptions of kinetic behavior into a fluid description by, for example, using a two-fluid formalism [33,34] or by invoking higher order closures of the fluid equations such as the Chapman-Enskog method [9], or the moment method of Grad [31]. Perturbation methods, however, have serious limitations. For example, the Chapman-Enskog method does not converge in low collisional regimes or in situations where large gradients or pressure anisotropies are important. Although the Grad approach can include both the FLR and Chew-Goldberger-Low [10] approximations, it works well only for nearly Maxwellian distributions, and is not well suited for describing nearly-collisionless turbulent plasmas. Ideally, one would like to use kinetic theory to construct physically realistic dissipation operators for transport processes driven by wave-particle interactions, but that is beyond present computational capabilities. Progress can be made, however, by limiting the kinetic effects to, *e.g.*, ion cyclotron damping.

The evolution of the pressure tensor as determined by the exact Vlasov moment equations has been described by Siregar and Goldstein [52] using microscopic information about the cyclotron interaction to calculate the transport coefficients and off-diagonal pressure elements that represent wave-particle momentum transfers. The resulting equations of state incorporate coupling of the mean magnetic field to particle motions in the field-aligned and perpendicular directions. The longitudinal action adiabatic invariant, J, is destroyed by the nonconservation of linear momentum in that direction, while the magnetic moment adiabatic invariant, μ, remains conserved when the magnetic field varies slowly during a cyclotron period.

To apply the resulting equations of state to macroscopic (fluid) phenomena, Siregar *et al.* [54] used kinetic information from hybrid simulations of the interaction of parallel propagating ion cyclotron waves with a non-Maxwellian plasma to define a coarse-graining process that averaged the kinetic interactions to temporal and spatial scales appropriate to a fluid description. A one-dimensional hybrid code and quasifluid model were used to study the evolution of temperature anisotropy in weak and strong resonances. The coarse fluid variables were constructed by taking velocity moments and by requiring that all processes contained within hundreds of proton-inertial lengths (the scales of nonlocal processes) be taken into account in a single coarse fluid particle. In this way, nonlocal processes for the standard fluid are scale-renormalized in the coarse fluid particles so that they become "local" in the coarse proton fluid.

A one-dimensional hybrid simulation with $m_e = 0$ was used to study low frequency cyclotron waves in the frequency range $\omega(k) \ll \omega_{ce}$ such that $\omega(k)/k_{\parallel} V_e \ll 1$ where V_e is the electron thermal velocity. At the fundamental cyclotron resonance

$$\mathbf{k} \cdot \mathbf{V} = \omega_{res} - \omega_{cp} < 0 \tag{3}$$

the resonant phase speed V_{res} satisfies

$$\frac{\omega_{res}}{k_{res}} = \frac{\omega_{cp}}{k_{res}} + V_{res} \tag{4}$$

where the resonance wave number k_{res} of protons moving with speed $V_{res}(p)$ at a fraction f of the thermal speed a is given by

$$k_{res} = \frac{\omega_{res} - \omega_c}{fa}$$

The strength of the resonance at $V = a$ can be defined as

$$\zeta^p(a) = |\frac{1}{k_{res}a}| \, |\omega_{res} - k_{res}a - \omega_c| \tag{5}$$

where ζ is the argument of the plasma dispersion function $Z(\zeta)$ [17].

Weak resonance ($\zeta^p(a) > 8$) is satisfied for large parallel proton velocities $v_{\parallel} \approx |\omega_c/k| > v_A$. For weak resonance, the number of resonant protons is small compared to the nonresonant thermal protons, and only a fraction of the nonresonant protons participate in the mechanical motion of the waves. As the resonance strength increases, the number of resonant protons also increases and the phase speeds of the cyclotron waves approach the speed of sound c_s and nonlocal effects become important.

The 1-1/2-D hybrid simulations were initialized with an (isotropic) Kappa proton velocity distribution with $\kappa = 1$ and $\beta = 0.1$ to which various distributions of parallel propagating proton cyclotron waves were added [55]. The two coarse variables $\bar{I}_1(t) = <\mu>$ and $\bar{I}_2(t) = <J>$ show strong anticorrelations in time for weak to intermediate resonance strengths (see Fig. 11) at hundreds of proton-cyclotron periods. Anticorrelation remains significant even in a random-phased spectrum of waves but disappears in the strongest resonance cases studied. The anticorrelations of the coarse variables reflect the (nonlocal) energy transfers between the cyclotron-damped wave(s) and the parallel and perpendicular velocities of the resonant and nonresonant protons. Fig. 11 shows the anticorrelation of $\bar{I}_1$ and $\bar{I}_2$, and the quasi-invariance of $\bar{I} = m_1 \bar{I}_1(t) + m_2 \bar{I}_2(t)$ ($m_1 = 3/5$ and $m_2 = 2/5$ for all cases studied). The significance of the values of m_1 and m_2 is not known [54,55].

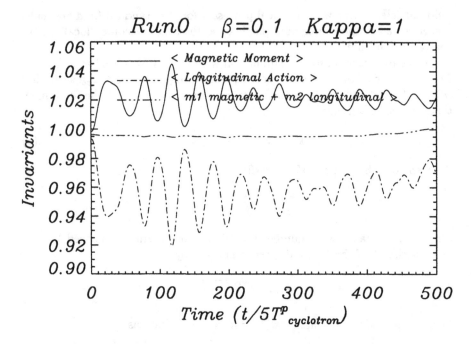

Fig. 11. Behavior of the coarse invariants for a single wave in weak resonance with a nonMaxwellian proton distribution. (From [55]).

3   Summary and conclusions

This review has discussed modifications to the magnetofluid equations that attempt to describe the transition between the inertial and dissipation ranges more physically than does the standard Navier-Stokes approach. The simplest approach is to add higher order derivatives in the form of hyperresistivity and viscosity (eq. 1) to the Navier-Stokes terms. While this simple change does improve resolution in the inertial range, the resulting equations have not been derived from first principles in a physically motivated way. Thus, it is difficult to determine the extent to which the evolution of the energy containing scales has been modified, although the simulations themselves suggest that any modifications that have occurred are small because the overall results of the simulations resemble closely results obtained without the higher order dissipation term.

The use of nonlinear operators [44,45] and [51] is more physically motivated. These operators are self-adjusting and become important only near sharp gradients. In studies of magnetic reconnection and shear-flow instabilities, the nonlinear operators produce dissipation that limits the current near the X-point of magnetic reconnection regions and confines dissipation to regions of strong velocity gradients in the vortex street studies.

By generalizing the MHD equations to include a more general Ohm's Law, physical processes dependent on the proton cyclotron scale can be incorporated. Neither the Hall nor FLR terms are dissipative, so additional modifications must be made to ensure numerical stability. (These modifications can include either the bi-Laplacian or nonlinear operators.) One limitation of this approach is that one may need to consider other contributions to Ohm's Law, such as the electron pressure gradient, that may be as important or more important than either the Hall or FLR terms (J. D. Scudder, private communication, 1995). Such two-fluid models are beyond the scope of the present review, but have been considered in other contexts by, *e.g.*, [33,34,37].

Particle simulations can be combined with fluid codes and theory to study how finite amplitude fluctuations evolve and dissipate. Vlasov theory can be used to construct dissipation operators that are both physically realistic and which permit accurate numerical calculation of small scale turbulent features. To merge the kinetic and fluid scales some averaging of kinetic processes is necessary. Coarse-graining of the fluid equations has shown the existence of a quasi-invariant [54,55] that relates the quasifluid parallel and perpendicular temperatures. For weak resonances, there is a strong anticorrelation between those components of the pressure tensor. The situation for strongly resonance plasmas remains to be explored. The quasi-invariant formulation predicts an inverse plasma $\beta_{||}$ dependence for the quasi-steady anisotropic state as observed and predicted [19].

References

1. Agim YZ, Viñas AF, Goldstein ML (1995) Magnetohydrodynamic and hybrid simulations of broadband fluctuations near interplanetary shocks. J. Geophys. Res. 100: 17,081
2. Ambrosiano J, Matthaeus WH, Goldstein ML (1988) Test-particle studies of acceleration by turbulent reconnection fields. J. Geophys. Res. 93: 14,383
3. Belcher JW, Davis L (1971) Large-amplitude Alfvén waves in the interplanetary medium, 2. J. Geophys. Res. 76: 3534
4. Biskamp D (1982) Effect of secondary tearing instability on the coalescence of magnetic islands. Phys. Lett. A 87: 357
5. Biskamp D, Welter H (1989) Dynamics of decaying two-dimensional magnetohydrodynamic turbulence. Phys. Fluids B 1: 1964
6. Buneman O, Neubert T, Nishikawa K-I (1992) Solar wind-magnetosphere interaction as simulated by a 3D, EM particle code. IEEE Trans. Plasma Sci.
7. Buneman O, Nishikawa K-I, Neubert T (1995) Solar wind-magnetosphere interaction as simulated by a 3D, EM particle code. In: Ashour-Abdalla M, Chang T and Dusenbery P (eds) Space plasmas: Coupling between small and medium scale processes. AGU, Washington DC Geophys. Monogr. Ser., vol 86, pp 347-355
8. Burlaga LF, Ness NF, Belcher J, Szabo A, Isenberg P, Lee M (1994) Pickup protons and pressure balanced structures: Voyager 2 observations in MIRs near 35 AU. J. Geophys. Res. 99: 21511

9. Chapman S, Cowling TG (1939) The Mathematical Theory of Non-uniform Gases. Cambridge University Press

10. Chew FG, Goldberger ML, Low FE (1956) The Boltzmann equation and the one-fluid hydromagnetic equations in the absence of particle collisions. Proc. Roy. Soc. A236: 112

11. Coleman PJ (1966) Hydromagnetic waves in the interplanetary medium. Phys. Rev. Lett. 17: 207

12. Coleman PJ (1967) Wave-like phenomena in the interplanetary plasma: Mariner 2. Planet. Space Sci. 15: 953

13. Coleman PJ (1968) Turbulence, viscosity, and dissipation in the solar wind plasma. Astrophys. J. 153: 371

14. Deane A, Roberts DA, Goldstein ML, Zalesak S, Spicer D (1998) A 3D compressible MHD code in spherical coordinates for solar wind studies. Comp. Phys. Comm. in preparation

15. Freeman JW (1988) Estimates of solar wind heating inside 0.3 AU. Geophys. Res. Lett. 15: 88

16. Freeman JW, Totten T, Arya S (1992) A determination of the polytropic index of the free streaming solar wind using improved temperature and density radial power-law indices. EOS, Trans. Am. Geophys. U. 73: 238

17. Fried BD, Conte SD (1961) The Plasma Dispersion Function. Academic Press, San Diego, Calif.

18. Frisch U, Pouquet A, Léorat J, Mazure A (1975) Possibility of an inverse cascade of magnetic helicity in magnetohydrodynamic turbulence. J. Fluid Mech. 68: 769

19. Gary SP, Lee MA (1994) The ion cyclotron anisotropy instability and the inverse correlation between proton anisotropy and proton beta. J. Geophys. Res. 99: 11,297-11,302

20. Gazis PR, Barnes A, Mihalov JD, Lazarus AJ (1994) Solar wind velocity and temperature in the outer heliosphere. J. Geophys. Res. 99: 6561

21. Gazis PR, Lazarus AJ (1982) Voyager observations of solar wind proton temperature: 1-10 AU. Geophys. Res. Lett. 9: 431

22. Gazis PR, Mihalov JD, Barnes A, Lazarus AJ, Smith EJ (1989) Pioneer and Voyager observations of the solar wind at large heliocentric distances and latitudes. Geophys. Res. Lett. 16: 223

23. Ghosh S, Goldstein ML (1997) Anisotropy in Hall MHD turbulence due to a mean magnetic field. J. Plasma Phys. 57: 129-154

24. Ghosh S, Siregar E, Roberts DA, Goldstein ML (1996) Simulation of high-frequency solar wind power spectra using Hall magnetohydrodynamics. J. Geophys. Res. 101: 2493-2504

25. Goldstein ML (1996) Turbulence in the Solar Wind: Kinetic Effects. In: Solar Wind 8, Dana Point, p 239

26. Goldstein ML, Roberts DA, Deane AE, Ghosh S, Wong HK (1999) Numerical simulation of Alfvénic turbulence in the solar wind. J. Geophys. Res. in press

27. Goldstein ML, Roberts DA, Fitch CA (1991) The structure of helical interplanetary magnetic fields. Geophys. Res. Lett. 18: 1505

28. Goldstein ML, Roberts DA, Fitch CA (1994) Properties of the fluctuating magnetic helicity in the inertial and dissipation ranges of solar wind turbulence. J. Geophys. Res. 99: 11,519

29. Goldstein ML, Roberts DA, Matthaeus WH (1995) Magnetohydrodynamic turbulence in the solar wind. Annual Review of Astronomy and Astrophysics 33: 283

30. Goldstein ML, Roberts DA, Matthaeus WH (1997) Magnetohydrodynamic turbulence in cosmic winds. In: Jokipii JR, Sonett CP and Giampapa MS (eds) Cosmic Winds and the Heliosphere. Univ. of Arizona Press, Tucson, p 521

31. Grad H (1949) On the kinetic theory of rarefied gases. Commun. Pure Appl. Math. 2: 231

32. Grappin R, Pouquet A, Léorat J (1983) Dependence of MHD turbulence spectra on the velocity field-magnetic field correlation. Astron. and Astrophys. 126: 51

33. Hesse M, Winske D (1998a) Electron dissipation in collisionless magnetic reconnection. Journal of Geophysical Research-Space Physics 103: 26,479-26,486

34. Hesse M, Winske D, Birn J (1998b) On the ion-scale structure of thin current sheets in the magnetotail. Physica Scripta T74: 63-66

35. Jayanti VB, Ghosh S, Goldstein ML (1999) Simulations of Hall and Finite Larmor Radius corrections on the development of MHD turbulence. J. Geophys. Res. in preparation

36. Kolmogorov AN (1941) The local structure of turbulence in incompressible viscous fluid for very large Reynolds numbers. C. R. Acad. Sci. URSS 30: 301

37. Kuznetsova MM, Hesse M, Winske D (1998) Kinetic quasi-viscous and bulk flow inertia effects in collisionless magnetotail reconnection. Journal of Geophysical Research-Space Physics 103: 199-213

38. Leamon RJ, Smith CW, Ness NF, Matthaeus WH, Wong HK (1998) Observational constraints on the dynamics of the interplanetary magnetic field dissipation range. J. Geophys. Res. 103: 4775-4788

39. Marsch E (1991) MHD turbulence in the solar wind. In: Schwenn R and Marsch E (eds) Physics of the Inner Heliosphere. Springer-Verlag, Heidelberg, vol 2, p 159

40. Matthaeus WH, Goldstein ML (1982) Measurement of the rugged invariants of magnetohydrodynamic turbulence. J. Geophys. Res. 87: 6011

41. Moffatt HK (1978) Magnetic Field Generation in Electrically Conducting Fluids. Cambridge University Press, New York

42. Neubert T, Miller RH, Buneman O, Nishikawa K-I (1992) The dynamics of low-plasma clouds as simulated by a three-dimensional, electromagnetic particle code. J. Geophys. Res. 97: 12,057-12,072

43. Parker EN (1958) Dynamics of the interplanetary gas and magnetic fields. Astrophys. J. 128: 664

44. Passot T, Politano H, Pouquet A, Sulem PL (1990) Comparative study of dissipation modeling in two-dimensional MHD turbulence. Theor. Comput. Fluid Dyn 1: 47

45. Passot T, Pouquet A (1988) Hyperviscosity for compressible flows using spectral methods. J. Comp. Phys. 75: 300

46. Pouquet A (1993) Magnetohydrodynamic Turbulence. In: Les Houches Summer School on Astrophysical Fluid Dynamics, Les Houches, France. Amsterdam: Elsevier, pp 139 ff.

47. Roberts DA, Ghosh S, Goldstein ML (1996) Nonlinear evolution of interplanetary Alfvénic fluctuations with convected structures. Geophys. Res. Lett. 23: 591

48. Roberts DA, Goldstein ML, Klein LW, Matthaeus WH (1987b) Origin and evolution of fluctuations in the solar wind: Helios observations and Helios-Voyager comparisons. J. Geophys. Res. 92: 12,023
49. Roberts DA, Klein LW, Goldstein ML, Matthaeus WH (1987a) The nature and evolution of magnetohydrodynamic fluctuations in the solar wind: Voyager observations. J. Geophys. Res. 92: 11,021
50. Schwenn R (1983) The 'average' solar wind in the inner heliosphere: Structure and slow variations. In: Solar Wind Five. NASA Conf. Publ., p 485
51. Siregar E, Ghosh S, Goldstein ML (1995) Nonlinear entropy production operators for magnetohydrodynamic plasmas. Physics of Plasmas 2(5): 1480
52. Siregar E, Goldstein ML (1996) A Vlasov moment description of cyclotron wave particle interactions. Phys. Plasmas 3(4): 3
53. Siregar E, Stribling WT, Goldstein ML (1994) On the dynamics of a plasma vortex street and its topological signatures. Physics of Plasmas 1: 2125
54. Siregar E, Viñas AF, Goldstein ML (1998a) Coarse graining and nonlocal processes in proton cyclotron resonant interactions. Phys. Plasmas 5: 333
55. Siregar E, Viñas AF, Goldstein ML (1998b) Topological invariants in cyclotron turbulence. In: The 1998 Conference on Multi-Scale Phenomena in Space Plasmas II, Cascais, Portugal. MIT Center for Theoretical Geo/Cosmo Plasma Physics, Cambridge, Mass., p. 307
56. Ting AC, Matthaeus WH, Montgomery C (1986) Turbulent relaxation processes in magnetohydrodynamics. Phys. Fluids: 3261
57. Tu C-Y (1988) The damping of interplanetary Alfvénic fluctuations and the heating of the solar wind. J. Geophys. Res. 93: 7
58. Turner L (1986) Hall effects on magnetic relaxation. IEEE Trans. Plasma Sci. PS-14: 849
59. Umeki H, Terasawa T (1992) Decay instability of incoherent Alfvén waves in the solar wind. J. Geophys. Res. 97: 3113
60. Verma MK, Roberts DA, Goldstein ML (1995) Turbulent heating and temperature evolution of the solar wind plasma. J. Geophys. Res. 100: 19,839
61. Viñas AF, Wong HK, Klimas AJ (1999) Generation of electron suprathermal tails in the upper solar atmosphere: Implications for coronal heating. Astrophys. J. submitted
62. Zalesak ST (1979) Fully multidimensional flux-corrected transport algorithms for fluids. J. Comp. Phys. 31: 335

A Weak Turbulence Theory
for Incompressible Magnetohydrodynamics

S. Galtier[1,2], S. Nazarenko[1], A.C. Newell[1], and A. Pouquet[2]

[1] Mathematics Institute, University of Warwick, Coventry, CV4 7AL, UK
[2] Observatoire de la Côte d'Azur, CNRS UMR 6529, BP 4229, 06304, France

Abstract. We derive a weak turbulence formalism for incompressible MHD. Three–wave interactions lead to a system of kinetic equations for the spectral densities of energy and helicity. The kinetic equations conserve energy in all wavevector planes normal to the applied magnetic field $B_0 \, \hat{e}_{\parallel}$. Numerically and analytically, we find energy spectra $E^{\pm} \sim k_{\perp}^{n_{\pm}}$, such that $n_{+} + n_{-} = -4$, where $E^{\pm}$ are the spectra of the Elsässer variables $z^{\pm} = v \pm b$ in the two-dimensional case ($k_{\parallel} = 0$). The constants of the spectra are computed exactly and found to depend on the amount of correlation between the velocity and the magnetic field. Comparison with several numerical simulations and models is also made.

1 Introduction and general discussion

Magnetohydrodynamic (MHD) turbulence plays an important role in many astrophysical situations [34], ranging from the solar wind [23], to the Sun [35], the interstellar medium [14] and beyond [50], as well as in laboratory devices such as tokamaks (see *e.g.* [46] [43] [12] [44]). A very instrumental step in recognizing some of the features that distinguished MHD turbulence from hydrodynamic turbulence was taken independently in the early sixties by Iroshnikov and Kraichnan in 1965 (IK; [19] [17]). They argued that the destruction of phase coherence by Alfvén waves traveling in opposite directions along local large eddy magnetic fields introduces a new time scale and a slowing down of energy transfer to small scales. They pictured the scattering process as being principally due to three wave interactions. Assuming 3D isotropy, dimensional analysis then leads to the prediction of a $k^{-3/2}$ Kolmogorov finite energy flux spectrum.

However, it is clear, and it has been a concern to Kraichnan and others throughout the years, that the assumption of local three dimensional isotropy is troublesome. Indeed numerical simulations and experimental measurements both indicate that the presence of strong magnetic fields make MHD turbulence strongly anisotropic. Anisotropy is manifested in a two dimensionalization of the turbulence spectrum in a plane transverse to the locally dominant magnetic field and in inhibiting spectral energy transfer along the direction parallel to the field [27] [29] [30]. Replacing the 3D isotropy assumption by a 2D one, and retaining of the rest of the IK picture, leads to the dimensional analysis prediction of a $k_{\perp}^{-2}$ spectrum ($\mathbf{B_0} = B_0\,\hat{\mathbf{e}}_{\parallel}$, the applied magnetic field, $k_{\parallel} = \mathbf{k} \cdot \hat{\mathbf{e}}_{\parallel}$, $\mathbf{k}_{\perp} = \mathbf{k} - k_{\parallel}\hat{\mathbf{e}}_{\parallel}$, $k_{\perp} = |\mathbf{k}_{\perp}|$) although recent numerical simulations by Ng and Bhattacharjee [31] [32] to confirm this spectrum were inconclusive. Indeed, it is also difficult to determine the exact spectral slope from experimental data.

A major controversy in the debate of the universal features of MHD turbulence was introduced by Sridhar and Goldreich (SG [40]) in 1994. They challenged that part of IK thinking which viewed Alfvén wave scattering as a three wave interaction process, an assumption implicit in the IK derivation of the $k^{-3/2}$ spectrum. SG argue that, in the inertial range where amplitudes are small, significant energy exchange between Alfvén waves can only occur for resonant three wave interactions. Moreover, their argument continues, because one of the fluctuations in such a resonant triad has zero Alfvén frequency, the three wave coupling is empty. They conclude therefore that the long time dynamics of weak MHD fields are determined by four wave resonant interactions.

This conclusion is false. In this paper, we will show that resonant three wave interactions are non empty and lead to a relaxation to universal behavior and significant spectral energy redistribution. Moreover, weak turbulence theory provides a set of closed kinetic equations for the long time evolution of the eight power spectra, corresponding to total energy $e^s(\mathbf{k})$, parallel energy $\Phi^s(\mathbf{k})$, magnetic and pseudo magnetic helicities $R^s(\mathbf{k})$, $I^s(\mathbf{k})$ constructed from the Elsässer fields $\mathbf{z}^s = \mathbf{v} + s\mathbf{b}$, $s = \pm1$, where $\mathbf{v}$ and $\mathbf{b}$ are the fluctuating velocity and Alfvén velocity respectively. The latter is defined such that $\mathbf{b} = \mathbf{B}/\sqrt{\mu_0\rho_0}$, where ρ_0 is the uniform density and μ_0 the magnetic permeability. We will also show that a unique feature of Alfvén wave weak turbulence is the existence of additional conservation laws. One of the most important is the conservation of energy on all wavevector planes perpendicular to the applied field $\mathbf{B_0}$. There is no energy transfer between planes. This extra symmetry means that relaxation to universal behavior only takes place as function of $k_{\perp}$ so that, in the inertial range (or window of transparency), $e^s(\mathbf{k}) = f(k_{\parallel})k_{\perp}^{-p}$ where $f(k_{\parallel})$ is non universal.

Because weak turbulence theory for Alfvén waves is not straightforward and because of the controversy raised by SG, it is important to discuss carefully and understand clearly some of the key ideas before outlining the main results. We therefore begin by giving an overview of the theory for the statistical initial value problem for weakly nonlinear MHD fields.

1.1   Alfvén weak turbulence: the kinematics, the asymptotic closure and some results

The starting point is a kinematic description of the fields. We assume that the Elsässer fields $\mathbf{z}^s(\mathbf{x}, t)$ are random, homogeneous, zero mean fields in the three spatial coordinates **x**. This means that the n-point correlation functions between combinations of these variables estimated at $x_1, ..., x_n$ depend only on the relative geometry of the spatial configuration. We also assume that for large separation distances $|x_i - x_j|$ along **any** of the three-spatial directions, fluctuations are statistically independent. We will also discuss the case of strongly two dimensional fields for which there is significant correlation along the direction of the applied magnetic field. We choose to use cumulants rather than moments, to which the cumulants are related by a one-to-one map. The choice is made for two reasons. The first is that they are exactly those combinations of moments which are asymptotically zero for all large separations. Therefore they have well defined and, at least initially before long distance correlations can be built up by nonlinear couplings, smooth Fourier transforms. We will be particularly interested in the spectral densities

$$q_{jj'}^{ss'}(\mathbf{k}) = \frac{1}{(2\pi)^3} \int_{-\infty}^{+\infty} \langle z_j^s(\mathbf{x}) z_{j'}^{s'}(\mathbf{x}+\mathbf{r}) \rangle e^{-i\mathbf{k}\cdot\mathbf{r}} \, d\mathbf{r} \tag{1}$$

of the two point correlations. (Remember, $z_j^s(\mathbf{x})$ has zero mean so that the second order cumulants and moments are the same.) The second reason for the choice of cumulants as dependent variables is that, for joint Gaussian fields, all cumulants above second order are identically zero. Moreover, because of linear wave propagation, initial cumulants of order three and higher decay to zero in a time scale $(b_0 k_\parallel)^{-1}$ where $\mathbf{b_0} = \mathbf{B_0}/\sqrt{\mu_0 \rho_0}$ is the Alfvén velocity $(b_0 = |\mathbf{b_0}|)$ and $k_\parallel^{-1}$ a dominant parallel length scale in the initial field. This is a simple consequence of the Riemann-Lebesgue lemma; all Fourier space cumulants become multiplied by fast nonvanishing oscillations because of linear wave properties and these oscillations give rise to cancelations upon integration. Therefore, the statistics approaches a state of joint Gaussianity. The amount by which it differs, and the reason for a nontrivial relaxation of the dynamics, is determined by the long time cumulative response generated by nonlinear couplings of the waves. The special manner in which third and higher cumulants are regenerated by nonlinear processes leads to a natural asymptotic closure of the statistical initial value problem.

Basically, because of the quadratic interactions, third order cumulants (equal to third order moments) are regenerated by fourth order cumulants and binary products of second order ones. But the only long time contributions arise from a subset of the second order products which lie on certain resonant manifold defined by zero divisors. It is exactly these terms which appear in the kinetic equations which describe the evolution of the power spectra of second order moments over time scales $(\epsilon^2 b_0 k_\parallel)^{-1}$. Here, ϵ is a measure of the strength of the nonlinear coupling. Likewise, higher order cumulants are nonlinearly regenerated principally by products of lower order cumulants. Some of these small divisor terms contribute to frequency renormalization and others contribute to further (*e.g.* four wave resonant interactions) corrections of the kinetic equations over longer times.

What are the resonant manifolds for three wave interactions and, in particular, what are they for Alfvén waves? They are defined by the divisors of a system of weakly coupled wavetrains $a_j e^{i(\mathbf{k}\cdot\mathbf{x}-\omega_s(k_j)t)}$, with $\omega_s(k_j)$ the linear wave frequency, s its level of degeneracy, which undergo quadratic coupling. One finds that triads $\mathbf{k}$, $\boldsymbol{\kappa}$, $\mathbf{L}$ which lie on the resonant manifold defined for some choice of s, s', s'', by

$$\mathbf{k} = \boldsymbol{\kappa} + \mathbf{L}\,,$$
$$\omega_s(\mathbf{k}) = \omega_{s'}(\boldsymbol{\kappa}) + \omega_{s''}(\mathbf{L})\,, \tag{2}$$

interact strongly (cumulatively) over long times $(\epsilon^2 \omega_0)^{-1}$, ω_0 a typical frequency. For Alfvén waves, $\omega_s(\mathbf{k}) = s b_0 \cdot \mathbf{k} = s b_0 k_\parallel$ when $s = \pm 1$ (Alfvén waves of a given wavevector can travel in one of two directions) and b_0, the Alfvén velocity, is the strength of the applied field. Given the dispersion relation, $\omega = s b_0 \cdot \mathbf{k}$, one might ask why there is any weak turbulence for Alfvén waves at all because for, $s = s' = s''$, (2) is satisfied for all triads. Furthermore, in that case, the fast oscillations multiplying the spectral cumulants of order $N + 1$ in the evolution equation for the spectral cumulant of order N disappear so that there is no cancelation (phase mixing) and therefore no natural asymptotic closure. However, the MHD wave equations have the property that the coupling coefficient for this interaction is identically zero and therefore the only interactions of importance occur between oppositely traveling waves where $s' = -s$, $s'' = s$. In this case, (2) becomes

$$2 s b_0 \cdot \boldsymbol{\kappa} = 2 s b_0 \kappa_\parallel = 0\,. \tag{3}$$

The third wave in the triad interaction is a fluctuation with zero Alfvén frequency. SG incorrectly conclude that the effective amplitude of this zero mode is zero and that therefore the resonant three wave interactions are null.

Although some of the kinetic equations will involve principal value integral (PVI) with denominator $s\omega(\mathbf{k}) + s\omega(\boldsymbol{\kappa}) - s\omega(\mathbf{k} - \boldsymbol{\kappa}) = 2 s b_0 \kappa_\parallel$, whose meaning we discuss later, the majority of the terms contain the Dirac delta functions of this quantity. The equation for the total energy density contains only the latter implying that energy exchange takes place by resonant interactions. Both the resonant delta functions and PVI arise from taking long time limits $t \to \infty$, $\epsilon^2 t$ finite, of integrals of the form

$$\int F(k_\parallel, \epsilon^2 t) \left(e^{(2 i s b_0 k_\parallel t)} - 1 \right) (2 i s b_0 k_\parallel)^{-1}\, dk_\parallel \sim$$
$$\int F(k_\parallel, \epsilon^2 t) \left(\pi\, \mathrm{sgn}(t)\, \delta(2 s b_0 k_\parallel) + i\mathcal{P}\left(\frac{1}{2 s b_0 k_\parallel}\right) \right) dk_\parallel\,. \tag{4}$$

Therefore, implicit in the derivation of the kinetic equations is the assumption that $F(k_\parallel, \epsilon^2 t)$ is relatively smooth near $k_\parallel = 0$ so that $F(k_\parallel, \epsilon^2 t)$ remains nearly constant for $k_\parallel \sim \epsilon^2$. In particular, the kinetic equation for the total energy density

$$e^s(\mathbf{k}_\perp, k_\parallel) = \Sigma_{j=1}^3 q_{jj}^{ss}(\mathbf{k}_\perp, k_\parallel) \tag{5}$$

is the integral over $\kappa_\perp$ of a product of a combination of $q_{jj'}^{ss}(\mathbf{k}_\perp - \kappa_\perp, k_\parallel)$ with $Q^{-s}(\kappa_\perp, 0) = \Sigma_{p,m} k_p k_m\, q_{pm}^{-s-s}(\kappa_\perp, 0)$. Three observations (O1,2,3) and two questions (Q1,2) arise from this result.

O1– Unlike the cases for most systems of dispersive waves, the resonant manifolds for Alfvén waves **foliate** wavevector space. For typical dispersion relations, a wavevector κ, lying on the resonant manifold of the wavevector $\mathbf{k}$, will itself have a different resonant manifold, and members of that resonant manifold will again have different resonant manifolds. Indeed the union of all such manifolds will fill $\mathbf{k}$ space so that energy exchange occur throughout all of $\mathbf{k}$ space.

O2– In contrast, for Alfvén waves, the kinetic equations for the total energy density contains $k_\parallel$ as a parameter which identifies which wavevector plane perpendicular to $\mathbf{B}_0$ we are on. Thus the resonant manifolds for all wavevectors of a given $\bar{k}_\parallel$ is the plane $k_\parallel = \bar{k}_\parallel$. The resonant manifolds foliate $\mathbf{k}$-space.

O3– Further, conservation of total energy holds for each $k_\parallel$ plane. There is energy exchange between energy densities having the same $k_\parallel$ value but not between those having different $k_\parallel$ values. Therefore, relaxation towards a universal spectrum with constant transverse flux occurs in wavevector planes perpendicular to the applied magnetic field. The dependence of the energy density on $k_\parallel$ is nonuniversal and is inherited from the initial distribution along $k_\parallel$.

Q1– If the kinetic equation describes the evolution of power spectra for values of $k_\parallel$ outside of a band of order ϵ^ξ, $\xi < 2$, then how does one define the evolution of the quantities contained in $Q^{-s}(\kappa_\perp, 0)$ so as to close the system in $k_\parallel$?

Q2– Exactly what is $Q^{-s}(\kappa_\perp, 0)$? Could it be effectively zero as SG surmise? Could it be possibly singular with singular support located near $k_\parallel = 0$ in which case the limit (4) is suspect? To answer the crucially important question 2, we begin by considering the simpler example of a one dimensional, stationary random signal $u(t)$ of zero mean. Its power spectrum is $f(\omega)$ the limit of the sequence $f_L(\omega) = \frac{1}{2\pi} \int_{-L}^{L} \langle u(t)u(t + \tau)\rangle e^{-i\omega\tau}\, d\tau$ which exists because the integrand decays to zero as $\tau \to \pm\infty$. Ergodicity and the stationarity of $u(t)$ allows us to estimate the average $R(\tau) = \langle u(t)u(t + \tau)\rangle$ by the biased estimator

$$R_L(\tau) = \frac{1}{2L} \int_{-L+|\tau|/2}^{L-|\tau|/2} u(t - \tau/2)u(t + \tau/2)\, dt$$

with mean $E\{R_L(\tau)\} = (1 - \tau/2L)R(\tau)$. Taking L sufficiently large and assuming a sufficiently rapid decay so that we can take $R_L(\tau) = 0$ for $|\tau| > 2L$ means that $R_L(\tau)$ is simply the convolution of the signal with itself. Furthemore the Fourier transform $S_L(\omega)$ can then be evaluated as

$$S_L(\omega) = \int_{-2L}^{2L} R_T(\tau)e^{-i\omega\tau}\, d\tau = \frac{1}{2L} \left| \int_{-L}^{L} u(t)e^{-i\omega t}\, dt \right|^2.$$

For sufficiently large L, the expected value of $S_L(\omega)$ is $S(\omega)$, the Fourier transform of $R(\tau)$ although the variance of this estimate is large. Nevertheless $S_L(\omega)$, and in particular $S_L(0)$, is generally non zero and measures the power in the low frequency modes. To make the connection with Fourier space, we can think of replacing the signal $u(t)$ by the periodic extension of the truncated signal $\tilde{u}_L(t) = u(t), |t| < L ; \tilde{u}_L(t+2L)$ for $|t| > L$. The zero mode of the Fourier transform $a_L(0) = \frac{1}{2L} \int_{-L}^{L} \tilde{u}(t)\, dt$ is a nonzero random variable and, while its expected value (for large L) is zero, the expected value of its square is certainly not zero. Indeed the expected value of $S_L(0) = 2L\, a_L^2(0)$ has a finite nonzero value which, as $L \to \infty$, is independent of L as $a_L(0)$ has mean zero and a standard deviation proportional to $(2L)^{-1/2}$. Likewise for Alfvén waves, the power associated with the zero mode $Q^{-s}(\kappa_\perp, 0)$ is nonzero and furthermore, for the class of three dimensional fields in which correlations decay in all directions, $Q^{-s}(\kappa_\perp, k_\parallel)$ is smooth near $k_\parallel = 0$. Therefore, for these fields, we may consider $Q^{-s}(\kappa_\perp, 0)$ as a limit of $Q^{-s}(\kappa_\perp, k_\parallel)$ as $\frac{k_\parallel}{k_{\perp 0}} \to 0$ and $\frac{1}{\epsilon^2}\frac{k_\parallel}{k_{\perp 0}} \to \infty$. Here $k_{\perp 0}$ is some wavenumber near the energy containing part of the inertial range. Therefore, in this case, we solve first the nonlinear kinetic equation for $\lim_{k_\parallel \to 0} e_k^s(k_\perp, k_\parallel)$, namely for very oblique Alfvén waves, and having found the asymptotic time behavior of $e^s(k_\perp, 0)$, then return to solve the equation for $e^s(k_\perp, k_\parallel)$ for finite $k_\parallel$.

Assuming isotropy in the transverse $k_\perp$ plane, we find universal spectra $c_n^s k_\perp^{n_s}$ for $E^s(k_\perp)$ ($\int E^s(k_\perp, 0)\, dk_\perp = \int e^s(k_\perp)\, d\mathbf{k}_\perp$), corresponding to finite fluxes of energy from low to high transverse wavenumbers. Then $e^s(k_\perp, k_\parallel) = 2\pi f^s(k_\parallel) c_n^s k_\perp^{n_s - 1}$ where $f^s(k_\parallel)$ is not universal. These solutions each correspond to energy conservation. We find that convergence of all integrals is guaranteed for $-3 < n_s, n_{-s} < -1$ and that

$$n_+ + n_- = -4 \qquad (6)$$

which means, that for no directional preference, $n_+ = n_- = -2$. These solutions have finite energy, i.e. $\int E\, dk_\perp$ converges. If we interpret them as being set up by a constant flux of energy from a source at low $k_\perp$ to a sink at high $k_\perp$, then, since they have finite capacity and can only absorb a finite amount of energy, they must be set up in finite time. When we searched numerically for the evolution of initial states to the final state, we found a remarkable result which we yet do not fully understand. Each $E^s(k_\perp)$ behaves as a propagating front in the form $E^s(k_\perp) = (t_0 - t)^{1/2} E_0(k_\perp(t_0 - t)^{3/2})$ and $E_0(l) \sim l^{-7/3}$ as $l \to +\infty$. This means that for $t < t_0$, the $E^s(k_\perp)$ spectrum had a tail for $k_\perp < (t_0 - t)^{-3/2}$ with stationary form $k_\perp^{-7/3}$ joined to $k_\perp = 0$ through a front $E_0(k_\perp(t_0 - t)^{3/2})$. The 7/3 spectrum is steeper than the $+2$ spectrum. Amazingly, as t approached very closely to t_0, disturbances in the high $k_\perp$ part of the $k_\perp^{-7/3}$ solution propagated back along the spectrum, rapidly turning it into the finite energy flux spectrum $k_\perp^{-2}$. We neither understand the origin nor the nature of this transition solution, nor do we understand the conservation law involved with the second equilibrium solution of the kinetic equations. Dimensional analysis suggests that it is associated with the conservation of the spectral density $k_\perp^{-2/3} E^s(k)$ and that would seem to connote a nonlocal quantity in physical space. However, it is more likely determined as a condition on the front solution of the integro-differential equation. The almost constant $k_\perp^{-7/3}$ spectrum would appear to allow a nonlocal interaction whereby the energy at the developing front can be instantaneously supplied by the reservoir at small k. Once the connection to infinity is made, however the circuit between source and sink is closed and the finite flux energy spectrum takes over.

To this point we have explained how MHD turbulent fields for which correlations decay in all directions relax to quasiuniversal spectra via the scattering of high frequency Alfvén waves with very oblique, low frequency ones. But there is another class of fields that it is also important to consider. There are homogeneous, zero mean random fields which have the anisotropic property that correlations in the direction of applied magnetic field do not decay with increasing separation $\mathbf{B_0} \cdot (\mathbf{x_1} - \mathbf{x_2})$. For this case, we may think of decomposing the Elsässer fields as

$$z_j^s(\mathbf{x_\perp}, x_\|) = \bar{z}_j^s(\mathbf{x_\perp}) + \hat{z}_j^s(\mathbf{x_\perp}, x_\|) \tag{7}$$

where the $\hat{z}_j^s(\mathbf{x_\perp}, x_\|)$ have the same properties of the fields considered heretofore but where the average of $z_j^s(\mathbf{x_\perp}, x_\|)$ over $x_\|$ is nonzero. The total average of z_j^s is still zero when one averages $\bar{z}_j^s(\mathbf{x_\perp})$ over $\mathbf{x_\perp}$. In this case, it is not hard to show that correlations

$$\langle z_j^s(\mathbf{x_\perp}, x_\|) \, z_{j'}^{s'}(\mathbf{x_\perp} + \mathbf{r_\perp}, x_\| + r_\|) \rangle$$

divide into two parts

$$\langle z_j^s(\mathbf{x_\perp}) \, z_{j'}^{s'}(\mathbf{x_\perp} + \mathbf{r_\perp}) \rangle + \langle \hat{z}_j^s(\mathbf{x_\perp}, x_\|) \, \hat{z}_{j'}^{s'}(\mathbf{x_\perp} + \mathbf{r_\perp}, x_\| + r_\|) \rangle$$

with Fourier transforms,

$$q_{jj'}^{ss'}(\mathbf{k}) = \delta(k_\|) \, \bar{q}_{jj'}^{ss'}(k_\perp) + \hat{q}_{jj'}^{ss'}(k_\perp, k_\|), \tag{8}$$

when $\bar{q}$ is smooth in $\mathbf{k_\perp}$ and $\hat{q}$ is smooth in both $\mathbf{k_\perp}$ and $k_\|$. The former is simply the transverse Fourier of the two point correlations of the $x_\|$ averaged field. Likewise all higher order cumulants have delta function multipliers $\delta(k_\|)$ for each $\mathbf{k}$ dependence. For example

$$\bar{q}_{jj'j''}^{ss's''}(\mathbf{k}, \mathbf{k}') = \delta(k_\|) \, \delta(k_\|') \, \bar{q}_{jj'j''}^{ss's''}(\mathbf{k_\perp}, \mathbf{k_\perp}')$$

is the Fourier transform of

$$\langle \bar{z}_j^s(\mathbf{x_\perp}) \, \bar{z}_{j'}^{s'}(\mathbf{x_\perp} + \mathbf{r_\perp}) \, \bar{z}_{j''}^{s''}(\mathbf{x_\perp} + \mathbf{r_\perp}') \rangle.$$

Such singular dependence of the Fourier space cumulants has a dramatic effect on the dynamics especially since the singularity is supported precisely on the resonant manifold. Indeed the hierarchy of cumulant equations for $\bar{q}^{(n)}$ simply loses the fast (Alfvén) time dependence altogether and becomes fully nonlinear MHD turbulence in two dimensions with time t replaced by ϵt. Let us imagine, then, that the initial fields are dominated by this two dimensional component and that the fields have relaxed on the time scale $t \sim \epsilon^{-1}$ to their equilibrium solutions of finite energy flux for which $\bar{E}(k_\perp)$ is the initial Kolmogorov finite energy flux spectrum $k_\perp^{-5/3}$ for $k_\perp > k_0$, k_0 some input wavenumber and $\bar{E}(k_\perp) \sim k_\perp^{-1/3}$ corresponding to the inverse flux of the squared magnetic vector potential $(\mathcal{A}\,;\mathbf{B} = \nabla \times \mathcal{A})$. $\hat{\mathcal{A}}(\mathbf{k})$, the spectral density of $\langle \mathcal{A}^2 \rangle$, bahaves as $k_\perp^{-7/3}$. These are predicted from phenomenological arguments and supported by numerical simulations.

Let us then ask: how do Alfvén waves (Bragg) scatter off this two dimensional turbulent field? To answer this question, one should of course redo all the analysis taking proper account of the $\delta(k_\parallel)$ factors in $\bar{q}^{(n)}$. However, there is a simpler way. Let us imagine that the power spectra for the $\hat{z}_j^s$ fields are supported at finite $k_\parallel$ and have much smaller integrated power over an interval $0 \leq k_\parallel < \beta \ll 1$ than do the two dimensional fields. Let us replace the $\delta(k_\parallel)$ multiplying $q_{jj'}^{ss'}(\mathbf{k}_\perp)$ by a function of finite width β and height β^{-1}. Then the kinetic equation is linear and describes how the power spectra, and in particular $\hat{e}^s(\mathbf{k})$, of the $\hat{z}_j^s$ fields interact with the power spectra of the two dimensional field $\bar{z}_j^s$. Namely, the $Q^{-s}(\mathbf{k}_\perp, 0)$ field in the kinetic equation is determined by the two dimensional field and taken as known. The time scale of the interaction is now $\beta\epsilon^{-2}$, because the strength of the interaction is increased by β^{-1} and, is faster than that of pure Alfvén wave scattering. But the equilibrium of the kinetic equation will retain the property that $n_+ + n_- = -4$ where now n_{-s} is the phenomenological exponent associated with two dimensional MHD turbulence and n_s the exponent of the Alfvén waves. Note that when $n_{-s} = -5/3$, n_s is $-7/3$, which would agree with the temporary spectrum observed in the finite time transition to the $k_\perp^{-2}$ spectrum.

We now proceed to a detailed presentation of our results.

2   The derivation of the kinetic equations

The purpose in this section is to obtain closed equations for the energy and helicity spectra of weak MHD turbulence, using the fact that, in the presence of a strong uniform magnetic field, only Alfvén waves of opposite polarities propagating in opposite directions interact.

2.1 The basic equations

We will use the weak turbulence approach, the ideas of which are described in great detail in the book of Zakharov, L'vov and Falkovich (1992). There are several different ways to derive the weak turbulence kinetic equations. We follow here the technique that can be found in Benney and Newell (1969). We write the 3D incompressible MHD equations for the velocity $\mathbf{v}$ and the Alfvén velocity $\mathbf{b}$

$$(\partial_t + \mathbf{v} \cdot \nabla)\mathbf{v} = -\nabla P_* + \mathbf{b} \cdot \nabla \mathbf{b} + \nu \nabla^2 \mathbf{v}, \tag{9}$$

$$(\partial_t + \mathbf{v} \cdot \nabla)\mathbf{b} = \mathbf{b} \cdot \nabla \mathbf{v} + \eta \nabla^2 \mathbf{b}, \tag{10}$$

where P_* is the total pressure, ν the viscosity, η the magnetic diffusivity and $\nabla \cdot \mathbf{v} = 0$, $\nabla \cdot \mathbf{b} = 0$. In the absence of dissipation, these equations have three quadratic invariants in dimension three, namely the total energy $E^T = \frac{1}{2}\langle v^2 + b^2 \rangle$, the cross–correlation $E^C = \langle \mathbf{v} \cdot \mathbf{b} \rangle$ and the magnetic helicity $H^M = \langle \mathcal{A} \cdot \mathbf{B} \rangle$ [47].

The Elsässer variables $\mathbf{z}^s = \mathbf{v} + s\mathbf{b}$ with $s = \pm 1$ give these equations a more symmetrized form, namely:

$$(\partial_t + \mathbf{z}^{-s} \cdot \nabla)\mathbf{z}^s = -\nabla P_*, \tag{11}$$

where we have dropped the dissipative terms which pose no particular closure problems. The first two invariants are then simply written as $2E^s = \langle |\mathbf{z}^s|^2 \rangle$.

We now assume that there is a strong uniform magnetic induction field $\mathbf{B_0}$ along the unit vector $\hat{\mathbf{e}}_\parallel$ and non dimensionalize the equations with the corresponding magnetic induction $\mathbf{B_0}$, where the z^s fields have an amplitude proportional to ϵ ($\epsilon \ll 1$) assumed small compared to b_0. Linearizing the equations leads to

$$(\partial_t - sb_0\partial_\parallel)z_j^s = -\epsilon\partial_{x_m} z_m^{-s} z_j^s - \partial_{x_j} P_*, \tag{12}$$

where $\partial_\parallel$ is the derivative along $\hat{\mathbf{e}}_\parallel$. The frequency of the modes at a wavevector $\mathbf{k}$ is $\omega(\mathbf{k}) = \omega_k = \mathbf{b_0} \cdot \mathbf{k} = b_0 k_\parallel$. We Fourier transform the wave fields using the interaction representation,

$$z_j^s(\mathbf{x}, t) = \int A_j^s(\mathbf{k}, t)\, e^{i\mathbf{k} \cdot \mathbf{x}}\, d\mathbf{k} = \int a_j^s(\mathbf{k}, t)\, e^{i(\mathbf{k} \cdot \mathbf{x} + s\omega_k t)}\, d\mathbf{k}, \tag{13}$$

where $a_j^s(\mathbf{k}, t)$ varies slowly in time because of the weak nonlinearities; hence

$$\partial_t a_j^s(\mathbf{k}, t) = -i\epsilon k_m P_{jn} \int a_m^{-s}(\boldsymbol{\kappa})\, a_n^s(\mathbf{L})\, e^{i(-s\omega_k - s\omega_\kappa + s\omega_L)t}\delta_{\mathbf{k}, \boldsymbol{\kappa}\mathbf{L}}\, d_{\boldsymbol{\kappa}\mathbf{L}} \tag{14}$$

with $d_{\boldsymbol{\kappa}\mathbf{L}} = d\boldsymbol{\kappa}d\mathbf{L}$ and $\delta_{\mathbf{k}, \boldsymbol{\kappa}\mathbf{L}} = \delta(\mathbf{k} - \boldsymbol{\kappa} - \mathbf{L})$; finally, $P_{jn}(k) = \delta_{jn} - k_j k_n k^{-2}$ is the usual projection operator keeping the $\mathbf{A}^s(\mathbf{k})$ fields transverse to $\mathbf{k}$ because of incompressibility. The exponentially oscillating term in (14) is essential:

its exponent should not vanish when $(\mathbf{k} - \boldsymbol{\kappa} - \mathbf{L}) = 0$, *i.e.* the waves should be dispersive for the closure procedure to work. In that sense, incompressible MHD can be coined "pseudo"–dispersive because although $\omega_k \sim k$, the fact that waves of one s–polarity interact *only* with the opposite polarity has the consequence that the oscillating factor is non–zero except at resonance ; indeed with $\omega_k = b_0 k_\parallel$, one immediately sees that $-s\omega_k - s\omega_\kappa + s\omega_L = s(-k_\parallel - \kappa_\parallel + L_\parallel) = -2s\kappa_\parallel$ using the convolution constraint between the three waves in interaction. In fact, Alfvén waves may have a particularly weak form of interactions since such interactions take place only when two waves propagating in opposite directions along the lines of the uniform magnetic field meet. As will be seen later (see §3), this has the consequence that the transfer in the direction parallel to $\mathbf{B}_0$ is zero, rendering the dynamics two–dimensional, as is well known (see *e.g.* [29] [39]). Technically, we note that there are two types of waves that propagate in opposite directions, so that the classical criterion [48] for resonance to occur, *viz.* $\omega'' > 0$ does not apply here.

2.2   Toroidal and poloidal fields

The divergence–free condition implies that only two scalar fields are needed to describe the dynamics ; following classical works in anisotropic turbulence, they are taken as [10] [15] [36]

$$\mathbf{z}^s = \mathbf{z}^s_1 + \mathbf{z}^s_2 = \nabla \times (\psi^s \hat{\mathbf{e}}_\parallel) + \nabla \times (\nabla \times (\phi^s \hat{\mathbf{e}}_\parallel)) , \qquad (15)$$

which in Fourier space gives

$$A^s_j(\mathbf{k}) = i\mathbf{k} \times \hat{\mathbf{e}}_\parallel \, \hat{\psi}^s(\mathbf{k}) - \mathbf{k} \times (\mathbf{k} \times \hat{\mathbf{e}}_\parallel) \, \hat{\phi}^s(\mathbf{k}) . \qquad (16)$$

We elaborate somewhat on the significance of the $\hat{\psi}^s$ and $\hat{\phi}^s$ fields since they are the basic fields with which we shall deal. Note that $\mathbf{z}^s_1$ are two–dimensional fields with no parallel component and thus with only a vertical vorticity component (vertical means parallel to $\mathbf{B}_0$), whereas the $\mathbf{z}^s_2$ fields have zero vertical vorticity ; such a decomposition is used as well for stratified flows (see [22] and references therein). Indeed, rewriting the double cross product in (16) leads to :

$$\mathbf{A}^s(\mathbf{k}) = i\mathbf{k} \times \hat{\mathbf{e}}_\parallel \, \hat{\psi}^s(\mathbf{k}) - \mathbf{k} \, k_\parallel \hat{\phi}^s(\mathbf{k}) + \hat{\mathbf{e}}_\parallel \, k^2 \hat{\phi}^s(\mathbf{k}) \qquad (17)$$

or using $\mathbf{k} = \mathbf{k}_\perp + k_\parallel \hat{\mathbf{e}}_\parallel$:

$$\mathbf{A}^s(\mathbf{k}) = i\mathbf{k} \times \hat{\mathbf{e}}_\parallel \, \hat{\psi}^s(\mathbf{k}) - \mathbf{k}_\perp k_\parallel \, \hat{\phi}^s(\mathbf{k}) + \hat{\mathbf{e}}_\parallel \, k^2_\perp \hat{\phi}^s(\mathbf{k}) . \qquad (18)$$

The above equations indicate the relationships between the two orthogonal systems (with $\mathbf{p} = \mathbf{k} \times \hat{\mathbf{e}}_\parallel$ and $\mathbf{q} = \mathbf{k} \times \mathbf{p}$) made of the triads $(\mathbf{k}, \mathbf{p}, \mathbf{q})$,

$(\hat{\mathbf{e}}_\parallel, \mathbf{p}, \mathbf{k}_\perp)$ and the system $(\hat{\mathbf{e}}_\parallel, \mathbf{p}, \mathbf{k})$. In terms of the decomposition used in [45] with

$$\mathbf{h}_\pm = \mathbf{p} \times \mathbf{k} \pm i\mathbf{p} \tag{19}$$

and writing $\mathbf{z}^s = A_+^s \mathbf{h}_+ + A_-^s \mathbf{h}_-$, it can be shown easily that $\psi^s = A_+^s - A_-^s$ and $\phi^s = A_+^s + A_-^s$. In these latter variables, the s–energies E^s are proportional to $\langle |A_+^s|^2 + |A_-^s|^2 \rangle$ and the s–helicities $\langle \mathbf{z}^s \cdot \nabla \times \mathbf{z}^s \rangle$ are proportional to $\langle |A_+^s|^2 - |A_-^s|^2 \rangle$. Note that E^s are not scalars: when going from a right-handed to a left-handed frame of reference, E^s changes into E^{-s}.

2.3 Moments and cumulants

We now seek a closure for the energy tensor $q_{jj'}^{ss'}(\mathbf{k})$ defined as

$$\langle a_j^s(\mathbf{k})\, a_{j'}^{s'}(\mathbf{k}') \rangle \equiv q_{jj'}^{ss'}(\mathbf{k}')\, \delta(\mathbf{k}+\mathbf{k}') \tag{20}$$

in terms of second order moments of the two scalars fields $\hat{\psi}^s(\mathbf{k})$ and $\hat{\phi}^s(\mathbf{k})$. Simple manipulations lead, with the restriction $s = s'$ (it can be shown that correlations with $s' = -s$ have no long time influence and therefore are, for convenience of exposition, omitted), to:

$$
\begin{aligned}
q_{11}^{ss}(\mathbf{k}') &= k_2^2 \Psi^s(\mathbf{k}) - k_1 k_2 k_\parallel I^s(\mathbf{k}) + k_\parallel^2 k_1^2 \Phi^s(\mathbf{k}), \\
q_{22}^{ss}(\mathbf{k}') &= k_1^2 \Psi^s(\mathbf{k}) + k_1 k_2 k_\parallel I^s(\mathbf{k}) + k_\parallel^2 k_2^2 \Phi^s(\mathbf{k}), \\
q_{12}^{ss}(\mathbf{k}') + q_{21}^{ss}(\mathbf{k}') &= -2k_1 k_2 \Psi^s(\mathbf{k}) + k_\parallel(k_1^2 - k_2^2) I^s(\mathbf{k}) + 2k_1 k_2 k_\parallel^2 \Phi^s(\mathbf{k}), \\
q_{1\parallel}^{ss}(\mathbf{k}') + q_{\parallel 1}^{ss}(\mathbf{k}') &= k_2 k_\perp^2 I^s(\mathbf{k}) - 2k_1 k_\parallel k_\perp^2 \Phi^s(\mathbf{k}), \\
q_{2\parallel}^{ss}(\mathbf{k}') + q_{\parallel 2}^{ss}(\mathbf{k}') &= -k_1 k_\perp^2 I^s(\mathbf{k}) - 2k_2 k_\parallel k_\perp^2 \Phi^s(\mathbf{k}), \\
q_{\parallel\,\parallel}^{ss}(\mathbf{k}') &= k_\perp^4 \Phi^s(\mathbf{k}), \\
\tfrac{1}{k_1}[q_{2\parallel}^{ss}(\mathbf{k}') - q_{\parallel 2}^{ss}(\mathbf{k}')] = \tfrac{1}{k_2}[q_{\parallel 1}^{ss}(\mathbf{k}') &- q_{1\parallel}^{ss}(\mathbf{k}')] = \tfrac{1}{k_\parallel}[q_{12}^{ss}(\mathbf{k}') - q_{21}^{ss}(\mathbf{k}')] \\
&= -ik_\perp^2 R^s(\mathbf{k}),
\end{aligned}
\tag{21}
$$

where the following correlators involving the toroidal and poloidal fields have been introduced:

$$
\begin{aligned}
\langle \hat{\psi}^s(\mathbf{k})\hat{\psi}^s(\mathbf{k}') \rangle &= \delta(\mathbf{k}+\mathbf{k}')\Psi^s(\mathbf{k}'), \\
\langle \hat{\phi}^s(\mathbf{k})\hat{\phi}^s(\mathbf{k}') \rangle &= \delta(\mathbf{k}+\mathbf{k}')\Phi^s(\mathbf{k}'), \\
\langle \hat{\psi}^s(\mathbf{k})\hat{\phi}^s(\mathbf{k}') \rangle &= \delta(\mathbf{k}+\mathbf{k}')\Pi^s(-\mathbf{k}), \\
\langle \hat{\phi}^s(\mathbf{k})\hat{\psi}^s(\mathbf{k}') \rangle &= \delta(\mathbf{k}+\mathbf{k}')\Pi^s(\mathbf{k}), \\
R^s(\mathbf{k}) &= \Pi^s(-\mathbf{k}) + \Pi^s(\mathbf{k}), \\
I^s(\mathbf{k}) &= i[\Pi^s(-\mathbf{k}) - \Pi^s(\mathbf{k})],
\end{aligned}
\tag{22}
$$

and where $k_\perp^2 = k_1^2 + k_2^2$, $k^2 = k_\perp^2 + k_\parallel^2$. Note that $\Sigma_s R^s$ is the only pseudo-scalar, linked to the possible lack of symmetry of the equations under plane reversal, *i.e.* to a non–zero helicity.

The density energy spectrum writes

$$e^s(\mathbf{k}) = \Sigma_j \, q_{jj}^{ss}(\mathbf{k}) = \mathbf{k}_\perp^2 \left(\Psi^s(\mathbf{k}) + k^2 \Phi^s(\mathbf{k}) \right) . \tag{23}$$

Note that it can be shown easily that the kinetic and magnetic energies $\frac{1}{2}\langle u^2 \rangle$ and $\frac{1}{2}\langle b^2 \rangle$ are equal in the context of the weak turbulence approximation. Similarly, expressing the magnetic induction as a combination of $\mathbf{z}^\pm$ and thus of $\hat{\psi}^\pm$ and $\hat{\phi}^\pm$, the following symmetrized cross–correlator of magnetic helicity (where the Alfvén velocity is used for convenience) and its Fourier transform are found to be

$$\frac{1}{2}\langle \hat{A}_j(\mathbf{k}) \, b_j(\mathbf{k}') \rangle + \frac{1}{2}\langle \hat{A}_j(\mathbf{k}') \, b_j(\mathbf{k}) \rangle = \frac{1}{4}\mathbf{k}_\perp^2 \, \Sigma_s R^s(\mathbf{k}) \, \delta(\mathbf{k} + \mathbf{k}') , \tag{24}$$

where the correlations between the $+$ and $-$ variables are ignored because they are exponentially damped in the approximation of weak turbulence. Similarly to the case of energy, there is equivalence between the kinetic and magnetic helical variables in that approximation, hence the kinetic helicity defined as $\langle \mathbf{u} \cdot \boldsymbol{\omega} \rangle$ writes simply in terms of its spectral density $H^V(\mathbf{k})$:

$$H^V(\mathbf{k}) = k^2 H^M(\mathbf{k}) = \frac{1}{4} k^2 \mathbf{k}_\perp^2 \, \Sigma_s R^s(\mathbf{k}) . \tag{25}$$

In summary, the eight fundamental spectral density variables for which we seek a weak turbulence closure are the energy $e^s(\mathbf{k})$ of the three components of the $\mathbf{z}^s$ fields, the energy density along the direction of the uniform magnetic field $\Phi^s(\mathbf{k})$, the correlators related to the off–diagonal terms of the spectral energy density tensor $I^s(\mathbf{k})$ and finally the only helicity–related pseudo–scalar correlators, namely $R^s(\mathbf{k})$.

The main procedure that leads to a closure of weak turbulence for in-compressible MHD is outlined in the Appendix. It leads to the equations (82) giving the temporal evolution of the components of the spectral tensor $q_{jj'}^{ss}(\mathbf{k})$ just defined. The last technical step consists in transforming equations (82) of the Appendix in terms of the eight correlators we defined above. This leads us to the final set of equations, constituting the kinetic equations for weak MHD turbulence.

2.4   The kinetic equations

In the general case the kinetic equations for weak MHD turbulence are

$$\partial_t e^s(\mathbf{k}) = \frac{\pi \varepsilon^2}{b_0} \int \left[\left(L_\perp^2 - \frac{X^2}{k^2} \right) \Psi^s(\mathbf{L}) - \left(k_\perp^2 - \frac{X^2}{L^2} \right) \Psi^s(\mathbf{k}) \right.$$

$$+ \left(L_\perp^2 L^2 - \frac{k_\parallel^2 W^2}{k^2} \right) \Phi^s(\mathbf{L}) - \left(k_\perp^2 k^2 - \frac{k_\parallel^2 Y^2}{L^2} \right) \Phi^s(\mathbf{k})$$

$$\left. + \left(\frac{k_\parallel XY}{L^2} \right) I^s(\mathbf{k}) - \left(\frac{k_\parallel XW}{k^2} \right) I^s(\mathbf{L}) \right] Q_k^{-s}(\boldsymbol{\kappa}) \delta(\kappa_\parallel) \delta_{\mathbf{k},\boldsymbol{\kappa}\mathbf{L}} \, d_{\boldsymbol{\kappa}\mathbf{L}} \tag{26}$$

$$\partial_t \left[k_\perp^2 k^2 \Phi^s(\mathbf{k}) \right] = \frac{\pi \varepsilon^2}{b_0} \int \left[k_\parallel^2 X^2 \left(\frac{\Psi^s(\mathbf{L})}{k_\perp^2 k^2} - \frac{\Phi^s(\mathbf{k})}{L_\perp^2} \right) \right.$$

$$+ \left(k_\parallel^2 Z + k_\perp^2 L_\perp^2 \right)^2 \left(\frac{\Phi^s(\mathbf{L})}{k_\perp^2 k^2} - \frac{\Phi^s(\mathbf{k})}{L_\perp^2 L^2} \right) + \frac{k_\parallel X}{k_\perp^2 k^2} \left(k_\parallel^2 Z + k_\perp^2 L_\perp^2 \right) I^s(\mathbf{L})$$

$$\left. + \left(\frac{k_\parallel XY}{2L^2} \right) I^s(\mathbf{k}) \right] Q_k^{-s}(\boldsymbol{\kappa}) \, \delta(\kappa_\parallel) \, \delta_{\mathbf{k},\boldsymbol{\kappa}\mathbf{L}} \, d_{\boldsymbol{\kappa}\mathbf{L}}$$

$$- \frac{\varepsilon^2}{b_0} s R^s(\mathbf{k}) \mathcal{P} \int \frac{X}{2\kappa_\parallel L^2} \left(k_\parallel Z - L_\parallel k_\perp^2 \right) Q_k^{-s}(\boldsymbol{\kappa}) \, \delta_{\mathbf{k},\boldsymbol{\kappa}\mathbf{L}} \, d_{\boldsymbol{\kappa}\mathbf{L}} \tag{27}$$

$$\partial_t \left[k_\perp^2 R^s(\mathbf{k}) \right] = -\frac{\pi \varepsilon}{b_0} \int \left[L_\perp^2 \left(\frac{Z + k_\parallel^2}{k^2} \right) R^s(\mathbf{L}) \right.$$

$$\left. + \frac{k_\perp^2}{2} \left(1 + \frac{(Z + k_\parallel^2)^2}{k^2 L^2} \right) R^s(\mathbf{k}) \right] Q_k^{-s}(\boldsymbol{\kappa}) \delta(\kappa_\parallel) \, \delta_{\mathbf{k},\boldsymbol{\kappa}\mathbf{L}} \, d_{\boldsymbol{\kappa}\mathbf{L}}$$

$$+ \frac{\varepsilon^2}{b_0} s \mathcal{P} \int \left[2X \left(k_\parallel Z - L_\parallel k_\perp^2 \right) \left(\Psi^s(\mathbf{k}) + k^2 \Phi^s(\mathbf{k}) \right) \right.$$

$$\left. + \left(\left(k_\parallel Z - L_\parallel k_\perp^2 \right)^2 - k^2 X^2 \right) I^s(\mathbf{k}) \right] \frac{Q_k^{-s}(\boldsymbol{\kappa})}{2\kappa_\parallel k^2 L^2} \, \delta_{\mathbf{k},\boldsymbol{\kappa}\mathbf{L}} d_{\boldsymbol{\kappa}\mathbf{L}} \tag{28}$$

$$\partial_t \left[k_\perp^2 k^2 I^s(\mathbf{k}) \right] = \frac{\pi \varepsilon^2}{b_0} \int \left[\left(L_\perp^2 Z + \frac{k_\parallel^2}{k_\perp^2} (Z^2 - X^2) \right) I^s(\mathbf{L}) \right.$$

$$+ \left(\frac{k_\parallel^2 Y^2}{2L^2} - k_\perp^2 k^2 + \frac{k^2 X^2}{2L^2} \right) I^s(\mathbf{k}) \left(\frac{k_\parallel XY}{L^2} \right) \left(\Psi^s(\mathbf{k}) + k^2 \Phi^s(\mathbf{k}) \right)$$

$$\left. + \frac{2k_\parallel X}{k_\perp^2} \left(Z\Psi^s(\mathbf{L}) - \left(k_\parallel^2 Z + k_\perp^2 L_\perp^2 \right) \Phi^s(\mathbf{L}) \right) \right] Q_k^{-s}(\boldsymbol{\kappa}) \delta(\kappa_\parallel) \, \delta_{\mathbf{k},\boldsymbol{\kappa}\mathbf{L}} \, d_{\boldsymbol{\kappa}\mathbf{L}}$$

$$- \frac{\varepsilon^2}{b_0} s R^s(\mathbf{k}) \mathcal{P} \int \frac{1}{2\kappa_\parallel L^2} \left((k_\parallel Z - L_\parallel k_\perp^2)^2 - k^2 X^2 \right) Q_k^{-s}(\boldsymbol{\kappa}) \, \delta_{\mathbf{k},\boldsymbol{\kappa}\mathbf{L}} \, d_{\boldsymbol{\kappa}\mathbf{L}} \tag{29}$$

with

$$\delta_{\mathbf{k},\boldsymbol{\kappa}\mathbf{L}} = \delta(\mathbf{L} + \boldsymbol{\kappa} - \mathbf{k}), \quad d_{\boldsymbol{\kappa}\mathbf{L}} = d\boldsymbol{\kappa} \, d\mathbf{L},$$

and

$$Q_k^{-s}(\kappa) = k_m k_p \, q_{p \; m}^{-s-s}(\kappa)$$
$$= X^2 \Psi^{-s}(\kappa) + X(k_3 \kappa_\perp^2 - \kappa_3 Y) I^{-s}(\kappa) + (\kappa_3 Y - k_3 \kappa_\perp^2)^2 \phi^{-s}(\kappa).$$

Note that Q_k^{-s} does not involve the spectral densities $R^s(\mathbf{k})$, because of symmetry properties of the equations. The geometrical coefficients appearing in the kinetic equations are

$$X = (\mathbf{k}_\perp \wedge \boldsymbol{\kappa}_\perp)_z = k_\perp \kappa_\perp \sin \theta, \tag{30a}$$
$$Y = \mathbf{k}_\perp \cdot \boldsymbol{\kappa}_\perp = k_\perp \kappa_\perp \cos \theta, \tag{30b}$$
$$Z = \mathbf{k}_\perp \cdot \mathbf{L}_\perp = k_\perp^2 - k_\perp \kappa_\perp \cos \theta = k_\perp^2 - Y, \tag{30c}$$
$$W = \boldsymbol{\kappa}_\perp \cdot \mathbf{L}_\perp = \kappa_\perp^2 - L_\perp^2 - k_\perp \kappa_\perp \cos \theta = Z - L_\perp^2, \tag{30d}$$

where θ is the angle between $\mathbf{k}_\perp$ and $\boldsymbol{\kappa}_\perp$, and with

$$d\boldsymbol{\kappa}_\perp = \kappa_\perp d\kappa_\perp d\theta = \frac{L_\perp}{k_\perp \sin \theta} \, d\kappa_\perp dL_\perp, \tag{31}$$

$$\cos \theta = \frac{\kappa_\perp^2 + k_\perp^2 - L_\perp^2}{2\kappa_\perp k_\perp}. \tag{32}$$

In (27), (28) and (29) $\mathcal{P} \int$ means the Cauchy Principal value of the integral.

3   General properties of the kinetic equations

3.1   Dynamical decoupling in the direction parallel to $\mathbf{B}_0$

The integral on the right-hand side of the kinetic equation (26) contains a delta function of the form $\delta(\kappa_\parallel)$, the integration variable corresponding to the parallel component of one of the wavenumbers in the interacting triad. This delta function arises because of the three-wave frequency resonance condition. Thus, in any resonantly interacting wave triad $(\mathbf{k}, \boldsymbol{\kappa}, \mathbf{L})$, there is always one wave that corresponds to a purely 2D motion – having no dependence on the direction parallel to the uniform magnetic field – whereas the other two waves have equal parallel components of their corresponding wavenumbers, viz. $L_\parallel = k_\parallel$. This means that the parallel components of the wavenumber enter in the kinetic equation of the total energy $e^s(\mathbf{k})$ as an external parameter and that the dynamics is decoupled at each level of $k_\parallel$. In other words, there is no transfer associated with the three-wave resonant interaction along the $k_\parallel$-direction in $\mathbf{k}$-space for the total energy. This result, using the exact kinetic equations developed here, corroborates what has already been found in [29] using a phenomenological analysis of the basic MHD equations, and in [31] [32] in the framework of a model of weak MHD turbulence using individual wave packets.

As for the kinetic equation (26), the other kinetic equations (27) to (29) have integrals containing delta functions of the form $\delta(\kappa_\parallel)$. But, in addition, they have PVIs which can, *a priori*, contribute to a transfer in the parallel direction. The eventual contributions of these PVIs are discussed in §3.4.

3.2   Detailed energy conservation

Detailed conservation of energy for each interacting triad of waves is a usual property in weak turbulence theory. This property is closely related with the frequency resonance condition

$$\omega_{\mathbf{k}} = \omega_{\mathbf{L}} + \omega_{\boldsymbol{\kappa}},$$

because ω can be interpreted as the energy of one wave "quantum". For Alfvén waves, the detailed energy conservation property is even stronger because one of the waves in any resonant triad belongs to the 2D state with frequency equal to zero,

$$\omega_{\boldsymbol{\kappa}} \propto \kappa_\parallel = 0.$$

Thus, for every triad of Alfvén waves $\mathbf{k}, \mathbf{L}$ and $\boldsymbol{\kappa}$ (such that $\kappa_\parallel = 0$) the energy is conserved within two co-propagating waves having wavevectors $\mathbf{k}$ and $\mathbf{L}$. Mathematically, this corresponds to the symmetry of the integrand in the equation for e^s with respect to changing $\mathbf{k} \leftrightarrow \mathbf{L}$ (and correspondingly $\boldsymbol{\kappa} = \mathbf{k} - \mathbf{L} \to -\boldsymbol{\kappa}$).

As we have said, energy is conserved $k_\parallel$ plane by $k_\parallel$ plane so that, for each $k_\parallel$, it can be shown from (26)

$$\frac{\partial}{\partial t} \int e^s(\mathbf{k}_\perp, k_\parallel)\, d\mathbf{k}_\perp = 0. \tag{33}$$

3.3   Properties of R^s and I^s

Although R^s and I^s evolve according to their own kinetic equations (29) and (28), the range of values they can take on is bounded by Ψ^s and Φ^s, with the bounds being a simple consequence of the definition of these quantities. The correlators $R^s(\mathbf{k})$ and $I^s(\mathbf{k})$ have been defined in the previous section as the real part and the imaginary part of $\Pi^s(\mathbf{k})$, the cross–correlator of the toroidal field $\hat{\psi}^s(\mathbf{k})$ and of the poloidal field $\hat{\phi}^s(\mathbf{k})$. Then $\mathbf{k}_\perp^2 \Sigma_s R^s(\mathbf{k})$ appears as the spectral density of the magnetic helicity, the integral of which is a conserved quantity in three dimensions (see [42] for the case with a mean magnetic field). On the other hand $I^s(\mathbf{k})$, which we will call the anisotropy correlator (or pseudo-helicity), is not a conserved quantity. Neither is it positive definite.

Two realizability conditions (see also [8]) between the four correlators Ψ^s, Φ^s, I^s and R^s can be obtained from

$$\langle\, |\hat{\psi}^s(\mathbf{k}) \pm k\hat{\phi}^s(\mathbf{k})|^2\,\rangle \geq 0\,, \tag{34}$$

and

$$\langle\, |\hat{\psi}^s(\mathbf{k})|^2\,\rangle\langle\, |\hat{\phi}^s(\mathbf{k})|^2\,\rangle \geq |\langle\, \hat{\psi}^s(\mathbf{k})\hat{\phi}^s(-\mathbf{k})\,\rangle|^2\,. \tag{35}$$

These conditions are found to be respectively

$$\Psi^s(\mathbf{k}) + k^2\Phi^s(\mathbf{k}) \geq |kR^s(\mathbf{k})|\,, \tag{36}$$

and

$$4\Psi^s(\mathbf{k})\Phi^s(\mathbf{k}) \geq R^{s\,2}(\mathbf{k}) + I^{s\,2}(\mathbf{k})\,. \tag{37}$$

Note that the combination

$$Z = (1/2)k_\perp^2[k^2\Phi(\mathbf{k}) - \Psi(\mathbf{k}) - i|k|I(\mathbf{k})]$$

is named polarization anisotropy in [8].

3.4   Purely 2D modes and two-dimensionalisation of 3D spectra

The first consequence of the fact that there is no transfer of the total energy in the $k_\parallel$ direction in $\mathbf{k}$-space is an asymptotic two-dimensionalisation of the energy spectrum $e^s(\mathbf{k})$. Namely, the 3D initial spectrum spreads over the transverse wavenumbers, $\mathbf{k}_\perp$, but remains of the same size in the $k_\parallel$ direction, and the support of the spectrum becomes very flat (pancake-like) for large time. The two-dimensionalisation of weak MHD turbulence has been observed in laboratory experiments [37] [49], and in the the solar wind data [4] [3] [16] [7] and in many direct numerical simulations of the three–dimensional MHD equations [33].

From the mathematical point of view, the two-dimensionalisation of the total energy means that, for large time, the energy spectrum $e^s(\mathbf{k})$ is supported on a volume of wavenumbers such that for most of them $k_\perp \gg k_\parallel$. This implies $\Psi^s(\mathbf{k})$ and $\Phi^s(\mathbf{k})$ are also supported on the same anisotropic region of wavenumbers because both of them are non-negative. This, in turn, implies that both R^s and I^s will also be non-zero only for the same region in the $\mathbf{k}$-space as $e^s(\mathbf{k})$, $\Psi^s(\mathbf{k})$ and $\Phi^s(\mathbf{k})$, as it follows from the bound (36) and (37). This fact allows one to expand the integrands in the kinetic equations in powers of small $k_{\parallel}/k_{\perp}$. At the leading order in $k_{\parallel}/k_{\perp}$, one obtains

$$\partial_t \left[k_{\perp}^2 \Psi^s(\mathbf{k}) \right] =$$

$$\frac{\pi \varepsilon^2}{b_0} \int \left[\left(L_{\perp}^2 - \frac{X^2}{k_{\perp}^2} \right) \Psi^s(\mathbf{L}) - \left(k_{\perp}^2 - \frac{X^2}{L_{\perp}^2} \right) \Psi^s(\mathbf{k}) \right] X^2 \Psi^{-s}(\boldsymbol{\kappa})\ \delta(\kappa_{\parallel})\ \delta_{\mathbf{k},\boldsymbol{\kappa}\mathbf{L}}\ d_{\boldsymbol{\kappa}\mathbf{L}}\ ,$$

$$(38)$$

$$\partial_t \left[k_{\perp}^4 \Phi^s(\mathbf{k}) \right] = \frac{\pi \varepsilon^2}{b_0} \int \left[L_{\perp}^4 \Phi^s(\mathbf{L}) - k_{\perp}^4 \Phi^s(\mathbf{k}) \right] X^2 \Psi^{-s}(\boldsymbol{\kappa})\ \delta(\kappa_{\parallel})\ \delta_{\mathbf{k},\boldsymbol{\kappa}\mathbf{L}}\ d_{\boldsymbol{\kappa}\mathbf{L}}\ ,$$

$$(39)$$

$$\partial_t \left[k_{\perp}^2 R^s(\mathbf{k}) \right] =$$

$$-\frac{\pi \varepsilon^2}{b_0} \int \left[\left(\frac{L_{\perp}^2 Z}{k_{\perp}^2} \right) R^s(\mathbf{L}) + \left(\frac{k_{\perp}^2}{2} + \frac{Z^2}{2L_{\perp}^2} \right) R^s(\mathbf{k}) \right] X^2 \Psi^{-s}(\boldsymbol{\kappa})\ \delta(\kappa_{\parallel})\ \delta_{\mathbf{k},\boldsymbol{\kappa}\mathbf{L}}\ d_{\boldsymbol{\kappa}\mathbf{L}}\ ,$$

$$(40)$$

$$\partial_t \left[k_{\perp}^2 I^s(\mathbf{k}) \right] =$$

$$\frac{\pi \varepsilon^2}{b_0} \int \left[\left(\frac{L_{\perp}^2 Z}{k_{\perp}^2} \right) I^s(\mathbf{L}) - \left(k_{\perp}^2 - \frac{X^2}{2L_{\perp}^2} \right) I^s(\mathbf{k}) \right] X^2 \Psi^{-s}(\boldsymbol{\kappa})\ \delta(\kappa_{\parallel})\ \delta_{\mathbf{k},\boldsymbol{\kappa}\mathbf{L}}\ d_{\boldsymbol{\kappa}\mathbf{L}}\ .$$

$$(41)$$

Note that the principal value terms drop out of the kinetic equations at leading order. This property means that there is no transfer of any of the eight correlators in the $k_{\parallel}$ direction in $\mathbf{k}$-space.

One can see from the above that the equations for the transverse and parallel energies decouple for large time. These equations describe the shear-Alfvén and pseudo-Alfvén waves respectively. Moreover, in the large time limit, the magnetic helicity $\Sigma_s R^s$ and the polarisation correlator I^s are also described by equations which are decoupled from each other and from the parallel and transverse energies. It is interesting that the kinetic equation for the shear-Alfvén waves (*i.e.* for the transverse energy) can be obtained also from the reduced MHD equations (or RMHD) which have been derived under the same conditions of quasi two-dimensionality (see *e.g.* [41]).

An important consequence of the dynamical decoupling at different $k_{\parallel}$'s within the kinetic equation formalism is that the set of purely 2D modes (corresponding to $k_{\parallel} = 0$) evolve independently of the 3D part of the spectrum (with $k_{\parallel} \neq 0$) and can be studied separately. One can interpret this fact as a neutral stability of the purely 2D state with respect to 3D perturbations. As we mentioned in the Introduction, the kinetic equations themselves are applicable to a description of $k_{\parallel} = 0$ modes only if the correlations of the dynamical fields decay in all directions, so that their spectra are sufficiently smooth for all wavenumbers including the ones with $k_{\parallel} = 0$. To be precise, the characteristic $k_{\parallel}$ over which the spectra can experience significant changes must be greater than ϵ^2. Study of such 2D limits of 3D spectrum will be presented in the next section. It is possible, however, that in some physical situations the correlations decay slowly along the magnetic field due to a (hypothetical) energy condensation at the $k_\parallel = 0$ modes. In this case, the modes with $k_\parallel = 0$ should be treated as a separate component, a condensate, which modifies the dynamics of the 3D modes in a manner somewhat similar to the superfluid condensate, as described by Bogoliubov [20]. We leave this problem for future study, but we give some thoughts about it in the conclusion.

4   Asymptotic solution of the 3D kinetic equations

The parallel wavenumber $k_\parallel$ enters equations (38)-(41) only as an external parameter. In other words, the wavenumber space is foliated into the dynamically decoupled planes $k_\parallel = 0$. Thus, the large-time asymptotic solution can be found in the following form,

$$\Phi^s(\mathbf{k}_\perp, k_\parallel) = f_1(k_\parallel)\Phi^s(\mathbf{k}_\perp, 0), \tag{42}$$

$$\Psi^s(\mathbf{k}_\perp, k_\parallel) = f_2(k_\parallel)\Psi^s(\mathbf{k}_\perp, 0), \tag{43}$$

$$R^s(\mathbf{k}_\perp, k_\parallel) = f_3(k_\parallel)R^s(\mathbf{k}_\perp, 0), \tag{44}$$

$$I^s(\mathbf{k}_\perp, k_\parallel) = f_4(k_\parallel)I^s(\mathbf{k}_\perp, 0), \tag{45}$$

where f_i, $(i = 1, 2, 3, 4)$ are some arbitrary functions of $k_\parallel$ satisfying the conditions $f_i(0) = 1$ (and such that the bounds (37) and (36) are satisfied). Substituting these formulae into (38)-(41), one can readily see that the functions f_i drop out of the problem, and the solution of the 3D equations is reduced to solving a 2D problem for $\Phi^s(\mathbf{k}_\perp, 0), \Psi^s(\mathbf{k}_\perp, 0), R^s(\mathbf{k}_\perp, 0)$ and $I^s(\mathbf{k}_\perp, 0)$, which will be described in the next section.

5   Two-dimensional problem

Let us consider Alfven wave turbulence which is axially symmetric with respect to the external magnetic field. Then $I^s(k_\perp, 0) = 0$ because of the condition $I^s(-\mathbf{k}) = I^s(\mathbf{k})$. In the following, we will consider only solutions with $R^s = 0$. (One can easily see that R^s will remain zero if it is zero initially.) The remaining equations to be solved are

$$\partial_t E_\perp^s(k_\perp, 0) =$$

$$\frac{\pi\varepsilon^2}{b_0} \int (\hat{\mathbf{e}}_L \cdot \hat{\mathbf{e}}_k)^2 \sin\theta \, \frac{k_\perp}{\kappa_\perp} \, E_\perp^{-s}(\kappa_\perp, 0) \, [k_\perp E_\perp^s(L_\perp, 0) - L_\perp E_\perp^s(k_\perp, 0)]d\kappa_\perp dL_\perp, \tag{46}$$

$$\partial_t E_\parallel^s(k_\perp, 0) =$$

$$\frac{\pi\varepsilon^2}{b_0} \int \sin\theta \, \frac{k_\perp}{\kappa_\perp} \, E_\perp^{-s}(\kappa_\perp, 0) \, [k_\perp E_\parallel^s(L_\perp, 0) - L_\perp E_\parallel^s(k_\perp, 0)]d\kappa_\perp dL_\perp, \tag{47}$$

where $\hat{\mathbf{e}}_k$ and $\hat{\mathbf{e}}_L$ are the unit vectors along $\mathbf{k}_\perp$ and $\mathbf{L}_\perp$ respectively and

$$E_\perp^s(k_\perp, 0) = k_\perp^3 \Psi^s(k_\perp, 0) \ , \tag{48}$$

$$E_\parallel^s(k_\perp, 0) = k_\perp^5 \Phi^s(k_\perp, 0) \ , \tag{49}$$

are the horizontal and the vertical components of the energy density. The equation (46) corresponds to the evolution of the shear-Alfvén waves for which the energy fluctuations are transverse to $\mathbf{B}_0$ whereas equation (47) describes the pseudo-Alfvén waves for which the fluctuations are along $\mathbf{B}_0$. Both waves propagate along $\mathbf{B}_0$ at the same Alfvén speed. Equation (46) describes interaction between two shear-Alfvén waves, $E_\perp^s$ and $E_\perp^{-s}$, propagating in opposite directions. On the other hand, the evolution of the pseudo-Alfvén waves depend on their interactions with the shear-Alfvén waves. The detailed energy conservation of the equation (46) implies that there is no exchange of energy between the two different kinds of waves. The physical picture in this case is that the shear-Alfvén waves interact only among themselves and evolve independently of the pseudo-Alfvén waves. The pseudo-Alfvén waves scatter from the shear-Alfvén waves without amplification or damping and they do not interact with each other.

Using a standard two–point closure of turbulence (see *e.g.* [22]) in which the characteristic time of transfer of energy is assumed known and written *a priori*, namely the EDQNM closure, Goldreich and Sridhar [13] derived a variant of the kinetic equation (46) but for strong MHD turbulence. In their analysis, the ensuing energy spectrum, which depends (as it is well known) on the phenomenological evaluation of the characteristic transfer time, thus differs from our result where the dynamics is self–consistent, closure being obtained through the assumption of weak turbulence.

It can be easily verified that the geometrical coefficient appearing in the closure equation in [13] is identical to the one we find for the $E_\perp^s(\mathbf{k}_\perp, k_\parallel)$ spectrum in the two–dimensional case. However, the two formulations, beyond the above discussion on characteristic time scales, differ in a number of ways: (i) We choose to let the flow variables to be non mirror–symmetric, whereas helicity is not taken into account in [13] where they have implicitly assumed $R^s \equiv 0$; (ii) However, because of the anisotropy introduced by the presence of a uniform magnetic field, one must take into account the coupled dynamics of the energy of the shear Alfvén waves, the pseudo–Alfvén wave and the anisotropy correlator I^s ; indeed, even if initially $I^s \equiv 0$, it is produced by wave coupling and is part of the dynamics. (iii) In three dimensions, all geometrical coefficients that depend on $k^2 = k_\perp^2 + k_\parallel^2$ have a $k_\parallel$–dependence which is a function of initial conditions and again is part of the dynamics.

5.1   Kolmogorov spectra

The Zakharov transformation. The symmetry of the previous equations allows us to perform a conformal transformation, called the Zakharov transformation (also used in modeling of strong turbulence, see [22]), in order to find the exact stationary solutions of the kinetic equations as power laws [48]. This operation (see Figure 1) consists of writing the kinetic equations in dimensionless variables $\omega_1 = \kappa_\perp/k_\perp$ and $\omega_2 = L_\perp/k_\perp$, setting $E_\perp^+$ and $E_\perp^-$ by $k_\perp^{n_\bullet}$ and $k_\perp^{n_{-\bullet}}$ respectively, and then rearranging the collision integral by the transformation

$$\omega_1' = \frac{\omega_1}{\omega_2} , \tag{50}$$

$$\omega_2' = \frac{1}{\omega_2} . \tag{51}$$

The new form of the collision integral, resulting from the summation of the integrand in its primary form and after the Zakharov transformation, is

$$\partial_t E_\perp^s \sim \int \left(\frac{\omega_2^2 + 1 - \omega_1^2}{2\omega_2}\right)^2 \left(1 - (\frac{\omega_1^2 + 1 - \omega_2^2}{2\omega_1})^2\right)^{1/2} \omega_1^{n_{-\bullet}-1}\omega_2$$
$$(\omega_2^{n_\bullet-1} - 1)(1 - \omega_2^{-n_\bullet-n_{-\bullet}-4}) \, d\omega_1 \, d\omega_2 .$$

The collision integral can be null for specific values of n_s and n_{-s}. The exact solutions, called the Kolmogorov spectra, correspond to these values which satisfy

$$n_+ + n_- = -4 . \tag{52}$$

It is important to understand that Zakharov transformation is not an identity transformation, and it can lead to spurious solutions. The necessary and sufficient condition for a spectrum obtained by Zakharov transformation to be a solution of the kinetic equation is that the right hand side integral in (46) (*i.e.* before the Zakharov transformation) equation converges. This condition is called the locality of the spectrum and leads to the following restriction on the spectral indices in our case:

$$-3 < n_\pm < -1 . \tag{53}$$

In the particular case of a zero cross–correlation one has $E_\perp^+ = E_\perp^- = E_\perp \sim k_\perp^n$ with only one solution

$$n = -2 .$$

Note that the thermodynamic equilibrium, corresponding to the equipartition state for which the flux of energy is zero instead of being finite as in the above spectral forms, corresponds to the choices $n_s = n_{-s} = 1$ for both the perpendicular and the parallel components of the energy.

The Kolmogorov constants $C_K(n_s)$ and $C_K'(n_s)$. The final expression of the Kolmogorov–like spectra found above as a function of the Kolmogorov

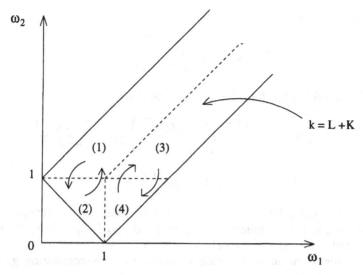

Fig. 1. Geometrical representation of the Zakharov transformation. The rectangular region, corresponding to the triad interaction $\mathbf{k}_\perp = \mathbf{L} + \boldsymbol{\kappa}$, is decomposed into four different regions called (1), (2), (3) and (4); ω_1 and ω_2 are respectively the dimensionless variables $\kappa_\perp/k_\perp$ and $L_\perp/k_\perp$. The Zakharov transformation applied to the collision integral consists to exchange regions (1) and (2), and regions (3) and (4).

constant (generalised to MHD) $C_K(n_s)$ and of the flux of energy $P_\perp^s(k_\perp)$ can be obtained in the following way. For a better understanding, the demonstration will be done in the simplified case of a zero cross–correlation. The generalization to the correlated case $(E^+ \neq E^-)$ is straightforward. Using the definition of the flux,

$$\partial_t E_\perp(k_\perp, 0) = -\partial_{k_\perp} P_\perp(k_\perp) , \qquad (54)$$

one can write the flux of energy as a function of the collision integral (with the new form of the integrand) depending on n_s. Then the limit $n_s \to -2$ is taken in order to have a constant flux $P_\perp$ with no more dependence in $k_\perp$, as it is expected for a stationary spectrum in the inertial range. Here we have considered an infinite inertial range to use the Zakharov transformation. Whereas the collision integral tends to zero when $n_s \to -2$, the limit with which we are concerned is not zero because of the presence of a denominator proportional to $2n_s + 4$, and which is a signature of the dimension in wavenumber of the flux. Finally the "L'Hospital's rule" gives the value of $P_\perp$ from which it is possible to write the Kolmogorov spectrum of the shear-Alfvén waves

$$E_\perp(k_\perp, 0) = P_\perp^{1/2} C_K(-2) k_\perp^{-2} , \qquad (55)$$

with the Kolmogorov constant

$$C_K(n_s) = \sqrt{\frac{-2b_0}{\pi \epsilon^2 J_1(n_s)}} , \qquad (56)$$

and with the following form for the integral $J_1(n_s)$

$$J_1(n_s) = 2^{n_s+3} \int_{x=1}^{+\infty} \int_{y=-1}^{1} \frac{\sqrt{(x^2-1)(1-y^2)}\,(xy+1)^2}{(x-y)^{n_s+6}\,(x+y)^{2-n_s}}$$

$$\left[2^{1-n_s} - (x+y)^{1-n_s}\right] \ln\left(\frac{x+y}{2}\right) dx\,dy . \qquad (57)$$

As expected, the calculation gives a negative value for the integral $J_1(n_s)$ and for the particular value $n_s = -2$, we obtain $C_K(-2) \simeq 0.585$. Note that the integral $J_1(n_s)$ converges only for $-3 < n_s < -1$.

The generalization to the case of non–zero cross–correlation gives the relations

$$E_\perp^+(k_\perp,0)\,E_\perp^-(k_\perp,0) = P_\perp^+\,C_K^2(n_s)\,k_\perp^{-4} = P_\perp^-\,C_K^2(-n_s-4)\,k_\perp^{-4}$$

$$= \sqrt{P_\perp^+ P_\perp^-}\,C_K(n_s)\,C_K(-n_s-4)\,k_\perp^{-4} , \qquad (58)$$

where the second formulation is useful to show the symmetry with respect to s. The computation of the Kolmogorov constant C_K as a function of $-n_s$ is given in Figure 2. An asymmetric form is observed which means the ratio $P_\perp^+/P_\perp^-$ is not constant, as we can see in Figure 3 where we plot this ratio as a function of $-n_s$. We see that for any ratio $P_\perp^+/P_\perp^-$ there corresponds a unique value of n_s, between the singular ratios $P_\perp^+/P_\perp^- = +\infty$ for $n_s = -3$ and $P_\perp^+/P_\perp^- = 0$ for $n_s = -1$. Thus, a larger flux of energy P^+ corresponds to a steeper slope of the energy spectra $E_\perp^+(k_\perp,0)$ in agreement with the physical image that a larger flux of energy implies a faster energy cascade.

In the zero cross–correlation case, a similar demonstration for the pseudo–Alfvén waves $E_\parallel^s(k_\perp,0)$ leads to the relation

$$E_\parallel(k_\perp,0) = P_\parallel P_\perp^{-1/2}\,C_K'(-2)\,k_\perp^{-2} , \qquad (59)$$

with the general form of the Kolmogorov constant

$$C_K'(n_s) = \sqrt{\frac{-2b_0 J_1(n_s)}{\pi \epsilon^2 J_2(n_s) J_2(-n_s-4)}} , \qquad (60)$$

where the integral $J_2(n_s)$ is

$$J_2(n_s) = 2^{n_s+3} \int_{x=1}^{+\infty} \int_{y=-1}^{1} \frac{\sqrt{(x^2-1)(1-y^2)}}{(x-y)^{n_s+6}\,(x+y)^{2-n_s}} \qquad (61)$$

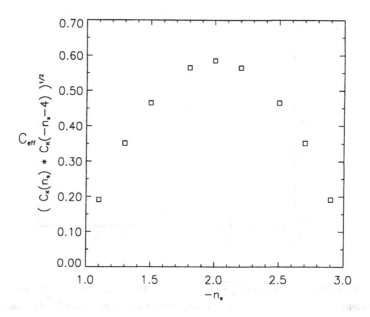

Fig. 2. Variation of $\sqrt{C_K(n_s)\,C_K(-n_s-4)}$ as a function of $-n_s$. Notice the symmetry around the value $-n_s$ corresponding to the case of zero velocity-magnetic field correlation.

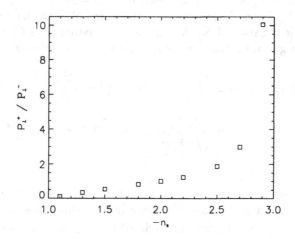

Fig. 3. Variation of $P_\perp^+/P_\perp^-$, the ratio of fluxes of energy, as a function of $-n_s$. For the zero cross correlation case the ratio is 1.

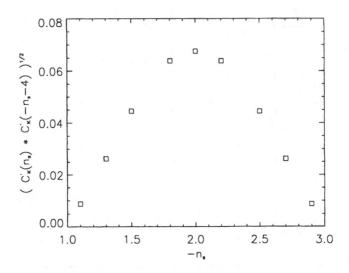

Fig. 4. Variation of $\sqrt{C'_K(n_s)\, C'_K(-n_s-4)}$ as a function of $-n_s$. Notice the symmetry around the value $-n_s$ corresponding to the zero cross correlation case.

$$\left[2^{1-n_s} - (x+y)^{1-n_s}\right] ln\left(\frac{x+y}{2}\right) dx dy .$$

Note that the integral $J_2(n_s)$ converges only for $-3 < n_s < -1$. The presence of the flux $P_\perp$ in the Kolmogorov spectrum is linked to the presence of $E_\perp$ in the kinetic equation of $E_\parallel$. A numerical evaluation gives $C'_K(-2) \simeq 0.0675$ whereas the generalization for the non-zero cross-correlation is

$$E^+_\parallel(k_\perp, 0)\, E^-_\parallel(k_\perp, 0) = \frac{P^+_\parallel P^-_\parallel}{P^+_\perp} C'^2_K(n_s)\, k^{-4}_\perp = \frac{P^+_\parallel P^-_\parallel}{P^-_\perp} C'^2_K(-n_s-4)\, k^{-4}_\perp$$

$$= \frac{P^+_\parallel P^-_\parallel}{\sqrt{P^+_\perp P^-_\perp}} C'_K(n_s)\, C'_K(-n_s-4)\, k^{-4}_\perp , \qquad (62)$$

where the last formulation shows the symmetry with respect to s. The power laws of the spectra $E^s_\parallel$ have the same indices than those of $E^s_\perp$ and the Kolmogorov constant C' is in fact related to C by the relation

$$\frac{C'_K(n_s)}{C'_K(-n_s-4)} = \frac{C_K(-n_s-4)}{C_K(n_s)} . \qquad (63)$$

Therefore the choice of the ratio $P^+_\perp/P^-_\perp$ determines not only $C_K(n_s)$ but also $C'_K(n_s)$, allowing for free choices of the dissipative rates of energy $P^\pm_\parallel$.

The result of the numerical evaluation of $C'_K(n_s)$ is shown in Figure 4. An asymmetrical form form is also visible; notice also that the values of $C'_K(n_s)$ (*i.e.* the constant in front of the parallel energy spectra) are smaller by an order of magnitude than those of $C_K(n_s)$ for the perpendicular spectra.

5.2   Temporal evolution of the kinetic equations

Numerical method. Equations (46) and (47) can be integrated numerically with a standard method, as for example presented in [21]. Since that the energy spectrum varies smoothly with k, it is convenient to use a logarithmic subdivision of the k axis

$$k_i = \delta k \; 2^{i/F} , \tag{64}$$

where i is a non–negative integer; δk is the minimum wave number in the computation and F is the number of wave numbers per octave. F defines the refinement of the "grid", and in particular it is easily seen that a given value of F introduces a cut–off in the degree of non–locality of the nonlinear interactions included in the numerical computation of the kinetic equations. But since the solutions are local, a moderate value of F can be used (namely, we take $F = 4$). Tests have nevertheless been performed with $F = 8$ and we show that no significant changes occur in the results to be described below.

This technique allows us to reach Reynolds numbers much greater than in direct numerical simulations. In order to regularize the equations at large k, we have introduced dissipative terms which were omitted in the derivation of the kinetic equations. We take the magnetic Prandtl number (ν/η) to be unity. For example, with $\delta k = 2^{-3}$, $F = 8$, $i_{max} = 225$; this corresponds to a ratio of scales $2^{28}/2^{-3}$. Taking a wave energy U_0^2 and an integral scale L_0 both of order one initially, and a kinematic viscosity of $\nu = 3.3 \times 10^{-8}$, the Reynolds number of such a computation is $\mathcal{R}_e = U_0 L_0/\nu \sim 10^8$. All numerical simulations to be reported here have been computed on an Alpha Server 8200 located at the Observatoire de Nice (SIVAM).

Shear-Alfvén waves. In this paper, we only consider decaying turbulence. As a first numerical simulation we have integrated the equation (46) in the zero cross–correlation case $(E^+ = E^-)$ and without forcing. Figure 5 (top) shows the temporal evolution of the total energy $E_\perp(t)$ with by definition

$$E_\perp(t) = \int_{k_{min}}^{k_{max}} E_\perp(k_\perp, 0) \, dk_\perp , \tag{65}$$

where k_{min} and k_{max} have the values given in the previous section. The total energy is conserved up to a time $t_* \simeq 1.6$ after which it decreases because of the dissipative effects linked to mode coupling, whereas the enstrophy $\int k^2 E_\perp(k) \, d^2\mathbf{k}$ increases sharply (bottom of Figure 5). The energy spectra

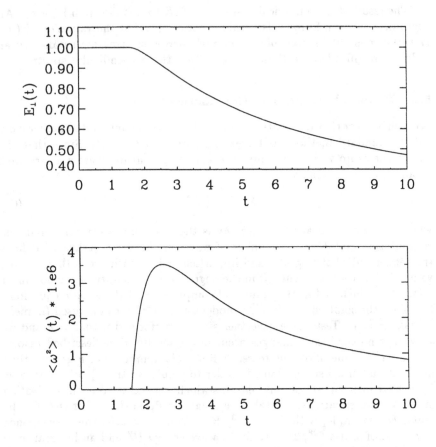

Fig. 5. Temporal evolution of the energy $E_\perp(t)$ (top) and the enstrophy $<\omega^2>(t)$ normalized to 1.10^6. Notice the conservation of the energy up to the time t_*.

at different times are displayed in Figure 6. As we approach the time t_*, the spectra spread out to reach the smallest scales (*i.e.* the largest wavenumbers). For $t > t_*$, constant energy flux spectrum $k_\perp^{-2}$ is obtained (indicated by the straight line). For times t significantly greater than t_*, we have a self-similar energy decay, in what constitutes the turbulent regime.

Shear-Alfvén *versus* pseudo-Alfvén waves. In a second numerical computation we have studied the system (46) (47) with an initial normalised cross-correlation of 80%. The following parameters have been used : $\delta k = 2^{-3}$, $F = 4$, $i_{max} = 105$ and $\nu = 6.4\,10^{-8}$. Figure 7 (top) shows the temporal evolution of energies for the four different waves ($E_\perp^\pm$ and $E_\parallel^\pm$). The same behavior as that of Figure 5 (top) is observed, with a conservation of energy up to the

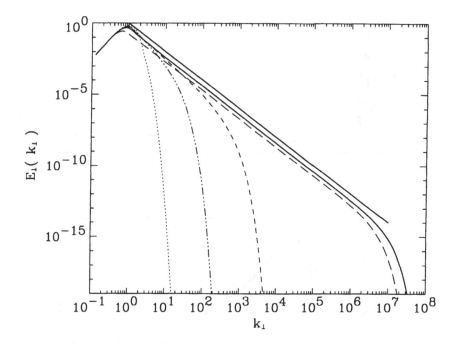

Fig. 6. Energy spectra $E_\perp(k_\perp, 0)$ of the shear-Alfvén waves in the zero cross correlation case for the times $t = 0$ (dot), $t = 1.0$ (dash-dot), $t = 1.5$ (small-dash), $t = 1.6$ (solid) and $t = 10.0$ (long-dash); the straight line follows a $k_\perp^{-2}$.

time $t'_* \simeq 5$, and a decay afterwards; this decay is nevertheless substantially weaker than when the correlation is zero, since in the presence of a significant amount of correlation between the velocity and the magnetic field. It is easily seen from the primitive MHD equations that the nonlinearities are strongly reduced. On the other hand the temporal evolution of enstrophies (bottom) displays that the maxima for these four types of waves are reached at different times: the pseudo-Alfvén waves are the fastest to reach their maxima at $t \simeq 5.5$ vs. $t \simeq 7.5$ for the shear-Alfvén waves. Figure 8 corresponds to the temporal evolution of another conserved quantity, the cross correlation ρ_x defined as

$$\rho_x = \frac{E_x^+ - E_x^-}{E_x^+ + E_x^-} , \qquad (66)$$

where x symbolizes either $\perp$ or $\parallel$. As expected, ρ_x is constant during an initial period (till $t = t'_*$) and then tends asymptotically to one, but in a faster way for the pseudo-Alfvén waves. This growth of correlation is well documented in the isotropic case [24] and is seen to hold as well here in the weak turbulence regime. Figures 9 and 10 give the compensated spectra

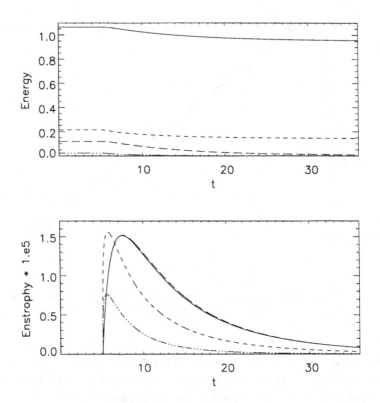

Fig. 7. Temporal evolution of energies (top) $E_\perp^+$ (solid), $E_\perp^-$ (long-dash), $E_\parallel^+$ (small-dash) and $E_\parallel^-$ (dash-dot); the same notation is used for the enstrophies (bottom) which are normalized to 1.10^5. Notice that energies are conserved till the time t'_*.

$E_\perp^+ E_\perp^- k_\perp^4$ and $E_\parallel^+ E_\parallel^- k_\perp^4$ respectively at different times. In both cases, from $t = 6$ onward, a plateau is observed over almost four decades and remains flat for long times; this confirms nicely the theoretical predictions (58) and (62).

As a final note, we remark that the analysis of the complete set of eight equations in dimension three as derived in section §2.4 remains to be done and is left for future work.

6   Front propagation

The numerical study of the transition between the initial state and the final state, where the $k_\perp^{-2}$-spectrum is reached, shows two remarkable properties illustrated by Figure 11 and 12.

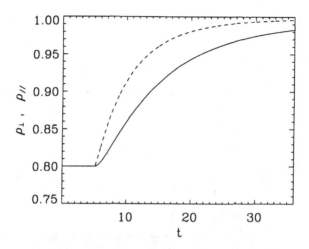

Fig. 8. Temporal evolution of the cross correlations $\rho_\perp$ (solid) and $\rho_\parallel$ (dash). These quantities are conserved up to the time t'_*.

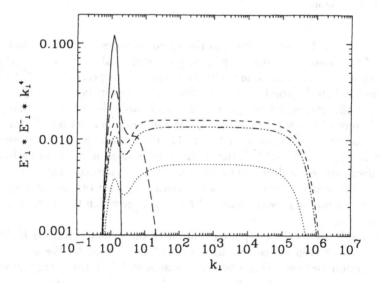

Fig. 9. Compensated spectra $E_\perp^+ E_\perp^- k_\perp^4$ at times $t = 0$ (solid), $t = 4$ (long-dash), $t = 6$ (small-dash), $t = 8$ (dash-dot) and $t = 20$ (dot).

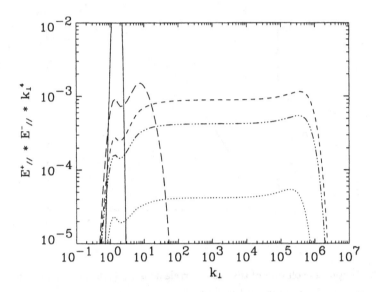

Fig. 10. Compensated spectra $E_{\parallel}^{+} E_{\parallel}^{-} k_{\perp}^{4}$ for the same times and with the same legend as in Figure 9.

We show in Figure 11 (top), in lin-log coordinates, the progression with time of the front of energy propagating to small scales; more precisely, we give the wavenumber at time t with an energy of, respectively, 10^{-25} (dash-dot line) and 10^{-16} (solid line). Note that all curves display an abrupt change at $t_0 \simeq 1.55$, after which the growth is considerably slowed down. Using this data, Figure 11 (bottom) gives $log(k_{\perp})$ as a function of $log(1.55-t)$; the lines having the same meaning as in Figure 11 (top); the large dash represents a power law $k_{\perp} \sim (1.55-t)^{-1.5}$. Hence, the small scales, in this weak turbulence formalism, are reached in a finite time *i.e.* in a catastrophic way. This is also seen on the temporal evolution of the enstrophy (see bottom of Figure 5), with a catastrophic growth ending at $t \simeq 2.5$, after which the decay of energy begins.

Figure 12 shows the temporal evolution of the energy spectrum $E_{\perp}(k_{\perp}, 0)$ of the shear-Alfvén waves around the catastrophic time t_0. We see that before t_0, evaluated here with a better precision to 1.544, the energy spectrum propagates to small scales following a stationary $k_{\perp}^{-7/3}-$ spectrum and not a $k_{\perp}^{-2}-$ spectrum. It is only when the dissipative scale is reached, at t_0, that a remarkable effect is observed: in a very fast time the $k_{\perp}^{-7/3}$ solution turns into the finite energy flux spectrum $k_{\perp}^{-2}$ with a change of the slope propagating from small scales to large scales.

Note that this picture is different from the scenario proposed by Falkovich and Shafarenko (FS) (1991) for the finite capacity spectra. According to FS,

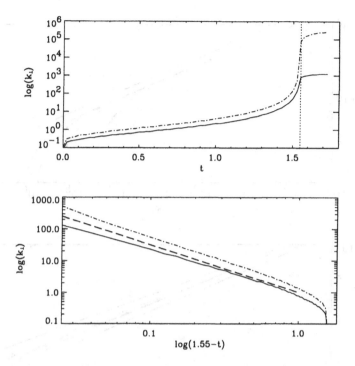

Fig. 11. Temporal evolution (top), in lin-log coordinates, of the front of energy propagating to small scales. The solid line and the dash-dot line correspond respectively to an energy of 10^{-16} and 10^{-25}. An abrupt change is visible at time $t_0 \simeq 1.55$ (vertical dotted line). The $log(k_\perp)$ as a function of $log(1.55 - t)$ (bottom) displays a power law in $k_\perp \sim (1.55 - t)^{-1.5}$ (large dash line).

the Kolmogorov spectrum forms right behing the propagating front, whereas in our case it only after the front reaches infinite wavenumbers (*i.e.* dissipative region). The front propagation can be described in terms of self-similar solutions having a form [11] [48]

$$E_\perp(k_\perp, 0) = \frac{1}{\tau^a} E_0\left(\frac{k_\perp}{\tau^b}\right), \tag{67}$$

where $\tau = t_0 - t$. Substituting (67) into the kinetic equation (46) we have

$$\partial_\tau \left(\frac{1}{\tau^a} E_0\left(\frac{k_\perp}{\tau^b}\right)\right) \sim \tau^{-a} E_0\left(\frac{\kappa_\perp}{\tau^b}\right) \left(\tau^{b-a} E_0\left(\frac{L_\perp}{\tau^b}\right) - \tau^{b-a} E_0\left(\frac{k_\perp}{\tau^b}\right)\right) \tau^{2b}.$$

which leads to the relation

$$1 + 3b = a. \tag{68}$$

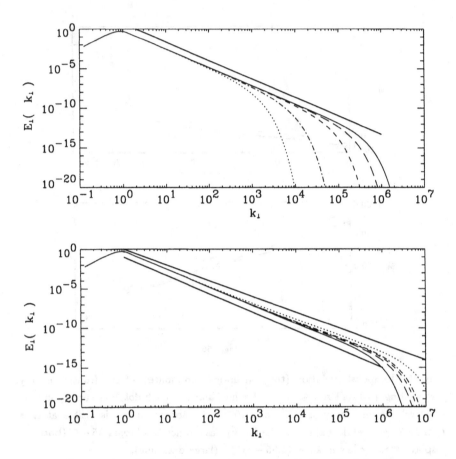

Fig. 12. Temporal evolution of the energy spectrum $E_\perp(k_\perp, 0)$ of the shear-Alfvén waves around the catastrophic time $t_0 \simeq 1.544$. For $t < t_0$ (top) ($t = 1.50$ (dot), $t = 1.53$ (dash-dot), $t = 1.54$ (dash), $t = 1.542$ (long dash) and $t = 1.543$ (solid)) a $k_\perp^{-7/3}$– spectrum is observed. For $t \geq t_0$ (bottom) ($t = 1.544$ (solid), $t = 1.546$ (long dash), $t = 1.548$ (dash), $t = 1.55$ (dash-dot) and $t = 1.58$ (dot)) a fast change of the slope appears to give finally a $k_\perp^{-2}$– spectrum. Note that this change propagates from small scales to large scales. In both cases straight lines follow either a $k_\perp^{-7/3}$ or a $k_\perp^{-2}$.

If E_0 is stationary and has a power-law form $E_0 \sim k^m$, then we have another relation between a and b

$$a + mb = 0 \tag{69}$$

Excluding a from (68) and (69) we have $1 + (3 + m)b = 0$. In our case this condition is satisfied because $b = -3/2$ and $m = -7/3$ which confirms that the front solution is of self-similar type.

7   Nonlocal isotropic 3D MHD turbulence

We have considered until now a turbulence of Alfvén waves that arises in the presence of a strong uniform magnetic field. Following Kraichnan [19], one can assume that the results obtained for turbulence in a strong external magnetic field are applicable to MHD turbulence at small scales which experience the magnetic field of the large-scale component as a quasi-uniform external field. Furthermore, the large-scale magnetic field is much stronger than the one produced by the small-scales themselves because most of the MHD energy is condensed at large scales due to the decreasing distribution of energy among modes as the wavenumbers grow. In this case therefore, the small-scale dynamics consists again of a large number of weakly interacting Alfvén waves. Using such a nonlocality hypothesis and applying a dimensional argument, Kraichnan derived the $k^{-3/2}$ energy spectrum for MHD turbulence. However, Kraichnan did not take into account the local anisotropy associated with the presence of this external field. In [32], the dimensional argument of Kraichnan is modified in order to take into account the anisotropic dependence of the characteristic time associated with Alfvén waves on the wavevector by simply writing

$$\tau \sim \frac{1}{b_0 k_{\parallel}} . \tag{70}$$

In that way, one obtains a $k_{\perp}^{-2}$ energy spectrum, which agrees with the analytical and numerical results of the present paper for the spectral dependence on $k_{\perp}$. On the other hand, the dependence of the spectra on $k_{\parallel}$, as we showed before in this paper, is not universal because of the absence of energy transfer in the $k_{\parallel}$ direction, although it is shown in [18] that for a quasi–uniform field as considered in this section, there is some transfer in the quasi–parallel direction. In the strictly uniform case, this spectral dependence is determined only by the dependence on $k_{\parallel}$ of the driving and/or initial conditions.

For large time, the spectrum is almost two-dimensional. The characteristic width of the spectrum in $k_{\parallel}$ (described by the function $f_1(k_{\parallel})$) is much less than its width in $k_{\perp}$, so that approximately one can write

$$e^s(\mathbf{k}) = C\, k_{\perp}^{-3}\, \delta(k_{\parallel}), \tag{71}$$

where C is a constant. The $k_\perp^{-3}$ factor corresponds to the $E_\perp \propto k_\perp^{-2}$ Kolmogorov–like spectrum found in this paper (the physical dimensions of e^s and $E_\perp^s$ being different). In the context of MHD turbulence, this spectrum is valid only locally, that is for distances smaller than the length–scale of the magnetic field associated with the energy–containing part of MHD turbulence. Let us average this spectrum over the large energy containing scales, that is over all possible directions of $\mathbf{B_0}$. Writing $k_\perp = |\mathbf{k} \times \mathbf{B_0}|/|B_0|$ and $k_\parallel = |\mathbf{k} \cdot \mathbf{B_0}|/|B_0|$ and assuming that $\mathbf{B_0}$ takes all possible directions in 3D space with equal probability, we have for the averaged spectrum

$$\langle e^s \rangle = \int e^s(\mathbf{k}, \mathbf{x})\, d\sigma(\zeta) = \int C\delta(\zeta \cdot \mathbf{k})\, |\zeta \times \mathbf{k}|^{-3} d\sigma(\zeta), \qquad (72)$$

where $\zeta = (\sin\theta \cos\phi, \sin\theta \sin\phi, \cos\theta)$ is a unit vector in the coordinate space and $d\sigma = \sin\theta\, d\theta\, d\phi$ is the surface element on the unit sphere. Choosing θ to be the angle between $\mathbf{k}$ and $\mathbf{B_0}$ and ϕ to be the angle in the transverse to $\mathbf{k}$ plane, we have

$$\langle e^s \rangle = C \int_0^{2\pi} d\phi \int_0^\pi \frac{\delta(\cos\theta)}{|k|} \frac{(\sin\theta)^{-3}}{|k|^3} \sin\theta\, d\theta = 2\pi C k^{-4}. \qquad (73)$$

This isotropic spectrum represents the averaged energy density in 3D wavevector space. By averaging over all possible directions of the wavevector, we get the following density of the energy distribution over the absolute value of the wavevector,

$$E_k = 8\pi^2 C\, k^{-2}. \qquad (74)$$

As we see, taking into account the local anisotropy and subsequent averaging over the isotropic energy containing scales results in an isotropic energy spectrum k^{-2}. This result is different from the $k^{-3/2}$ spectrum derived by Kraichnan without taking into account the local anisotropy of small scales.

The difference in spectral indices may also arise from the fact that the approach here is that of weak turbulence, whereas in the strong turbulence case, isotropy is recovered on average and a different spectrum – that of Kraichnan – obtains. Solar wind data [25] indicates that the isotropic spectrum scales as $k^{-\alpha}$ with $\alpha \sim 1.67$, close to the Kolmogorov value for neutral fluids (without intermittency corrections which are known to occur) ; hence, it could be interpreted as well as being a Kraichnan–like spectrum steepened by intermittency effects which are known to take place in strong MHD turbulence as well in the form of current and vorticity filaments and magnetic flux tubes.

8   Conclusion

We have obtained in this paper the kinetic equations for weak Alfvénic turbulence in the presence of correlations between the velocity and the magnetic field, and taking into account the non–mirror invariance of the MHD equations leading to non–zero helical terms. These equations, contrary to what is stated in [40], obtain at the level of three–wave interactions.

In this anisotropic medium, a new spectral tensor must be taken into account in the formalism when compared to the isotropic case (which can include terms proportional to the helicity); this new spectral tensor I^s is linked to the anisotropy induced by the presence of a strong uniform magnetic field, and we can also study its dynamics. This purely anisotropic correlator was also analysed in the case of neutral fluids in the presence of rotation [8].

We obtain an asymptotic two-dimensionalisation of the spectra: indeed, the evolution of the turbulent spectra at each $k_\parallel$ is determined only by the spectra at the same $k_\parallel$ and by the purely 2D state characterized by $k_\parallel = 0$. This property of bi–dimensionalisation was previously obtained theoretically from an analysis of the linearised MHD equations [29] and using phenomeno-logical models [31], and numerically as well [33], whereas they are obtained in our paper from the rigorously derived kinetic equations. Note that the strong field $\mathbf{B_0}$ has no structure (it is a $\mathbf{k} = 0$ field), whereas the analysis performed in [18] considers a strong quasi–uniform magnetic field of characteristic wavenumber $k_L \neq 0$, in which case the authors find that bi-dimensionalisation obtains as well for large enough wavenumbers.

A $k_\perp^{-2}$ energy spectrum is obtained for the inertial range turbulence, well verified numerically, and reached in a singular fashion with small scales developing in a finite time. We also obtain a family of Kolmogorov solutions with different values of spectra for different wave polarities and we show that the sum of the spectral exponents of these spectra is equal to -4. The dynamics of both the shear-Alfvén waves and the pseudo-Alfvén waves is obtained. Finally, the small-scale spectrum of nonlocal 3D MHD turbulence in the case when there is no external field is also derived.

In ongoing work, we are considering three–dimensional turbulence in the asymptotic regime of large time when the spectrum tends to a quasi-2D form. This is the same regime for which RMHD approach is valid [41]. However, in addition to the shear-Alfvén waves described by the RMHD equations, the kinetic equations describe also the dynamics of the so-called pseudo-Alfvén waves which are decoupled from the shear-Alfvén waves in this case, from the magnetic helicity and from the polarisation covariance. Finding the Kolmogorov solution for the 3D case is technically very similar to the case of 2D turbulence. Preliminary results indicate that we obtain for the energy spectrum, $f(k_\parallel) k_\perp^{-2}$ where $f(k_\parallel)$ is an arbitrary function which is to be fixed by matching in the forcing region at small wavenumbers. The $k_\parallel$ dependence is non-universal and depends on the form of the forcing because of the property that there is no energy transfer between different $k_\parallel$'s.

The weak turbulence regime remains valid as long as the Alfvén characteristic time $(k_\parallel b_0)^{-1}$ remains small compared to the eddy turn–over time of the waves $(k_\perp z_l)^{-1}$ where z_l is the characteristic Elsässer field at scale l. Using the exact scaling law found in this paper $(E(\mathbf{k}) \sim k_\perp^{-2})$ together with a local analysis $(z_l^2 \sim k E(\mathbf{k}))$ and recalling that $k_\parallel$ is negligible compared to $k_\perp$, one arrives at a scaling $k_\perp \sim B_0^2$ for this to happen.

The dynamo problem in the present formalism reduces to its simplest expression: in the presence of a strong uniform magnetic field $\mathbf{B_0}$, to a first approximation (closing the equations at the level of second–order correlation tensors), one obtains immediate equipartition between the kinetic and the magnetic wave energies, corresponding to an instantaneous efficiency of the dynamo. Of course, one may ask about the origin of $\mathbf{B_0}$ itself, in which case one may resort to standard dynamo theories (see e.g. [34]). We see no tendency towards condensation.

In view of the ubiquity of turbulent conducting flows embedded in strong quasi–uniform magnetic fields, the present derivation should be of some use when studying the dynamics of such media, even though compressibility effects have been ignored. However, it can be argued [13] that this incompressible approximation may be sufficient (see also [6]), because of the damping of the fast magnetosonic wave by plasma kinetic effects. Finally, the wave energy may not remain negligible for all times, in which case resort to phenomenological models for strong MHD turbulence is required. Is desirable as well an exploration of such complex flows through analysis of laboratory and numerical experiments, and through detailed observations like those stemming from satellite data for the solar wind, from the THEMIS instrument for the Sun looking at the small–scale magnetic structures of the photosphere, and the planned large array instrumentation (LSA) to observe in detail the interstellar medium.

Acknowledgements

This work has been performed using the computing facilities provided by the program "Simulations Interactives et Visualisation en Astronomie et Mécanique (SIVAM)" at OCA. Grants from CNRS (PNST and PCMI) and from EC (FMRX-CT98-0175) are gratefully acknowledged.

Appendix

From the dynamical equations, one writes successively for the second and third–order moments of the $\mathbf{z}^s$ fields:

$$\partial_t \left\{ a_j^s(\mathbf{k}) a_{j'}^{s'}(\mathbf{k'}) \right\} = -i\epsilon k_m P_{jn}(\mathbf{k}) \int \left\{ a_m^s(\boldsymbol{\kappa}) a_n^s(\mathbf{L}) a_{j'}^{s'}(\mathbf{k'}) \right\} e^{-2is\omega_\kappa t} \delta_{\kappa \mathbf{L}, \mathbf{k}} d_{\kappa \mathbf{L}}$$

$$-i\epsilon k_m' P_{j'm}(\mathbf{k'}) \int \left\{ a_m^{-s'}(\boldsymbol{\kappa}) a_n^{s'}(\mathbf{L}) a_j^s(\mathbf{k}) \right\} e^{-2is'\omega_\kappa t} \delta_{\kappa \mathbf{L}, \mathbf{k'}} d_{\kappa \mathbf{L}} \qquad (75)$$

and

$$\partial_t \left\{ a_j^s(\mathbf{k}) a_{j'}^{s'}(\mathbf{k'}) a_{j''}^{s''}(\mathbf{k''}) \right\} =$$

$$-i\epsilon k_m P_{jn}(\mathbf{k}) \int \left\{ a_m^{-s}(\boldsymbol{\kappa}) a_n^s(\mathbf{L}) a_{j'}^{s'}(\mathbf{k'}) a_{j''}^{s''}(\mathbf{k''}) \right\} e^{-2is\omega_\kappa t} \delta_{\kappa \mathbf{L}, \mathbf{k}} d_{\kappa \mathbf{L}}$$

$$+ \{[\mathbf{k}, s, j] \to [\mathbf{k'}, s', j'] \to [\mathbf{k''}, s'', j''] \to [\mathbf{k}, s, j]\} \qquad (76)$$

The fourth–order moment in the above equation, $\langle \kappa Lk'k'' \rangle$ in short–hand notation, decomposes into the sum of three products of second–order moments, and a fourth–order cumulant $\{mnj'j''\}$. The latter does not contribute to secular behavior, and of the remaining terms one is absent as well in the kinetic equations because it involves the combination of wavenumbers $\langle \kappa L \rangle \langle k'k'' \rangle$: it introduces, because of homogeneity, a $\delta(\kappa + L)$ factor which combined with the convolution integral leads to a zero contribution for $k = 0$. Hence, the third–order cumulant leads to six terms that read:

$$\partial_t \left\{ a_j^s(\mathbf{k}) a_{j'}^{s'}(\mathbf{k}') a_{j''}^{s''}(\mathbf{k}'') \right\} = -i\epsilon k_m P_{jn}(\mathbf{k}) q_{mj'}^{-ss'}(\mathbf{k}') q_{nj''}^{ss''}(\mathbf{k}'') e^{2is\omega't}$$

$$-i\epsilon k_m P_{jn}(\mathbf{k}) q_{mj''}^{-ss''}(\mathbf{k}'') q_{nj'}^{ss'}(\mathbf{k}') e^{2is\omega''t}$$

$$-i\epsilon k_m' P_{j'n}(\mathbf{k}') q_{mj''}^{-s's''}(\mathbf{k}'') q_{nj}^{s's}(\mathbf{k}) e^{2is'\omega''t}$$

$$-i\epsilon k_m' P_{j'n}(\mathbf{k}') q_{mj}^{-s's}(\mathbf{k}) q_{nj''}^{s's''}(\mathbf{k}'') e^{2is'\omega t}$$

$$-i\epsilon k_m'' P_{j''n}(\mathbf{k}'') q_{mj}^{-s''s}(\mathbf{k}) q_{nj'}^{s''s'}(\mathbf{k}') e^{2is''\omega t}$$

$$-i\epsilon k_m'' P_{j''n}(\mathbf{k}'') q_{mj'}^{-s''s'}(\mathbf{k}') q_{nj}^{s''s}(\mathbf{k}) e^{2is''\omega't} \quad (77)$$

It can be shown that, of these six terms, only the fourth and fifth ones give non–zero contributions to the kinetic equations. Defining

$$\omega_{k,\kappa L} = \omega_k - \omega_\kappa - \omega_L \quad (78)$$

and integrating equation (77) over time, the exponential terms will lead to

$$\Delta(\omega_{k,\kappa L}) = \int_0^t \exp\left[it\omega_{k,\kappa L}\right] dt = \frac{\exp\left[i\omega_{k,\kappa L}\right] - 1}{i\omega_{k,\kappa L}}. \quad (79)$$

Substituting these expressions in (77), only the terms which have an argument in the Δ functions that cancel exactly with the arguments in the exponential appearing in (75) will contribute. We then obtain the fundamental kinetic equations for the energy tensor, viz. :

$$\partial_t q_{jj'}^{ss'}(\mathbf{k}')\delta(\mathbf{k} + \mathbf{k}') =$$

$$-\epsilon^2 k_m P_{jn}(\mathbf{k}) \int k_{2p} P_{nq}(\mathbf{L}) q_{pm}^{-s-s}(\kappa) q_{qj'}^{ss'}(\mathbf{k}') \Delta(-2s\omega_1) \delta_{\kappa L, k} d\kappa L$$

$$-\epsilon^2 k_m P_{jn}(\mathbf{k}) \delta_{ss'} \int k_p' P_{j'q}(\mathbf{k}') q_{pm}^{-s'-s}(\kappa) q_{qn}^{s's}(\mathbf{L}) \Delta(-2s\omega_1) \delta_{\kappa L, k} d\kappa L$$

$$-\epsilon^2 k_m' P_{j'n}(\mathbf{k}') \int k_{2p} P_{nq}(\mathbf{L}) q_{pm}^{-s'-s}(\kappa) q_{qj}^{s's}(\mathbf{k}) \Delta(-2s'\omega_1) \delta_{\kappa L, k'} d\kappa L$$

$$-\epsilon^2 k_m' P_{j'n}(\mathbf{k}') \delta_{ss'} \int k_p P_{nq}(\mathbf{k}) q_{pm}^{-s-s'}(\kappa) q_{qn}^{ss'}(\mathbf{L}) \Delta(-2s'\omega_1) \delta_{\kappa L, k'} d\kappa L \quad (80)$$

We now perform an integration over the delta and taking the limit $t \to +\infty$ we find

$$\partial_t[q_{jj'}^{ss'}(\mathbf{k'})\delta(\mathbf{k}+\mathbf{k'})] =$$

$$-\epsilon^2 \int \int \int \left\{ Q_k^{-s}(\boldsymbol{\kappa}) P_{jn}(\mathbf{k}) P_{nl}(\mathbf{L})[q_{jj'}^{ss'}(\mathbf{k'})\left[\frac{\pi}{2}\delta(\kappa_\parallel) - i\mathcal{P}(\frac{1}{2s\kappa_\parallel})\right]\right.$$

$$+Q_k^{-s'}(\boldsymbol{\kappa}) P_{j'n}(\mathbf{k}) P_{nl}(\mathbf{L})[q_{lj}^{s's}(\mathbf{k})\left[\frac{\pi}{2}\delta(\kappa_\parallel) + i\mathcal{P}(\frac{1}{2s'\kappa_\parallel})\right] K$$

$$\left. -\pi\delta_{ss'} Q_k^{-s}(\boldsymbol{\kappa}) P_{j'l}(\mathbf{k}) P_{jn}(\mathbf{k'}) q_{ln}^{ss}(\mathbf{L})\delta(\kappa_\parallel) \right\} d\kappa_1 d\kappa_2 d\kappa_\parallel. \tag{81}$$

where $\mathcal{P}$ stands for the principal value of the integral.

In the case where $s = s'$ of interest here because the cross–correlators between z–fields of opposite polarities decay to zero in that approximation, the above equations simplify to:

$$\frac{2}{\pi}\partial_t[q_{jj'}^{ss}(\mathbf{k'}) \pm q_{j'j}^{ss}(\mathbf{k'})] =$$

$$2\int P_{jn}(\mathbf{k}) P_{j'q}(\mathbf{k})[q_{qn}^{ss}(\mathbf{L}) \pm q_{nq}^{ss}(\mathbf{L})]Q_k^{-s}(\boldsymbol{\kappa})\delta(\kappa_\parallel) d\kappa_1 d\kappa_2 d\kappa_\parallel$$

$$-\int P_{jn}(\mathbf{k}) P_{nq}(\mathbf{L})[q_{qj'}^{ss}(\mathbf{k'}) \pm q_{j'q}^{ss}(\mathbf{k'})]Q_k^{-s}(\boldsymbol{\kappa})\delta(\kappa_\parallel) d\kappa_1 d\kappa_2 d\kappa_\parallel$$

$$-\int P_{j'n}(\mathbf{k}) P_{nq}(\mathbf{L})[q_{jq}^{ss}(\mathbf{k'}) \pm q_{qj}^{ss}(\mathbf{k'})]Q_k^{-s}(\boldsymbol{\kappa})\delta(\kappa_\parallel) d\kappa_1 d\kappa_2 d\kappa_\parallel. \tag{82}$$

When developping the above expression in terms of the correlators defined in (22), one arrives at the kinetic equations for weak MHD turbulence given in Section 2.4.

References

1. Achterberg, A. 1979 *Astron. & Astrophys.* **76**, 276.

2. Akhiezer, A., Akhiezer, I.A., Polovin, R.V., Sitenko, A.G., Stepanov, K.N. 1975, *Plasma Electrodynamics II: Non-linear Theory and Fluctuations*, Pergamon Press.

3. Bavassano, B., Dobrowolny, M., Fanfoni, G., Mariani, F., Ness, N.F. 1982 *Solar Phys* **78**, 373.

4. Belcher, J.W., Davis, L. 1971 *J. Geophys. Res.* **76**, 3534.

5. Benney J., Newell A. 1969 *Stud. Appl. Math.* **48**, 29.

6. Bhattacharjee A. 1998 Puebla Conference in Astrophysics, Cambridge University Press Astrophysical Series, J. Franco & A. Carraminama Ed..

7. Bieber, J.W., Wanner, W., Matthaeus, W.H. 1996 *J. Geophys. Res.* **101**, 2511.

8. Cambon, C., Jacquin, L 1989 *J. Fluid Mech.* **202**, 295.

9. Cambon, C., Mansour, N.N., Godeferd, F.S. 1997 *J. Fluid Mech.* **337**, 303.

10. Craya A. 1958 *P.S.T. Ministère de l'Air* **345**.

11. Falkovich, G.E., Shafarenko, A.V. 1991 *J. Nonlinear Sci.* **1**, 457.

12. Gekelman, W., Pfister, H. 1988 *Phys. Fluids* **31**, 2017.

13. Goldreich P., Sridhar S. 1995 *ApJ.* **438** 763.

14. Heiles, C., Goodman, A.A., Mc Kee, C.F., Zweibel, E.G. 1993 in Protostars and Planets **IV** (Univ. Arizona Press, Tucson), 279.

15. Herring J. 1974 *Phys. Fluids* **17**, 859.

16. Horbury, T., Balogh, A., Forsyth, R.J., Smith, E.J. 1995 *Geophys. Rev. Lett.* **22**, 3405.

17. Iroshnikov P. 1963 *Sov. Astron.* **7**, 566.

18. Kinney R., McWilliams J. 1997 *J. Plasma Phys.* **57**, 73.

19. Kraichnan R. 1965 *Phys. Fluids* **8**, 1385.

20. Landau, L.D., Lifshitz, E.M *Statistical Physics II* **9**, Oxford Pergamon Press.

21. Leith, C.E., Kraichnan, R.H. 1972 *J. Atmos. Sci.* **29**, 1041.

22. Lesieur M. 1990 *Turbulence in Fluids*, Second Edition, Kluwer.

23. Marsch, E., C.Y. Tu 1994 *Ann. Geophys.*, *12*, 1127.

24. Matthaeus, W.H., Montgomery, D. 1980 *Ann. N.Y. Acad. Sci.* **357**, 203.

25. Matthaeus, W.H., Goldstein, M.L. 1982 *J. Geoph. Res.* **87A**, 6011.

26. Matthaeus, W.H., Goldstein, M.L., Roberts, D.A. 1990 *J. Geophys. Res.* **95**, 20673.

27. Matthaeus, W.H., Ghosh, S. Oughton, S., Roberts, D.A. 1996 *J. Geophys. Res.* **101**, 7619.

28. McIvor, I 1977 *Mon. Not. R. astr. Soc.* **178**, 85.

29. Montgomery D., Turner L. 1981 *Phys. Fluids* **24**, 825.

30. Montgomery, D.C., Matthaeus, W.H. 1995 *ApJ* **447**, 706.

31. Ng C.S., Bhattacharjee A. 1996 *ApJ.* **465** 845.

32. Ng C.S., Bhattacharjee A. 1997 *Phys. Plasma* **4** 605.

33. Oughton S., Priest E.R., Matthaeus W.H 1994 *J. Fluid Mech.* **280**, 95.

34. Parker, E.N. 1994 *Spontaneous current sheets in magnetic fields with applications to stellar X-rays*, Oxford University Press.

35. Priest, E.R. 1982 *Solar Magnetohydrodynamics*, D. Reidel Pub. Comp..

36. Riley J., Metcalfe R., Weissman M. 1981 in *Nonlinear properties of internal waves*, AIP Conf. Proc. **76**, 79, B. West Ed..

37. Robinson, D., Rusbridge, M. 1971 *Phys. Fluids* **14**, 2499.

38. Sagdeev, R., Galeev, A. 1969 *Nonlinear plasma theory*, Benjamin Inc.

39. Shebalin, J.V., Matthaeus, W.H., Montgomery, D. 1983 *J. Plasma Phys.* **29**, 525.

40. Sridhar S., Goldreich P. 1994 *ApJ.* **432** 612.

41. Strauss, H.R. 1976 *Phys. Fluids* **19**, 134.

42. Stribling, T., Matthaeus, W.H., Oughton, S. 1995 *Phys. Plasmas* **2**, 1437.

43. Taylor, J.B. 1986 *Rev. Mod. Phys.* **58**, 741.

44. Taylor, J.B. 1993 *Phys. Fluids B* **5**, 4378.

45. Waleffe F. 1992 *Phys. Fluids A* **4**, 350.

46. Wild, N., Gekelman, W., Stenzel, R.L. 1981 *Phys. Rev. Lett.* **46**, 339.

47. Woltjer L. 1958 *Proc. Natl. Acad. Sci. U.S.A.* **44**, 489.

48. Zakharov E., L'vov V., Falkovich G. 1992 *Kolmogorov Spectra of Turbulence I : Wave Turbulence*, Springer–Verlag.

49. Zweben, S., Menyuk, C., Taylor, R. 1979 *Phys. Rev. Lett.* **42**, 1270.

50. Zweibel, E.G., Heiles, C. 1997 *Nature* **385**, 131.

Shell Models for MHD Turbulence

P. Giuliani

Departement of Physics, University of Calabria, 87036 Rende (CS), Italy

Abstract. We review the main properties of shell models for magnetohydrodynamic (MHD) turbulence. After a brief account on shell models with nearest neighbour interactions, the paper focuses on the most recent results concerning dynamical properties and intermittency of a model which is a generalization to MHD of the Gledzer-Yamada-Okhitani (GOY) model for hydrodynamic. Applications to astrophysical problems are also discussed.

1 Introduction

Shell models are dynamical systems (ordinary differential equations) representing a simplified version of the spectral Navier–Stokes or MHD equations for turbulence. They were originally introduced and developed by Obukhov [1], Desnyansky and Novikov [2] and Gledzer [3] in hydrodynamic turbulence and constitute nowdays a consistent and relevant alternative approach to the analytical and numerical study of fully developed turbulence (see [4] for a complete review).

Shell models are built up by dividing wave-vector space (k–space) in a discrete number of shells whose radii grow exponentially like $k_n = k_0 \lambda^n$, $(\lambda > 1)$, $n = 1, 2, \ldots, N$. Each shell is assigned a scalar dynamic variable, $u_n(t)$, (real or complex) which takes into account the averaged effects of velocity modes between k_n and k_{n+1}. The equation for $u_n(t)$ is then written in the form

$$\frac{du_n}{dt} = k_n C_n + D_n + F_n \qquad (1)$$

where $k_n C_n$, D_n and F_n are respectively quadratic nonlinear coupling terms (involving nearest or next-nearest shell interactions), dissipation terms and forcing terms, the last generally restricted to the first shells. Nonlinear terms are chosen to satisfy scale-invariance and conservation of ideal invariants. The main advantage shell models offer is that they can be investigated by means of rather easy numerical simulations at very high Reynolds (Re) numbers. The degrees of freedom of a shell model are $N \sim \ln Re$, to be compared with $N \sim Re^{9/4}$ for a three dimensional hydrodynamic turbulence following the Kolmogorov scaling.

The paper is organized as follows. In section 2 shell models with nearest neighbour interactions are briefly reviewed. In section 3 equations for MHD models with nearest and next-nearest neighbour interactions are presented and conservations laws for the ideal case are discussed. Section 4 is devoted to dynamo action in shell models and section 5 to spectral properties in forced stationary state and intermittency. In section 6 conclusions are drawn and a brief mention to astrophysical applications is made.

2   Models with nearest neighbour interactions

The simplest hydrodynamic shell model is the Obukhov–Novikov model, which is a linear superposition of the Obukhov equation [1] and the Novikov equation [2]. The model involves real variables $u_n(t)$ and conserves the energy $1/2 \sum_{n=1}^{N} u_n^2$ in absence of forcing and dissipation. It does not conserve phase space volume nor other quadratic invariants exist. The extension of the Obukhov-Novikov model to MHD is due to Gloaguen et al. [5]. We write down the equations for clarity (u_n and b_n represent respectively the velocity and the magnetic field in dimensionless units)

$$\frac{du_n}{dt} = -\nu k_n^2 u_n + \alpha \left(k_n u_{n-1}^2 - k_{n+1} u_n u_{n+1} - k_n b_{n-1}^2 + k_{n+1} b_n b_{n+1} \right) \quad (2)$$

$$+ \beta \left(k_n u_{n-1} u_n - k_{n+1} u_{n+1}^2 - k_n b_n b_{n-1} + k_{n+1} b_{n+1}^2 \right)$$

$$\frac{db_n}{dt} = -\eta k_n^2 b_n + \alpha k_{n+1} \left(u_{n+1} b_n - u_n b_{n+1} \right) + \beta k_n (u_n b_{n-1} - u_{n-1} b_n) \quad (3)$$

Here ν is the kinematic viscosity, η is the magnetic diffusivity, α and β are two arbitrary coupling coefficients. The ideal invariants of the system are the total energy, $1/2 \sum_{n=1}^{N} \left(u_n^2 + b_n^2 \right)$ and the cross-correlation, $\sum_{n=1}^{N} u_n b_n$ which are two ideal invariants of the MHD equations [6]. When written in terms of the Elsässer variables $Z_n^+ = u_n + b_n$, $Z_n^- = u_n - b_n$, the equations assume a symmetric form and the conservation of the two previous invariants is equivalently expressed as the conservation of the pseudo-energies $E^\pm = (1/4) \sum_{n=1}^{N} Z_n^{\pm \, 2}$. It is remarkable to note that, unlike the hydrodynamic model, the MHD version satisfies a Liouville theorem $\sum_{n=1}^{N} \partial(dZ_n^\pm/dt)/\partial Z_n^\pm = 0$, impling phase-space volume conservation. The MHD equations conserve a third ideal invariant which is the magnetic helicity in three dimensions (3D) and the mean square potential in two dimensions (2D), but no further ideal quadratic invariant can be imposed to this shell model.

A detailed bifurcation analysis for a three-mode system was performed in [5] for different values of α and β. The low Reynolds (kinetic and magnetic) numbers, used as control parameters, allowed to identify a great variety of regions in the parameter space. Turbulence was investigated with a nine-mode system which produces an inertial range with spectra following approximately the Kolmogorov scaling $E(k) \sim k^{-5/3}$. Temporal intermittency was also observed and then reconsidered in more details by Carbone [7] who calculated the scaling exponents of the structure functions for the Elsässer variables and for the pseudo-energy transfer rates, showing consistency with the usual multifractal theory. Other interesting MHD phenomena were also observed in [5] such as dynamo effect and the growth of correlation between velocity and magnetic field in an unforced simulation. These phenomena will be treated in more details in the next paragraphs.

The complex version of (2) and (3) was thoroughly investigated by Biskamp [8]. The complex model allows to include the Alfvèn effect [9], [10], [11], that is the interaction of a constant large scale magnetic field with small scale turbulent eddies. The main consequence of this effect should be a reduction of the spectral energy transfer rate and a consequent change of the spectra from the Kolmogorov scaling, $E(k) \sim k^{-5/3}$, to the Iroshnikov-Kraichnan one, $E(k) \sim k^{-3/2}$. In this paper the Alfvèn effect will not be furtherly treated. The reader is referred to [8] for a complete discussion concerning the inclusion of Alfvènic terms in shell models.

3   Models with nearest and next-nearest interactions

Shell models with nearest and next-nearest neighbour interactions were introduced by Gledzer [3]. In particular the so called GOY (Gledzer-Yamada-Ohkitani) model has been extensively both numerically and analitically investigated [12], [13], [14]. The GOY model allows to conserve another quadratic invariant besides energy which was identified with the kinetic helicity [15]. A generalization of the GOY model to MHD can be found in Biskamp [8]. All the parameters of the model are now fixed by imposing the conservation of another quadratic invariant that can be chosen to distinguish between a 3D and a 2D model. A more refined version was then considered by Frick and Sokoloff [16] to take into account the fact that the magnetic helicity is a quantity not positive definite. The situation can be summarized as follows [17].

Let us consider the following set of equations (u_n and b_n are now complex variables representing the velocity and the magnetic field in dimensionless units)

$$\frac{du_n}{dt} = -\nu k_n^2 u_n - \nu' k_n^{-2} u_n + ik_n \Big\{ (u_{n+1}u_{n+2} - b_{n+1}b_{n+2})$$
$$-\frac{\delta}{\lambda}(u_{n-1}u_{n+1} - b_{n-1}b_{n+1}) - \frac{1-\delta}{\lambda^2}(u_{n-2}u_{n-1} - b_{n-2}b_{n-1}) \Big\}^* + f_n \quad (4)$$

$$\frac{db_n}{dt} = -\eta k_n^2 b_n + ik_n \Big\{ (1 - \delta - \delta_m)(u_{n+1}b_{n+2} - b_{n+1}u_{n+2})$$
$$+\frac{\delta_m}{\lambda}(u_{n-1}b_{n+1} - b_{n-1}u_{n+1}) + \frac{1-\delta_m}{\lambda^2}(u_{n-2}b_{n-1} - b_{n-2}u_{n-1}) \Big\}^* + g_n \quad (5)$$

or, in terms of the complex Elsässer variables $Z_n^\pm(t) = v_n(t) \pm b_n(t)$, particularly useful in some solar–wind applications,

$$\frac{dZ_n^\pm}{dt} = -\nu^+ k_n^2 Z_n^\pm - \nu^- k_n^2 Z_n^\mp - \frac{\nu'}{2} k_n^{-2} Z_n^+ - \frac{\nu'}{2} k_n^{-2} Z_n^- + ik_n T_n^{\pm*} + f_n^\pm \quad (6)$$

where

$$T_n^\pm = \left\{ \frac{\delta + \delta_m}{2} Z_{n+1}^\pm Z_{n+2}^\mp + \frac{2 - \delta - \delta_m}{2} Z_{n+1}^\mp Z_{n+2}^\pm \right.$$
$$+ \frac{\delta_m - \delta}{2\lambda} Z_{n+1}^\pm Z_{n-1}^\mp - \frac{\delta + \delta_m}{2\lambda} Z_{n+1}^\mp Z_{n-1}^\pm$$
$$\left. - \frac{\delta_m - \delta}{2\lambda^2} Z_{n-1}^\pm Z_{n-2}^\mp - \frac{2 - \delta - \delta_m}{2\lambda^2} Z_{n-1}^\mp Z_{n-2}^\pm \right\} \quad (7)$$

Here $\nu^\pm = (\nu \pm \eta)/2$, being ν the kinematic viscosity and η the resistivity, $-\nu' k_n^{-2} u_n$, eq. (4), is a drag term specific to 2D cases (see below), $f_n^\pm = (f_n \pm g_n)/2$ are external driving forces, δ and δ_m are real coupling coefficients to be determined. In the inviscid unforced limit, equations (6) conserve both pseudoenergies $E^\pm(t) = (1/4) \sum_n |Z_n^\pm(t)|^2$ for any value of δ and δ_m (the sum is extended to all the shells), which corresponds to the conservation of both the total energy $E = E^+ + E^- = (1/2) \sum_n (|v_n(t)|^2 + |b_n(t)|^2)$ and the cross-helicity $h_C = E^+ - E^- = \sum_n Re(v_n b_n^*)$. As far as the third ideal invariant is concerned, we can define a generalized quantity as

$$H_B^{(\alpha)}(t) = \sum_{n=1}^N (\text{sign}(\delta - 1))^n \frac{|b_n(t)|^2}{k_n^\alpha} \quad (8)$$

whose conservation implies $\delta = 1 - \lambda^{-\alpha}$, $\delta_m = \lambda^{-\alpha}/(1 + \lambda^{-\alpha})$ for $\delta < 1$, $0 < \delta_m < 1$ and $\delta = 1 + \lambda^{-\alpha}$, $\delta_m = -\lambda^{-\alpha}/(1 - \lambda^{-\alpha})$ for $\delta > 1$, $\delta_m < 0$, $\delta_m > 1$. Thus two classes of MHD GOY models can be defined with respect to the values of δ: 3D–like models for $\delta < 1$, where $H_B^{(\alpha)}$ is not positive definite and represents a generalized magnetic helicity; 2D–like models where $\delta > 1$ and $H_B^{(\alpha)}$ is positive definite. This situation strongly resembles what happens in the hydrodynamic case where 2D–like ($\delta > 1$) and 3D–like ($\delta < 1$) models are conventionally distinguished with respect to a second generalized conserved quantity $H_K^{(\alpha)}(t) = \sum_n (\text{sign}(\delta - 1))^n k_n^\alpha |v_n(t)|^2$. Here the 3D and 2D cases are recovered for $\alpha = 1, 2$ where the ideal invariants are identified respectively with kinetic helicity and enstrophy. It should be noted that, although the hydrodynamic invariants are not conserved in the magnetic case, the equations which link α and δ are exactly the same for hydrodynamic and MHD models. Thus, once fixed α and δ, it is a simple matter to find out which GOY model the MHD GOY one reduces to when $b_n = 0$ [17]. To summarize we have that (with $\lambda = 2$) the model introduced in [16] for the 3D case will be called, hence on, 3D MHD GOY model or simply 3D model. It is recovered for $\alpha = 1$, $\delta = 1/2$, $\delta_m = 1/3$ and reduces to the usual 3D GOY model for $b_n = 0$. The Biskamp's 3D model [8] is actually a 2D–like model and will be called pseudo 3D model. It is obtained for $\alpha = 1$, $\delta = 3/2$, $\delta_m = -1$ and reduces to a 2D–like GOY model that conserves a quantity which has the same dimensions as kinetic helicity but is positive definite. The 2D models introduced in [8] and in [16] coincide, they are recovered for $\alpha = 2$, $\delta = 5/4$, $\delta_m = -1/3$ and reduce to the usual 2D GOY model for $b_n = 0$. In the following the properties of the 3D model will be mainly investigated.

4   Dynamo action in MHD shell models

The problem of magnetic dynamo, that is the amplification of a seed of magnetic field and its maintenance against the losses of dissipation in an electrically conducting flow, is of great interest by itself and for astrophysical applications (see for example [18] for an excellent introduction to the problem). Shell models offer the opportunity to test with relative simplicity whether a small value of the magnetic field can grow in absence of forcing terms on the magnetic field. Previous considerations about dynamo action in shell models can be found in [5]. In that case numerical study of bifurcations in the three-mode system revealed instabilities of kinetic fixed points to magnetic ones or magnetic chaos. The existence of a sort of dynamo effect in MHD GOY models was put forward by Frick and Sokoloff [16]. The authors investigate the problem of the magnetic field generation in a free-decaying turbulence, thus showing that: 1) in the $3D$ case magnetic energy grows and reaches a value comparable with the kinetic one, in a way that the magnetic field growth is unbounded in the kinematic case; 2) in the $2D$ case magnetic energy slowly decays in the nonlinear as well as in the kinematic case. These results have been interpreted as a 3D "turbulent dynamo effect" and seem to be in agreement with well-known results by which dynamo effect is not possible in two dimensions [19]. The problem was then reexamined in [17] in a forced situation looking at a comparison between the 3D MHD GOY model and the pseudo 3D model.

Starting from a well developed turbulent velocity field, a seed of magnetic field is injected and the growth of the magnetic spectra monitored. System is forced on the shell $n = 4$ ($k_0 = 1$), setting $f_4^+ = f_4^- = (1 + i) \, 10^{-3}$, which corresponds to only inject kinetic energy at large scales. Method of integration is a modified fourth order Runge-Kutta scheme. In fig. 1 we plot $\log_{10}\langle |b_n|^2 \rangle$ and $\log_{10}\langle |v_n|^2 \rangle$ versus $\log_{10} k_n$ for the 3D model. Angular brackets $\langle \ \rangle$ stand for time averages. It can be seen that the magnetic energy grows rapidly in time and forms a spectrum where the amplitude of the various modes is, at small scale, of the same order as the kinetic energy spectrum. (The subsequent evolution of magnetic and kinetic spectra will be considered in the next section). The spectral index is close to $k^{-2/3}$ which is compatible with a

Kolmogorov scaling of the second order structure function. For a comparison

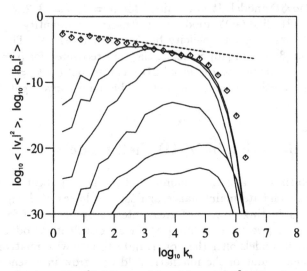

Fig. 1. 3D model: $\log_{10}\langle|v_n|^2\rangle$ (*diamonds*) and $\log_{10}\langle|b_n|^2\rangle$ (*lines*) versus $\log_{10}k_n$. The averages of $|b_n|^2$ are made over intervals of 3 large scale turnover times. Time proceeds upwards. The kinetic spectrum is averaged over 30 large scale turnover times. The straight line has slope $-2/3$. Parameters used: N=24, $\nu = \eta = 10^{-8}$, $\nu' = 0$

we integrated the pseudo 3D model and it can be seen (fig. 2) that a magnetic spectrum is formed, but it slowly decays in time. Notice that, because of the smallness of b_n, its back-reaction on the velocity field is negligible, thus the kinematic part of the model evolves independently from the magnetic one. Now the scaling $|v_n|^2 \sim k_n^{-4/3}$ follows, as a cascade of generalized enstrophy is expected for 2D–like hydrodynamic GOY models when $\alpha < 2$ (see [20] for details). The question now arises whether it is correct the interpretation of the growth of the magnetic field in the 3D model as the corresponding dynamo effect expected in the real 3D magnetohydrodynamics. First of all it should be noted that in the kinematic case an analogy with the vorticity equation predicts the following relations between velocity and magnetic energy spectra [6]: $|v_n|^2 \sim k^{-a}$, $|b_n|^2 \sim k^{2-a}$, so that if $a = 2/3$ it follows a magnetic energy spectrum growing with k. The kinematic case corresponds to the first stage of growth of our simulation where this behaviour is sometimes visible, at least qualitatively. Note however that the averages are made on very small time intervals because of the rapid growth of the magnetic energy. A similar, much more pronounced behaviour is found for the pseudo 3D model as well.

Let us stress that the sign of the third ideal invariant seems to play a crucial role as far as the growth of small magnetic fields is concerned. In ef-

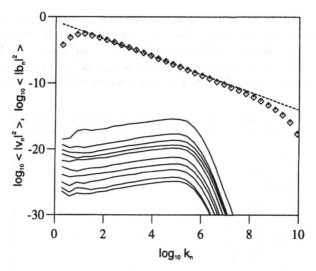

Fig. 2. Pseudo 3D model: $\log_{10}\langle |v_n|^2 \rangle$ (*diamonds*) and $\log_{10}\langle |b_n|^2 \rangle$ (*lines*) versus $\log_{10} k_n$. Averages are made over intervals of 100 large scale turnover times. Time proceeds downwards. The kinetic spectrum is only shown for the last interval. The straight line has slope $-4/3$, see text for explanation. Parameters used: N=33, $\nu = 10^{-16}$, $\eta = 0.5 \cdot 10^{-9}$, $\nu' = 1$

fect this sort of dynamo effect can also be considered under a different point of view. Let us consider the ideal evolution of the model $dZ_n^\pm/dt = ik_n T_n^{\pm*}$. We can build up the phase space S of dimension $D = 4N$, by using the Elsässer variables as axes, so that a point in S represents the system at a given time. A careful analysis of (6) shows that there exist some subspaces $I \subset S$ of dimension $D = 2N$ which remain invariant under the time evolution [21]. More formally, let $y(0) = (v_n, b_n)$ be a set of initial conditions such that $y(0) \in I$, I is time invariant if the flow T^t, representing the time evolution operator in S, leaves I invariant, that is $T^t[y(0)] = y(t) \in I$. The kinetic subspace $K \subset S$, defined by $y(0) = (v_n, 0)$ is obviously the usual fluid GOY model. Further subspaces are the Alfvénic subspaces $A^\pm$ defined by $y(0) = (v_n, \pm v_n)$, say $Z_n^+ \neq 0$ and $Z_n^- = 0$ (or vice versa). Each initial condition in these subspaces is actually a fixed point of the system. We studied the properties of stability of K and $A^\pm$. Following [21], let us define for each I the orthogonal complement P, namely $S = I \oplus P$. Let us then decompose the solution as $y(t) = (y_{int}(t), y_{ext}(t))$ where the subscripts refer to the I and P subspaces respectively. Finally we can define the energies $E_{int} = \|y_{int}\|^2$ and $E_{ext} = \|y_{ext}\|^2$. Note that the distance of a point $y = (y_{int}, y_{ext})$ from the subspace I is $d = \min_{\hat{y} \in I} \|y - \hat{y}\| = \|y_{ext}\|$. Then E_{ext} represents the square of the distance of the solution from the invariant subspace. At time $t = 0$, $E_{ext} = \epsilon E_{int}$ ($\epsilon \ll 1$) represents the energy of the perturbation. Since the

total energy is constant in the ideal case, two extreme situations can arise: 1)
The external energy remains of the same order of its initial value, that is the
solution is trapped near I which is then a stable subspace; 2) The external
energy assumes values of the same order as the internal energy, that is the
solution is repelled away from the subspace which is then unstable. Since the

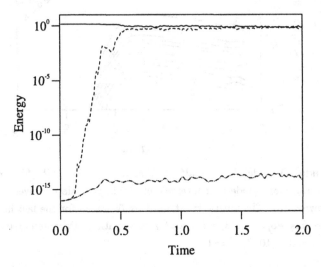

Fig. 3. Ideal case: kinetic energy (*continuous line*) and magnetic energy (*dashed
line*) versus time for the 3D model; magnetic energy (*dot-dashed line*) versus time
for the pseudo 3D model

external and internal energies for the Alfvénic subspaces are nothing but the
pseudoenergies E^+ and E^-, which are ideal invariants, the Alfvénic subspaces
are stable. As regards the kinetic subspace, E_{int} and E_{ext} represent respectively the kinetic and magnetic energies. Looking at the numerical solutions
of the ideal model (fig. 3) we can see the difference in the stability properties
between the pseudo 3D model and the 3D one. In the first case the external
energy remains approximately constant, while in the second case the system
fills up immediately all the available phase space. This striking difference is
entirely due to the nonlinear term, and in fact must be ascribed to the differences in sign of the third invariant. The effect of the unstable subspace, which
pushes away the solutions, is what in ref. [16] is called "turbulent dynamo
effect".

5   Spectral properties in stationary forced state

The main fundamental difference between hydrodynamic and MHD shell
models lies in the fact that the behaviour of the former is not so sensitive to

the type of forcing, at least as far as the main features are concerned. On the contrary in the magnetic case phase space is more complex because of the presence of invariant subspaces which can act as attractors of the dynamics of the system, hence the type of forcing becomes crucial in selecting the stationary state reached by the system. The spectral properties of the 3D model have been investigated by Frick and Sokolov in [16] under different choices of the forcing terms. In their simulations they observe that the spectral indexes of kinetic and magnetic spectra depend on the level of cross helicity and magnetic helicity. In particular spectra with spectral index $-5/3$ appear if the cross helicity vanishes. Even in this case results may be deceptive. In fact, defining the reduced cross helicity h_R as the cross helicity divided by the total energy, long runs [17] show that, in case of constant forcing on the velocity variables, even from an initial value $h_R = 0$ the system evolves inevitably towards a state in which the reduced cross helicity reaches either the value $+1$ or -1, corresponding to a complete correlation or anti-correlation between velocity and magnetic field. In terms of attractors the system is attracted towards one of the Alfvènic subspaces where velocity and magnetic field are completely aligned or anti-aligned. Due to the particular form of the nonlinear interactions in MHD (6), the nonlinear transfer of energy towards the small scales is stopped. In this case Kolmogorov-like spectra appear as a transient of the global evolution. This is shown in fig.4 where it is clearly seen a component (Z_n^-) which is completely vanishing while the Z_n^+ spectrum becomes steeper and steeper as energy is not removed from large scales. If an exponentially correlated in time gaussian random forcing on the velocity field is adopted the system shows a very interesting behaviour. It spends long periods (several large scale turnover times) around one of the Alfvènic attractors, jumping from one to the other rather irregularly (fig. 5). This behaviour assures the existence of a flux of energy to the small scales, modulates the level of nonlinear interactions and the consequent dissipation of energy at small scales, which is burstly distributed in time. What we want to stress is the fact that the Alfvènic attractors play a relevant role in the dynamics of the system. This fact should be taken into account especially when a stationary state is investigated in order to determine the scaling exponents of the structure functions (see below). Two regimes, the Kolmogorov transient and the completely aligned regime, could be mixed during the average procedure, thus leading to unreliable values of the scaling exponents.

6   Fluxes, inertial range and intermittency

The "four-fifth" relation $\langle \delta v(l)^3 \rangle = (-4/5)\epsilon l$, where ϵ is the mean rate of energy dissipation and l the separation, derived by Kolmogorov in [22], can be generalized to MHD flows [23],[24],[25]. A corresponding relation exists in MHD shell models, which can be derived following the considerations in [26].

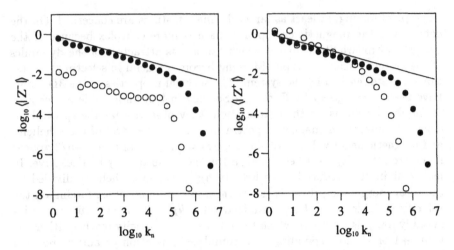

Fig. 4. *Left*: $\log_{10}\langle|Z_n^-|\rangle$ versus $\log_{10} k_n$ at different times (**a**) Black circles: The average of $|Z_n^-|$ are made over the first 30 large scale turnover times (**b**) White circles: The average is made after 300 large scale turnover times (**c**). The straight line has slope $-1/3$. *Right*: The same for $|Z_n^+|$

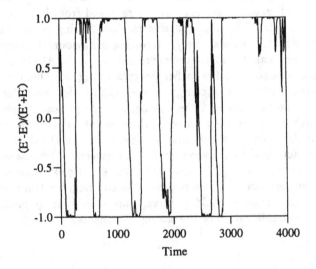

Fig. 5. Reduced cross helicity versus time in the case of an exponentially time-correlated random gaussian forcing on the velocity variables

Assuming for simplicity $\nu = \eta$, the scale–by–scale energy budget equation is:

$$\frac{d}{dt} \sum_{i=1,n} \frac{|Z_i^\pm|^2}{4} = k_n \, (Z^\pm Z^\pm Z^\mp)_n - \nu \sum_{i=1,n} k_i^2 \frac{|Z_i^\pm|^2}{2} + \sum_{i=1,n} \frac{1}{2} \mathrm{Re}\{Z_i^\pm (f_i^\pm)^*\}$$

where the quantities $(Z^\pm Z^\pm Z^\mp)_n$ are defined as

$$(Z^\pm Z^\pm Z^\mp)_n = \frac{1}{4}\mathrm{Im}\{(\delta + \delta_m)\} Z_n^\pm Z_{n+1}^\pm Z_{n+2}^\mp + \frac{(2 - \delta - \delta_m)}{\lambda} Z_{n-1}^\pm Z_n^\mp Z_{n+1}^\pm$$
$$+ (2 - \delta - \delta_m) Z_n^\pm Z_{n+1}^\mp Z_{n+2}^\pm + \frac{(\delta_m - \delta)}{\lambda} Z_{n-1}^\mp Z_n^\pm Z_{n+1}^\pm\} \quad (9)$$

Assuming that i) forcing terms only act at large scales; (ii) the system tends to a statistically stationary state; (iii) in the infinite Reynolds numbers limit ($\nu \to 0$) the mean energy dissipation tends to a finite positive limit $\varepsilon^\pm$, we obtain

$$\langle (Z^+ Z^+ Z^-)_n \rangle = -\varepsilon^+ \, k_n^{-1}$$

$$\langle (Z^- Z^- Z^+)_n \rangle = -\varepsilon^- \, k_n^{-1}$$

These are the equations that define the inertial range of the system and that can be easily checked and confirmed by numerical simulations (fig. 6). It is to be remarked that these are the appropriate combinations that are expected to scale exactly as k^{-1}. Let us finally remind that, as far as cascade properties of shell models are concerned, the major drawback lies in the difficulty to reproduce cascades of quantities that are expected to flow inversely, such as energy in 2D hydrodynamic [20] or magnetic helicity in MHD [8].

A deep understanding of intermittency in turbulence is nowdays one of the most challenging tasks from a theoretical point of view (see [27] for review). A lot of papers have been dedicated in the last years to investigate temporal intermittency in shell models. Deviations from the Kolmogorov scaling $\xi_p = p/3$ of the scaling exponents in the structure functions, $\langle |u_n|\rangle^p \sim k^{-\xi_p}$, have been observed and described in the context of a multifractal approach [13]. A precise calculation of the scaling exponents may have difficulties related to the presence of periodic oscillations superimposed to the power law. Another source of uncertainty is linked to the exact identification of the inertial range where the fit should be performed. These problems are at lenght discussed and investigated in [28] where a new shell model (called Sabra model) has been introduced in the context of hydrodynamic turbulence. The Sabra model is a slight modification of the standard GOY model and allows to eliminate spurious oscillations in the spectra. The same problems are in principle encountered in magnetohydrodynamic models thus a generalization of the Sabra model to MHD is required ([29]). An alternative approach to the determination of scaling exponents for the 3D MHD GOY model can be

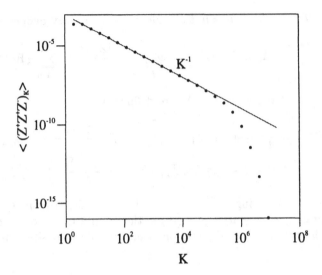

Fig. 6. Numerical check of the exact scaling relation involving the mixed third order moment

found in [30] where concepts and techniques related to ESS and GESS [31] are used.

We have determined the scaling exponents of the structure functions $\langle|u_n|^p\rangle \sim k^{-\xi_p^u}$, $\langle|b_n|^p\rangle \sim k^{-\xi_p^u}$, $\langle|Z_n^+|^p\rangle \sim k^{-\xi_p^+}$, $\langle|Z_n^-|^p\rangle \sim k^{-\xi_p^-}$ adopting a random forcing on the velocity variables (on shell n=1 and n=2) to assure the system does not "align". The forcing terms were calculated solving a Langevin equation $\dot{f}_n = -(1/\tau_0)\,f_n + \mu$, where τ_0 is a correlation time chosen equal to the large scale turnover time and μ is a gaussian delta-correlated noise. The total number of shells is 23 and the values of viscosity and resistivity are $\nu = 0.5 \cdot 10^{-9}$, $\eta = 0.5 \cdot 10^{-9}$. In fig. 7 the first three structure functions are plotted for the magnetic field, together with the best fit lines. From a comparison with spectra obtained in the standard GOY model [26], it should be remarked that the cross over region between the inertial range and the dissipative one is not so sharp as in the hydrodynamic case.

We then decided to perform a least-square fit in the range, determined visually, between the shell numbers $n = 3$ and $n = 12$. The values of ξ_p^u and ξ_p^b are reported in Table 1 together with the values of ξ_p, extracted from [26], for the hydrodynamic GOY model. The values of the scaling exponents of the other structure functions are compatible, within errors coming from the fit procedure, with those of the velocity variables. It can be seen that the values found are compatible with those obtained for the standard hydrodynamic GOY model.

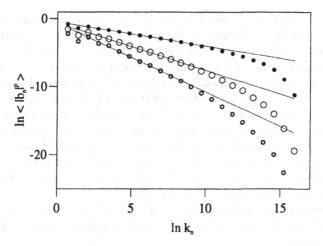

Fig. 7. Structure functions $\ln\langle|b_n|^p\rangle$ versus $\ln k_n$ for (a) $p = 1$ (*black circles*) (b) $p = 2$ (*large white circles*) (c) $p = 3$ (*small white circles*). The straight lines are the best fit in the range between $n = 3$ and $n = 12$

Table 1. Scaling exponents ξ_p (GOY model), ξ_p^u, ξ_p^b

p	ξ_p (GOY)	ξ_p^u	ξ_p^b
1		0.37 ± 0.01	0.36 ± 0.01
2	0.72	0.71 ± 0.01	0.70 ± 0.01
3	1.04	1.02 ± 0.02	1.02 ± 0.02
4	1.34	1.31 ± 0.03	1.32 ± 0.03
5	1.61	1.57 ± 0.04	1.59 ± 0.04
6	1.86	1.82 ± 0.05	1.84 ± 0.05
7	2.11	2.05 ± 0.06	2.07 ± 0.06
8	2.32	2.28 ± 0.08	2.30 ± 0.08

7 Conclusions

In this paper we have reported about the main properties concerning dynamical behaviour and intermittency of a shell model for MHD turbulence. The properties of the model reveal a complex structure of phase space in which invariant subspaces are present. The stability properties of the kinetic subspace are related to a dynamo action in the system while Alfvènic subspaces act as strong attractors which cause the system to evolve towards a state in which no energy cascade is present. A careful choice of forcing terms seems to be crucial in determining the stationary state reached by the system.

We want finally mention that shell models, as good candidates to reproduce the main features of MHD turbulence, can be used to check conjectures and ideas in astrophysical applications where very high Reynolds numbers are often present. We briefly remind two examples of applications. In [32] MHD shell models have been used to simulate magnetohydrodynamics in the early universe to investigate the effects of plasma viscosity on primordial magnetic fields. As second example, scaling laws found in the probability distribution functions of quantities connected with solar flares (eruption events in the solar corona) are at present matter of investigation by means of shell models. Results on this subject can be found in [33].

I am grateful to Vincenzo Carbone, Pierluigi Veltri, Leonardo Primavera, Guido Boffetta, Antonio Celani and Angelo Vulpiani for useful discussions and suggestions.

References

1. Obukhov A. M., Atmos. Oceanic Phys., **7** (1971) 41.

2. Desnyanski V. N. and Novikov E. A., prik. Mat. Mekh., **38** (1974) 507.

3. Gledzer E. B., Dokl. Akad. Nauk SSSR, **208** (1973) 1046.

4. Bohr T., Jensen M. H., Paladin G. and Vulpiani A., Dynamical Systems Approach to Turbulence (Cambridge University Press) 1998.

5. Gloaguen C., Leorat J., Pouquet A. and Grappin R., Physica D **17** (1985) 154.

6. Biskamp D., Nonlinear Magnetohydrodynamics (Cambridge University Press) 1993.

7. Carbone V., Europhys. Lett., **27** (1994) 581.

8. Biskamp D., Phys. Rev. E, **50** (1994) 2702; Chaos, Solitons & Fractals, **5** (1995) 1779.

9. Iroshnikov P. S., Astron. Zh., **40** (1963) 742.

10. Kraichnan R. H., Phys. Fluids, **8** (1965) 1385.

11. Dobrowolny M., Mangeney A. and Veltri P., Phys. Rev. Lett., **45** (1980) 1440.

12. Yamada M., Ohkitani K., J. Phys. Soc. Jpn., **56** (1987) 4810.

13. Jensen M. H., Paladin G. and Vulpiani A., Phys. Rev. A, **43** (1991) 798.

14. Biferale L., Lambert A., Lima R. and Paladin G., Physica D, **80** (1995) 105.

15. Kadanoff L., Lohse D., Wang J. and Benzi R., Phys. Fluids, **7**1995 617.

16. Frick P. and Sokoloff D. D., Phys. Rev. E., **57** (1998) 4195.

17. Giuliani P. and Carbone V., Europhys. Lett., **43** (1998) 527.

18. Moffatt H. K., Magnetic Field Generation in Electrical Conducting Fluids (Cambridge University Press) 1978.

19. Zeldovich Ya. B., Sov. Phys. JETP, **4** (1957) 460.

20. Ditlevsen P. and Mogensen I. A., Phys. Rev. E, **53** (1996) 4785.

21. Carbone V. and Veltri P., Astron. Astrophys., **259** (1992) 359.

22. Kolmogorov A. N., Dokl. Akad. Nauk SSSR, **32** (1941) 19.

23. Chandrasekhar S., Proc. R. Soc. London, Ser. A **204**, (1951) 435.

24. Chandrasekhar S., Proc. R. Soc. London, Ser. A **207**, (1951) 301.

25. Politano H. and Pouquet A., Phys. Rev. E, **57** (1998) R21.

26. Pisarenko D., Biferale L., Courvoisier D., Frisch U. and Vergassola M., Phys. Fluids A, **5** (1993) 2533.

27. Frisch U., Turbulence: the Legacy of A. N. Kolmogorov (Cambridge University Press) 1995.

28. L'vov V. S., Podivilov E., Pomyalov A., Procaccia I. and Vandembroucq D., Phys. Rev. E, **58** (1998) 1811.

29. Boffetta G., Carbone V., Giuliani P. and Veltri P., in preparation.

30. Basu A., Sain A., Dhar S. K. and Pandit R., Phys. Rev. Lett., **81** (1998) 2687.

31. Benzi R., Ciliberto S., Tripiccione R., Baudet C., Massaioli F. and Succi S., Phys. Rev. E, **48** (1993) R29.

32. Brandenburg A., Enqvist K. and Olesen P., Phys. Lett. B, **392** (1997) 395

33. Boffetta G., Carbone V., Giuliani P., Veltri P., and Vulpiani A., submitted to Phys. Rev. Lett.

Dynamics of Vortex and Magnetic Lines in Ideal Hydrodynamics and MHD

E.A. Kuznetsov and V.P. Ruban

Landau Institute for Theoretical Physics, Kosygin str., 117334 Moscow, Russia

Abstract. Vortex line and magnetic line representations are introduced for description of flows in ideal hydrodynamics and MHD, respectively. For incompressible fluids it is shown that the equations of motion for vorticity Ω and magnetic field with the help of this transformation follow from the variational principle. By means of this representation it is possible to integrate the system of hydrodynamic type with the Hamiltonian $\mathcal{H} = \int |\Omega| d\mathbf{r}$. It is also demonstrated that these representations allow to remove from the noncanonical Poisson brackets, defined on the space of divergence-free vector fields, degeneracy connected with the vorticity frozenness for the Euler equation and with magnetic field frozenness for ideal MHD. For MHD a new Weber type transformation is found. It is shown how this transformation can be obtained from the two-fluid model when electrons and ions can be considered as two independent fluids. The Weber type transformation for ideal MHD gives the whole Lagrangian vector invariant. When this invariant is absent this transformation coincides with the Clebsch representation analog introduced in [1].

1 Introduction

There are a large number of works devoted to the Hamiltonian description of the ideal hydrodynamics (see, for instance, the review [2] and the references therein). This question was first studied by Clebsch (a citation can be found in Ref. [3]), who introduced for nonpotential flows of incompressible fluids a pair of variables λ and μ (which later were called as the Clebsch variables). A fluid dynamics in these variables is such that vortex lines represent themselves intersection of surfaces $\lambda = $ const and $\mu = $ const and these quantities, being canonical conjugated variables, remain constant by fluid advection. However, these variables, as known (see, i.e.,[4]) describe only partial type of flows. If λ and μ are single-valued functions of coordinates then the linking degree of vortex lines characterizing by the Hopf invariant [5] occurs to be equal to zero. For arbitrary flows the Hamiltonian formulation of the equation for incompressible ideal hydrodynamics was given by V.I.Arnold [6,7]. The Euler equations for the velocity curl $\Omega = \text{curl } \mathbf{v}$

$$\frac{\partial \Omega}{\partial t} = \text{curl}\,[\mathbf{v} \times \Omega], \quad \text{div } \mathbf{v} = 0 \tag{1}$$

are written in the Hamiltonian form,

$$\frac{\partial \Omega}{\partial t} = \{\Omega, \mathcal{H}\}, \tag{2}$$

by means of the noncanonical Poisson brackets [4]

$$\{F, G\} = \int \left(\Omega \left[\text{curl}\,\frac{\delta F}{\delta \Omega} \times \text{curl}\,\frac{\delta G}{\delta \Omega} \right] \right) d\mathbf{r} \tag{3}$$

where the Hamiltonian

$$\mathcal{H}_h = -\frac{1}{2} \int \Omega \Delta^{-1} \Omega d^3\mathbf{r}, \tag{4}$$

coincides with the total fluid energy.

In spite of the fact that the bracket (3) allows to describe flows with arbitrary topology its main lack is a degeneracy. By this reason it is impossible to formulate the variational principle on the whole space S of divergence-free vector fields.

The cause of the degeneracy, namely, presence of Casimirs annulling the Poisson bracket, is connected with existence of the special symmetry formed the whole group - the relabeling group of Lagrangian markers (for details see the reviews [8,2]). All known theorems about the vorticity conservation (the Ertel's, Cauchy's and Kelvin's theorems, the frozenness of vorticity and conservation of the topological Hopf invariant) are a sequence of this symmetry. The main of them is the frozenness of vortex lines into fluid. This is related to the local Lagrangian invariant – the Cauchy invariant. The physical meaning of this invariant consists in that any fluid particle remains all the time on its own vortex line.

The similar situation takes place also for ideal magneto-hydrodynamics (MHD) for barotropic fluids:

$$\rho_t + \nabla(\rho \mathbf{v}) = 0, \tag{5}$$

$$\mathbf{v}_t + (\mathbf{v}\nabla)\mathbf{v} = -\nabla w(\rho) + \frac{1}{4\pi\rho}[\text{curl}\,\mathbf{h} \times \mathbf{h}], \tag{6}$$

$$\mathbf{h}_t = \text{curl}[\mathbf{v} \times \mathbf{h}]. \tag{7}$$

Here ρ is a plasma density, $w(\rho)$ plasma entalpy, $\mathbf{v}$ and $\mathbf{h}$ are velocity and magnetic fields, respectively. As well known (see, for instance, [9]-[13]), the

MHD equations possesses one important feature – frozenness of magnetic field into plasma which is destroyed only due to dissipation (by finite conductivity). For ideal MHD combination of the continuity equation (5) and the induction equation (7) gives the analog of the Cauchy invariant for MHD.

The MHD equations of motion (5-7) can be also represented in the Hamiltonian form,

$$\rho_t = \{\rho, \mathcal{H}\} \qquad \mathbf{h}_t = \{\mathbf{h}, \mathcal{H}\}, \qquad \mathbf{v}_t = \{\mathbf{v}, \mathcal{H}\}, \tag{8}$$

by means of the noncanonical Poisson brackets [14]:

$$\{F, G\} = \int \left(\frac{\mathbf{h}}{\rho} \cdot \left(\left[\operatorname{curl} \frac{\delta F}{\delta \mathbf{h}} \times \frac{\delta G}{\delta \mathbf{v}} \right] - \left[\operatorname{curl} \frac{\delta G}{\delta \mathbf{h}} \times \frac{\delta F}{\delta \mathbf{v}} \right] \right) \right) d^3 r + \tag{9}$$

$$+ \int \left(\frac{\operatorname{curl} \mathbf{v}}{\rho} \cdot \left[\frac{\delta F}{\delta \mathbf{v}} \times \frac{\delta G}{\delta \mathbf{v}} \right] \right) d^3 r + \int \left(\frac{\delta G}{\delta \rho} \nabla \left(\frac{\delta F}{\delta \mathbf{v}} \right) - \frac{\delta F}{\delta \rho} \nabla \left(\frac{\delta G}{\delta \mathbf{v}} \right) \right) d^3 r.$$

This bracket is also degenerated. For instance, the integral $\int (\mathbf{v}, \mathbf{h}) d r$, which characterizes mutual linkage knottiness of vortex and magnetic lines, is one of the Casimirs for this bracket.

The analog of the Clebsch representation in MHD serves a change of variables suggested in 1970 by Zakharov and Kuznetsov [1]:

$$\mathbf{v} = \nabla \phi + \frac{[\mathbf{h} \times \operatorname{curl} \mathbf{S}]}{\rho}. \tag{10}$$

New variables (ϕ, ρ) and $\mathbf{h}, \mathbf{S}$ represent two pairs canonically conjugated quantities with the Hamiltonian coinciding with the total energy

$$\mathcal{H} = \int \left(\rho \frac{\mathbf{v}^2}{2} + \rho \varepsilon(\rho) + \frac{\mathbf{h}^2}{8\pi} \right) dr.$$

In the present paper we suggest a new approach of the degeneracy resolution of the noncanonical Poisson brackets by introducing new variables, i.e., Lagrangian markers labeling each vortex lines for ideal hydrodynamics or magnetic lines in the MHD case.

The basis of this approach is the integral representation for the corresponding frozen-in field, namely, the velocity curl for the Euler equation and magnetic field for MHD. We introduce new objects, i.e., the vortex lines or magnetic lines and obtain the equations of motion for them. This description is a mixed Lagrangian-Eulerian description, when each vortex (or magnetic) line is enumerated by Lagrangian marker, but motion along the line is described in terms of the Eulerian variables. Such representation removes all degeneracy from the Poisson brackets connected with the frozenness, remaining the equations of motion to be gauge invariant with respect to reparametrization of each line. Important, that the equations for line motion, as the equations for curve deformation, are transverse to the line tangent.

It is interesting that the line representation also solves another problem - the equations of line motion follow from the variational principle, being Hamiltonian.

This approach allows also simply enough to consider the limit of narrow vortex (or magnetic) lines. For two-dimensional flows in hydrodynamics this "new" description corresponds to the well-known fact, namely, to the canonical conjugation of x and y coordinates of vortices (see, for instance, [3]).

The Hamiltonian structure introduced makes it possible to integrate the three-dimensional Euler equation (2) with Hamiltonian $\mathcal{H} = \int |\Omega| d\mathbf{r}$. In terms of the vortex lines the given Hamiltonian is decomposed into a set of Hamiltonians of noninteracting vortex lines. The dynamics of each vortex lines is, in turn, described by the equation of a vortex induction which can be reduced by the Hasimoto transformation [15] to the integrable one-dimensional nonlinear Schrodinger equation.

For ideal MHD a new representation - analog of the Weber transformation - is found. This representation contains the whole vector Lagrangian invariant. In the case of ideal hydrodynamics this invariant provides conservation of the Cauchy invariant and, as a sequence, all known conservation laws for vorticity (for details see the review [2]). It is important that all these conservation laws can be expressed in terms of observable variables. Unlike the Euler equation, these vector Lagrangian invariants for the MHD case can not be expressed in terms of density, velocity and magnetic field. It is necessary to tell that the analog of the Weber transformation for MHD includes the change of variables (10) as a partial case. The presence of these Lagrangian invariants in the transform provides topologically nontrivial MHD flows.

The Weber transform and its analog for MHD play a key role in constructing the vortex line (or magnetic line) representation. This representation is based on the property of frozenness. Just therefore by means of such transform the noncanonical Poisson brackets become non-degenerated in these variables and, as a result, the variational principle may be formulated. Another peculiarity of this representation is its locality, establishing the correspondence between vortex (or magnetic) line and vorticity (or magnetic field). This is a specific mapping, mixed Lagrangian-Eulerian, for which Jacobian of the mapping can not be equal to unity for incompressible fluids as it is for pure Lagrangian description.

2 General remarks

We start our consideration from some well known facts, namely, from the Lagrangian description of the ideal hydrodynamics.

In the Eulerian description for barotropic fluids, pressure $p = p(\rho)$, we have coupled equations - discontinuity equation for density ρ and the Euler equation for velocity:

$$\rho_t + \mathrm{div}\,\rho\mathbf{v} = 0, \tag{11}$$

$$\mathbf{v}_t + (\mathbf{v}\nabla)\mathbf{v} = -\nabla w(\rho), \qquad dw(\rho) = dp/\rho. \tag{12}$$

In the Lagrangian description each fluid particle has its own label. This is three-dimensional vector $\mathbf{a}$, so that particle position at time t is given by the function

$$\mathbf{x} = \mathbf{x}(\mathbf{a},t). \tag{13}$$

Usually initial position of particle serves the Lagrangian marker: $\mathbf{a} = \mathbf{x}(\mathbf{a},0)$.

In the Lagrangian description the Euler equation (12) is nothing more than the Newton equation:

$$\ddot{\mathbf{x}} = -\nabla w.$$

In this equation the second derivative with respect to time t is taken for fixed $\mathbf{a}$, but the r.h.s. of the equation is a function of t and $\mathbf{x}$. Excluding from the latter the x-dependence, the Euler equation takes the form:

$$\ddot{x}_i \frac{\partial x_i}{\partial a_k} = -\frac{\partial w(\rho)}{\partial a_k}, \tag{14}$$

where now all quantities are functions of t and $\mathbf{a}$.

In the Lagrangian description the continuity equation (11) is easily integrated and the density is given through the Jacobian of the mapping (13) $J = \det(\partial x_i/\partial a_k)$:

$$\rho = \frac{\rho_0(\mathbf{a})}{J}. \tag{15}$$

Now let us introduce a new vector,

$$u_k = \frac{\partial x_i}{\partial a_k} v_i, \tag{16}$$

which has a meaning of velocity in a new curvilinear system of coordinates or it is possible to say that this formula defines the transformation law for velocity components. It is worth noting that (16) gives the transform for the velocity $\mathbf{v}$ as a *co-vector*.

The straightforward calculation gives that the vector $\mathbf{u}$ satisfies the equation

$$\frac{du_k}{dt} = \frac{\partial}{\partial a_k}\left(\frac{\mathbf{v}^2}{2} - w\right). \tag{17}$$

In this equation the right-hand-side represents gradient relative to **a** and therefore the "transverse" part of the vector **u** will conserve in time. And this gives the Cauchy invariant:

$$\frac{d}{dt}\mathrm{curl}\,_a\,\mathbf{u} = 0, \tag{18}$$

or

$$\mathrm{curl}_a\,\mathbf{u} = \mathbf{I}. \tag{19}$$

If Lagrangian markers **a** are initial positions of fluid particles then the Cauchy invariant coincides with the initial vorticity: $\mathbf{I} = \Omega_0(\mathbf{a})$. This invariant is expressed through instantaneous value of $\Omega(\mathbf{x}, t)$ by the relation

$$\Omega_0(\mathbf{a}) = J(\Omega(\mathbf{x}, t)\nabla)\mathbf{a}(\mathbf{x}, t) \tag{20}$$

where $\mathbf{a} = \mathbf{a}(\mathbf{x}, t)$ is inverse mapping to (13). Following from (20) relation for $\mathbf{B} = \Omega/\rho$,

$$B_{0i}(a) = \frac{\partial a_i}{\partial x_k} B_k(x, t),$$

shows that, unlike velocity, **B** transforms as a vector.

By integrating the equation (17) over time t we arrive at the so-called Weber transformation

$$\mathbf{u}(\mathbf{a}, t) = \mathbf{u}_0(\mathbf{a}) + \nabla_a \Phi, \tag{21}$$

where the potential Φ obeys the Bernoulli equation:

$$\frac{d\Phi}{dt} = \frac{\mathbf{v}^2}{2} - w(\rho) \tag{22}$$

with the initial condition: $\Phi|_{t=0} = 0$. For such choice of Φ a new function $\mathbf{u}_0(\mathbf{a})$ is connected with the "transverse" part of **u** by the evident relation

$$\mathrm{curl}_a\,\mathbf{u}_0(\mathbf{a}) = \mathbf{I}.$$

The Cauchy invariant **I** characterizes the vorticity frozenness into fluid. It can be got by standard way considering two equations - the equation for the quantity $\mathbf{B} = \Omega/\rho$,

$$\frac{d\mathbf{B}}{dt} = (\mathbf{B}\nabla)\mathbf{v}, \tag{23}$$

and the equation for the vector $\delta\mathbf{x} = \mathbf{x}(\mathbf{a} + \delta\mathbf{a}) - \mathbf{x}(\mathbf{a})$ between two adjacent fluid particles:

$$\frac{d\delta\mathbf{x}}{dt} = (\delta\mathbf{x}\nabla)\mathbf{v}, \tag{24}$$

The comparison of these two equations shows that if initially the vectors $\delta\mathbf{x}$ are parallel to the vector $\mathbf{B}$, then they will be parallel to each other all time. This is nothing more than the statement of the vorticity frozenness into fluid. Each fluid particle remains all the time at its own vortex line. The combination of Eqs. (23) and (24) leads to the Cauchy invariant. To establish this fact it is enough to write down the equation for the Jacoby matrix $J_{ij} = \partial x_i/\partial a_j$ which directly follows from (24):

$$\frac{d}{dt}\frac{\partial a_i}{\partial x_k} = -\frac{\partial a_i}{\partial x_j}\frac{\partial v_j}{\partial x_k},$$

that in combination with Eq. (23) gives conservation of the Cauchy invariant (19).

If now one comes back to the velocity field $\mathbf{v}$ then by use of Eqs. (16) and (21) one can get that

$$\mathbf{v} = u_{0k}\nabla a_k + \nabla\Phi \tag{25}$$

where gradient is taken with respect to $\mathbf{x}$. Here the equation for potential Φ has the standard form of the Bernoulli equation:

$$\Phi_t + (\mathbf{v}\nabla)\Phi - \frac{\mathbf{v}^2}{2} + w(\rho) = 0.$$

It is interesting to note that relations (19), as equations for determination of $\mathbf{x}(\mathbf{a},t)$, unlike Eqs (17), are of the first order with respect to time derivative. This fact also reflects in the expression for velocity (25) which can be considered as a result of the partial integration of the equations of motion (17). Of course, the velocity field given by (25) contain two unknown functions: one is the whole vector $\mathbf{a}(\mathbf{x},t)$ and another is the potential Φ. For incompressible fluids the latter is determined from the condition $\mathrm{div}\,\mathbf{v} = 0$. In this case the Bernoulli equation serves for determination of the pressure.

Another important moment connected with the Cauchy invariant is that it follows from the variational principle (written in terms of Lagrangian variables) as a sequence of relabelling symmetry remaining invariant the action (for details, see the reviews [8,2]). Passing from Lagrangian to Hamiltonian in this description we have no any problems with the Poisson bracket. It is given by standard way and does not contain any degeneracy that the noncanonical Poisson brackets (3) and (9) have. One of the main purposes of this paper is to construct such new description of the Euler equation (as well as the ideal MHD) which, from one side, would allow to retain the Eulerian description, as maximally as possible, but, from another side, would exclude from the very beginning all remains from the gauge invariance of the complete Euler description connected with the relabeling symmetry.

As for MHD, this system in one point has some common feature with the Euler equation: it also possesses the frozenness property. The equation for $\mathbf{h}/\rho$

coincides with (23) and therefore dynamics of magnetic lines is very familiar to that for vortex lines of the Euler equation. However, this analogy cannot be continued so far because the equation of motion for velocity differs from the Euler equation by the presence of pondermotive force. This difference remains also for incompressible case.

3   Vortex line representation

Consider the Hamiltonian dynamics of the divergence-free vector field $\Omega(\mathbf{r}, t)$, given by the Poisson bracket (3) with some Hamiltonian $\mathcal{H}$ [1]:

$$\frac{\partial \Omega}{\partial t} = \mathrm{curl} \left[\mathrm{curl} \frac{\delta \mathcal{H}}{\delta \Omega} \times \Omega \right]. \tag{26}$$

As we have said, the bracket (3) is degenerate, as a result of which it is impossible to formulate the variational principle on the entire space $\mathcal{S}$ of solenoidal vector fields. It is known [2] that Casimirs f, annulling Poisson brackets, distinguish in $\mathcal{S}$ invariant manifolds $\mathcal{M}_f$ (symplectic leaves) on each of which it is possible to introduce the standard Hamiltonian mechanics and accordingly to write down a variational principle. We shall show that solution of this problem for the equations (26) is possible on the base of the property of frozenness of the field $\Omega(\mathbf{r}, t)$, which allows to resolve all constrains, stipulated by the Casimirs, and gives the necessary formulation of the variational principle.

To each Hamiltonian $\mathcal{H}$ - functional of $\Omega(\mathbf{r}, t)$ - we associate the generalized velocity

$$\mathbf{v}(\mathbf{r}) = \mathrm{curl} \frac{\delta \mathcal{H}}{\delta \Omega}. \tag{27}$$

However one should note that the generalized $\mathbf{v}(\mathbf{r})$ is defined up to addition of the vector parallel to Ω:

$$\mathrm{curl} \frac{\delta \mathcal{H}}{\delta \Omega} \rightarrow \mathrm{curl} \frac{\delta \mathcal{H}}{\delta \Omega} + \alpha \Omega,$$

that in no way does change the equation for Ω. Under the condition $(\Omega \cdot \nabla \alpha) = 0$ a new generalized velocity will have zero divergence and the frozenness equation (26) can be written already for the new $\mathbf{v}(\mathbf{r})$. A gauge changing of the generalized velocity corresponds to some addition of a Casimir to the Hamiltonian :

$$\mathcal{H} \rightarrow \mathcal{H} + f; \qquad \{f, ..\} = 0.$$

---

[1] The Hamiltonian (4) corresponds to ideal incompressible hydrodynamics.

Hence becomes clear that the transformation

$$\mathbf{x} = \mathbf{x}(\mathbf{a}, t)$$

of the initial positions of fluid particles $\mathbf{x}(\mathbf{a}, 0) = \mathbf{a}$ by the generalized velocity field $\mathbf{v}(\mathbf{r})$ through solution of the equation

$$\dot{\mathbf{x}} = \mathbf{v}(\mathbf{x}, t) \tag{28}$$

is defined ambiguously due to the ambiguous definition of $\mathbf{v}(\mathbf{r})$ by means of (27). Therefore using full Lagrangian description to the systems (26) becomes ineffective.

Now we introduce the following general expression for $\Omega(\mathbf{r})$, which is gauge invariant and fixes all topological properties of the system that are determined by the initial field $\Omega_0(\mathbf{a})$[16]:

$$\Omega(\mathbf{r}, t) = \int \delta(\mathbf{r} - \mathbf{R}(\mathbf{a}, t))(\Omega_0(\mathbf{a})\nabla_{\mathbf{a}})\mathbf{R}(\mathbf{a}, t)d^3\mathbf{a}. \tag{29}$$

Here now

$$\mathbf{r} = \mathbf{R}(\mathbf{a}, t) \tag{30}$$

does not satisfy any more the equation (28) and, consequently, the mapping Jacobian $J = \det\|\partial\mathbf{R}/\partial\mathbf{a}\|$ is not assumed to equal 1, as it was for full Lagrangian description of incompressible fluids.

It is easily to check that from condition $(\nabla_{\mathbf{a}}\Omega_0(\mathbf{a})) = 0$ it follows that divergence of (29) is identically equal to zero.

The gauge transformation

$$\mathbf{R}(\mathbf{a}) \to \mathbf{R}(\tilde{\mathbf{a}}_{\Omega_0}(\mathbf{a})) \tag{31}$$

leaves this integral unchanged if $\tilde{\mathbf{a}}_{\Omega_0}$ is arisen from $\mathbf{a}$ by means of arbitrary nonuniform translations along the field line of $\Omega_0(\mathbf{a})$. Therefore the invariant manifold $\mathcal{M}_{\Omega_0}$ of the space $\mathcal{S}$, on which the variational principle holds, is obtained from the space $\mathcal{R} : \mathbf{a} \to \mathbf{R}$ of arbitrary continuous one-to-one three-dimensional mappings identifying $\mathcal{R}$ elements that are obtained from one another with the help of the gauge transformation (31) with a fixed solenoidal field $\Omega_0(\mathbf{a})$.

The integral representation for Ω (29) is another formulation of the frozenness condition - after integration of the relation (29) over area σ, transverse to the lines of Ω, follows that the flux of this vector remains constant in time:

$$\int_{\sigma(t)} (\Omega, d\mathbf{S}_{\mathbf{r}}) = \int_{\sigma(0)} (\Omega_0, d\mathbf{S}_{\mathbf{a}}).$$

Here $\sigma(t)$ is the image of $\sigma(0)$ under the transformation (30).

It is important also that $\Omega_0(\mathbf{a})$ can be expressed explicitly in terms of the instantaneous value of the vorticity and the mapping $\mathbf{a} = \mathbf{a}(\mathbf{r}, t)$, inverse to (30). By integrating over the variables $\mathbf{a}$ in the relation (29),

$$\Omega(\mathbf{R}) = \frac{(\Omega_0(\mathbf{a})\nabla_\mathbf{a})\mathbf{R}(\mathbf{a})}{\det||\partial\mathbf{R}/\partial\mathbf{a}||}, \tag{32}$$

where $\Omega_0(\mathbf{a})$ can be represented in the form:

$$\Omega_0(\mathbf{a}) = \det||\partial\mathbf{R}/\partial\mathbf{a}||(\Omega(\mathbf{r})\nabla)\mathbf{a}. \tag{33}$$

This formula is nothing more than the Cauchy invariant (19). We note that according to Eq. (32) the vector $\mathbf{b} = (\Omega_0(\mathbf{a})\nabla_\mathbf{a})\mathbf{R}(\mathbf{a})$ is tangent to $\Omega(\mathbf{R})$. It is natural to introduce parameter s as an arc length of the initial vortex lines $\Omega_0(\mathbf{a})$ so that

$$\mathbf{b} = \Omega_0(\nu)\frac{\partial\mathbf{R}}{\partial s}.$$

In this expression Ω_0 depends on the transverse parameter ν labeling each vortex line. In accordance with this, the representation (29) can be written in the form

$$\Omega(\mathbf{r}, t) = \int \Omega_0(\nu)d^2\nu \int \delta(\mathbf{r} - \mathbf{R}(s, \nu, t))\frac{\partial\mathbf{R}}{\partial s}ds, \tag{34}$$

whence the meaning of the new variables becomes clearer: To each vortex line with index ν there is associated the closed curve

$$\mathbf{r} = \mathbf{R}(s, \nu, t),$$

and the integral (34) itself is a sum over vortex lines. We notice that the parametrization by introduction of s and ν is local. Therefore as global the representation (34) can be used only for distributions with closed vortex lines.

To get the equation of motion for $\mathbf{R}(\nu, s, t)$ the representation (34) (in the general case - (29)) must be substituted in the Euler equation (26) and then a Fourier transform with respect to spatial coordinates performed. As a result of simple integration one can obtain:

$$\left[\mathbf{k} \times \int \Omega_0(\nu)d^2\nu \int ds e^{-i\mathbf{k}\mathbf{R}}[\mathbf{R}_s \times \{\mathbf{R}_t(\nu, s, t) - \mathbf{v}(\mathbf{R}, t)\}]\right] = 0.$$

This equation can be resolved by putting integrand equal identically to zero:

$$[\mathbf{R}_s \times \mathbf{R}_t(\nu, s, t)] = [\mathbf{R}_s \times \mathbf{v}(\mathbf{R}, t)]. \tag{35}$$

With this choice there remains the freedom in both changing the parameter s and relabelling the transverse coordinates ν. In the general case of arbitrary topology of the field $\Omega_0(\mathbf{a})$ the vector $\mathbf{R}_s$ in the equation (35) must be replaced by the vector $\mathbf{b} = (\Omega_0(\mathbf{a})\nabla_\mathbf{a})\mathbf{R}(\mathbf{a}, t)$. Notice that, as it follows from (35) and (32), a motion of a point on the manifold $\mathcal{M}_{\Omega_0}$ is determined only by the transverse to $\Omega(\mathbf{r})$ component of the generalized velocity.

The obtained equation (35) is the equation of motion for vortex lines. In accordance with (35) the evolution of each vector $\mathbf{R}$ is principally transverse to the vortex line. The longitudinal component of velocity does not effect on the line dynamics.

The description of vortex lines with the help of equations (34) and (35) is a mixed Lagrangian-Eulerian one: The parameter ν has a clear Lagrangian origin whereas the coordinate s remains Eulerian.

4   Variational principle

The key observation for formulation of the variational principle is that the following general equality holds for functionals that depend only on Ω:

$$\left[\mathbf{b} \times \mathrm{curl}\left(\frac{\delta F}{\delta \Omega(\mathbf{R})}\right)\right] = \frac{\delta F}{\delta \mathbf{R}(\mathbf{a})}\bigg|_{\Omega_0}. \tag{36}$$

For this reason, the right-hand-side of (35) equals the variational derivative $\delta \mathcal{H}/\delta \mathbf{R}$:

$$[(\Omega_0(\mathbf{a})\nabla_\mathbf{a})\mathbf{R}(\mathbf{a}) \times \mathbf{R}_t(\mathbf{a})] = \frac{\delta \mathcal{H}\{\Omega\{\mathbf{R}\}\}}{\delta \mathbf{R}(\mathbf{a})}\bigg|_{\Omega_0}. \tag{37}$$

It is not difficult to check now that the equation (37) described dynamics of vortex line is equivalent to the requirement of extremum of the action ($\delta S = 0$) with the Lagrangian [16]

$$\mathcal{L} = \frac{1}{3}\int d^3\mathbf{a}([\mathbf{R}_t(\mathbf{a}) \times \mathbf{R}(\mathbf{a})] \cdot (\Omega_0(\mathbf{a})\nabla_\mathbf{a})\mathbf{R}(\mathbf{a})) - \mathcal{H}(\{\Omega\{\mathbf{R}\}\})). \tag{38}$$

Thus, we have introduced the variational principle for the Hamiltonian dynamics of the divergence-free vector field topologically equivalent to $\Omega_0(\mathbf{a})$.

Let us discuss some properties of the equations of motion (37), which are associated with excess parametrization of elements of $\mathcal{M}_{\Omega_0}$ by objects from $\mathcal{R}$. We want to pay attention to the fact that From Eq. (36) follows the property that the vector $\mathbf{b}$ and $\delta F/\delta \mathbf{R}(\mathbf{a})$ are orthogonal for all functionals defined on $\mathcal{M}_{\Omega_0}$. In other words the variational derivative of the gauge-invariant functionals should be understood (specifically, in (36)) as

$$\hat{P}\frac{\delta F}{\delta \mathbf{R}(\mathbf{a})},$$

where $\hat{P}_{ij} = \delta_{ij} - \tau_i\tau_j$ is a projector and $\tau = \mathbf{b}/|\mathbf{b}|$ a unit tangent (to vortex line) vector. Using this property as well the transformation formula

(36) it is possible, by a direct calculation of the bracket (3), to obtain the Poisson bracket (between two gauge-invariant functionals) expressed in terms of vortex lines:

$$\{F, G\} = \int \frac{d^3\mathbf{a}}{|\mathbf{b}|^2} \left(\mathbf{b} \cdot \left[\hat{P} \frac{\delta F}{\delta \mathbf{R}(\mathbf{a})} \times \hat{P} \frac{\delta G}{\delta \mathbf{R}(\mathbf{a})} \right] \right). \tag{39}$$

The new bracket (39) does not contain variational derivatives with respect to $\Omega_0(\mathbf{a})$. Therefore, with respect to the initial bracket the Cauchy invariant $\Omega_0(\mathbf{a})$ is a Casimir fixing the invariant manifolds $\mathcal{M}_{\Omega_0}$ on which it is possible to introduce the variational principle (38).

In the case of the hydrodynamics of a superfluid liquid a Lagrangian of the form (38) was apparently first used by Rasetti and Regge [17] to derive an equation of motion, identical to Eq. (35), but for a separate vortex filament. Later, on the base of the results [17], Volovik and Dotsenko Jr. [18] obtained the Poisson bracket between the coordinates of the vortices and the velocity components for a continuous distribution of vortices. The expression for these brackets can be extracted without difficulty from the general form for the Poisson brackets (39) . However, the noncanonical Poisson brackets obtained in [17,18] must be used with care. Their direct application gives for the equation of motion of the coordinate of a vortex filament an answer that is not gauge-invariant. For a general variation, which depends on time, additional terms describing flow along a vortex appear in the equation of motion. For this reason, the dynamics of curves (including vortex lines) is in principle "transverse" with respect to the curve itself.

We note that for two-dimensional (in the $x-y$ plane) flows the variational principle for action with the Lagrangian (38) leads to the well-known fact that $X(\nu, t)$- and $Y(\nu, t)$- coordinates of each vortex are canonically conjugated quantities (see [3]).

5   Integrable hydrodynamics

Now we present an example of the equations of the hydrodynamic type (26), for which transition to the representation of vortex lines permits to establish of the fact of their integrability [16].

Consider the Hamiltonian

$$\mathcal{H}\{\Omega(\mathbf{r})\} = \int |\Omega| d\mathbf{r} \tag{40}$$

and the corresponding equation of frozenness (26) with the generalized velocity

$$\mathbf{v} = \mathrm{curl}\,(\Omega/\Omega)\,.$$

We assume that vortex lines are closed and apply the representation (34). Then due to (32) the Hamiltonian in terms of vortex lines is decomposed as a sum of Hamiltonians of vortex lines:

$$\mathcal{H}\{\mathbf{R}\} = \int |\Omega_0(\nu)| d^2\nu \int \left|\frac{\partial \mathbf{R}}{\partial s}\right| ds. \qquad (41)$$

The standing here integral over s is the total length of a vortex line with index ν. According to (37), with respect to these variables the equation of motion for the vector $\mathbf{R}(\nu, s)$ is local, it does not contain terms describing interaction with other vortices:

$$\eta[\tau \times \mathbf{R}_t(\nu, s, t)] = [\tau \times [\tau \times \tau_s]]. \qquad (42)$$

Here $\eta = \text{sign}(\Omega_0)$, $\tau = \mathbf{R}_s/|\mathbf{R}_s|$ is the unit vector tangent to the vortex line.

This equation is invariant against changes $s \to \tilde{s}(s, t)$. Therefore the equation (42) can be resolved relative to $\mathbf{R}_t$ up to a shift along the vortex line – the transformation unchanged the vorticity Ω. This means that to find Ω it is enough to have one solution of the equation

$$\eta|\mathbf{R}_s|\mathbf{R}_t = [\tau \times \tau_s] + \beta \mathbf{R}_s, \qquad (43)$$

which follows from (42) for some value of β. Arisen from here equation for τ as a function of filament length l $(dl = |\mathbf{R}_s| ds)$ and time t (by choosing a new value $\beta = 0$) reduces to the integrable one-dimensional Landau-Lifshits equation for a Heisenberg ferromagnet:

$$\eta\frac{\partial \tau}{\partial t} = \left[\tau \times \frac{\partial^2 \tau}{\partial l^2}\right].$$

This equation is gauge-equivalent to the 1D nonlinear Schrödinger equation [19] and, for instance, can be reduced to the NLSE by means of the Hasimoto transformation [15]:

$$\psi(l, t) = \kappa(l, t) \cdot \exp(i \int^l \chi(\tilde{l}, t) d\tilde{l}),$$

where $\kappa(l, t)$ is a curvature and $\chi(l, t)$ the line torsion.

The considered system with the Hamiltonian (40) has direct relation to hydrodynamics. As known (see the paper [15] and references therein), the local approximation for thin vortex filament (under assumption of smallness of the filament width to the characteristic longitudinal scale) leads to the Hamiltonian (41) but only for one separate line. Respectively, the equation (26) with the Hamiltonian (40) can be used for description of motion of a few number of vortex filaments, thickness of which is small compared with a distance between them. In this case (nonlinear) dynamics of each filament is independent upon neighbor behavior. In the framework of this model singularity appearance (intersection of vortices) is of an inertial character very similar to the wave breaking in gas-dynamics. Of course, this approximation does not work on distances between filaments comparable with filament thickness.

It should be noted also that for the given approximation the Hamiltonian of vortex line is proportional to the filament line whence its conservation follows that, however, in no cases is adequate to behavior of vortex filaments in turbulent flows where usually process of vortex filament stretching takes place. It is desirable to have the better model free from this lack. A new model must necessarily describe nonlocal effects.

In addition we would like to say that the list of equations (26) which can be integrated with the help of representation (34) is not exhausted by (40). So, the system with the Hamiltonian

$$\mathcal{H}_\chi\{\Omega(\mathbf{r})\} = \int |\Omega|\chi d\mathbf{r} \tag{44}$$

is gauge equivalent to the modified KdV equation

$$\psi_t + \psi_{lll} + \frac{3}{2}|\psi|^2\psi_l = 0 \; -$$

the second one after NLSE in the hierarchy generated by Zakharov-Shabat operator. As against previous model (40) some physical application of (44) has not yet been found.

6   Lagrangian description of MHD

Consider now how the relabelling symmetry works in the ideal MHD. First, rewrite equations of motion (5-7) in the Lagrangian representation by introducing markers **a** for fluid particles

$$\mathbf{x} = \mathbf{x}(\mathbf{a}, t)$$

with

$$\mathbf{v}(\mathbf{x}, t) = \dot{\mathbf{x}}(\mathbf{a}, t).$$

In this case the continuity equation (5) and the equation for magnetic field (7) can be integrated. The density and the magnetic field are expressed in terms of the Jacoby matrix by means of Eq. (15) and by the equation

$$B_i(x, t) = \frac{\partial x_i}{\partial a_k} B_{0k}(a), \tag{45}$$

where $\mathbf{B} = \mathbf{h}/\rho$. In the latter transformation the Jacoby matrix serves the evolution operator for vector $\mathbf{B}$. The vector $\mathbf{B}$, in turn, transforms as a vector.

In terms of Lagrangian variables the equation of motion (6) is written as follows

$$\frac{\partial x_i}{\partial a_k}\ddot{x}_i = -\frac{\partial w(\rho)}{\partial a_k} + \frac{J}{4\pi\rho_0(a)}[\text{curl}\,\mathbf{h}\times\mathbf{h}]_i\frac{\partial x_i}{\partial a_k} \qquad (46)$$

With the help of relation (45) and Eq. (17) the vector $\mathbf{u}$ given by (16) will satisfy the equation

$$\frac{d\mathbf{u}}{dt} = \nabla\left(\frac{\mathbf{v}^2}{2} - w\right) - \frac{1}{4\pi}\left[\mathbf{B}_0(a)\times\text{curl}_a\mathbf{H}\right]. \qquad (47)$$

Here vector $\mathbf{B}_0(\mathbf{a}) = \mathbf{h}_0(\mathbf{a})/\rho_0(\mathbf{a})$ is a Lagrangian invariant and $\mathbf{H}$ represents the co-adjoint transformation of the magnetic field, analogous to (16):

$$H_i(a,t) = \frac{\partial x_m}{\partial a_i}h_m(x,t).$$

Now by analogy with (17) and (21), integration of Eq.(47) over time leads to the Weber type transformation:

$$\mathbf{u}(\mathbf{a},t) = \mathbf{u}_0(\mathbf{a}) + \nabla_a\Phi + \left[\mathbf{B}_0(\mathbf{a})\times\text{curl}_a\tilde{\mathbf{S}}\right]. \qquad (48)$$

Here $\mathbf{u}_0(\mathbf{a})$ is a new Lagrangian invariant which can be chosen as pure transverse, namely, with $\text{div}_a\,\mathbf{u}_0 = 0$. This new Lagrangian invariant cannot be expressed through the observed physical quantities such as magnetic field, velocity and density. In spite of this fact, as it will be shown in the next section, the vector Lagrangian invariant $\mathbf{u}_0(\mathbf{a})$ has a clear physical meaning. As far as new variables Φ and $\tilde{\mathbf{S}}$, they obey the equations:

$$\frac{d\Phi}{dt} = \frac{\mathbf{v}^2}{2} - w,$$

$$\frac{d\tilde{\mathbf{S}}}{dt} = -\frac{\mathbf{H}}{4\pi} + \nabla_a\psi.$$

The transformation (48) for velocity $\mathbf{v}(\mathbf{x},t)$ takes the form:

$$\mathbf{v} = u_{0k}(\mathbf{a})\nabla a_k + \nabla\Phi + \left[\frac{\mathbf{h}}{\rho}\times\text{curl}\,\mathbf{S}\right] \qquad (49)$$

where $\mathbf{S}$ is the vector $\tilde{\mathbf{S}}$ transformed by means of the rule (16):

$$S_i(x,t) = \frac{\partial a_k}{\partial x_i}\tilde{S}_k(a,t).$$

In Eulerian description Φ satisfies the Bernoulli equation

$$\frac{\partial\Phi}{\partial t} + (\mathbf{v}\nabla)\Phi - \frac{\mathbf{v}^2}{2} + w = 0 \qquad (50)$$

and equation of motion for $\mathbf{S}$ is of the form:

$$\frac{\partial \mathbf{S}}{\partial t} + \frac{\mathbf{h}}{4\pi} - [\mathbf{v} \times \text{curl} \mathbf{S}] + \nabla \psi_1 = 0. \tag{51}$$

For $\mathbf{u}_0 = 0$ the transformation (49) was introduced for ideal MHD by Za-kharov and Kuznetsov in 1970 [1]. In this case magnetic field $\mathbf{h}$ and vector $\mathbf{S}$ as well as Φ and ρ are two pairs of canonically conjugated variables. It is interesting to note that in the canonical case the equations of motion for $\mathbf{S}$ and Φ obtained in [1] coincide with (50) and (51). However, the canonical parametrization describes partial type of flows, in particular, it does not describe topological nontrivial flows for which mutual knottiness between magnetic and vortex lines is not equal to zero. This topological characteristics is given by the integral $\int (\mathbf{v}, \mathbf{h}) d\mathbf{x}$. Only when $\mathbf{u}_0 \neq 0$ this integral takes non-zero values.

7   Frozen-in MHD fields

To clarify meaning of new Lagrangian invariant $\mathbf{u}_0(\mathbf{a})$ we remind that the MHD equations (5-7) can be obtained from two-fluid system where electrons and ions are considered as two separate fluids interacting each other by means of self-consistent electromagnetic field. The MHD equations follow from two-fluid equations in the low-frequency limit when characteristic frequencies are less than ion gyro-frequency. The latter assumes i) neglecting by electron inertia, ii) smallness of electric field with respect to magnetic field, and iii) charge quasi-neutrality. We write down at first some intermediate system called often as MHD with dispersion [20]:

$$\text{curl curl} \mathbf{A} = \frac{4\pi e}{c}(n_1 \mathbf{v}_1 - n_2 \mathbf{v}_2), \tag{52}$$

$$(\partial_t + \mathbf{v}_1 \nabla)m\mathbf{v}_1 = \frac{e}{c}(-\mathbf{A}_t + [\mathbf{v}_1 \times \text{curl } \mathbf{A}]) - \nabla \frac{\partial \varepsilon}{\partial n_1}, \tag{53}$$

$$0 = -\frac{e}{c}(-\mathbf{A}_t + [\mathbf{v}_2 \times \text{curl } \mathbf{A}]) - \nabla \frac{\partial \varepsilon}{\partial n_2}. \tag{54}$$

In these equations $\mathbf{A}$ is the vector potential so that the magnetic field $\mathbf{h} = \text{curl} \mathbf{A}$ and electric field $\mathbf{E} = -\frac{1}{c}\mathbf{A}_t$. This system is closed by two continuity equations for ion density n_1 and electron density n_2:

$$n_{1,t} + \nabla(n_1 \mathbf{v}_1) = 0, \qquad n_{2,t} + \nabla(n_2 \mathbf{v}_2) = 0. \tag{55}$$

In this system $\mathbf{v}_{1,2}$ are velocities of ion and electron fluids, respectively. The first equation of this system is a Maxwell equation for magnetic field in static limit. The second equation is equation of motion for ions. The next one is equation of motion for electrons in which we neglect by electron inertia. By means of the latter equation one can obtain the equation of frozenness of magnetic field into electron fluid (this is another Maxwell equation):

$$\mathbf{h}_t = \mathrm{curl}[\mathbf{v}_2 \times \mathbf{h}].$$

Applying the operator div to (52) gives with account of continuity equations the quasi-neutrality condition: $n_1 = n_2 = n$. Next, by excluding n_2 and $\mathbf{v}_2$ we have finally the MHD equations with dispersion in its standard form [20]:

$$(\partial_t + \mathbf{v}\nabla)m\mathbf{v} = -\nabla w(n) + \frac{1}{4\pi n}[\mathrm{curl}\,\mathbf{h} \times \mathbf{h}], \qquad w(n) = \frac{\partial}{\partial n}\varepsilon(n,n),$$

$$n_t + \nabla(n\mathbf{v}) = 0, \qquad \mathbf{h}_t = \mathrm{curl}\left[\left(\mathbf{v} - \frac{c}{4\pi en}\mathrm{curl}\,\mathbf{h}\right) \times \mathbf{h}\right], \qquad (56)$$

where $\mathbf{v}_1 = \mathbf{v}$, and $\varepsilon(n,n)$ is internal energy density so that $w(n)$ is entalpy per one pair ion-electron. The classical MHD follows from this system in the limit when the last term $c/(4\pi en)\mathrm{curl}\,\mathbf{h}$ in equation (56) should be neglected with respect to $\mathbf{v}$. At the same time, the vector potential $\mathbf{A}$ must be larger characteristic values of $(mc/e)\mathbf{v}$ in order to provide inertia and magnetic terms in Eq. (53) being of the same order of magnitude. Both requirements are satisfied if $\epsilon = c/(\omega_{pi}L) \ll 1$ where L is a characteristic scale of magnetic field variation and $\omega_{pi} = \sqrt{4\pi ne^2/m}$ is ion plasma frequency.

Unlike MHD equations (5-7), the given system has two frozen-in fields. These are the field $\Omega_2 = -\frac{e}{mc}\mathbf{h}$ frozen into electron fluid and the field

$$\Omega_1 = \mathrm{curl}(\mathbf{v} + \frac{e}{mc}\mathbf{A}) = \Omega - \Omega_2$$

frozen into ion component:

$$\Omega_{1t} = \mathrm{curl}\left[\mathbf{v} \times \Omega_1\right],$$

$$\Omega_{2t} = \mathrm{curl}\left[\mathbf{v}_2 \times \Omega_2\right]$$

where

$$\mathbf{v}_2 = \mathbf{v} - \frac{c}{4\pi en}\mathrm{curl}\,\mathbf{h}.$$

Hence for both fields one can construct two Cauchy invariants by the same rule (19) as for ideal hydrodynamics:

$$\Omega_{10}(\mathbf{a}) = J_1(\Omega_1(\mathbf{x},t)\nabla)\mathbf{a}(\mathbf{x},t) \qquad (57)$$

where $\mathbf{a}(\mathbf{x},t)$ is inverse mapping to $\mathbf{x} = \mathbf{x}_1(\mathbf{a},t)$ which is solution of the equation $\dot{\mathbf{x}} = \mathbf{v}(\mathbf{x},t)$;

$$\Omega_{20}(\mathbf{a}_2) = J_2(\Omega_2(\mathbf{x},t)\nabla)\mathbf{a}_2(\mathbf{x},t) \qquad (58)$$

with $\mathbf{a}_2(\mathbf{x},t)$ inverse to the mapping $\mathbf{x} = \mathbf{x}_2(\mathbf{a}_2, t)$ and $\dot{\mathbf{x}} = \mathbf{v}_2(\mathbf{x},t)$.

In order to get the corresponding Weber transformation for MHD as a limit of the system it is necessary to introduce two momenta for ion and electron fluids:

$$\mathbf{p}_1 = m\mathbf{v} + \frac{e}{c}\mathbf{A} \tag{59}$$

$$\mathbf{p}_2 = -\frac{e}{c}\mathbf{A}. \tag{60}$$

In these expressions the terms containing the vector potential are greater sum of $\mathbf{p}_1$ and $\mathbf{p}_2$ in parameter ϵ. For each momentum in Lagrangian representation one can get equations, analogous to (14), (17):

$$\frac{\partial x_k}{\partial a_{1i}}\frac{dp_{1k}}{dt} = -p_{1k}\frac{\partial v_k}{\partial a_{1i}} + \frac{\partial}{\partial a_{1i}}\left(-\frac{\partial \varepsilon}{\partial n_1} + \frac{e}{c}(\mathbf{v}\cdot\mathbf{A}) + m\frac{v^2}{2}\right) \tag{61}$$

$$\frac{\partial x_k}{\partial a_{2i}}\frac{dp_{2k}}{dt} = -p_{2k}\frac{\partial v_{2k}}{\partial a_{2i}} + \frac{\partial}{\partial a_{2i}}\left(-\frac{\partial \varepsilon}{\partial n_2} - \frac{e}{c}(\mathbf{v}_2\cdot\mathbf{A})\right). \tag{62}$$

By introducing the vector $\tilde{\mathbf{p}}$ for each type of fluids, by the same rule as (16),

$$\tilde{p}_i = \frac{\partial x_k}{\partial a_i}p_k,$$

after integration over time of equations of motion for $\tilde{\mathbf{p}}$ one can arrive at two Weber transformations for each momentum:

$$\mathbf{p}_1 = \tilde{p}_{1i}(a_1)\nabla a_{1i} + \nabla \Phi_1, \tag{63}$$
$$\mathbf{p}_2 = \tilde{p}_{2i}(a_2)\nabla a_{2i} + \nabla \Phi_2. \tag{64}$$

In the limit $\epsilon \to 0$ the markers $\mathbf{a}_1$ and $\mathbf{a}_2$ can be put approximately equal. This means that their difference will be small:

$$\mathbf{a}_2 - \mathbf{a}_1 = \mathbf{d} \sim \epsilon.$$

Besides, due to charge quasi-neutrality, Jacobians with respect to a_1 and a_2 must be equal each other (here we put $n_{10}(\mathbf{a}_1) = n_{20}(\mathbf{a}_2) = 1$ without loss of generality):

$$\det\|\partial\mathbf{x}/\partial\mathbf{a}_1\| = \det\|\partial\mathbf{x}/\partial\mathbf{a}_2\|.$$

As a result, the infinitesimal vector $\mathbf{d}(\mathbf{a},t)$ relative to the argument $\mathbf{a}$ occurs divergence free: $\partial d_i/\partial a_i = 0$.

Then, summing (63) and (64) and considering the limit $\epsilon \to 0$, we obtain the Weber-type transformation coinciding with (48):

$$\mathbf{u}(\mathbf{a},t) = \mathbf{u}_0(\mathbf{a}) + \nabla_a\Phi + \left[\mathbf{B}_0(\mathbf{a}) \times \mathrm{curl}_a\,\tilde{\mathbf{S}}\right], \tag{65}$$

where vectors $u_0(a)$ and $\tilde{S}$ are expressed through the Lagrangian invariants $\tilde{p}_1(a)$ and $\tilde{p}_2(a)$ and displacement d between electron and ion by means of relations [21]:

$$u(a,t) = \frac{1}{m}(\tilde{p}_1(a) + \tilde{p}_2(a)),$$

$$d = -\frac{mc}{e}\mathrm{curl}_a\tilde{S}.$$

Important that in (65) all terms are of the same order of magnitude (zero order relative to ϵ). Curl of vectors $\tilde{p}_1(a)$ and $\tilde{p}_2(a_2)$ yield the corresponding Cauchy invariants (57) and (58).

8 Relabeling symmetry in MHD

Now let us show how existence of new Lagrangian invariants corresponds to the relabeling symmetry.

Consider the MHD Lagrangian [2],

$$\mathcal{L}_* = \int \left(\rho\frac{v^2}{2} - \rho\tilde{\varepsilon}(\rho) - \frac{h^2}{8\pi}\right) d\mathbf{r},$$

where we neglect by contribution from electric field in comparison with that from magnetic field. Here $\tilde{\varepsilon}(\rho)$ is specific internal energy.

In terms of mapping $x(a, t)$ the Lagrangian $\mathcal{L}_*$ is rewritten as follows [22]:

$$\mathcal{L}_* = \int \frac{\dot{x}^2}{2}d^3a - \int \tilde{\varepsilon}(J_x^{-1}(a))d^3a - \frac{1}{8\pi}\int \left(\frac{(h_0(a)\nabla_a)x}{J_x(a)}\right)^2 J_x(a)d^3a. \quad (66)$$

Here density and magnetic field are expressed by means of relations

$$\rho = 1/J_x, \qquad h = (h_0(a)\nabla_a)x/J_x,$$

and

$$J_x(a, t) = \det\|\partial x/\partial a\|$$

is the Jacobian of mapping $x = x(a,t)$ and initial density is put to equal 1. Notice, that variation of the action with by the Lagrangian (66) relative to $x(a)$ gives the equation of motion (46) (or the equivalent equation for vector u (47)).

Due to the presence of magnetic field in the Lagrangian (66), the relabeling symmetry, in comparison with ideal hydrodynamics, reduces. If the first two terms in (66) are invariant with respect to all incompressible changes $a \to a(b)$ with $J|_b = 1$, invariance of the last term, however, restricts the class of possible deformations up to the following class

$$(h_0(a)\nabla_a)b = h_0(b).$$

For infinitesimal transformations

$$\mathbf{a} \rightarrow \mathbf{a} + \tau \mathbf{g}(\mathbf{a})$$

where τ is a (small) group parameter the vector $\mathbf{g}$ must satisfy two conditions:

$$\text{div}_a \mathbf{g} = 0, \quad \text{curl}_a [\mathbf{g} \times \mathbf{h_0}] = 0. \tag{67}$$

The first condition is the same as for ideal hydrodynamics, the second one provides conservation of magnetic field frozenness.

The conservation laws generating by this symmetry, in accordance with Noether theorem, can be obtained by standard scheme from the Lagrangian (66). They are written through the infinitesimal deformation $\mathbf{g}(\mathbf{a})$ as integral over $\mathbf{a}$:

$$I = \int (\mathbf{u}, \mathbf{g}(\mathbf{a})) da \tag{68}$$

where the vector $\mathbf{u}$ is given by (16). Putting $\mathbf{g} = \mathbf{h_0}$ from this (infinite) family of integrals one gets the simplest one

$$I_{ch} = \int (\mathbf{v}, \mathbf{h}) d\mathbf{r}$$

which represents a cross-helicity characterizing degree of mutual knottiness of vortex and magnetic lines.

The conservation laws (68) are compatible with the Weber-type transformation. Really, substituting (48) into (68) and using (67) one leads to the relation

$$\int (\mathbf{u_0}(\mathbf{a}), \mathbf{g}(\mathbf{a})) da.$$

Hence conservation of (68) also follows. Note that if one would not suppose an independence of $\mathbf{u_0}$ on t then, due to arbitrariness of $\mathbf{g}(\mathbf{a})$, this could be considered as independent verification of conservation of solenoidal field $\mathbf{u_0}$:

$$\frac{d}{dt} \mathbf{u_0} = 0.$$

The MHD equations expressed in terms of Lagrangian variables become Hamiltonian ones, as in usual mechanics, for momentum $\mathbf{p} = \dot{\mathbf{x}}$ and coordinate $\mathbf{x}$. These variables assign the canonical Poisson structure.

In the Eulerian representation the MHD equations can be written also in the Hamiltonian form [14]:

$$\rho_t = \{\rho, H\}, \quad \mathbf{v}_t = \{\mathbf{v}, H\}, \quad \mathbf{h}_t = \{\mathbf{h}, H\},$$

where noncanonical Poisson bracket $\{F, G\}$ is given by the expression (9). As for ideal hydrodynamics, this Poisson bracket occurs to be degenerated. For example, the cross helicity I_{ch} serves a Casimir for the bracket (9). The reason of the Poisson bracket degeneracy is the same as for one-fluid hydrodynamics - it is connected with a relabeling symmetry of Lagrangian markers.

For incompressible case the Poisson bracket (9) reduces so that it can be expressed only through magnetic field $\mathbf{h}$ and vorticity Ω:

$$\{F,G\} = \int \left(\frac{\mathbf{h}}{\rho} \cdot \left(\left[\mathrm{curl}\frac{\delta F}{\delta \mathbf{h}} \times \mathrm{curl}\frac{\delta G}{\delta \Omega}\right] - \left[\mathrm{curl}\frac{\delta G}{\delta \mathbf{h}} \times \mathrm{curl}\frac{\delta F}{\delta \Omega}\right]\right)\right) d^3\mathbf{r} \quad (69)$$

$$+ \int \left(\Omega \left[\mathrm{curl}\frac{\delta F}{\delta \Omega} \times \mathrm{curl}\frac{\delta G}{\delta \Omega}\right]\right) d^3\mathbf{r}.$$

This bracket remains also degenerated.

9 Variational principle for incompressible MHD

By analogy with incompressible hydrodynamics, one can introduce magnetic line representation:

$$\mathbf{h}(\mathbf{r},t) = \int \delta(\mathbf{r} - \mathbf{R}(\mathbf{a},t))(\mathbf{h}_0(\mathbf{a})\nabla_\mathbf{a})\mathbf{R}(\mathbf{a},t)d^3\mathbf{a}. \quad (70)$$

For vorticity the analog of vortex line parametrization (29) can be obtained, for instance, as a limit $\epsilon \to 0$ of the corresponding representations for the two-fluid system. Calculations give [21]:

$$\Omega(\mathbf{r},t) = \int \delta(\mathbf{r} - \mathbf{R}(\mathbf{a},t))((\Omega_0(\mathbf{a}) + \mathrm{curl}_\mathbf{a}[\mathbf{h}_0(\mathbf{a}) \times \mathbf{U}(\mathbf{a},t)])\nabla_\mathbf{a})\mathbf{R}(\mathbf{a},t)d^3\mathbf{a}, \quad (71)$$

Here the field $\mathbf{U}(\mathbf{a},t)$ is not assumed solenoidal, as well as the Jacobian of mapping $\mathbf{r} = \mathbf{R}(\mathbf{a},t)$ is not equal to unity.

From the corresponding limit of the two-fluid system to incompressible MHD it is possible also to get the expression for Lagrangian

$$L = \int d^3\mathbf{a}([(\mathbf{h}_0\nabla_\mathbf{a})\mathbf{R} \times (\mathbf{U}\nabla_\mathbf{a})\mathbf{R}] \cdot \mathbf{R}_t) + \quad (72)$$

$$+1/3 \int d^3\mathbf{a}([\mathbf{R}_t \times \mathbf{R}] \cdot (\Omega_0\nabla_\mathbf{a})\mathbf{R}) - \mathcal{H}\{\Omega\{\mathbf{R},\mathbf{U}\},\mathbf{h}\{\mathbf{R}\}\}.$$

The Hamiltonian of the incompressible MHD $\mathcal{H}_{MHD}$ in terms of $\mathbf{U}(\mathbf{a},t)$ and $\mathbf{R}(\mathbf{a},t)$ takes the form

$$\mathcal{H}_{MHD} = \frac{1}{8\pi} \int \frac{((\mathbf{h}_0(\mathbf{a})\nabla_\mathbf{a})\mathbf{R}(\mathbf{a}))^2}{\det||\partial\mathbf{R}/\partial\mathbf{a}||} d^3\mathbf{a}+$$

$$+\frac{1}{8\pi} \int \int \frac{((\Omega(\mathbf{a}_1)\nabla_1)\mathbf{R}(\mathbf{a}_1) \cdot (\Omega(\mathbf{a}_2)\nabla_2)\mathbf{R}(\mathbf{a}_2))}{|\mathbf{R}(\mathbf{a}_1) - \mathbf{R}(\mathbf{a}_2)|} d^3\mathbf{a}_1 d^3\mathbf{a}_2, \qquad (73)$$

where we introduce the notation

$$\Omega(\mathbf{a}, t) = \Omega_0(\mathbf{a}) + \mathrm{curl}_{\mathbf{a}}[\mathbf{h}_0(\mathbf{a}) \times \mathbf{U}(\mathbf{a}, t)].$$

Equations of motion for $\mathbf{U}$ and $\mathbf{R}$ follow from the variational principle for action with Lagrangian (72):

$$[(\mathbf{h}_0\nabla_{\mathbf{a}})\mathbf{R} \times \mathbf{R}_t] \cdot (\partial \mathbf{R}/\partial a_\lambda) = -\delta \mathcal{H}/\delta U_\lambda, \qquad (74)$$

$$[(\Omega(\mathbf{a},t)\nabla_{\mathbf{a}})\mathbf{R} \times \mathbf{R}_t] - [(\mathbf{h}_0\nabla_{\mathbf{a}})\mathbf{R} \times (\mathbf{U}_t\nabla_{\mathbf{a}})\mathbf{R}] = \delta \mathcal{H}/\delta \mathbf{R}. \qquad (75)$$

These equations can be obtained also directly from the MHD system (5-7) by the same scheme as it was done for ideal hydrodynamics.

Thus, we have variational principle for the MHD-type equations for two solenoidal vector fields. Their topological properties are fixed by $\Omega_0(\mathbf{a})$ and $\mathbf{h}_0(\mathbf{a})$. These quantities represent Casimirs for the initial Poisson bracket (69). It is worth noting that the obtained equations of motion have the gauge invariant form. This gauge invariance is a remaining symmetry connected with relabeling of Lagrangian markers of magnetic lines in two-dimensional manifold which can be specified always locally. Coordinates of this manifold enumerate magnetic lines. This symmetry leads to conservation of volume of magnetic tubes including infinitesimally small magnetic tubes, namely, magnetic lines. This property explains why the Jacobian of the mapping $\mathbf{r} = \mathbf{R}(\mathbf{a},t)$ can be not equal identically to unity.

Acknowledgments

Authors thank A.B.Shabat for useful discussion of the connection between NLSE and equations (26), that resulted in integrability declaration for (44). This work was supported by the Russian Foundation of Basic Research under Grant no. 97-01-00093 and by the Russian Program for Leading Scientific Schools (grant no.96-15-96093). Partially the work of E.K. was supported by the Grant INTAS 96-0413, and the work of V.R. by the Grant of Landau Scholarship.

References

1. V.E.Zakharov and E.A.Kuznetsov, Doklady USSR Ac. Nauk. (Soviet Doklady), **194**, 1288 (1970).

2. V.E.Zakharov and E.A.Kuznetsov, Uspekhi fizicheskikh nauk (Physics Us-pekhi),**167**, 1137 (1997).

3. H.Lamb, Hydrodynamics, Cambridge Univ. Press, 1932.

4. E.A.Kuznetsov, A.V.Mikhailov, Phys. Lett., **77A**, 37 (1980).

5. J.J. Moreau, C.R.Acad. Sc. Paris, 252, 2810 (1961); H.K.Moffatt, J. Fluid Mech. 35, 117 (1969).

6. V.I.Arnold, Doklady Ac. Nauk SSSR, 162, 773-777 (1965) (in Russian).

7. V.I.Arnold, Uspekhi matematicheskikh nauk, 24, N3, 225 (1969); Mathematical methods of classical mechanics, Moscow, Nauka, 1974 (in Russian).

8. R.Salmon, Ann. Rev. Fluid Mech., **20**, 225 (1988).

9. L.D.Landau and E.M.Lifshits, *Electrodynamics of continuous media*, Moscow, Nauka (1982).

10. M.I. Monastyrskii and P.V.Sasorov, ZhETF (JETP), **93**, 1210, (1987).

11. V.V.Yan'kov, ZhETF (JETP), **107**, 414,(1995).

12. S.S.Moiseev, R.Z.Sagdeev, A,V.Tur, and V.V.Yanovskii, ZhETF (JETP), **83**, 215, (1982).

13. B.N.Kuvshinov, FIzika Plazmy (Russian Plasma Phys.), **22**, 971, (1996).

14. P.J.Morrison and J.M.Greene, Phys. Rev. Lett., **45**, 790 (1980)

15. R.Hasimoto, J. Fluid Mech., **51**, 477 (1972).

16. E.A.Kuznetsov and V.P.Ruban, Pis'ma v ZhETF, **67**, 1012 (1998) [JETP Letters, **67**, 1076 (1998)].

17. M.Rasetti and T.Regge, Physica **80A**, 217 (1975).

18. G.E.Volovik, V.S.Dotsenko (jun.), Pis'ma ZhETF (JETP Letters), **29**, 630 (1979).

19. V.E.Zakharov and L.A.Takhtajan, Teor. Mat. Fiz., **38**, 26 (1979) (in Russian).

20. V.I.Karpman, *Nonlinear waves in dispersive media*, Moscow, Nauka, (1973) (in Russian).

21. V.P.Ruban, *Magnetic line motion in MHD*, Preprint of Landau Institute, JETP, submitted (1999).

22. V.I.Il'gisonis and V.P.Pastukhov, Fizika Plazmy (Russian Plasma Phys.), **22**, 228 (1996).

Quasi-Two-Dimensional Hydrodynamics and Interaction of Vortex Tubes

V.E. Zakharov

Landau Institute for Theoretical Physics, 2 Kosygin str., Moscow 117334, Russia

1 Introduction

This paper is long overdue. Most of the results presented here were obtained in 1986-87. Just a small portion of them (the equations for the dynamics of a pair of counter-rotating vortices and their self-similar solutions) were published in time - in 1988 [1]. The publication was very brief and did not include any details of the calculations.

Nevertheless, it was noticed and then generously cited by R. Klein, A. Majda and K. Damodaran [2]. In this publication I would like to express my gratitude to these authors who reobtained an essential part of the results published below. In spite of the fact that some of the results published below (equations for the systems of almost parallel vortices) could be found in their article, I believe that my paper deserves to be published. Some of the results presented here are completely new, and, which is more important, the methodology published here is completely different from the one used in [2].

In this article we develop a systematic approach to study stationary and nonstationary flows of ideal incompressible fluid under assumption that the gradients in one preferred direction z are much less than the gradients in the orthogonal plane. Such flows could be called quasi-two dimensional. There are two motivations for paying a special attention to this class of fluid motion.

One is connected with the classical problem of the "blow-up" in the Euler equation. According to the most plausible scenario, (see, for instance, [3]), in the point of blow-up the vorticity becomes infinite. As far as vorticity is a vector, this assumption presumes that the flow near the blow-up point is almost two-dimensional and the velocity field is concentrated mostly in the plane orthogonal to the vorticity vector. An elaboration of this regular tool for description of this type of flow looks very timely.

Another motivation is the vortex dynamics. This is a subject which has a chance to become the backbone of the future theory of turbulence. Probably, there is no way to explain qualitatively and quantitatively the fundamental phenomenon of intermittency but a careful study of the dynamics of the vortex tubes or their systems in a real three-dimensional nonstationary flow. "Vortices are the sinews of turbulence" said K. Moffatt (look at his lecture on the Seventh European Turbulence Conference [4]).

2   Quasi-two-dimensional hydrodynamics

Let us consider the dynamics of incompressible and inviscid fluid. We will describe the fluid motion by means of two stream functions Ψ and Φ. We put:

$$
\begin{aligned}
v_x &= \Psi_y + \Phi_{xz} \\
v_y &= -\Psi_x + \Phi_{yz} \\
v_z &= -\Delta_\perp \Phi = -\left(\frac{\partial^2}{\partial x^2} + \frac{\partial^2}{\partial y^2} \right) \Phi.
\end{aligned}
\tag{1}
$$

The condition of incompressibility $\nabla \cdot v = 0$ is automatically satisfied in virtue of (1). The components of the vorticity are defined as follow:

$$
\operatorname{curl} \mathbf{v} = \Omega_1 \mathbf{i} + \Omega_2 \mathbf{j} + \Omega_3 \mathbf{k}
\tag{2}
$$

where

$$
\Omega_1 = \Psi_{xz} - \frac{\partial}{\partial y} \Delta \Phi
\tag{3}
$$

$$
\Omega_2 = \Psi_{yz} + \frac{\partial}{\partial x} \Delta \Phi
\tag{4}
$$

$$
\Omega_3 = -\Delta_\perp \Psi.
\tag{5}
$$

The components of the vorticity satisfies the equation

$$
\frac{\partial \Omega_i}{\partial t} + (v \nabla) \Omega_i = (\Omega \Delta) v_i.
\tag{6}
$$

From (6) with $i = 3$, one can easily obtain the equation for Ψ:

$$
-\frac{\partial}{\partial t} \Delta_\perp \Psi + \left\{ \Psi, \Delta_\perp \Psi \right\} = \operatorname{div}_\perp \left(\Delta_\perp \Psi \nabla_\perp \Phi_z \right) -
$$
$$
- \operatorname{div}_\perp \left(\Delta_\perp \Phi \nabla_\perp \Psi_z \right) + \left\{ \Delta_\perp \Phi, \Phi_{zz} \right\}.
\tag{7}
$$

Here and further

$$
\left\{ A, B \right\} = A_x B_y - A_y B_x.
$$

To find the equation for Φ one can notice that the expression

$$
S = \frac{\partial \Omega_1}{\partial y} - \frac{\partial \Omega_2}{\partial x} = -\Delta_\perp \Delta \Phi
\tag{8}
$$

does not include Ψ.

From (6) with $i = 1, 2$ one can calculate dS/dt. Cumbersome calculations lead to the following result:

$$
\Delta_\perp \left(-\Delta \Phi_t + \{\Psi, \Delta\Phi\} \right) = \frac{\partial}{\partial z} \left\{ \frac{1}{2} \Delta_\perp (\nabla_\perp \Psi)^2 - \operatorname{div} \Delta_\perp \Psi \nabla_\perp \Psi \right\}
$$
$$
+ \Delta_\perp \{\Phi_z, \Psi_z\} - \frac{\partial}{\partial z} \{\Delta_\perp \Phi, \Phi_z\} +
$$
$$
+ \Delta_\perp (\nabla_\perp \Phi_z, \nabla_\perp \Delta \Phi) - \frac{\partial}{\partial z} \operatorname{div} \Delta_\perp \Phi \nabla_\perp \Delta \Phi
\tag{9}
$$

The system of equations (7), (9) is just a little bit bizarre form of the Euler equations for incompressible fluid. They can be transformed to the Navier-Stokes system by the simple change

$$\frac{\partial}{\partial t} \to \frac{\partial}{\partial t} - \nu \Delta, \tag{10}$$

where ν is a viscosity coefficient.

Equations (7) and (9) preserve the integral of energy

$$E = \frac{1}{2} \int \left\{ (\nabla_\perp \Psi)^2 + (\Delta_\perp \Phi)^2 \right\} dr\, dz \; ; \qquad \frac{dE}{dt} = 0. \tag{11}$$

The new form of Euler's equation is good for the description of quasi-two dimensional stationary and nonstationary flows, when

$$\frac{\partial}{\partial z} \ll \frac{\partial}{\partial x}, \frac{\partial}{\partial y}.$$

Let us first put

$$\frac{\partial \Psi}{\partial z} = 0; \qquad \frac{\partial \Phi}{\partial z} = 0.$$

Then

$$\frac{\partial}{\partial t} \Delta_\perp \Psi = \left\{ \Psi, \Delta_\perp \Psi \right\} \tag{12}$$

$$\frac{\partial}{\partial t} \Delta_\perp \Phi = \left\{ \Psi, \Delta_\perp \Phi \right\}. \tag{13}$$

Equation (12) is a standard two-dimensional Euler equation for an incompressible fluid. The passive scalar equation (13) describes a transport of the z-independent vertical velocity.

In the next step we keep in (7), (9) the terms linear in $\partial/\partial z$. One obtains the following system:

$$\frac{\partial}{\partial t} \Delta_\perp \Psi - \{\Psi, \Delta_\perp \Psi\} = \mathrm{div}_\perp \left[\Delta_\perp \Phi \nabla_\perp \Psi_z - \Delta_\perp \Psi \nabla_\perp \Phi_z \right] \tag{14}$$

$$\Delta_\perp \left(- \Delta_\perp \Phi_t + \{\Psi, \Delta_\perp \Phi\} \right) = \frac{\partial}{\partial z} \left[\frac{1}{2} \Delta_\perp (\nabla_\perp \Psi)^2 - \mathrm{div}_\perp \Delta_\perp \Psi \nabla_\perp \Psi \right]$$

$$\Delta_\perp (\nabla_\perp \Phi_z, \nabla_\perp \Delta_\perp \Phi) - \frac{\partial}{\partial z} \mathrm{div}_\perp \Delta_\perp \Phi \nabla_\perp \Delta_\perp \Phi. \tag{15}$$

The system (14), (15) describes a generic almost two-dimensional nonstationary flow of an incompressible fluid. If one assumes that the vertical velocity is small ($v_z \ll v_x, v_y$), the equation (15) can be simplified into the form:

$$\Delta_\perp \left(- \Delta_\perp \Phi_t + \{\Psi, \Delta_\perp \Phi\} \right) = \frac{\partial}{\partial z} \left[\frac{1}{2} \Delta_\perp (\nabla_\perp \Psi)^2 - \mathrm{div}_\perp \Delta_\perp \Psi \nabla_\perp \Psi \right] \tag{16}$$

This is remarkable that both systems (14), (15) and (14), 16) preserve the energy integral (11). Both are Hamiltonian systems, having the same Hamiltonian (11), but different Poisson's structures. In this article we will not discuss this interesting question in more details.

It is important to explore how far the approximate systems (14), (15) and (14), (16) differ from the exact Euler equation. One can use the axial symmetric case as a test. In this case

$$v_\phi = -\frac{\partial \Psi}{\partial r}$$

$$v_z = -\frac{1}{r}\frac{\partial}{\partial r} r \frac{\partial \Phi}{\partial r} \tag{17}$$

$$v_r = \Phi_{rz}. \tag{18}$$

Plugging (17) into (14) we find that v_ϕ satisfies the exact equation

$$\frac{\partial v_\phi}{\partial t} + v_r \frac{\partial v_\phi}{\partial r} + v_z \frac{\partial v_\phi}{\partial z} + \frac{v_r\, v_\phi}{r} = 0. \tag{19}$$

Equations (15) and (16) transform into the following reduced equations:

$$\frac{\partial}{\partial r}\left(\frac{\partial v_z}{\partial t} + v_r \frac{\partial v_z}{\partial r} + v_z \frac{\partial v_z}{\partial z}\right) = \frac{1}{r}\frac{\partial}{\partial z} v_\phi^2 \tag{20}$$

$$\frac{\partial}{\partial r}\frac{\partial v_z}{\partial t} = \frac{1}{r}\frac{\partial}{\partial z} v_\phi^2. \tag{21}$$

At the same time the exact Euler equation after excluding the pressure takes the form

$$\frac{\partial}{\partial r}\left(\frac{\partial v_z}{\partial t} + v_r \frac{\partial v_z}{\partial r} + v_z \frac{\partial v_z}{\partial z}\right) - \frac{\partial}{\partial z}\left(\frac{\partial v_r}{\partial t} + v_r \frac{\partial v_r}{\partial r} + v_z \frac{\partial v_r}{\partial z}\right) = \frac{1}{r}\frac{\partial}{\partial z} v_\phi^2 \tag{22}$$

There is one more modification of the quasi-two dimensional hydrodynamics. Instead of (16) one can use a more exact equation

$$\Delta_\perp\left(-\Delta\Phi_t + \{\Psi, \Delta\Phi\}\right) = \frac{\partial}{\partial z}\left[\frac{1}{2}\Delta_\perp(\nabla_\perp\Psi)^2 - \mathrm{div}\Delta_\perp\Psi\nabla_\perp\Psi\right]. \tag{23}$$

In the axial symmetric case it leads to the following modification of equation (22):

$$\frac{\partial}{\partial r}\frac{\partial v_z}{\partial t} - \frac{\partial}{\partial z}\frac{\partial v_r}{\partial z} = \frac{1}{r}\frac{\partial}{\partial z} v_\phi^2. \tag{24}$$

The difference between $\Delta\Phi$ and $\Delta_\perp\Phi$

$$\Delta\Phi - \Delta_\perp\Phi = \frac{\partial^2\Phi}{\partial z^2}$$

is of the second order in $\partial/\partial z$ and looks to be negligible. We will see further that this is not quite right. In some cases the approximation (16) is too crude, and the more exact equation (23) should be used to obtain the correct results.

Comparison of (19) and (21) demonstrates the fact that in the quasi-two dimensional equation one takes into account only the lowest order in $\partial/\partial z$.

3   Dynamics of the isolated vortex tube

In this chapter we apply the derived quasi-two dimensional hydrodynamic equations for the description of the dynamics of a single vortex tube. We will assume that the core of the tube is axially symmetric and small with respect to its characteristic curvature radius.

Let the central line of the tube be given by the formula

$$\chi = \chi_0 = a(t, z)$$
$$y = y_0 = b(t, x). \tag{25}$$

One can introduce polar coordinates in the coordinate frame attached with the vortex line

$$\chi = a + r \cos \phi$$
$$y = b + r \sin \phi. \tag{26}$$

Now

$$\{A, B\} = \frac{1}{r} \left(A_r \, B_\phi - A_\phi \, B_r \right) \tag{27}$$

$$\frac{\partial}{\partial t} \to D_t = \frac{\partial}{\partial t} - (\dot{a} \sin \phi + \dot{b} \cos \phi)\frac{\partial}{\partial r} - \frac{1}{r}(\dot{a} \cos \phi - \dot{b} \sin \phi)\frac{\partial}{\partial \phi}$$
$$\frac{\partial}{\partial z} \to D_z = \frac{\partial}{\partial z} - (a' \sin \phi + b' \cos \phi)\frac{\partial}{\partial r} - \frac{1}{r}(a' \cos \phi - b' \sin \phi)\frac{\partial}{\partial \phi}. \tag{28}$$

In polar coordinates

$$v_\phi = -\frac{\partial \Psi}{\partial r} + \frac{1}{r}\frac{\partial}{\partial \phi}D_z \Phi$$

$$v_r = \frac{1}{r}\frac{\partial \Psi}{\partial \phi} + \frac{\partial}{\partial r}D_z \Phi$$

$$v_z = -U = -\frac{1}{r}\frac{\partial}{\partial r}r\Phi - \frac{1}{r^2}\frac{\partial^2 \Phi}{\partial \phi^2} \tag{29}$$

$$\Omega_z = -\Omega = \Delta_\perp \Psi = -\frac{1}{r}\frac{\partial}{\partial r}r\Psi_r - \frac{1}{r^2}\frac{\partial^2 \Psi}{\partial \phi^2}$$

$$\Omega_r = D_z \Psi_r - \frac{1}{r}\frac{\partial}{\partial \phi}\Delta_\perp \Phi$$

$$\Omega_\phi = -\frac{1}{r}D_z \Psi_\phi + \frac{\partial}{\partial r}\Delta_\perp \Phi. \tag{30}$$

In the new variables, equation (14) can be rewritten as follows:

$$\frac{1}{r}\left(\Psi_r \Omega_\phi - \Psi_\phi \Omega_r\right) = D_t \, \Delta_\perp \Psi - \frac{1}{r}\frac{\partial}{\partial r} r \left(U \frac{\partial}{\partial r} D_z \Psi - \Omega_z \frac{\partial}{\partial r} D_z \Phi\right)$$

$$- \frac{1}{r}\frac{\partial}{\partial r}\left(U \frac{\partial}{\partial \phi} D_z \Psi - \Omega_z \frac{\partial}{\partial \phi} D_z \Phi\right). \tag{31}$$

In this article we will use only simplified equation (16). In polar coordinates it reads

$$\Delta_\perp \left(\Psi_r \, U_\phi - \Psi_\phi \, U_r\right) = \Delta_\perp D_t \, \Phi +$$

$$D_z \left(\frac{1}{2r}\frac{\partial}{\partial r} \Psi_r^2 + \frac{1}{r^2}\left(\Psi_{rr}\Psi_{\phi\phi} - \Psi_{r\phi}^2\right) - \frac{1}{r^3}\Psi_\phi \Psi_{r\phi} - \frac{1}{r^4}\Psi_\phi^4\right). \tag{32}$$

If $a = b = 0$, systems (30), (31) and (30), (32) have a trivial solution

$$\Psi = \Psi_0(r); \quad \Phi = 0, \tag{33}$$

describing a solitary stationary vortex tube. In this case only the angular component of velocity exists

$$v_\phi = -\frac{\partial \Psi_0}{\partial r}, \tag{34}$$

while the vorticity reduces to its vertical component

$$\Omega_3 = -\Omega_0 = \frac{1}{r}\frac{\partial}{\partial r} r \, \Psi_{0r}, \tag{35}$$

where $\Omega(r)$ is an arbitrary function. We will assume that this function has a finite support. In another words:

$$\Omega_0(r) \neq 0 \quad if \ \ 0 < r < \rho$$

$$\Omega_0(r) = 0 \quad if \ \ r > \rho$$

Here ρ can be interpreted as the size of the tube's core.

Let us define

$$\Gamma = -2\pi \int_0^\infty r\Omega_0(r)\,dr. \tag{36}$$

This is the total vorticity of the tube. As $r \to \infty$,

$$\Psi_{0r} \to \frac{\Gamma}{2\pi r}, \quad \Psi_0 \simeq \frac{\Gamma}{2\pi}\ln r + c. \tag{37}$$

The constant $c = c(z,t)$ is indefinite so far. In presence of a, b (32) is not anymore an exact solution. One has to seek the solution in the form

$$\Psi = \Psi_0(r, z, t) + \Psi'(r, z, \phi, t) \tag{38}$$
$$\Phi = \Phi_0(r, z, t) + \Phi'(r, z, \phi, t). \tag{39}$$

Here Ψ', Φ' are periodic functions of ϕ:

$$\Psi' = \sum_{n \neq 0} \Psi_n\, e^{in\phi}, \quad \Psi_{-n} = \Psi_n^*$$
$$\Phi' = \sum_{n \neq 0} \Phi_n\, e^{in\phi}, \quad \Phi_{-n} = \Phi_n^*. \tag{40}$$

Let us introduce the complex coordinate of the vortex line (the center of the vortex tube) as

$$w = a + ib. \tag{41}$$

The derivative $w' = \partial w / \partial z$ is a dimensionless parameter characterizing the angle between the tangent to the vortex line and the vertical axis. We assume that this angle is small,

$$w' \simeq \epsilon \ll 1. \tag{42}$$

Another small parameter,

$$\mu \sim \rho w'', \tag{43}$$

is the ratio of the size of the vortex line to its curvature radius. We assume that this parameter is small too,

$$\mu \ll 1. \tag{44}$$

All components of the Fourier series (39) are small and can be expanded in powers of ϵ and μ. The question about order of magnitudes averaged in angle stream functions Ψ_0 and Φ_0 is more delicate. In principle, a shape of tube described by the function $\Psi_0(r, z, t)$ can essentially depend on z.

In a long run this dependence can cause the development of intensive vertical flow, described by a relative high value of Φ_0. But all these effects are out of a scope of this article. We will assume that Φ_0, as well as the difference $\Psi_0(r, z, t) - \Psi_0(r)$ are small, and in the first approximation can be neglected.

In this theory the leading terms in the expansions (39) are:

$$\Phi \simeq \Phi_1\, e^{i\phi} + \Phi_1^*\, e^{-i\phi} \tag{45}$$
$$\Psi' \simeq \Psi_1\, e^{i\phi} + \Psi_1^*\, e^{-i\phi}. \tag{46}$$

In the terms of w, w^* the derivatives (28) can be rewritten as follows:

$$D_t = \frac{\partial}{\partial t} + \frac{i}{2}\left(\dot{w}\,e^{i\phi} - \dot{w}^*\,e^{-i\phi}\right)\frac{\partial}{\partial r} - \frac{1}{2r}\left(\dot{w}\,e^{i\phi} + \dot{w}^*\,e^{-i\phi}\right)\frac{\partial}{\partial\phi}$$

$$D_z = \frac{\partial}{\partial z} + \frac{i}{2}\left(w'\,e^{i\phi} - w'^*\,e^{-i\phi}\right)\frac{\partial}{\partial r} - \frac{1}{2r}\left(w'\,e^{i\phi} + w'^*\,e^{-i\phi}\right)\frac{\partial}{\partial\phi}. \quad (47)$$

Using (44), (45) one finds:

$$U \simeq U_1\,e^{i\phi} + U_1^*\,e^{-i\phi}$$
$$\Omega' \simeq \Omega_1\,e^{i\phi} + \Omega_1^*\,e^{-i\phi}. \quad (48)$$

Here

$$U_1 = \left(\frac{1}{r}\frac{\partial}{\partial r}r\frac{\partial}{\partial r} - \frac{1}{r^2}\right)\Phi_1 = L\,\Phi_1$$

$$\Omega_1 = \left(\frac{1}{r}\frac{\partial}{\partial r}r\frac{\partial}{\partial r} - \frac{1}{r^2}\right)\Psi_1 = L\,\Psi_1 \quad (49)$$

and

$$L = \frac{1}{r}\frac{\partial}{\partial r}r\frac{\partial}{\partial r} - \frac{1}{r^2}.$$

At leading order in ϵ, μ, equation (31) can be rewritten as follows:

$$L\frac{1}{r}\Psi_{0r}\,U_1 = \frac{1}{4}w'\frac{1}{r}\frac{\partial}{\partial r}r\frac{\partial}{\partial r}\Psi_{0r}^2. \quad (50)$$

This equation can be integrated twice. The result is

$$U_1 = \frac{1}{2}w'\,\Psi_{0r}. \quad (51)$$

This result has a very simple physical explanation. To leading order the vertical component of the velocity is proportional to the local angular velocity. It is just a result of the tilting of the planes where the fluid rotates caused by bending of the vortex line.

This point need to be commented. Actually, formula (45) cannot be valid for very large values of r. It is correct only for $r \ll l$, where l is a characteristic scale along the vortex line. To find v_z at $r \geq l$ one should use the more exact equation (23). Now one obtains instead of (50):

$$-\Delta\Phi = \frac{1}{2}w'\,\Psi_{0r}. \quad (52)$$

This equation should be complemented by the boundary condition

$$\Phi \to 0 \quad \text{at}\quad r \to \infty.$$

The solution of (52) decays exponentially at $r \to \infty$:

$$U_1 \simeq l^{\frac{r}{l}} \qquad r \to \infty.$$

This fact makes possible to perform a choice of an unknown constant $c = c(z,t)$. The explicit formula for $c(z,t)$ is not important. One has to introduce a new stream function $\tilde{\Psi}_0(r)$ defined as follow:

$$\tilde{\Psi}_{0r}(r) = \Psi_{0r} \qquad r < l$$
$$\tilde{\Psi}_0(r) \to 0 \qquad r \to \infty. \tag{53}$$

A correct definition of u_1 is

$$u_1 = \frac{1}{2} w' \, \tilde{\Psi}_{0r}. \tag{54}$$

By integrating (52) one obtains

$$\Phi_1 = \frac{1}{2} w' \frac{1}{r} \int_0^r r \tilde{\Psi}_0(r) \, dr. \tag{55}$$

Hence in the leading order $\Phi_1 \sim \epsilon$.

For finding Ψ_1 one should use equation (31). After simple transformation in the leading order one obtains

$$-\frac{i}{r}\left[\Psi_{0r} L \Psi_1 - \Psi_1 L \Psi_{0r}\right] =$$
$$= \frac{i}{2} \dot{w} \frac{\partial}{\partial r} \Omega_0 + \frac{1}{2} w'' \left[-\frac{1}{r^2} \frac{\partial}{\partial r} r^2 \, \Omega_0 \, \Phi_1 + \frac{1}{r} \frac{\partial}{\partial r} r \, \Omega_0 \, \tilde{\Psi}_0 \right]. \tag{56}$$

One can multiply (52) to r^2 and integrate over r from zero to infinity. It is easy to check that the left hand vanishes. The right hand gives the equation describing the dynamics of the vortex line

$$i \, \dot{w} + \lambda \, w'' = 0. \tag{57}$$

Here

$$\lambda = -\frac{\pi}{\Gamma} \int_0^\infty r \, \Omega_0 \, \tilde{\Psi}_0(r) \, dr. \tag{58}$$

The vorticity Γ is defined by (35). Integrating by parts in (56) leads to the result

$$\lambda = \frac{E}{\Gamma}$$
$$E = 2\pi \int_0^\infty r \, \tilde{\Psi}_{0r}^2 \, dr. \tag{59}$$

Integral (57) diverges logarithmically

$$E \simeq \ln L/\rho. \tag{60}$$

Equation (55) is not a new one, of course. It can be obtained from so called "local induction approximation".

Equation (55) can be solved analytically. The correction $\Psi_1 \sim \mu$, but the explicit expression for Ψ_1 is cumbersome and will not be used further in the paper.

Assuming $w \simeq e^{-i\Omega t + ikx}$ one finds that (56) describes the waves with the dispersion relation

$$\Omega = \lambda k^2, \tag{61}$$

with $\lambda \simeq \Gamma \ln k\rho$. These waves are circularly polarized and rotate in the direction opposite to rotation of the fluid in the vortex tube. Equation (56) is just the leading term in expansion of a more accurate nonlinear equation describing the dynamics of a free vortex line. The developed procedure makes possible to find this equation up to any desirable accuracy. These calculations will be published separately.

4   Vortex tube in an external flow

In this chapter we study the behavior of a vortex tube in a two-dimensional flow. We will assume that the characteristic scale of the flow is much larger that the size of the tube core ρ. In this case one can separate

$$\Psi = \Psi_0 + \Psi' + \tilde{\Psi}. \tag{62}$$

Here Ψ_0 and Ψ' are defined from (37), while $\tilde{\Psi}$ is a stream function of the external flow.

At leading order the most important contribution of the external flow appears in equation (14), which should be replaced by the equation

$$\frac{\partial}{\partial t} \Delta_\perp \Psi - \{\Psi, \Delta_\perp \Psi\} = \{\tilde{\Psi}, \Delta_\perp \Psi\} +$$
$$\text{div}_\perp \left[\Delta_\perp \Phi \Delta_\perp \Psi_z - \Delta_\perp \Psi \nabla_\perp \Phi_z \right]. \tag{63}$$

Omitting the details, we just present the result of influence of the large scale flow. Equation (56) should be replaced by the nonlinear Schrödinger equation

$$i\dot{w} + \lambda w'' = i F(w, \bar{w}). \tag{64}$$

Here

$$F = \frac{\partial \tilde{\Psi}}{\partial y} - i \frac{\partial \tilde{\Psi}}{\partial x}\bigg|_{x - iy = w} = \frac{\partial \tilde{\Psi}}{\partial \bar{w}}\bigg|_{x - iy = w}. \tag{65}$$

Equation (64) has a very clear physical meaning : it just describes the transport of the point vortex by the external flow having the velocity components

$$v_x = \frac{\partial \tilde{\Psi}}{\partial y}, \quad v_y = -\frac{\partial \tilde{\Psi}}{\partial x}.$$

The system (64) is Hamiltonian. It can be written as follow

$$i \dot{w} = \frac{\delta H}{\delta \bar{\omega}}. \tag{66}$$

Here

$$H = \int \left\{ \lambda |\omega'|^2 + \tilde{\Psi}(\omega, \bar{\omega}) \right\} dz. \tag{67}$$

The function $\tilde{\Psi}$ is a stream function of the external flow. One can admit that it depends slowly on z and t.

We can now mention that

$$F = \bar{U}.$$

Here $U = \tilde{v}_x - i \tilde{v}_y$ is a complex velocity of the external flow. Suppose, that the external flow is potential

$$\frac{\partial \tilde{v}_x}{\partial y} = \frac{\partial \tilde{v}_y}{\partial x}.$$

In this case U is an analytic function on w. Hence, F is anti-analytic. In a potential case

$$F = F(\bar{w}). \tag{68}$$

If the external flow is axially symmetric, then

$$\tilde{\Psi} = A(|w|^2)$$
$$F = -i\, w\, A'(|w|^2). \tag{69}$$

Equation (64) takes the form, which is standard in nonlinear optics,

$$i \dot{w} + \lambda w'' = w\, A'(|w|^2). \tag{70}$$

Equations (64), (70) may have solitonic solutions, which are very interesting from the physical view-point. Suppose, $\Gamma > 0$, $\lambda < 0$. By the rescaling

$$\lambda \frac{\partial^2}{\partial z^2} \to \frac{\partial^2}{\partial z^2}$$

one can transform the equation to the standard form, neglecting the logarithmic difference,

$$i \dot{w} + w'' - w\, A'(|w|^2) = 0. \tag{71}$$

Let us consider the Taylor expansion

$$-A'(\xi) = \alpha + \beta\xi + \cdots \qquad \xi \to 0. \tag{72}$$

Then we obtain the cubic Nonlinear Schrödinger equation

$$i\dot{w} + w'' + \left(\alpha + \beta|w|^2\right)w = 0. \tag{73}$$

The vorticity of the external flow

$$\Omega_3 = -\frac{1}{r}\frac{\partial}{\partial r}r\frac{\partial\tilde{\Psi}}{\partial r}$$

can be expanded at small r as follow

$$\Omega_3 \simeq 2(\alpha + 2\beta\,r^2). \tag{74}$$

Equation (73) is focusing, if $\beta > 0$, and defocusing, if $\beta < 0$. The sign of α defines the direction of rotation of the external flow. If $\alpha < 0$, this direction is opposite to the rotation of the vortex tube (the case of counter-rotation). For $\alpha > 0$ the signs of the rotations are the same (the case of co-rotation). In a typical case the absolute maximum of external vorticity is located at the center $r = 0$. It assumes $\alpha\beta < 0$.

Assuming that this combination is satisfied one can conclude that the vortex is described by the focusing NLS in the co-rotational case, while in the counter-rotational case one should use the defocusing NLS. This leads to another important conclusion: in the counter-rotational case the position of the vortex tube outside of the center is stable, while in the co-rotational case it is unstable. Let us recall again that this is correct only if the absolute value of the external vorticity reaches the maximum at $r = 0$. If the absolute value of the vorticity grows with r, the situation is quite opposite.

One can easily study solitonic solutions of the equation (70).

5   Interaction of vortex tubes

Let us study the interaction of a system of n almost parallel vortex tubes with vorticities $\Gamma_1, \dots, \Gamma_n$. The system of vortex tubes can be described by the following systems of Nonlinear Schrödinger equations

$$i\dot{w}_k + \lambda_k w_k'' = -\sum_{l \neq k}\frac{\Gamma_l}{\bar{w}_k - \bar{w}_l} \tag{75}$$

(See [2]). The right hand of (75) is anti-analytic in all w_k.

It is interesting to study the interaction of two identical vortices. We can put $n = 2$, $\Gamma_1 = \Gamma_2 = \Gamma$ and assume that the vortices $\lambda_1 = \lambda_2$ are located antisymmetrically with respect to the origin of coordinate. Thus, $\bar{w}_2 = -\bar{w}_1$, and equation (75) can be transformed into the following NLS equation

$$i\dot{w} + w_{zz} + \frac{w}{|w|^2} = 0. \tag{76}$$

Another interesting problem is interaction of two antiparallel vortices. In this case $\Gamma_2 = -\Gamma_1 = -\Gamma$, $\lambda_1 = -\lambda_2$, and one can assume

$$w_2 = \bar{w}_1 = w.$$

Equation (75) reads now (see [1])

$$i\dot{w} + w'' = \frac{1}{w - \bar{w}} \tag{77}$$

or

$$\dot{x} = -y'' + \frac{1}{2y}$$
$$\dot{y} = x''. \tag{78}$$

One can study also the continuous limit of system (75). Suppose, that we have a congruency of almost parallel vortices. Each vortex is marked by two Lagrangian labels p, q. We can introduce $s = p + iq$ and consider

$$w = w(t, z, s, \bar{s}), \qquad \lambda = \lambda(s, \bar{s}), \qquad \Gamma = \Gamma(s, \bar{s})$$

Equation (75) can be replaced by the system

$$i\frac{\partial w}{\partial t} + \lambda(s, \bar{s})\frac{\partial^2 w}{\partial z^2} + \int \frac{\Gamma(s', \bar{s}')ds'\, d\bar{s}'}{\bar{w}(s, \bar{s}) - \bar{w}(s', \bar{s}')} = 0. \tag{79}$$

The integral in (79) is understood as a principle value. We can interpret w as a displacement of a vortex from its initial position

$$w\Big|_{t=0} = s. \tag{80}$$

6   Instability and collapse of traveling vortex pair

Equation (78) has a trivial solution

$$y = a = \text{const}$$
$$x = \frac{1}{2a}t \tag{81}$$

describing the pair of antiparallel vortex tubes moving along the x axis.

In the moving frame

$$x' = x - \frac{1}{2a}t.$$

System (78) reads

$$\dot{x} = -y'' + \frac{1}{2y} - \frac{1}{2a}x'$$

$$\dot{y} = x''. \tag{82}$$

Linearization of (82)

$$w = x' + iy = a + xe^{i\Omega t + ipz}$$

leads to the result

$$\Omega^2 = -\frac{1}{2a^2}p^2 + p^4. \tag{83}$$

The counter-rotating pair of vortices is unstable if

$$p^2 < \frac{1}{2a^2}. \tag{84}$$

This is the so-called "Crow instability" [5]. It is interesting to study the nonlinear stage of this instability. One can look for the self-similar solutions of the system (78).

$$x = \sqrt{t_0 - t}A\left(\frac{z}{\sqrt{t_o - t}}\right)$$

$$y = \sqrt{t_0 - t}B\left(\frac{z}{\sqrt{t_o - t}}\right) \tag{85}$$

where A and B satisfy the equations

$$\frac{1}{2}(-A + xA') = -B'' + \frac{1}{2B}$$

$$\frac{1}{2}(-B + xB') = A''. \tag{86}$$

Asymptotically as $z \to \infty$

$$A \to \alpha z \qquad\qquad B \to \beta z \tag{87}$$

at $z \to -\infty$

$$A \to -\alpha z \qquad\qquad B \to \beta z.$$

Here α, β denote some constants. Solution (85) describes the collapse. Vortex lines converge ($\|w\| \sim \sqrt{t_0 - t}$) and meet at $z = 0$ in the moment of time $t = t_0$. At this moment the asymptotic formula (87) is valid for all z, and the vortex tubes are straight lines angled at $z = 0$. It is natural to make a conjecture that the collapse leads to the reconnection of the vortex tubes. But the system (84) is applicable only if the distance between the tubes is much greater than the size of their cores ($\|\omega\| \gg \rho$). Numerical experiments [6] show that at $|w| \sim \rho$ vortex cores deform and lose their round shape. They become flat, and the process of collapse slows down.

The system (78) has another class of approximate solutions. Let us suppose that

$$y'' \ll \frac{1}{2y}. \tag{88}$$

Then y satisfies the equation

$$\ddot{y} - \left(\frac{1}{y}\right)'' = 0. \tag{89}$$

This is an elliptic equation and the Cauchy problem for this equation is ill-posed. Anyway, it has a family of self-similar solutions:

$$x = (t_0 - t)^{1-\alpha} F\left(\frac{z}{(t_0 - t)^{1-\alpha}}\right)$$

$$y = (t_0 - t)^{\alpha} G\left(\frac{z}{(t_0 - t)^{1-\alpha}}\right). \tag{90}$$

To satisfy the condition (88) one has to put

$$1 > \alpha > \frac{1}{2}. \tag{91}$$

Solution (90) describes a slower collapse than (85).

7   Solitons on the co-rotating vortex pair

It was shown in section 5 that the co-rotating pair of identical vortices is described by equation (76). The solution

$$W = A e^{\frac{i}{A^2}t} \tag{92}$$

corresponds to the uniform rotation. This solution is stable. But equation (76) has interesting solitonic solutions ("dark solutions"). By separating the phase and the amplitude

$$W = A(z,t)e^{\frac{i}{2}\Phi} \tag{93}$$

One obtains the system

$$\frac{\partial}{\partial t}A^2 + \frac{\partial}{\partial z}A^2\phi_z = 0$$

$$\frac{A}{2}\left(\frac{\partial\Phi}{\partial t} + \frac{1}{2}(\frac{\partial\Phi}{\partial z})^2\right) = A_{zz} + \frac{1}{A}. \tag{94}$$

In the long-wave limit $A_{zz} \ll \frac{1}{A^2}$ system (94) transforms into the hyperbolic system

$$\frac{\partial}{\partial t}A^2 + \frac{\partial}{\partial z}A^2\Phi_z = 0$$

$$\frac{\partial\Phi}{\partial t} + \frac{1}{2}(\frac{\partial\Phi}{\partial z})^2 - \frac{1}{A^2} = 0. \tag{95}$$

This is the system of gas-dynamic equations with the exotic dependence of pressure on density $(P = \ln\rho)$. It was derived independently by V. Rulan [7].

To find the solitonic solutions one has to put

$$\frac{\partial}{\partial t} = c\frac{\partial}{\partial z}$$

$$\Phi \simeq e^{\frac{it}{A_0^2}}\Phi_0(z - ct).$$

Then

$$\Phi_{0z} = c(-1 + \frac{A_0^2}{A^2}) \tag{96}$$

$$A_{zz} + \frac{\bar{c}^2}{4}(A - \frac{A_0^4}{A^3}) + \frac{1}{A} - \frac{A}{A_0^2} = 0. \tag{97}$$

Equation (97) has solitonic solutions if

$$c^2 < \frac{2}{A_0^2}. \tag{98}$$

The solitons are the "necks" - moving domains of "closing in" of the vortex tubes. Equation (97) has the integral

$$\frac{1}{2}A_z^2 + \frac{\bar{c}^2}{8}(A^2 + \frac{A_0^4}{A^2}) + \ln\frac{A}{A_0} + \frac{A_0^2 - A^2}{2A_0^2} = 0. \tag{99}$$

The maximum of the amplitude A, $A_{max} = A_0$, is reached at infinity. The minimum of A is the second solution of the transcendent equation

$$\frac{\bar{c}^2}{8}(A_{min}^2 + \frac{A_0^4}{A_{min}^2}) + \ln\frac{A_{min}}{A_0} + \frac{A_0^2 - A_{min}^2}{2A_0^2} = 0. \tag{100}$$

If $\bar{\bar{c}}A_0 \ll 1$,

$$A_{min} \simeq \bar{\bar{c}}A_0^2 \ll A_0.$$

In the limiting case of $c = 0$ equation (99) describes the merging of the vortex pair into a single vortex.

One should remember that if $\bar{\bar{c}} \to 0$ $A_z \to \infty$ at $A \to A_{min}$, and the conditions of applicability of the used model are no longer valid.

The author is happy to express a deep gratitude to Dr. E. Kuznetsov for some very fruitful discussions.

References

1. V.E.Zakharov, Wave Collapse, *Uspehi Fizicheskoi Nauki* **155**, 529–533 (1988).

2. R.Klein, A.Majda and K. Damodaran, Simplified analysis of nearly parallel vortex filaments, *J. Fluid Mech.* **228**, 201–248 (1995).

3. A.Pumir, E.D.Siggia, Collapsing solutions in the 3-D Euler equations in *Topological Fluid Mechanics* 469–477, eds H.K. Moffatt, A. Tsinober, Cambridge University Press, Cambridge (1990); A. Pumir, E.D. Siggia, Development of singular solutions to the axisymmetric Euler equations *Phys. Fluids A* **4**, 1472–1491 (1992).

4. H.K. Moffatt, S.Kida and K.Okhitani, Stretched vortices - the sinews of turbulence; high Reynolds number asymptotics *J. Fluid Mech.* **259**, 241–264 (1994).

5. S.C.Crow, Stability theory for a pair of trailing vortices, Amer. Inst. Aeronaut. Astronaut. J. **8**, 2172–2179 (1970).

6. M.V.Melander and F. Hussain, Cross-linking of two antiparallel vortex tubes. *Phys. Fluids A* **1**, 633–636 (1989); M.J. Shelley, D.I. Meiron and S.A. Orszag, Dynamic aspects of vortex reconnection on perturbed anti-parallel tubes *Journal of Fluid Mechanics* **246**, 613-652 (1993).

7. V. Ruban, Hamiltonian dynamics of frozen fluids in ideal fluid *(in Russian) Dissertation.* Landau Institute for Theoretical Physics. Chernogolovka.

Lecture Notes in Physics

For information about Vols. 1–497
please contact your bookseller or Springer-Verlag

Monographs

For information about Vols. 1–16
please contact your bookseller or Springer-Verlag